D0461774

APPLIED STATISTICS
A First Course

APPLIED STATISTICS
A First Course

Mark L. Berenson
Department of Statistics and Computer Information Systems
Baruch College

David M. Levine
Department of Statistics and Computer Information Systems
Baruch College

David Rindskopf
Departments of Educational Psychology and Psychology
City University of New York, Graduate Center

PRENTICE HALL, *Englewood Cliffs, New Jersey 07632*

Library of Congress Cataloging-in-Publication Data

Berenson, Mark L.
Applied statistics.

Bibliography: p.
Includes index.
1. Statistics. I. Levine, David M.
II. Rindskopf, David. III. Title.
QA276.12.B46 1988 519.5 87-7233
ISBN 0-13-041476-X

Editorial/production supervision: Eleanor Perz
Interior design: Levavi & Levavi
Cover design: Maureen Eide
Manufacturing buyer: Barbara Kittle

Printed in the United States of America

10 9 8 7 6 5 4 3 2 1

ISBN 0-13-041476-X 01

PRENTICE-HALL INTERNATIONAL (UK) LIMITED, *London*
PRENTICE-HALL OF AUSTRALIA PTY. LIMITED, *Sydney*
PRENTICE-HALL CANADA INC., *Toronto*
PRENTICE-HALL HISPANOAMERICANA, S.A., *Mexico*
PRENTICE-HALL OF INDIA PRIVATE LIMITED, *New Delhi*
PRENTICE-HALL OF JAPAN, INC., *Tokyo*
SIMON & SCHUSTER ASIA PTE. LTD., *Singapore*
EDITORA PRENTICE-HALL DO BRASIL, LTDA., *Rio de Janeiro*

To my parents, the memory of my father, and to my wife, Rhoda, and daughters, Kathy and Lori.

M.L.B.

To my parents, to my wife, Marilyn, and my daughter, Sharyn.

D.M.L.

To my parents, the memory of my mother, and to my wife, Toni.

D.R.

Contents

Preface

When planning an introductory text in statistics that is intended to transcend a wide range of academic disciplines, the authors must decide how the text will differ from those already available and what contribution it will make to the field of study. In developing *Applied Statistics: A First Course*, our goal was to present a practical, data-analytic approach to statistical applications so that fundamental methods and concepts could be taught to students in all disciplines in a one-semester, noncalculus-based introductory course. Hence, the distinguishing characteristics of this text are its innovative approaches combined with its internal pedagogical devices.

Innovations

- *Using Case Studies as Chapter Motivators*: Detailed case studies involving extensive analysis based on actual or realistic data are used to motivate the discussions throughout the text.
- *Using a Survey/Database for Integrating Course Topics*: Perhaps the most difficult aspect for a student taking a statistics course is to perceive its continuity (i.e., how one set of topics interrelates with another). The text attempts to overcome this problem by utilizing the results from an alumni association membership survey (Case Study A) in 139 end-of-chapter Database Problems to show the interrelationships between descriptive statistics, cross-tabulations, statistical inference, ANOVA, nonparametrics, and regression and correlation.
- *Using Computer Packages in a Statistics Course*: The rapid expansion in the use of computer packages to assist in data analysis has posed a dilemma for statistics educators in terms of how to best employ software for a statistics course. This text goes beyond merely illustrating computer output. Rather, it devotes an entire chapter (Chapter 15—Using Computers for Statistical Analysis) to demonstrate how computer software is actually used to assist in data analysis. Utilizing questions from the alumni association membership survey (Case Study A) that cover the entire range of topics presented in Chapters 3 through 14 of the text, three widely used commercially available packages (SPSS^X, Minitab, and STAT-

GRAPHICS) are described and the resulting (annotated) computer outputs are interpreted. Hence, this chapter is not intended to make the reader proficient at any particular package; it is, however, intended to familiarize the reader with the use of statistical software and to illustrate that learning the syntax of a command-driven package (SPSS^X or Minitab) or learning to manipulate the displays from a menu-driven package (STATGRAPHICS) are not difficult. Moreover, since the chapter utilizes statistical procedures that had been covered throughout the text, it serves to demonstrate how the various topics are interwoven when one is performing a statistical analysis of a large data set.

- *End-of-Section Problems and End-of-Chapter Problems*: 725 additional problems are presented both at the end of sections and at the end of chapters for purposes of pedagogy. Most problems contain, on average, four or five parts. However, emphasis is not on "number crunching" but on understanding and interpretation. It is essential that students be able to express what they have learned. *Literacy* is enhanced by numerous thought-provoking questions involving discussion, letter-writing, memos, and reports. Problems that are either especially thought-provoking or have no "exact solution" are flagged with a "↯" symbol.

Internal Pedagogical Devices

- *Writing Style*: The material is written in a conversational, narrative type of style in order to emphasize both the concepts and methods of applied statistics. The basic philosophy is to "write for the student, not for the professor." Each particular concept is thoroughly developed with detailed examples. Moreover, to reduce student anxiety in dealing with end-of-section Problems or Supplementary Problems, humorous names are interjected as the situation merits.
- *Realistic Applications*: To motivate student interest, the material in the text is developed by referring to realistic type applications, to applications with actual data, and to applications selected from the alumni association membership survey (Case Study A).
- *Pedagogical Tools*: Numerous important pedagogical tools (boxed formulas, boxed examples and solutions, drawings, tabular interpretations, annotated computer output) have been utilized throughout the text to enhance the student's ability to comprehend the concepts and methods of applied statistics. Use of color is given for emphasis.
- *Reading and Interpreting Statistical Tables*: Each of the statistical tables shown in Appendix C is examined in depth when it is initially presented. Detailed explanations and illustrations are provided to enable the student to read and properly interpret the particular tables.
- *Highlighted Examples with Solutions*: Throughout the text several examples are provided with solutions "boxed off" in color for emphasis. This pedagogical feature permits the student to quickly review a particular topic.

- *End-of-Text Material*: To assist the student, the following end-of-text material is included:

a. Appendix A lists the rules for arithmetic and algebraic operations and discusses summation notation in depth.
b. Appendix B provides a list of statistical symbols and notation.
c. Appendix C provides an extensive set of statistical tables.
d. Answers to Selected Problems—To provide students with immediate feedback, answers to specific problems (designated with a ● symbol) selected from the end of sections and the end of chapters throughout the text are given.
e. Index—a detailed index consisting of more than 450 entries is provided along with extensive cross-referencing.

It is our hope and anticipation that the unique approaches taken in this text will make the study of introductory statistics more meaningful, rewarding, and comprehensible for all readers.

We are extremely grateful to the many organizations and companies that generously allowed us to use their actual data for developing problems and examples throughout our text. In particular, we would like to cite Time Inc. (publisher of *Fortune*), The American Association of University Professors (publisher of *Academe*), and CBS Inc. (publisher of *Road & Track*). Moreover, we would like to thank the Biometrika Trustees, American Cyanamid Company, The Rand Corporation, and the Institute of Mathematical Statistics for their kind permission to publish various tables in Appendix C. Furthermore, we would like to acknowledge SPSS Inc., Minitab Inc., and STSC Inc. for their permission to describe syntax and present computer output in this text.

We wish to express our thanks to Dennis Hogan, Carol Sobel, Kate Moore, and Eleanor Perz of the editorial staff at Prentice Hall for their continued encouragement. We also wish to thank Professors Robert F. Brown, UCLA; Larry J. Ringer, Texas A & M University; Thomas A. O'Connor, University of Louisville; and Phillip McGill, Illinois Central College for their constructive comments during the development of this textbook. Finally, we wish to thank our wives and children for their patience, understanding, love, and assistance in making this project a reality. It is to them that we dedicate this book.

MARK L. BERENSON
DAVID M. LEVINE
DAVID RINDSKOPF

Chapter 1

Introduction

1.1 WHAT IS MODERN STATISTICS?

A century ago H. G. Wells commented that "statistical thinking will one day be as necessary for efficient citizenship as the ability to read and write." Each day of our lives we are exposed to a wide assortment of numerical information pertaining to such phenomena as stock market activity, unemployment rates, medical research findings, opinion poll results, weather forecasts, and sports data. Frequently, such information has a profound effect on our lives.

The subject of modern statistics encompasses the collection, presentation, and characterization of information to assist in both data analysis and the decision-making process.

1.2 THE GROWTH AND DEVELOPMENT OF MODERN STATISTICS

The growth and development of modern statistics can be traced to two separate phenomena: the needs of government to collect information on its citizenry (see References 1 and 2) and the development of the mathematics of probability theory.

The collection of data began at least as early as recorded history. During the Egyptian, Greek, and Roman civilizations information was obtained primarily for the purposes of taxation and military conscription. In the Middle Ages, church institutions often kept records concerning births, deaths, and marriages. In America, although various records were kept back to colonial times (see Reference 3), the Federal Constitution required the taking of a census every ten years beginning in 1790. Today, these data are used for many purposes including congressional apportionment and the allocation of federal funds.

1.2.1 Descriptive Statistics

These and other needs for data on a nationwide basis were closely intertwined with the development of the subject of descriptive statistics.

> **Descriptive statistics can be defined as those methods involving the collection, presentation, and characterization of a set of data in order to properly describe the various features of that set of data.**

Although descriptive statistical methods are important for presenting and characterizing information (see Chapters 2–5), it has been the development of inferential statistical methods as an outgrowth of probability theory that has led to the great expansion in the application of statistics in all fields of research today.

1.2.2 Inferential Statistics

The initial impetus for the formulation of the mathematics of probability theory came from the investigation of games of chance during the Renaissance. The foundations of the subject of probability can be traced back to the middle of the seventeenth century in the correspondence between the mathematician Pascal and the gambler Chevalier de Mere (see Reference 1). These and other developments by such mathematicians as Bernoulli, DeMoivre, and Gauss were the forerunners of the subject of inferential statistics. However, it has only been since the turn of this century that statisticians such as Pearson, Fisher, Gosset, Neyman, Wald, and Tukey have pioneered in the development of the methods of inferential statistics that are so widely applied in so many fields today.

Inferential statistics can be defined as those methods that make possible the estimation of a characteristic of a population or the making of a decision concerning a population based only on sample results.

To clarify this, a few definitions are necessary.

A population (or universe) is the totality of items or things under consideration.

A sample is the portion of the population that is selected for analysis.

A parameter is a summary measure that is computed to describe a characteristic of an entire population.

A statistic is a summary measure that is computed to describe a characteristic from only a sample of the population.

Thus, one major aspect of inferential statistics is the process of using sample statistics to draw conclusions about the true population parameters.

The need for inferential statistical methods derives from the need for sampling. As a population becomes large, it is usually too costly, too time consuming, and too cumbersome to obtain our information from the entire population. Decisions pertaining to the characteristics of the population have to be based on the information contained in but a sample of that population. Probability theory provides the link by ascertaining the likelihood that the results from the sample reflect the results in the population.

These ideas can be clearly seen in the example of a political poll. If the pollster wishes to estimate the percentage of the votes a candidate will receive in a particular election, he or she will not interview each of the thousands (or even millions) of voters. Instead, a sample of voters will be selected. Based on the outcome from the sample, conclusions will be drawn concerning the entire population of voters. Appended to these conclusions will be a probability statement specifying the likelihood or confidence that the results from the sample reflect the true voting behavior in the population.

1.3 WHY STUDY MODERN STATISTICS?

With the expansion in the use of both large-scale mainframe computers and minicomputer systems as well as the advent of the personal computer, the use of statistical methods as an aid to data analysis and decision making has grown dramatically over the past decade and will continue to grow in the future.

By studying the subject of modern statistics, we will obtain an appreciation for and understanding of those techniques which are used on the numerical information we encounter in both our professional and nonprofessional lives. The concepts and methods described in this text are intended to provide a fundamental background in the subject of modern statistics and its applications in a wide variety of disciplines.

1.4 EXAMPLES OF APPLICATIONS OF STATISTICS

- *Agriculture*: A university's Agriculture College develops a new type of corn which it hopes will increase yield. But it knows that the number of bushels per acre it will get from the seed will vary, depending on factors such as the weather, the type of soil, and the equipment used by the farmer. How can it get a good estimate of what the average yield will be for the new corn, and whether the new corn is better than types already in use?
- *Business*: An airline needs to decide how many reservations to take for its flights. If it takes too many, then it will alienate customers who show up and are told there is no room for them on the flight. If it takes only as many reservations as there are seats, it will lose money because some people who reserve seats will not show up. How can the airline estimate the number of ''no-shows'' in order to plan the optimal number of reservations to allow—that is, the number which will minimize the number of people with reservations who get denied a seat, while maximizing the number of seats filled on the flights?
- *Health*: A toothpaste manufacturer wants to show that the use of its product reduces the incidence of tooth decay. The number of cavities which people have depends on many factors, such as what they eat, their genetic characteristics, how well they take care of their teeth, and the amount of fluorine in their water. How might studies be designed to prove that one toothpaste is better than another, given that there would be variability among people in the number of cavities they have, even if they all used the same toothpaste, let alone different ones?
- *Education*: Developers of a new computer language, PORKY, claim that students who learn to program in PORKY will also show improvement in academic subjects, reasoning skills, and IQ. Students, of course, vary tremendously in all of these areas, even when they are in the same schools and classes. Given this variability among students, how might we investigate whether students learning PORKY also improve in other areas?
- *Physics*: Scientists measuring the speed of light set up an apparatus which measures how long it takes a beam of light to return after being reflected in a mirror. Although they use instruments as precise as can be made, they find that the length of time varies slightly from trial to trial. Given that they know, from theory, that the measurement should be the same each time if the experiment is done perfectly, how should they estimate the speed of light from the many (different) measurements they have? How can they express the level of accuracy of their estimate?

What the above applications have in common is that they ask for a conclusion about a situation in which there is uncertainty because not all of the relevant factors can be measured or controlled, and because we cannot study the whole population. This causes the outcome to be variable, and thus not completely predictable. For example, the yield of corn, the number of passengers actually showing up, the number of cavities, the number of items correct on a test, and the measurement of the speed of light all involve measurements of quantities which are uncertain. If we knew how many cavities each person should have under "normal" conditions, then testing the effect of using a new toothpaste would be easy.

Problems

For Problems 1.1 to 1.6, specify the general problem to be solved, the specific inference to be made, what the population is, and (if you are describing the results of an actual published study) what the weaknesses of the study might be. Where appropriate, tell what parameters are of primary interest and what statistics are used to arrive at a conclusion.

1.1 Describe three applications of statistics to business.

1.2 Describe three applications of statistics to sports.

1.3 Describe three applications of statistics to political science or public administration.

1.4 Describe three applications of statistics to psychology.

1.5 Describe three applications of statistics to education.

1.6 Describe three applications of statistics to medicine or medical research.

1.7 Political polls taken before elections have some aspects which are descriptive, and others which are inferential. What is descriptive, and what is inferential, about the results reported in such polls? What assumptions are usually made when inferences are made from these polls?

1.8 We often hear or read statements such as "One-third of the children in America go hungry each night." Presuming that these statements have some factual basis (that is, they are not just invented), tell how you might count or estimate the number of children going hungry each night.

1.9 When a doctor gives you a blood test, or takes your temperature or blood pressure, he or she makes a decision about whether the results are abnormal. How do you think the criteria for normality are arrived at for such tests? What are some possible problems which might arise in establishing what is normal?

References

1. PEARSON, E. S., ed., *The History of Statistics in the Seventeenth and Eighteenth Centuries* (New York: Macmillan, 1978).
2. PEARSON, E. S., AND M. G. KENDALL, eds., *Studies in the History of Statistics and Probability* (Darien, Conn.: Hafner, 1970).
3. WATTENBERG, B. E., ed., *Statistical History of the United States: From Colonial Times to the Present* (New York: Basic Books, 1976).

Data Collection

2.1 INTRODUCTION: THE NEED FOR RESEARCH

Throughout history human inquisitiveness has led to experimentation and research in order to aid in the decision-making process. This process is a common denominator to all fields of endeavor. A few examples were listed in Chapter 1; some others are:

1. The personnel director wants to hire the right person for the right job.
2. The market researcher looks for the characteristics that distinguish a product from its competitors.
3. Potential investors want to determine which firms within which industries are likely to have accelerated growth in a period of economic recovery.
4. Public officials want information pertaining to the attitudes and desires of their constituents for legislative purposes (and for being retained in office).

5. A college administrator wishes to know information pertaining to the attributes, attitudes, and interests of the student body in order to implement decisions regarding the quality of campus life.

Defining one's goals leads to considering various courses of action. The rational decision maker seeks to evaluate information in order to select the course of action that maximizes objectives. To the statistician or researcher, the needed information is the **data.** For a statistical analysis to be useful in the decision-making process, the input data must be appropriate. If the data are flawed by biases, ambiguities, or other types of errors, even the fanciest and most sophisticated tools would not be enough to compensate for the deficiencies.

The goals of this chapter are:

1. To develop an understanding for formulating a research problem and conducting research.
2. To examine various sources of data for research.
3. To obtain an understanding of different types of data and measurement levels.
4. To obtain an appreciation for the art of questionnaire design and its problems.
5. To obtain an appreciation for and a knowledge of selecting a random probability sample.
6. To obtain an appreciation for the problems involved in data collection with respect to editing, coding, and data entry as well as response and nonresponse.

2.2 SOURCES OF DATA FOR RESEARCH

There are many methods by which researchers may obtain needed data. First, they may seek data already published by governmental, industrial, or individual sources. Second, they may design an experiment to obtain the necessary data. Third, they may conduct a survey. Fourth, they may make observations of the behavior of people or animals in whom they are interested.

The federal government is a major collector of data for both public and private purposes. The Bureau of Labor Statistics is responsible for collecting data on employment as well as establishing the well-known monthly Consumer Price Index. In addition to its constitutional requirement for conducting a decennial census, the Bureau of the Census is also concerned with a variety of ongoing surveys regarding population, housing, and manufacturing and from time to time undertakes special studies on topics such as crime, travel, and health care.

A second method of obtaining needed data is through experimentation. In an experiment, strict control is exercised over the treatments given to participants. For example, in a study testing the effectiveness of toothpaste, the researcher would determine which participants in the study would use the new brand and which would not, instead of leaving the choice to the subjects. Proper experimental designs are usually the subject matter of more advanced texts, since they often involve sophisticated statistical procedures. However, in order to develop a feeling for testing and experimentation, the fundamental experimental design concepts will be considered in Chapters 10 through 12.

A third method of obtaining data is to conduct a survey. Here no control is exercised over the behavior of the people being surveyed. They are merely asked questions about their beliefs, attitudes, behaviors, and other characteristics.

In an observational study, the researcher observes the behavior of interest directly, usually in its natural setting. Most knowledge of animal behavior was originally developed this way, as is the case in astronomy and geology, where experimentation and surveys are impractical. Observational studies also play a major role in anthropology and sociology, because they can often provide a richness of description which is absent from more structured situations such as surveys.

If a study is to be useful, the data gathered must be *valid*; that is, the "right" responses must be assessed, and in a manner that will elicit meaningful measurements. In order to design a survey or experiment, one must understand the different types of data and measurement levels. To demonstrate some of the issues involved in obtaining data, we will present them in the context of a survey, although most of the same issues will arise in other types of research.

2.3 OBTAINING DATA THROUGH SURVEY RESEARCH

A survey statistician will most likely want to develop an instrument that asks several questions and deals with a variety of phenomena or characteristics. These phenomena or characteristics are called **variables.** The data, which are the observed outcomes of these variables, may differ from response to response.

2.3.1 Types of Data

As outlined in Figure 2.1, there are basically two types of variables yielding two types of data: qualitative and quantitative. The difference between them is that

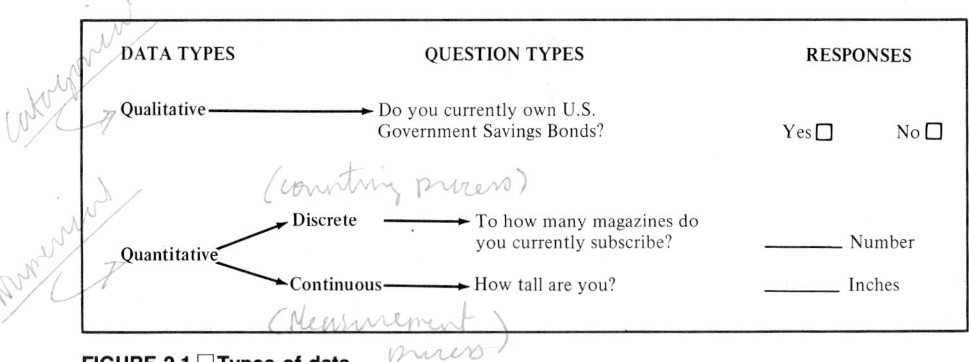

FIGURE 2.1 □ Types of data.

qualitative variables yield **categorical** responses while **quantitative** variables yield **numerical** responses. For example, the response to the question "Do you currently own United States Government Savings Bonds?" is categorical. The choices are clearly "yes" or "no." On the other hand, responses to questions such as "To how many magazines do you currently subscribe?" or "How tall are you?" are clearly numerical. In the first case the quantitative variable may be considered as **discrete** and in the second case as **continuous.**

Discrete quantitative data are numerical responses which arise from a counting process, while continuous quantitative data are numerical responses which arise from a measuring process.

"The number of magazines subscribed to" is an example of a discrete quantitative variable, since the response values are all integers. The individual currently subscribes to either no magazines, one magazine, two magazines, etc. On the other hand, "height of an individual" is an example of a continuous quantitative variable, since the response can take on any value within a continuum or interval, depending on the precision of the measuring instrument. For example, a person whose height is reported as 67 inches may be measured as $67\frac{1}{4}$ inches, $67\frac{7}{32}$ inches, or $67\frac{58}{250}$ inches, depending on the precision of the instrumentation available. Therefore, we can see that height is a continuous phenomenon which can take on any value within an interval.

It is interesting to note that theoretically no two persons could have exactly the same height, since the finer the measuring device used, the greater the likelihood of detecting differences between them. However, measuring devices used by researchers may not be sophisticated enough to detect small differences, and hence "tied" observations are often found in experimental data even though the variable is truly continuous.

Some variables are treated as continuous even when they are really discrete, if we think that the approximation to a continuous variable is close enough. Consider the case of the results of laboratory tests such as blood counts. As the name implies, these are typically counts of the number of cells of a particular type in a sample of blood. But because these numbers are often so large and can have so many possible values, they are very close to being continuous, and are usually treated as such.

2.3.2 Levels of Measurement and Types of Measurement Scales

From the above discussion, then, we see that our resulting data may also be described in accordance with the "level of measurement" attained.

In the broadest sense, all collected data are "measured" in some form. For example, even discrete quantitative data can be thought of as arising by a process of "measurement through counting." The four widely recognized levels of measurement are—from weakest to strongest level of measurement—the nominal, ordinal, interval, and ratio scales.

NOMINAL AND ORDINAL SCALES

Data obtained from a qualitative variable are said to have been measured either on a nominal scale or on an ordinal scale. If the observed data are merely classified into various distinct categories in which no ordering is implied, a **nominal** level of measurement is achieved. On the other hand, if the observed data are classified into distinct categories in which ordering is implied, an **ordinal** level of measurement is attained. These distinctions are depicted in Figures 2.2 and 2.3, respectively.

Qualitative Variable	*Categories*
Automobile ownership	Yes No
Type of life insurance owned	Term Endowment Straight-Life Other None
Political party affiliation	Democrat Republican Independent Other

FIGURE 2.2 □ Examples of nominal scaling.

Nominal scaling is the weakest form of measurement because no attempt can be made to account for differences within a particular category or to specify any ordering or direction across the various categories. Ordinal scaling is a somewhat stronger form of measurement, because an observed value classified into one category is said to possess more of a property being scaled than does an observed value classified into another category. Nevertheless, within a particular category no attempt is made to account for differences between the classified values. Moreover, ordinal scaling is still a weak form of measurement, because no meaningful numerical statements can be made about differences between the categories. That is, the ordering implies

Qualitative Variable	*Ordered Categories*
	(Lowest-Highest)
Student class designation	Freshman Sophomore Junior Senior
Product satisfaction	Very Unsatisfied Fairly Unsatisfied Neutral Fairly Satisfied Very Satisfied
Movie classification	G PG PG-13 R X
	(Highest-Lowest)
Faculty rank	Professor Associate Professor Assistant Professor Instructor
Standard & Poor's bond ratings	AAA AA A BBB BB B CCC CC C DDD DD D
Restaurant ratings	***** **** *** ** *
Student grades	A B C D F

FIGURE 2.3 □ Examples of ordinal scaling.

only *which* category is "greater," "better," or "more preferred"—not *how much* "greater," "better," or "more preferred."[1]

INTERVAL AND RATIO SCALES

An **interval scale** is an ordered scale in which the difference between measurements is a meaningful quantity. For example, a person who is 67 inches tall is 2 inches taller than someone who is 65 inches tall. In addition, that 2 inches is the same quantity that would be obtained if the two people were 76 and 74 inches tall, so the difference has the same meaning anywhere on the scale.

If, in addition to differences being meaningful and equal at all points on a scale, there is a true zero point so that ratios of measurements are sensible to consider, then the scale is a **ratio scale.** A person who is 76 inches tall is twice as tall as someone who is 38 inches tall; in general, then, measurements of length are ratio scales. Temperature is a trickier case: Fahrenheit and centigrade (Celsius) scales are interval but not ratio; the "0" demarcation is arbitrary, not real. No one should say that 76 degrees Fahrenheit is twice as hot as 38 degrees Fahrenheit. But when measured from absolute zero, as in the Kelvin scale, temperature is on a ratio scale, since a doubling of temperature is really a doubling of the average speed of the molecules making up the substance. Figure 2.4 gives examples of interval- and ratio-scaled variables.

Quantitative Variable	*Level of Measurement*
Temperature (in degrees Celsius or Fahrenheit)	Interval
Calendar time (Gregorian, Hebrew, or Islamic)	Interval
Height (in inches or centimeters)	Ratio
Weight (in pounds or kilograms)	Ratio
Age (in years or days)	Ratio
Salary (in American dollars or Japanese yen)	Ratio

FIGURE 2.4 □ Examples of interval and ratio scaling.

Data obtained from a quantitative variable are usually assumed to have been measured either on an interval scale or on a ratio scale. These scales constitute the highest levels of measurement. They are stronger forms of measurement than an ordinal scale because we are able to discern not only which observed value is the largest but also by how much.

Problems

2.1 Explain the differences between qualitative and quantitative variables and give three examples of each.

[1] College basketball and college football rankings are other examples of ordinal scaling. The differences in ability between the teams ranked first and second might not be the same as the differences in ability between the teams ranked second and third, or those ranked sixth and seventh, and so on.

2.2 Explain the differences between discrete and continuous variables and give three examples of each.

2.3 **(a)** If two students both score a 90 on the same examination, what arguments could be used to show that the underlying variable (phenomenon of interest)—knowledge of the subject—is continuous?
(b) If the test were a multiple-choice test scored by counting the number of items answered correctly, give reasons why this might be treated as an interval-scaled variable, even though it is not continuous. Further, give reasons why you might treat such test scores (or any other count variable) as a ratio-scaled variable.

• 2.4 Determine whether each of the following variables is qualitative or quantitative. If quantitative, determine whether the phenomenon of interest is discrete or continuous.
(a) Number of telephones per household.
(b) Type of telephone.
(c) Number of long-distance calls made per month.
(d) Length (in minutes) of long-distance calls.
(e) Color of telephone.
(f) Monthly charges (in dollars and cents) for long-distance calls made.

2.5 Suppose the following information is obtained for Georgina Steinbrenner on admission to the Stephan College infirmary:

(1) Sex: Female
(2) Residence or Dorm: Mogelever Hall
(3) Class: Sophomore
(4) Temperature: 102.2 F (oral)
(5) Pulse: 70 beats per minute
(6) Blood Pressure: 130/80 mg/mm
(7) Blood Type: B Positive
(8) Known Allergies to Medicines: None
(9) Preliminary Diagnosis: Influenza
(10) Estimated Length of Stay: 3 days

Classify each of the ten responses by type of data and level of measurement. [*Hint*: Be careful with blood pressure; it's tricky.]

2.6 Give three examples of variables which are actually discrete but might be considered continuous.

2.7 For each variable in the examples of applications of statistics you mentioned in the answers to Problems 1.1–1.6 on page 5, tell whether the variable is quantitative or qualitative; whether it is discrete or continuous; what level of measurement it has; and, if it is not continuous, whether it could be treated as such.

2.8 List three examples in an area of interest to you where data are useful for decision making. What data are useful? How might they be obtained? How might the data be used in the process of making a decision?

2.9 Why is temperature an interval-scaled variable when measured in degrees Fahrenheit or centigrade, but a ratio-scaled variable when measured from absolute zero?

2.10 One of the variables most often included in surveys is income. Sometimes the question is phrased: "What is your income (in thousands of dollars)?" In other surveys, the respondent is asked: "Is your income over $10,000? Over $20,000? Over $30,000?"
(a) For each of these cases, tell whether the variable is nominal, ordinal, interval, or ratio.
(b) In the first case, tell why income might be considered either discrete or continuous.

Note: Bullet● indicates those problems whose solutions are included in the Answers to Selected Problems.

2.4 DESIGNING THE QUESTIONNAIRE INSTRUMENT

The general procedure for designing a questionnaire will involve:

1. Choosing the broad topics which are to reflect the theme of the survey.
2. Deciding on a mode of response.
3. Formulating the questions.
4. Pilot testing and making final revisions.

To better understand this process let us consider an illustrative case.

Case A—The Alumni Association President's Study

Suppose that the president of the alumni association of a well-known college wishes to sample the membership in order to develop an "alumni profile." The purpose is to use the survey results for public relations/publicity campaigns, recruiting, fund raising, etc., in order to enhance the reputation and image of the college in the public's eye and to stimulate further contributions on the part of the membership to pay for the development of collegewide resources, construction of buildings, and scholarships for needy and outstanding students.

In order to obtain the needed (profile) information on the past achievements, attitudes, and interests, as well as current activities of its membership, the president of the alumni association hires a statistician to assist in developing a questionnaire and conducting the survey.

2.4.1 Selection of Broad Topics

Working together, the president and the statistician choose the broad topics which are to reflect the theme of the survey. The topics considered are:

1. Socioeconomic and demographic characteristics.
2. Past educational achievements, attitudes, and interests.
3. Past and present employment activities.

A first effort is then made to develop a series of questions within each of these broad topics.

2.4.2 Length of Questionnaire

Before very long a large number of questions will have been created. Unfortunately, however, there is an inverse relationship between the length of a questionnaire and the rate of response to the survey. That is, the longer the questionnaire, the lower will be the rate of response; the shorter the questionnaire, the higher will be the rate of response. It is, therefore, imperative that we carefully determine the merits of each question in terms of whether the question is really necessary, and if so, how to word it optimally. Questions should be as short as possible. The response categories for qualitative questions should be nonoverlapping and complete.

2.4.3 Mode of Response

The particular questionnaire format to be selected and the specific question wording are affected by the intended mode of response. There are essentially three modes through which survey work is accomplished: personal interview, telephone interview, and mail. The personal interview and the telephone interview usually produce a higher response rate than does the mail survey—but at a higher cost.

After careful consideration, our president and statistician determined that the mail survey would be the most practical means of obtaining the desired information from the membership.

2.4.4 Formulating the Questions

Because of the inverse relationship between the length of a questionnaire and the rate of response to the survey, each question must be clearly presented in as few words as possible, and each question should be deemed essential to the survey. In addition, questions must be free of ambiguities. For example, consider the following two questions:

1 Do you smoke? Yes _____ No _____

2 How old are you? ____________ years

Question 1 has several possible ambiguities. It is not clear if the desired response pertains to cigarettes, to cigars, to pipes, or to combinations thereof. It is also not clear whether occasional smoking or habitual smoking was the primary concern of the question. If we were interested only in current cigarette consumption, perhaps it would be better to ask:

About how many cigarettes do you currently smoke each day? _____

When replying to question 2, the respondent may be confused as to whether to base the answer on the *last* birthday or the *nearest* birthday. This problem may be avoided if the respondent is merely asked to

State your date of birth: __________ __________ __________
month day year

ALUMNI ASSOCIATION MEMBERSHIP PROFILE OF BACCALAUREATE DEGREE RECIPIENTS WHO ARE CURRENTLY EMPLOYED FULL-TIME

Codes — *(To answer the questions below, please circle the appropriate number or fill in where applicable)*

Codes	Question
____ 1, 2	1. How many years has it been since your graduation from the college?________
____ 4	2. In what area was your undergraduate major? Business [1] Humanities & Social Science [2] Science & Mathematics [3]
____ 6	3. What was the *primary* reason that attracted you to the college? Convenient location [1] Course offerings [3] Cost [2] Reputation [4] Other______________________ [5] *(specify)*
____ 8	4. When compared to your undergraduate days at the college, would you say that your current philosophical ideology is: More conservative [1] About the same [2] More liberal [3]
____ 10	5. Have you ever attended graduate school? Yes [1] No [2]
____ 12	6. On a scale from one to five, how closely related is your present job to your undergraduate major? (Circle choice.) Completely Unrelated 1 2 3 4 5 Completely Related
____ 14, 15, 16	7. What is your current annual salary (in thousands)?________
____ 18	8. In what type of organization are you presently employed? Corporation/Company with over 100 employees [1] Corporation/Company with under 100 employees [2] Education [3] Nonprofit or government agency [4] Self-employed [5]
____ 20, 21	9. Which *one* of the following terms *best* describes your current job activity? Accounting [01] Administration/Management [02] Banking/Finance [03] Computers [04] Marketing & Retailing [05] Personnel [06] Public Relations [07] Real Estate/Insurance [08] Sales [09] Statistics or Research [10] Teaching [11] Other______________________ [12] *(specify)*
____ 23, 24	10. How many years have you been working in this area?

FIGURE 2.5 □ Questionnaire.

____ 26, 27	11. Which *one* of the following areas *best* describes your first full-time job after graduation from college?

Accounting [01]
Administration/Management [02]
Banking/Finance [03]
Computers [04]
Marketing & Retailing [05]
Personnel [06]
Public Relations [07]
Real Estate/Insurance [08]
Sales [09]
Statistics or Research [10]
Teaching [11]
Other____________ (*specify*) [12]

____ 29, 30

12. In how many of the different areas listed in Question 11 have you worked on a full-time basis during your professional career? ______

____ 32

13. How many different full-time jobs at companies or organizations other than your present employer have you held in the past 10 years? ______

14. The following is a list of subject areas taught at the college:

(01) Accountancy
(02) Art
(03) Black & Hispanic Studies
(04) Computers
(05) Economics & Finance
(06) Education
(07) English Composition
(08) English Literature
(09) Foreign Languages
(10) History
(11) Law
(12) Management
(13) Marketing
(14) Mathematics
(15) Music
(16) Natural Sciences
(17) Philosophy & Religion
(18) Physical & Health Education
(19) Political Science
(20) Psychology
(21) Public Administration
(22) Secretarial Studies
(23) Sociology & Anthropology
(24) Speech
(25) Statistics
(26) Other ____________ (*specify*)

____ 34, 35

A. Which subject area taken at the college have your found to be *most valuable in your career*? ____________

____ 37, 38

B. Which subject area taken at the college have you found to be *most valuable in your daily nonworking activities*? ____________

____ 40, 41, 42

15. What was your cumulative college grade-point index? ______

____ 44

16. In retrospect, if you had to make the choice again, would you attend the college?
Yes [1] No [2] Not Sure [3]

____ 46

17. Assume a close friend or relative came to you for college advice today. Would you:
Highly advise against the college? [1]
Not recommend the college? [2]
Be undecided? [3]
Recommend the college? [4]
Highly recommend the college? [5]

____ 48

18. What is your sex? Male [1] Female [2]

____ 50, 51

19. What is your height (in inches)? ______

____ 53, 54

20. What is your date of birth? ____/____/____
Month Day Year

____ 56, 57, 58

(Respondent Number)

2.4.5 Testing the Questionnaire Instrument

Once the president and the statistician have discussed the pros and cons of each question, the instrument is properly organized and made ready for **pilot testing** so that it may be examined for clarity and length. Pilot testing on a small group of subjects is an essential phase in conducting a survey. Not only will this group of individuals be providing an estimate of the time needed for responding to the survey, but they will also be asked to comment on any perceived ambiguities in each question and to recommend additional questions. Once the president and the statistician have evaluated these results, changes are made. If time and budget permit, a second pilot study can be undertaken on a fresh sample of respondents to further improve the document.

Figure 2.5 on pages 15 and 16 depicts the questionnaire devised by the president and the statistician in its final form.

Problems

2.11 Why might you expect a greater response rate from a survey conducted by personal or telephone interview than from one conducted using a mailed questionnaire instrument?

2.12 Why would you expect a survey conducted by personal or telephone interview to be more costly than one conducted using a mailed questionnaire instrument?

2.13 Suppose that the A. C. Nielsen Company wanted to conduct a survey throughout the New York metropolitan area primarily to determine the amount of television viewing per household on Mondays. Describe both the population and the sample of interest to the A. C. Nielsen Company and describe the type of data that it primarily wishes to collect.

2.14 Develop a first draft of the questionnaire needed in Problem 2.13 by writing a series of three qualitative questions and three quantitative questions that you feel would be appropriate for this survey.

2.15 Write a survey which could be used to measure political orientation. Include not only questions about the person's political beliefs and preferences, but also personal characteristics (such as gender) which might be related to them. (Since this is for practice, don't make the survey too long—under ten questions should do it. Concentrate on writing good items.)

2.16 Write a brief survey which could be used to measure attitudes toward environmental issues (nuclear power, water purity, and so on).

2.17 Describe the strong and weak points of the following ways of wording a questionnaire item.

(i) What is your religion? ____________

(ii) Do you consider yourself to be a(n)
(a) Christian (b) Jew (c) other

(continued on page 18)

(Problem 2.17 continued)

(iii) What is your religion?

(a)	Episcopalian	(i)	Lutheran (L.C.A.)
(b)	Evangelical	(j)	Anglican
(c)	Roman Catholic	(k)	Hindu
(d)	Baptist	(l)	Buddhist
(e)	Jewish (Reform)	(m)	Moslem
(f)	Jewish (Conservative)	(n)	Shinto
(g)	Jewish (Orthodox)	(o)	none (atheist)
(h)	Lutheran (Missouri Synod)	(p)	none (agnostic)
		(q)	other

(a) Does the level of detail desired vary with the purpose of the survey? Illustrate your answer using the above questions.

(b) Of (i), (ii), and (iii), to which do you feel the most comfortable responding? Explain.

2.18 Write a question asking how much education a person has; write three versions of the question which give different levels of detail. Describe situations in which each might be appropriate or inappropriate to use.

2.5 CHOOSING THE SAMPLE SIZE FOR THE SURVEY

Since the president and the statistician had already determined that a mail survey was the most economical way of obtaining their desired information, it was then necessary for them to determine the appropriate sample size for the study. Rather than taking a complete census, statistical sampling procedures have become the preferred tool in most survey situations. There are three main reasons for drawing a sample. First of all, it is usually too time consuming to perform a complete census. Second, it is too costly to do a complete census. Third, it is just too cumbersome and inefficient to obtain a complete count of the target population. (See Section 2.7.) Thus the president and the statistician decided that it was their goal to make inferences about the entire population of full-time employed alumni members based on the results obtained from the sample.

After we determine the most essential quantitative and qualitative questions in a survey, the sample size needed will be based on satisfying the question with the most stringent requirements. In our case, the president and the survey statistician have determined that Questions 7 and 17 are the most essential quantitative and qualitative questions, respectively. However, calculation of the sample size required for a given survey is a matter that will be examined more appropriately in Chapter 9. As we shall see, the required sample size is 200 graduates out of a population of 9,660 full-time currently employed members of the college alumni association. However, because not everyone will respond when mailed the questionnaire, we must send it to a larger number of alumni. If we expect that only two out of three alumni will return it (i.e., a rate of return of 67%), then we will mail the survey to 300 alumni so that we can expect 200 responses.

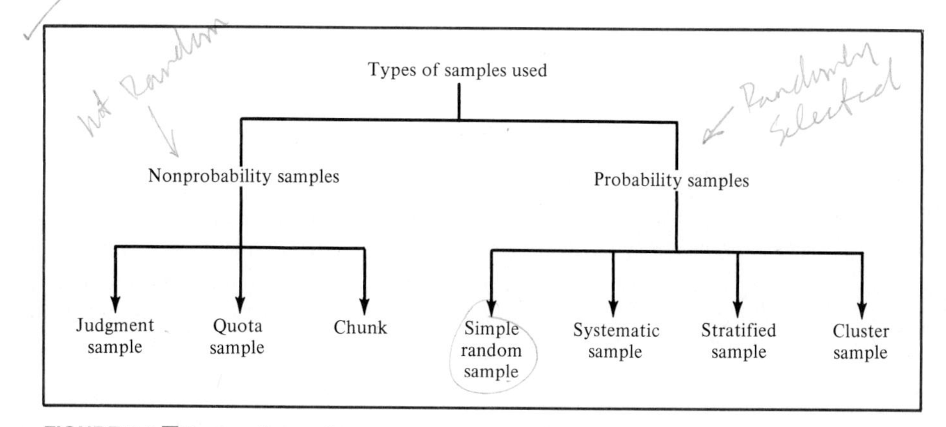

FIGURE 2.6 □ Types of samples.

2.6 TYPES OF SAMPLES

As depicted in Figure 2.6, there are basically two kinds of samples: the nonprobability sample and the probability sample. **Nonprobability samples** are not selected randomly, so there is no way to know how well they represent the overall population. They are used because they are less costly and time consuming to obtain than probability samples, and all too often it is assumed that the data are representative. Some typical procedures of nonprobability sampling are judgment sampling, quota sampling, and chunk sampling; these are discussed in detail in specialized books on sampling methods.

The only correct way in which the researcher can make statistical inferences from a sample to a population and interpret the results probabilistically is through the use of a probability sample. A **probability sample** is one in which the subjects of the sample are chosen on the basis of known probabilities. The four types most commonly used are the **simple random sample,** the **systematic sample,** the **stratified sample,** and the **cluster sample.** The simple random sample is one in which every subject has the same chance of selection as every other subject, and in which the selection of one subject does not affect the chances that any other subject is chosen. Moreover, a simple random sample may also be construed as one in which each possible sample that is drawn had the same chance of selection as any other sample. A detailed discussion of systematic sampling, stratified sampling, and cluster sampling procedures can be found in References 1 through 4.

2.7 DRAWING THE SIMPLE RANDOM SAMPLE

In this section we will be concerned with the process of selecting a simple random sample. Although it is not the most economical or efficient of the probability sampling procedures, it provides the base from which the more sophisticated sampling procedures have evolved.

The key to proper sample selection is obtaining and maintaining an up-to-date list of all the subjects from which the sample will be drawn. Such a list is known as the **population frame.** This population listing will serve as the **target population.** If many different probability samples were to be drawn from such a list, hopefully each sample would be a miniature representation of the population and yield reasonable estimates of its characteristics. If the listing is inadequate because certain groups of subjects in the population were not properly included, the random probability samples will only be providing estimates of the characteristics of the *target population*—not the *actual population*—and biases in the results will occur.

To draw a simple random sample of size 300 one could conceivably record the names of each of the 9,660 members in the population on a separate index card, place the index cards in a large fish bowl, thoroughly mix the cards, and then randomly select the 300 sample subjects from the fish bowl. Naturally, in order to ensure that each person (as well as each sample) has the same chance of being selected, all the index cards used must be of the same dimensions.

2.7.1 Sampling With or Without Replacement from Finite Populations

Two basic methods could be used for selecting the sample: the sample could be obtained *with replacement* or *without replacement* from the finite population. The method employed must be clearly stated by the survey statistician, since various formulas subsequently used for purposes of statistical inference are dependent upon the selection method.[2]

Let N represent the population size. When sampling **with replacement,** the chance that any particular subject in the population, say Susan Rubin, is selected on the first draw from the fish bowl is $1/N$ (or 1/9,660 for our example). Regardless of whoever is actually selected on the first draw, pertinent information is recorded on a master file, and then the particular index card is *replaced* in the bowl (sampling *with* replacement). The N cards in the bowl are then well shuffled and the second card is drawn. Since the first card had been replaced, the chance for selection of any particular subject, including Susan Rubin, on the second draw—regardless of whether or not that individual has been previously selected—is still $1/N$. Again the pertinent information is recorded on a master file and the index card is replaced in order to prepare for the third draw. Such a process is repeated until the desired sample size is obtained. Thus, when sampling with replacement, every subject on every draw will always have the same 1 out of N chance of being selected.

But should the researcher want to have the same individual subject possibly selected

[2] It is interesting to note that whether sampling with replacement from finite populations or sampling without replacement from infinite populations (such as some continuous, ongoing production process), the formulas used are the same.

more than once? When sampling from human populations, it is generally more appropriate to have a sample of different subjects than to permit repeated measurements of the same subject. Thus the researcher would employ the method of sampling **without replacement,** whereby once a subject is drawn, the same subject cannot be selected again. As before, when sampling without replacement, the chance that any particular subject in the population, say Susan Rubin, is selected on the first draw from the fish bowl is $1/N$. Whoever is selected, the pertinent information is recorded on a master file and then the particular index card is set aside rather than replaced in the bowl (sampling *without* replacement). The remaining $N - 1$ cards in the bowl are then well shuffled, and the second card is drawn. The chance that any individual not previously selected will be selected on the second draw is now 1 out of $N - 1$. This process of selecting a card, recording the information on a master file, shuffling the remaining cards, and then drawing again continues until the desired sample is obtained.

"Fish bowl" methods of sampling, while easily understandable, are not very efficient. Less cumbersome and more scientific methods of selection are desirable. One such method utilizes a **table of random numbers** (see Table C.1) for obtaining the sample.

2.7.2 Using a Table of Random Numbers

Such a table consists of a series of digits randomly generated (by electronic impulses) and listed in the sequence in which the digits were generated (see Reference 5). Since our numeric system uses 10 digits (0, 1, 2, . . . , 9), the chance of randomly generating any particular digit is equal to the probability of generating any other digit. This probability is 1 out of 10. Hence if a sequence of 500 digits were generated, we would expect about 50 of them to be the digit 0, 50 to be the digit 1, and so on. In fact, researchers who use tables of random numbers usually test such generated digits for randomness before employing them. Table C.1 has met all such criteria for randomness. Since every digit or sequence thereof in the table is random, we may use the table by reading either horizontally or vertically. The margins of the table designate row numbers and column numbers. The digits themselves are grouped into sequences of five for the sole purpose of facilitating the viewing of the table.

To use such a table in lieu of a fish bowl for selecting the sample, it is first necessary to assign code numbers to the individual members of the population. Suppose that the president's file contained a listing, in alphabetical sequence, of all $N = 9{,}660$ currently enrolled, full-time employed, dues-paying members. Since the population size (9,660) is a four-digit number, each assigned code number must also be *four* digits, so that every individual has an equal chance for selection. Thus a code of 0001 is given to the first individual in the population listing, a code of 0002 is given to the second individual in the population listing, . . . , a code of 0875 is given to the 875th individual in the population listing, and so on until a code of 9660 is given to the Nth individual in the listing.

In order to select the random sample, a random starting point for the table of random numbers (Table C.1) must be established. One such method is to close one's eyes and strike the table of random numbers with a pencil. Suppose the statistician utilized such a procedure and thereby selected row 06, column 01, as the starting point. Reading from left to right in sequences of four digits without skipping, groupings are obtained from Table 2.1.

Since $N = 9{,}660$ is the largest possible coded value, all four-digit code sequences greater than N (9661 through 9999 and 0000) are discarded. Hence the individual with code number 0233 is the first person in the sample (row 06 and columns 05

TABLE 2.1 □ Using a table of random numbers

		Column						
	00000	00001	11111	11112	22222	22223	33333	33334
Row	**12345**	**67890**	**12345**	**67890**	**12345**	**67890**	**12345**	**67890**
1	49280	88924	35779	00283	81163	07275	89863	02348
2	61870	41657	07468	08612	98083	97349	20775	45091
3	43898	65923	25078	86129	78496	97653	91550	08078
4	62993	93912	30454	84598	56095	20664	12872	64647
5	33850	58555	51438	85507	71865	79488	76783	31708
Begin Selection⟶ (Row 06, Column 01) 6	97340	23364	88472	04334	63919	36394	11095	92470
7	70543	29776	10087	10072	55980	64688	68239	20461
8	89382	93809	00796	95945	34101	81277	66090	88872
9	37818	72142	67140	50785	22380	16703	53362	44940
10	60430	22834	14130	96593	23298	56203	92671	15925
11	82975	66158	84731	19436	55790	69229	28661	13675
12	39087	71938	40355	54324	08401	26299	49420	59208
13	55700	24586	93247	32596	11865	63397	44251	43189
14	14756	23997	78643	75912	83832	32768	18928	57070
15	32166	53251	70654	92827	63491	04233	33825	69662
16	23236	73751	31888	81718	06546	83246	47651	04877
17	45794	26926	15130	82455	78305	55058	52551	47182
18	09893	20505	14225	68514	46427	56788	96297	78822
19	54382	74598	91499	14523	68479	27686	46162	83554
20	94750	89923	37089	20048	80336	94598	26940	36858
21	70297	34135	53140	33340	42050	82341	44104	82949
22	85157	47954	32979	26575	57600	40881	12250	73742
23	11100	02340	12860	74697	96644	89439	28707	25815
24	36871	50775	30592	57143	17381	68856	25853	35041
25	23913	48357	63308	16090	51690	54607	72407	55538
⋮	⋮	⋮	⋮	⋮	⋮	⋮	⋮	⋮

SOURCE: Reprinted from *A Million Random Digits with 100,000 Normal Deviates* by The RAND Corporation (New York: The Free Press, 1955). Copyright 1955 and 1983 by The RAND Corporation. Used by permission.

through 08) because the first four-digit code sequence (9734) in row 06 was discarded. The second individual selected has code number 6488 (row 06 and columns 09 through 12). Individuals with code numbers 4720, 4334, 6391, 9363, 9411, 0959, 2470, and 7054 are selected third through tenth, respectively.

The selecting process continues in a similar manner until the desired sample size of 300 individuals is obtained. During the selection process, if any four-digit coded sequence repeats, the person corresponding to that coded sequence would be included again as part of the sample if sampling with replacement; however, the repeating coded sequence is merely discarded if sampling without replacement. This was noted to first happen when the coded sequence 4205 appeared in row 12, columns 33 through 36, and then again in row 21, columns 21 through 24. Since the survey statistician is sampling without replacement, the repeating sequences were discarded and a sample of 300 unique individuals was obtained.

Problems

2.19 Draw a simple random sample of one-tenth of the people living in your dormitory (or neighborhood or apartment building, etc.). Explain the procedure you used.

2.20 For a study which would involve doing personal interviews with participants (rather than mail or phone surveys) tell why a simple random sample might be less practical than some other methods.

2.21 If you wanted to determine what proportion of novels published in the United States last year had themes based on sex or violence, how might you get a random sample to answer your question?

2.22 In a political poll to try to predict the outcome of an election, what is the population to which we usually want to generalize? How might we get a random sample from that population? From what you know about how such polls are actually conducted, what might be some problems with the sampling in these polls?

2.23 Suppose that I want to select a random sample of size 1 from a population of 3 items (which we can call A, B, and C). My rule for drawing the sample is: Flip a coin; if it is heads, pick item A; if it is tails, flip the coin again; this time, if it is heads choose B, if tails, choose C. Tell why this is a random sample, but not a simple random sample.

2.24 Suppose that a population has 4 members (call them A, B, C, and D). I would like to draw a random sample of size 2, which I decide to do in the following way: Flip a coin; if it is heads, my sample will be items A and B; if it is tails, the sample will be items C and D. While this is a random sample, it is not a simple random sample. Tell why. (If you did Problem 2.23, compare the procedure described there with the procedure described in this problem.)

2.25 For a population list containing $N = 902$ individuals, what code number would you assign for:

(a) The first person on the list?
(b) The fortieth person on the list?
(c) The last person on the list?

2.26 For a population of $N = 902$, verify that by starting in row 05 of the table of random numbers (Table C.1) only six rows are needed to draw a sample of size $n = 60$ *without* replacement.

2.8 OBTAINING THE RESPONSES

Now that the sample of 300 individuals has been selected, their responses must be obtained. For a mail survey sufficient time must be permitted for initial response. The questionnaire and any set of instructions should be mailed with a covering letter and with a self-addressed stamped envelope for the convenience of the respondent.

The covering letter should be brief and to the point. It should state the goal or purpose of the survey, how it will be used, and why it is important that selected subjects promptly respond. Moreover, it should give any necessary assurance of respondent anonymity and offer a note of appreciation for the respondent's time.

2.8.1 The Covering Letter

A draft of the covering letter used by the president of the alumni association is displayed in Figure 2.7 on the next page. The final version, of course, would be typed on the (official) alumni association stationery.

Problems

2.27 Write a draft of the cover letter needed for the television viewing survey developed in Problems 2.13 and 2.14 on page 17.

2.28 Write a draft of the cover letter needed for the political orientation survey developed in Problem 2.15 on page 17.

2.29 Suppose the political orientation survey developed in Problem 2.15 was not to be delivered by mail. Write a script for use in person or over the phone.

2.30 Repeat Problem 2.29 for the environmental issues survey you wrote for Problem 2.16 on page 17.

2.9 DATA PREPARATION: EDITING, CODING, AND TRANSCRIBING

Once the set of data is collected, it must be carefully prepared for tabular and chart presentation, analysis, and interpretation. The editing, coding, and transcribing processes are extremely important. Responses are scrutinized for completeness and for errors. If necessary, response validation is obtained by recontacting individuals

April 10, 1987

Dear Alumnus:

I am writing this letter to ask your cooperation and participation in a survey being conducted by the College Alumni Association.

The purpose of the survey is to provide alumni profile information concerning both your academic experience at the college and your employment history after graduation. You have been randomly selected as one of the participants in this survey. YOUR opinions and YOUR answers—both good and bad—are extremely important in making these results statistically valid. You can be assured that the responses obtained shall remain strictly anonymous.

I would greatly appreciate it if you could find the few minutes needed to complete this questionnaire and send it back as soon as possible in the enclosed self-addressed, stamped envelope. The results of the survey will be published in Alumni Reports.

Sincerely,

Robert Shurgot
President
College Alumni Association

FIGURE 2.7 □ A draft of the covering letter.

whose answers appear inconsistent or unusual. In addition, responses to open-ended questions are properly classified or scored, while responses to both quantitative and qualitative questions are coded for data entry.

Figure 2.8 represents the responses of Jackie Daniels, file code identification number 0233, who was the first subject selected in the sample.[3]

[3] To facilitate data entry, each individual's four-digit identification number (obtained from the alphabetical listing in the alumni file) is replaced by his/her corresponding respondent number, which specifies position in the sample-selection process. For example, the first person selected, Jackie Daniels (file code number 0233), has respondent number 1.

Question	*Type of Question*	*Computer Code*	*Columns Allocated for Data Entry*	*Jackie Daniels' Responses*	*Coded Response*
1	Years since graduation	YEARS	1, 2	29	29
2	Major	MAJOR	4	Science & Mathematics	3
3	Primary reason for attending	REASON	6	Other (Athletics)	5
4	Current ideology	PHIL	8	More conservative	1
5	Graduate school	GRAD	10	Yes	1
6	Relationship of job to major	JOBREL	12	2	2
7	Current annual salary	SALARY	14, 15, 16	$91,000	091
8	Type of organization	ORGANIZ	18	Under 100 employees	2
9	Current job description	JOBCAR	20, 21	Administration/ Management	02
10	No. years in area	YRSWRK	23, 24	25	25
11	First full-time job description	FIRSTJOB	26, 27	Sales	09
12	No. of different full-time areas worked in career	AREAS	29, 30	3	03
13	No. of other full-time jobs in past 10 years	DIFFJOBS	32	1	1
14A	Most valuable course in career	SUBJCAR	34, 35	Other (Business Ethics)	26
14B	Most valuable course in daily life	SUBDAILY	37, 38	Art	02
15	Grade-point index	GPI	40, 41, 42	2.7	2.7
16	Choosing college again	ATTEND	44	No	2
17	Recommendation level	ADVICE	46	Not recommend	2
18	Sex	SEX	48	Female	2
19	Height	HEIGHT	50, 51	63 inches	63
20	Age	AGE	53, 54	Nov. 13, 1935	51
—	Respondent Number	RESPNO	56, 57, 58	—	001

FIGURE 2.8 □ Coding the questionnaire responses of Jackie Daniels, file code identification number 0233.

Notice how the questionnaire responses are coded for data entry. Most qualitative questions require a one-digit code, such as observed with Question 2, "In what area was your undergraduate major?" Jackie Daniels majored in science and mathematics, and this response is given a code of 3. For quantitative questions, however, the number of spaces to allocate for a response must be based on the most extreme answers possible. For example, in Question 12 two spaces are required, because an alumnus could have worked in more than nine different areas. However, since Jackie Daniels has worked in only three, a coded value of 03 is recorded.

Whether we are entering our data on a portable terminal or on a personal computer, it is often desirable to leave one blank space between responses to the various questions. Not only is the data format more esthetically pleasing, but, even more important, it is easier to spot typographical errors. In our particular example, the coded responses for each subject take 58 spaces (including blanks). The responses

```
29 3 5 1 1 2 091 2 02 25 09 03 1 26 02 2.7 2 2 2 63 51 001
```

FIGURE 2.9 □ Data entered on a computer terminal for the responses of Jackie Daniels, #0233.

from Jackie Daniels are entered on a computer terminal, as depicted in Figure 2.9.[4]

Figure 2.10 on pages 28–31 is a printout of the data. This printout corresponds to the responses collected from the 200 alumni members who returned the questionnaire. We note that Jackie Daniels' responses appear first, since she was the first person selected into the sample.

Problems

2.31 Code the following responses for data entry:

(a) Height: 5 feet 2 inches	________	inches
(b) Weight: 97.5 pounds	________	pounds
(c) Date of birth: June 27, 1958	________	years

2.32 For each case in Problem 2.31 describe the rules you used for coding. What alternatives could you have considered?

2.10 DATA COLLECTION: A REVIEW AND A PREVIEW

Once data have been collected—whether from a published source, observational study, designed experiment, or from a survey such as we have just described—they must be organized and prepared to aid the researcher in making various analyses. In the following three chapters, various summary measures useful for data analysis and interpretation will be described, methods of tabular and chart presentation will be demonstrated, and various exploratory methods of evaluating the results will be considered.

[4] As we shall observe in Chapter 15, when using certain statistical software packages, data may be entered in either a free-format or fixed-format style. In the former, only one blank space is needed between the responses to the various questions; in the latter, specific columns are allocated for the responses to each of the questions.

YEARS	MAJOR	REASON	PHIL	GRAD	JOBREL	SALARY	ORGANIZ	JOBCAR	YRSWRK	FIRSTJOB	AREAS	DIFFJOBS	SUBJCAR	SUBDAILY	GPI	ATTEND	ADVICE	SEX	HEIGHT	AGE	RESPNO
29	3	5	1	1	2	91	2	2	25	9	3	1	26	2	2.7	2	2	2	63	51	1
21	3	5	3	1	1	23	3	11	10	12	2	1	11	11	2.6	1	3	1	72	56	2
10	1	3	1	1	3	59	1	12	6	1	2	5	1	1	2.7	2	1	1	68	31	3
25	2	5	2	2	3	69	1	9	30	5	2	1	7	8	2.7	1	3	1	72	53	4
6	1	4	2	2	5	23	2	1	9	1	5	4	1	19	3.1	1	5	2	62	30	5
27	3	3	2	1	5	52	3	10	27	10	3	1	25	1	2.6	2	3	2	64	48	6
10	2	2	2	2	3	32	1	3	19	3	1	1	1	11	2.6	1	4	1	71	38	7
42	1	4	1	2	5	61	5	1	43	1	1	0	1	8	2.4	1	4	1	74	62	8
32	1	2	1	1	5	67	1	1	25	1	1	3	1	15	3.2	1	4	1	69	54	9
10	1	2	2	1	2	42	2	9	1	3	5	2	1	12	3.0	1	4	2	64	38	10
12	1	2	2	2	5	46	2	1	11	1	1	1	1	8	2.9	3	3	1	67	35	11
34	2	4	1	2	1	34	1	5	34	5	1	1	13	8	2.7	3	2	1	72	59	12
16	1	2	2	2	5	53	1	2	5	1	5	3	1	7	3.4	1	4	1	68	38	13
17	1	2	1	1	5	36	1	1	29	1	1	3	1	8	3.2	1	5	1	69	48	14
9	2	2	2	2	4	163	1	6	15	6	2	1	20	20	2.3	1	3	2	62	40	15
11	3	4	1	1	1	29	1	4	3	12	2	3	12	19	3.2	1	5	2	64	37	16
13	3	3	3	1	5	31	4	10	13	2	2	2	25	15	3.3	1	3	1	66	35	17
5	1	4	1	1	5	25	1	1	5	1	1	3	1	8	2.9	1	5	1	71	25	18
15	3	1	2	2	2	32	4	2	15	6	1	0	20	20	3.9	1	4	1	68	55	19
13	1	4	2	1	4	36	4	1	13	1	1	1	1	20	3.0	1	4	2	65	36	20
10	1	3	1	2	5	35	5	1	11	1	1	1	1	4	2.5	1	3	1	70	34	21
22	3	2	2	2	1	44	5	9	20	9	1	1	1	8	3.2	3	3	1	69	62	22
5	2	3	2	2	1	20	1	12	1	9	1	4	13	26	2.4	2	4	1	72	30	23
22	3	3	1	1	4	57	1	10	29	10	10	1	5	20	2.8	1	4	2	64	54	24
34	1	5	2	2	5	49	2	1	33	1	1	2	1	7	3.5	1	4	1	66	58	25
7	2	4	2	1	5	65	2	9	7	9	2	1	13	11	3.1	1	4	1	74	29	26
18	1	2	1	1	3	82	1	2	8	1	2	4	1	10	3.4	1	3	1	66	40	27
15	2	2	2	2	4	82	2	3	15	3	2	3	5	11	3.2	2	2	1	72	37	28
13	1	3	2	2	1	43	1	1	13	1	1	4	1	19	2.7	1	4	2	66	36	29
13	1	4	2	2	4	27	4	1	9	1	1	2	1	20	2.5	1	4	1	67	35	30
15	3	4	1	2	1	20	5	9	2	12	3	2	7	19	2.9	1	5	1	70	36	31
6	2	2	2	2	1	25	1	3	3	5	2	2	1	3	2.1	1	4	2	61	36	32
30	1	2	1	1	5	52	1	1	37	1	3	2	1	9	2.5	1	4	1	68	64	33
44	2	2	2	1	5	47	5	12	29	12	6	1	5	8	2.5	1	5	1	70	64	34
12	1	4	1	1	4	76	5	2	5	1	4	3	1	24	2.8	1	5	2	63	37	35
12	2	2	2	2	1	24	4	12	5	12	3	2	1	26	2.3	2	2	1	74	33	36
6	2	4	2	2	1	28	1	1	6	1	1	2	1	12	2.8	1	5	1	70	29	37
11	1	4	1	2	4	42	4	2	12	1	2	2	1	26	2.0	1	5	1	69	45	38
13	2	3	1	1	5	49	4	2	7	2	1	1	5	5	2.9	3	3	2	65	35	39
44	1	2	2	1	2	56	3	11	38	1	1	1	1	10	3.0	1	5	1	71	69	40
19	2	2	2	2	5	185	5	5	23	5	1	1	13	13	2.7	1	5	1	66	42	41
24	2	4	1	2	3	34	1	2	32	2	2	0	5	10	2.6	1	5	1	70	64	42
29	1	5	1	2	5	76	1	1	26	1	1	1	1	8	3.9	1	3	1	71	51	43
5	2	4	2	2	5	42	2	5	5	5	2	4	13	20	2.9	1	4	2	63	28	44
5	2	3	2	2	4	27	4	3	5	3	1	1	5	15	2.3	2	3	2	63	26	45
6	2	5	1	2	4	31	1	3	18	3	1	2	5	14	2.5	1	5	1	71	41	46
34	2	2	2	2	5	67	1	5	34	5	5	2	13	8	3.4	2	1	1	74	57	47
27	1	2	1	2	5	70	2	1	24	1	1	1	1	10	2.3	1	2	1	73	50	48
19	2	4	1	1	4	61	1	5	17	5	1	5	5	26	2.2	1	4	2	69	41	49
23	1	4	2	2	3	97	1	2	10	1	3	1	1	1	2.4	1	3	1	70	49	50
20	3	4	1	1	5	56	5	12	18	12	2	4	20	13	2.8	1	4	1	65	42	51
24	1	3	2	2	3	81	1	2	5	1	2	3	1	20	2.4	1	5	1	72	46	52
46	1	2	1	1	3	46	2	5	40	1	4	1	1	8	3.4	1	5	2	62	67	53

FIGURE 2.10 □ Computer listing of responses to questionnaire from a sample of 200 currently enrolled, full-time employed, dues-paying members of the alumni association.

YEARS	MAJOR	REASON	PHIL	GRAD	JOBREL	SALARY	ORGANIZ	JOBCAR	YRSWRK	FIRSTJOB	AREAS	DIFFJOBS	SUBJCAR	SUBDAILY	GPI	ATTEND	ADVICE	SEX	HEIGHT	AGE	RESPNO
36	1	2	1	1	4	92	2	2	35	1	2	2	1	10	2.8	2	2	1	69	61	54
37	1	2	2	1	3	95	5	8	25	1	2	1	1	26	2.5	1	3	1	69	61	55
33	1	4	1	1	1	54	4	12	28	12	12	2	1	16	2.7	2	2	1	65	54	56
8	1	2	1	2	5	40	2	1	8	1	1	3	1	1	3.1	2	3	1	70	31	57
43	1	4	2	2	1	58	2	2	38	1	1	0	1	10	3.0	1	3	2	65	67	58
28	3	1	3	2	2	34	1	8	15	6	2	1	7	20	2.6	1	4	2	66	50	59
19	3	4	2	1	4	58	4	12	15	12	1	2	7	10	2.6	1	4	1	69	40	60
31	2	2	1	2	5	74	1	9	24	10	4	2	13	20	2.7	3	3	1	66	52	61
5	2	3	2	1	5	17	3	11	3	12	1	3	22	14	2.3	1	5	2	61	28	62
9	1	4	2	2	5	30	5	1	9	1	1	4	1	5	2.8	1	5	1	72	30	63
35	1	2	2	2	3	63	1	2	30	1	6	2	1	26	3.1	1	3	1	70	67	64
39	1	2	2	2	5	46	2	1	35	1	1	2	1	26	2.8	2	2	2	66	59	65
36	1	3	1	1	4	84	1	2	32	1	2	1	1	8	3.0	3	5	1	71	57	66
12	1	4	1	1	2	72	2	11	1	1	3	3	1	14	2.8	1	4	1	70	36	67
31	3	2	1	1	5	81	1	10	32	10	1	1	13	20	2.6	2	3	1	70	55	68
17	1	3	1	1	1	79	2	3	14	1	2	3	1	26	2.6	1	2	1	71	38	69
41	2	2	1	1	4	74	2	8	38	8	1	1	1	2	3.1	3	4	2	65	66	70
37	1	2	2	1	2	21	5	9	32	1	2	2	1	10	2.6	3	3	1	70	60	71
34	1	2	1	1	5	61	1	2	20	1	3	5	1	26	2.3	2	3	1	73	55	72
35	1	2	2	1	5	68	5	1	35	1	1	1	1	20	2.6	1	3	1	72	56	73
13	2	4	3	2	4	178	5	2	9	3	2	2	1	20	2.7	1	5	1	70	41	74
42	1	3	1	1	5	71	5	1	30	1	1	1	1	8	2.0	3	4	2	63	62	75
30	1	2	2	2	5	72	1	1	30	1	1	1	1	17	3.1	1	5	1	70	51	76
46	1	4	3	2	5	53	5	1	40	1	1	1	1	8	2.1	3	3	1	59	69	77
8	1	2	2	1	5	47	5	1	8	1	1	4	1	15	2.7	1	4	1	67	29	78
5	2	1	2	2	3	30	4	2	5	2	2	1	12	12	2.8	1	4	1	70	53	79
7	1	4	2	2	5	33	2	1	17	1	7	4	1	2	3.2	1	5	2	68	30	80
22	3	2	2	1	4	51	5	12	15	10	2	1	20	9	2.5	1	5	2	67	44	81
30	1	2	2	2	3	97	1	2	30	2	4	1	12	15	2.6	1	2	1	74	51	82
13	2	3	1	2	2	88	2	2	22	2	1	1	12	24	3.4	1	4	2	67	40	83
61	2	2	2	2	4	13	1	6	1	1	2	2	1	5	3.5	1	5	1	77	29	84
51	1	5	1	1	5	43	5	1	50	1	3	1	1	7	3.2	1	5	1	74	73	85
41	1	2	2	2	1	90	1	1	38	1	1	1	1	20	3.1	3	3	2	66	62	86
47	1	2	2	2	1	40	4	2	35	1	2	2	1	1	3.4	1	4	1	71	68	87
5	2	5	2	1	4	10	4	12	3	12	1	3	20	20	2.8	1	4	1	66	28	88
40	1	2	1	2	5	40	1	1	21	1	2	1	1	15	3.0	1	5	1	66	61	89
9	1	2	3	1	5	34	1	2	1	1	2	3	1	10	2.9	1	5	1	75	48	90
38	3	2	1	1	3	85	3	11	28	12	3	1	5	10	2.8	2	3	2	66	59	91
6	2	4	2	1	3	21	1	5	4	5	3	3	13	10	3.2	1	5	1	71	31	92
26	1	4	1	1	5	112	5	1	30	1	1	1	1	6	3.1	1	4	1	64	54	93
25	3	4	1	1	3	79	2	12	22	12	1	2	26	17	2.5	1	2	1	66	46	94
6	2	1	1	2	1	63	1	5	32	5	5	5	10	10	3.1	1	4	1	70	57	95
33	2	2	2	2	4	51	2	8	32	8	1	1	1	9	2.6	1	5	2	63	55	96
19	1	2	2	2	1	67	2	2	25	2	2	3	1	20	2.9	1	3	1	70	31	97
22	1	2	2	1	5	61	1	1	20	1	1	2	1	11	3.5	1	4	1	72	45	98
27	1	2	2	2	5	52	5	1	24	1	2	1	1	5	3.4	1	5	1	71	50	99
31	1	2	1	2	4	21	4	1	35	1	4	1	1	26	2.7	1	4	1	69	64	100
8	1	4	3	1	3	59	1	3	13	3	1	2	1	10	3.7	1	5	2	60	36	101
31	2	2	1	1	1	55	1	2	26	8	1	2	12	2	2.5	2	3	1	72	53	102
48	1	2	1	2	5	59	4	1	47	1	1	1	1	9	2.8	1	5	1	69	69	103
44	3	2	1	2	3	76	1	9	38	10	2	1	25	8	2.9	1	3	1	70	66	104
10	1	1	1	2	2	60	1	3	15	3	5	2	1	4	2.9	1	4	1	67	38	105
36	1	1	1	2	5	48	5	1	36	1	2	1	1	7	3.2	1	4	2	66	61	106

YEARS	MAJOR	REASON	PHIL	GRAD	JOBREL	SALARY	ORGANIZ	JOBCAR	YRSWRK	FIRSTJOB	AREAS	DIFFJOBS	SUBJCAR	SUBDAILY	GPI	ATTEND	ADVICE	SEX	HEIGHT	AGE	RESPNO
36	1	5	1	1	5	88	2	1	30	1	2	1	1	7	3.0	1	4	1	71	59	107
28	2	2	2	2	5	83	1	5	30	5	3	2	26	20	3.2	1	4	1	66	56	108
9	1	2	1	2	5	30	1	1	20	1	2	2	1	11	3.0	1	5	1	71	47	109
41	1	2	1	2	5	27	4	1	15	5	2	1	1	15	3.8	1	5	2	65	70	110
37	1	2	2	1	4	31	4	1	31	1	1	1	1	2	3.4	1	4	1	70	62	111
37	1	2	2	2	5	34	2	4	40	1	2	1	1	9	2.6	1	3	1	68	64	112
44	1	2	3	1	5	34	2	1	30	1	1	1	1	14	2.8	1	4	1	68	66	113
5	1	5	3	1	5	17	1	1	5	1	1	5	1	11	2.5	1	5	2	60	43	114
41	1	2	1	2	5	71	5	1	41	1	1	1	1	26	2.2	1	2	2	63	62	115
31	1	2	1	1	4	81	1	2	16	1	2	1	1	5	3.3	1	3	1	65	51	116
18	3	2	2	1	2	43	1	4	18	4	4	1	25	10	3.0	3	3	1	71	50	117
22	2	2	2	2	4	44	4	2	20	2	2	2	12	17	2.7	1	4	1	67	43	118
20	1	2	1	1	5	71	5	1	20	1	1	4	1	24	3.4	3	3	1	70	42	119
36	1	2	1	1	5	71	2	1	35	1	1	1	1	9	3.4	1	3	2	64	56	120
32	2	2	2	1	5	44	1	5	40	9	2	1	13	1	3.2	1	5	1	70	62	121
23	1	4	1	1	1	98	1	1	15	1	1	1	1	26	3.1	3	3	1	67	45	122
28	1	3	2	1	5	98	2	3	34	3	3	1	1	8	2.2	1	5	1	70	60	123
19	1	1	1	1	5	75	1	3	26	3	1	1	1	24	2.8	1	4	1	63	46	124
11	1	4	2	2	4	59	1	3	15	3	1	3	5	19	2.7	1	4	2	60	45	125
17	2	2	3	1	4	47	3	2	20	2	4	1	12	26	2.8	3	3	1	74	62	126
21	1	2	1	1	3	78	5	12	10	12	1	1	1	26	2.9	1	2	1	67	44	127
44	1	2	1	2	3	61	2	2	30	1	3	1	1	1	3.7	3	2	1	66	71	128
5	2	3	2	2	3	27	1	7	4	7	1	3	12	5	3.1	1	4	2	64	34	129
30	2	2	1	2	5	78	1	12	30	5	1	0	13	7	3.3	3	2	2	67	53	130
7	2	4	1	1	2	11	1	8	1	10	2	2	13	20	2.2	3	4	1	73	29	131
15	1	3	2	1	5	77	1	1	15	1	1	1	1	17	3.1	3	3	1	70	37	132
14	1	4	2	2	5	28	4	1	26	1	3	2	1	8	2.8	1	4	1	64	62	133
8	3	1	1	2	3	26	1	2	7	2	2	0	7	20	2.4	1	4	1	66	31	134
36	1	2	1	1	5	51	5	1	36	1	1	0	1	10	2.0	2	2	2	62	59	135
11	2	3	3	1	4	21	4	2	10	1	2	3	21	8	3.0	1	3	2	66	36	136
41	1	2	1	2	2	98	1	3	14	1	1	3	1	5	2.7	3	3	1	67	61	137
37	1	2	3	1	5	47	4	1	38	1	1	2	1	8	2.9	3	3	1	72	62	138
13	1	4	1	1	4	34	1	1	6	1	4	3	1	12	2.5	1	4	1	75	35	139
8	1	4	1	2	5	47	2	1	8	1	1	2	1	26	2.4	1	5	1	68	30	140
27	2	5	2	1	4	51	1	3	25	3	3	3	1	10	3.0	3	3	2	61	56	141
14	1	4	2	1	5	41	1	1	14	1	1	0	1	15	3.0	1	4	1	68	35	142
18	2	4	3	1	5	123	1	9	9	3	2	1	5	5	2.2	2	4	1	70	39	143
7	1	4	3	1	4	31	4	1	20	1	1	4	1	20	3.2	3	4	1	71	46	144
5	2	4	3	1	3	21	2	9	2	12	3	3	12	7	3.0	2	2	1	71	28	145
9	2	4	2	2	3	36	1	2	9	2	2	3	12	9	2.4	1	5	2	63	34	146
23	2	4	1	1	1	89	5	9	6	3	2	4	5	13	2.6	1	5	1	74	51	147
44	1	2	1	1	5	76	5	1	46	1	1	0	7	24	2.4	2	2	1	71	69	148
5	1	4	1	1	3	25	4	1	5	1	1	1	1	10	2.9	1	4	1	71	27	149
26	1	5	1	2	5	62	5	1	20	1	4	1	1	8	3.0	1	4	1	72	54	150
36	1	2	1	2	5	75	5	1	38	1	1	1	1	10	3.3	3	3	2	63	59	151
36	1	2	2	2	5	58	2	1	38	1	1	2	1	5	3.1	1	4	1	69	62	152
35	1	3	2	1	5	47	5	12	26	1	2	0	1	10	2.8	1	2	1	67	58	153
43	2	2	2	2	4	53	5	12	26	5	5	2	13	8	2.7	1	3	1	71	64	154
25	3	2	1	1	1	41	1	1	20	6	4	2	1	19	2.7	3	3	1	71	53	155
15	1	3	3	1	5	68	4	1	15	1	3	3	1	1	3.0	1	3	2	69	35	156
8	2	3	3	1	5	53	1	5	16	5	3	2	13	19	2.5	1	1	1	74	38	157
15	1	4	2	1	3	48	3	11	4	12	2	3	1	5	2.9	1	4	1	73	36	158
42	1	2	2	2	5	42	5	1	39	1	1	1	1	5	3.5	1	4	1	72	63	159

YEARS	MAJOR	REASON	PHIL	GRAD	JOBREL	SALARY	ORGANIZ	JOBCAR	YRSWRK	FIRSTJOB	AREAS	DIFJOBS	SUBJCAR	SUBDAILY	GPI	ATTEND	ADVICE	SEX	HEIGHT	AGE	RESPNO
43	1	1	1	2	5	59	5	1	43	1	2	1	1	11	2.1	1	2	1	74	65	160
22	2	2	2	2	5	63	1	2	20	5	3	2	12	10	2.8	2	2	2	65	44	161
12	1	4	1	2	5	55	1	1	15	1	2	3	1	5	2.7	1	5	1	71	40	162
38	1	2	1	2	5	62	5	1	38	1	1	1	1	24	2.2	1	3	1	71	58	163
7	2	2	1	1	5	33	4	10	6	12	3	5	13	22	2.3	2	3	2	68	29	164
32	1	2	2	2	5	70	2	2	15	1	2	1	1	26	2.8	3	3	1	66	57	165
50	1	2	2	1	4	56	5	2	20	11	3	2	1	7	2.7	1	5	2	61	71	166
35	1	2	1	2	2	73	1	2	28	1	2	1	1	20	2.6	3	3	1	71	57	167
10	3	4	1	1	1	77	5	3	10	3	1	4	26	26	2.3	1	5	1	65	32	168
8	2	5	2	1	5	34	4	6	4	2	2	3	12	11	3.3	1	5	2	63	30	169
38	1	5	1	2	5	45	2	1	40	1	1	2	1	5	2.9	1	4	1	71	59	170
10	2	4	1	2	4	31	1	9	10	9	1	1	13	2	3.6	1	5	1	70	32	171
14	1	3	1	1	5	41	3	11	12	11	11	3	1	20	3.0	1	4	2	66	35	172
6	1	2	1	2	4	24	2	1	6	1	1	3	1	26	3.0	2	2	1	71	30	173
48	1	4	2	1	4	32	2	2	30	1	2	1	1	8	3.0	1	4	1	68	68	174
56	1	4	1	1	1	70	5	1	55	5	2	1	1	11	2.7	2	3	1	71	77	175
8	3	4	2	2	3	25	1	2	8	2	2	3	9	10	2.7	1	3	2	65	42	176
45	2	2	2	1	2	74	5	12	32	10	3	1	1	20	3.1	3	3	1	71	67	177
33	1	2	1	2	5	99	1	1	30	12	1	1	1	10	2.8	1	3	2	66	54	178
25	3	3	3	1	4	17	5	2	4	12	5	2	1	20	2.7	1	5	2	68	46	179
20	1	2	2	2	5	91	5	1	20	1	1	1	1	10	2.8	2	5	1	71	42	180
14	1	3	2	2	4	37	4	2	2	1	2	2	1	20	2.4	1	5	1	73	58	181
34	1	2	1	1	5	91	1	1	30	1	1	1	1	8	2.3	2	2	1	69	55	182
16	1	2	1	2	5	75	2	1	14	1	1	2	1	20	2.6	1	5	2	58	39	183
11	1	2	2	2	5	36	4	1	13	1	1	1	1	2	2.9	3	3	2	62	62	184
35	2	5	1	1	1	71	5	2	30	5	4	1	1	9	3.0	3	2	1	75	57	185
11	1	2	1	1	4	64	1	1	8	1	1	5	1	26	3.3	3	2	1	68	37	186
11	1	4	2	1	5	36	5	1	11	1	1	5	1	17	3.1	1	5	1	70	33	187
8	2	5	2	1	2	52	5	12	5	12	1	2	1	17	2.7	1	4	2	65	40	188
42	1	2	1	2	5	39	4	1	37	1	1	1	1	15	3.3	2	1	1	69	63	189
41	1	2	2	2	5	79	1	1	40	1	1	1	1	2	2.7	1	3	1	67	61	190
21	2	3	1	2	5	57	1	9	21	9	3	1	1	5	2.6	1	5	2	63	47	191
26	2	2	3	2	1	43	2	1	9	2	2	2	1	20	3.5	2	2	1	69	48	192
9	2	3	1	2	2	25	4	1	7	1	3	3	1	5	2.9	1	4	1	73	38	193
42	1	2	1	1	4	34	4	1	35	1	1	2	1	17	2.4	1	4	1	72	63	194
13	1	3	1	1	5	52	2	1	13	1	2	5	1	10	2.7	1	4	1	67	35	195
5	2	3	1	2	1	44	2	4	6	2	1	4	12	26	2.5	1	4	2	66	30	196
6	1	4	2	2	5	38	2	1	5	1	1	1	1	20	2.6	2	4	1	71	28	197
10	1	1	2	2	3	36	2	8	30	8	8	2	1	5	2.7	1	1	1	69	50	198
6	2	4	1	1	5	21	2	1	5	3	2	2	1	24	2.7	1	5	2	66	31	199
36	2	3	2	1	5	63	1	5	30	10	2	1	25	20	3.6	1	4	1	73	58	200

Supplementary Problems

2.33 Determine whether each of the following random variables is qualitative or quantitative. If quantitative, determine whether the phenomenon of interest is discrete or continuous.

(a) Automobile ownership by students.
(b) Net weight (in grams) of packaged dry cereal.
(c) Political party affiliation of civil service workers.
(d) Number of bankrupt corporations per month in the United States.
(e) Useful lifetimes (in hours) of 100-watt light bulbs.
(f) Number of on-time arrivals per hour at a large airport.

2.34 Suppose the following information is obtained from Edson Wolfe on his application for a personal loan at the Donovan County Savings & Loan Association:

(1) Place of Residence: New York City
(2) Type of Residence: Condominium
(3) Monthly Payments: $1,875
(4) Date of Birth: October 30, 1947
(5) Occupation: Geo-chemist
(6) Employer: Jon Schwarz Builders, Inc.
(7) No. Years at Job: 7
(8) No. Jobs in Past Ten Years: 2
(9) Annual Salary: $55,000
(10) Other Income: $12,000
(11) Martial Status: Married
(12) No. Dependents: 3
(13) Collateral Offered: Auto
(14) Collateral Value: $7,500
(15) Other Loans: None
(16) Period of Desired Loan: 24 months

Classify each of the 16 responses by type of data and level of measurement.

2.35 Suppose that a plastic surgeon was interested primarily in studying the amount of satisfaction that former patients had with elective facial cosmetic surgery. A survey of 400 such patients was being planned.

(a) Describe both the population and sample of interest to the plastic surgeon.
(b) Describe the type of data that the plastic surgeon primarily wishes to collect.
(c) Develop a first draft of the questionnaire by writing a series of three qualitative questions and three quantitative questions that you feel would be appropriate for this survey.
(d) Write a draft of the cover letter needed for this survey.

2.36 Suppose that the manager of the Customer Services Division of RCA was interested primarily in determining whether customers who had purchased a video cassette recorder over the past twelve months were satisfied with their products. Using the warranty cards submitted after the purchase, the manager was planning to survey 1,425 of these customers.

(a) Describe both the population and sample of interest to the manager.
(b) Describe the type of data that the manager primarily wishes to collect.
(c) Develop a first draft of the questionnaire by writing a series of three qualitative questions and three quantitative questions that you feel would be appropriate for this survey.
(d) Write a draft of the cover letter needed for this survey.

• 2.37 Given a population of $N = 93$, draw a sample of size $n = 15$ *without* replacement by starting in row 29 of the table of random numbers (Table C.1). Reading across the row, list the 15 coded sequences obtained.

2.38 For a population of $N = 1{,}250$, a *two-stage* usage of the table of random numbers (Table C.1) may be recommended to avoid wasting time and effort. To obtain the sample by a two-stage approach, list the four-digit coded sequences after adjusting the first digit in each sequence as follows:
If the first digit is a 0, 2, 4, 6, or 8, change the digit to 0. If the first digit is a 1, 3, 5, 7, or 9, change the digit to 1. Thus, starting in row 07 of the random numbers table (Table C.1), the sequence 7054 becomes 1054, while the sequence 3297 becomes 1297, etc. Verify that only ten rows are needed to draw a sample of size $n = 60$ *without* replacement.

2.39 For a population of $N = 2{,}202$, a *two-stage* usage of the table of random numbers (Table C.1) may be recommended to avoid wasting time and effort. To obtain the sample by a two-stage approach, list the four-digit coded sequences after adjusting the first digit in each sequence as follows:
If the first digit is a 0, 3, or 6, change the digit to 0; if the first digit is a 1, 4, or 7, change the digit to 1; if the first digit is a 2, 5, or 8, change the digit to 2; if the first digit is a 9, discard the sequence. Thus, starting in row 07 of the random numbers table (Table C.1), the sequence 7054 becomes 1054, while the sequence 3297 becomes 0297, etc. Verify that only ten rows are needed to draw a sample of size $n = 60$ *without* replacement.

2.40 For a population of $N = 4{,}202$, what adjustment can you suggest for a two-stage usage of the random numbers table so that each of the 4,202 individuals has the same chance of selection? (See Problems 2.38 and 2.39.)

2.41 Develop a set of five quantitative and five qualitative questions that were not included in the questionnaire (Figure 2.5) and which might also have been of interest to the president of the alumni association.

• 2.42 The following computerized output is extracted from a set of data similar to those collected in the alumni association survey (Figure 2.10). However, each of the five lines, representing the respective responses of five particular individuals, has *at least one error* in it. Use the coded statements shown in Figures 2.5 and 2.8 to determine the particular recording errors in each of the five responses.

```
15  6  1  1  1  3  034  1  04  07  04  01  2  01  10  3.2  1  1  1  70  30  201
11  3  2  2  1  2  027  3  11  05  11  01  1  02  09  2.8  2  1  1  08  29  202
07  2  7  1  2  1  048  1  03  05  03  01  2  03  08  3.5  3  1  1  71  33  203
02  1  3  1  1  2  039  2  09  02  09  02  2  04  07  5.5  2  1  2  67  36  204
04  1  4  4  2  5  055  1  01  13  01  01  2  05  06  2.9  1  1  2  64  37  205
```

2.43 Suppose that the American Kennel Club (AKC) was planning to survey 1,500 of its club members primarily to determine the percentage of its membership that currently own more than one dog. Describe both the population and sample of interest to the AKC and describe the type of data that the AKC primarily wishes to collect.

2.44 Develop a first draft of the questionnaire needed in Problem 2.43 by writing a series of three qualitative questions and three quantitative questions that you feel would be appropriate for this survey.

References

1. Cochran, W. G., *Sampling Techniques*, 3d ed. (New York: Wiley, 1977).
2. Deming, W. E., *Sample Design in Business Research* (New York: Wiley, 1960).

3. FRANKEL, M. R., AND L. R. FRANKEL, "Some Recent Developments in Sample Survey Design," *Journal of Marketing Research*, vol. 14, August 1977, pp. 280–293.
4. HANSEN, M. H., W. N. HURWITZ, AND W. G. MADOW, *Sample Survey Methods and Theory*, Vols. I and II (New York: Wiley, 1953).
5. RAND CORPORATION, *A Million Random Digits with 100,000 Normal Deviates* (New York: Free Press, 1955).

Describing and Summarizing Data

3.1 INTRODUCTION: WHAT'S AHEAD

In the preceding chapter we learned how to obtain data through survey research. How do we make sense out of such collected information? What are the data telling us? How can the results ultimately be used by us? By management? By government? It is obvious that collecting data is only one aspect of the field of descriptive statistics. In this and the following two chapters we shall examine the other aspects: the description, summarization, presentation, analysis, and interpretation of the data.

The goals of this chapter are:

1. To develop an understanding of the properties of data and the various summary measures computed from them.
2. To be able to appropriately distinguish between the use of the various summary measures.

In order to introduce the relevant ideas of this chapter let us consider an illustrative case.

Case B—The Heldman Township Comptroller's Problem

Suppose that to obtain revenue information for the proposed state budget, the governor's office has requested the comptroller for Heldman Township to prepare a preliminary report, which must project the economic activities of small businesses in the township over the next year. Heldman Township contains (a population of) 50 small-type stores, establishments, shops, and restaurants—businesses which are affiliated with the local branch of the Small Business Association (SBA).

With the assistance of the president of the local SBA the comptroller takes a random sample of 7 such small businesses and, among other things, requests from the owners of these businesses a projection (in thousands of dollars) of the sales tax revenues they expect to be collecting during the present quarterly period. The results follow:

Business #	Projected Quarterly Sales Tax Receipts (in $000)
1	9.0
2	13.0
3	12.0
4	7.0
5	6.0
6	11.0
7	12.0

What can be learned from such data that will assist the comptroller with the preliminary report?

3.2 EXPLORING THE DATA

Based on this sample, we should be able to agree immediately on four points.

1. The data are in **raw form.** That is, the collected data seem to be in a random sequence with no apparent pattern to the manner in which the individual observations are listed.
2. The value of 12.0 thousands of dollars in sales tax receipts was more frequently projected for the current quarterly period for such small businesses than any other value.
3. The spread in the projections from store to store ranges from 6.0 to 13.0 thousands of dollars.

4. There do not appear to be any unusual or extraordinary sales tax values projected by any of the stores in the sample. The projected values are 6.0, 7.0, 9.0, 11.0, 12.0, 12.0, 13.0—not, for instance 6.0, 7.0, 9.0, 11.0, 12.0, 12.0, 31.0. Extreme observations (such as the 31.0 given in the example above) are called **outliers.**[1]

If the comptroller had asked us to examine the data and present a short summary of our findings, then comments similar to the four above are basically all that we could be expected to produce without formal statistical training. But what we have done is to provide both an elementary descriptive statistical analysis and interpretation of the given data. An *analysis* is **objective;** we should all agree with these findings. An *interpretation* is **subjective;** we may form different conclusions when interpreting our analytical findings. On this basis, points 2 through 4 are based on analysis, while point 1 is an interpretation. With respect to the latter, no formal analytical test (such as the runs test described in Chapter 14) was made—it is simply our conjecture that there is no pattern to the sequence of collected data. Moreover, our conjecture would seem to be appropriate if the sample of seven businesses was randomly and independently drawn from the population listing. Through survey research, whether collected by mail, by telephone, or by personal interview, the observed responses are usually obtained in a random order—as would be the case here.

Let us now see how we could add to our understanding of what the data are telling us by more formally examining the three properties of quantitative data.

3.3 PROPERTIES OF QUANTITATIVE DATA

The three major properties which describe a set of numerical data are:

1. Central tendency
2. Dispersion.
3. Shape.

In any analysis and/or interpretation, a variety of descriptive measures representing the properties of central tendency, dispersion, and shape may be used to extract and summarize the salient features of the data set. If these descriptive summary measures are computed from a sample of data, they are called **statistics;** if they are computed from an entire population of data, they are called **parameters.** Since statisticians usually take samples rather than use entire populations, our primary emphasis in this text is on statistics rather than parameters.

[1] Outliers, which may occur as true data values or as a consequence of errors, can have an undue influence on a statistical analysis. Because of this, we should always be aware of whether there are any outliers in a data set we are analyzing. Later in the book, we will see that one strategy for dealing with outliers is to use statistical methods which are less sensitive to their effects.

3.4 MEASURES OF CENTRAL TENDENCY

Most sets of data show a distinct tendency to group or cluster about a certain "central" point. Thus for any particular set of data, it usually becomes possible to select some typical value to describe the entire set. Such a descriptive typical value is a measure of central tendency or "location."

Four common measures of central tendency are the arithmetic mean, the median, the mode, and the midrange.

3.4.1 The Arithmetic Mean

The **arithmetic mean** (also called the **mean** or **average**) is the most commonly used measure of central tendency. It is calculated by summing all the observations in a set of data and then dividing the total by the number of items involved.

INTRODUCING ALGEBRAIC NOTATION

Thus, for a sample containing a set of n observations $X_1, X_2, \ldots, X_n$, the arithmetic mean (given by the symbol $\overline{X}$—called "X bar") can be written as

$$\overline{X} = \frac{X_1 + X_2 + \cdots + X_n}{n}$$

To simplify the notation, for convenience the term ΣX (meaning the sum of all the X values) is conventionally used whenever we wish to add together a series of observations.[2] That is

$$\Sigma X = X_1 + X_2 + \cdots + X_n$$

Using this summation notation, the arithmetic mean of the sample can be more simply expressed as

$$\overline{X} = \frac{\Sigma X}{n} \tag{3.1}$$

where
$\overline{X}$ = sample arithmetic mean
n = sample size
ΣX = sum of all of the X values in the sample

[2] See Appendix A, Section A.3, for a discussion of rules pertaining to summation notation.

For our comptroller's example,

$X_1 = 9.0$ (projected quarterly sales tax revenues from Business #1)
$X_2 = 13.0$ (projected quarterly sales tax revenues from Business #2)
$X_3 = 12.0$ (projected quarterly sales tax revenues from Business #3)
$X_4 = 7.0$ (projected quarterly sales tax revenues from Business #4)
$X_5 = 6.0$ (projected quarterly sales tax revenues from Business #5)
$X_6 = 11.0$ (projected quarterly sales tax revenues from Business #6)
$X_7 = 12.0$ (projected quarterly sales tax revenues from Business #7)

The arithmetic mean for this sample is calculated as

$$\bar{X} = \frac{\Sigma X}{n} = \frac{9.0 + 13.0 + \cdots + 12.0}{7} = 10.0$$

Note that the mean is computed as 10.0, even though not one individual business in the sample actually projected that value. In fact, we see from the "dot scale" of Figure 3.1 that for this set of data four observations are larger than the mean and three are smaller. In the dot scale, each observation is thought of as a "dot" weighing some amount. The mean is the balancing point for the scale, where observations which are larger balance out those which are smaller.

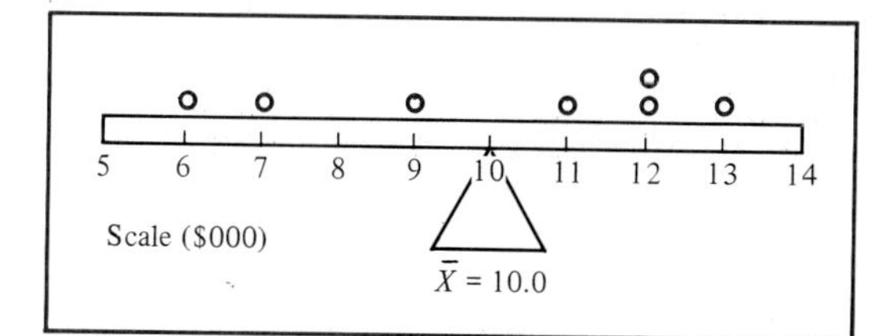

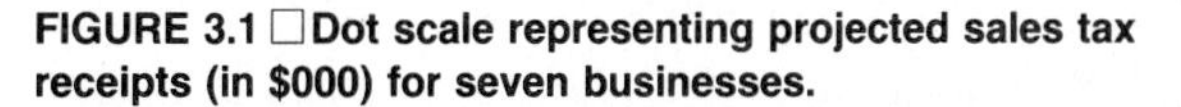
FIGURE 3.1 □ Dot scale representing projected sales tax receipts (in $000) for seven businesses.

CHARACTERISTICS OF THE MEAN

One important characteristic of the mean is that it may be used to estimate a **total amount** in a population when such a computation is appropriate.[3] The total can be estimated from the following equation:

$$\text{estimated total} = N\bar{X} \qquad \textbf{(3.2)}$$

where N = population size
$\bar{X}$ = sample arithmetic mean

[3] The estimation of the total amount is not always appropriate. Suppose, for example, a particular elementary school had (a population of) $N = 50$ fifth-grade students. If a sample of $n = 7$ such students was taken and the average age was computed to be $\bar{X} = 10.0$ years, there would be no meaningful reason to estimate the "total amount of years."

This is precisely what our comptroller of Heldman Township would want to compute for the preliminary report. Since the sample average (based on $n = 7$ small businesses) is 10.0 thousands of dollars, the total amount of sales tax receipts projected for the entire population of $N = 50$ small businesses during the current quarterly period is

$$\text{estimated total} = N\overline{X} = 50(10.0) = 500.0 \text{ thousands of dollars}$$

If it can be assumed that economic conditions during the current quarterly period will represent conditions over the entire year, the comptroller will obtain the desired annual projections by simply multiplying the quarterly projection by four.

$$\begin{aligned}\text{estimated yearly total} = 4N\overline{X} = 4(50)(10.0) &= 2{,}000 \text{ thousands of dollars}\\ &= 2.0 \text{ millions of dollars}\end{aligned}$$

Another important characteristic of the mean is that its calculation is based on all the observations ($X_1, X_2, \ldots, X_n$) in the set of data. No other commonly used measure of central tendency enjoys this "democratic feature." In particular, it is precisely this feature that makes the mean the appropriate measure to use when calculating the total amount. On the other hand, when it is simply desirable to select a *typical value* to describe an entire set of data, this "democratic feature" can be detrimental. Since its computation is based on every observation, the arithmetic mean is greatly affected by any extreme value or values (outliers). In such instances, the arithmetic mean will not be the best measure to use for describing or summarizing a set of data (although we would still want to use it in order to estimate the total amount).

3.4.2 The Median

The **median** is the middle value in an ordered sequence of data. If there are no tied (equal) values, half of the observations will be smaller and half will be larger than the median. The median is unaffected by any extreme observations in a set of data. Thus whenever any extreme observation is present, it is usually more appropriate to use the median than the mean to describe the typical value in a set of data.

To calculate the median from a set of data collected in its raw form, we must first arrange the data in rank order, from the smallest to the largest observation. Such an ordered sequence of data is called an **ordered array.** These ordered values are statistics, since they are calculated from the original observations. They are called, not surprisingly, the **order statistics** of the sample. The first order statistic is the smallest value, the second order statistic is the second smallest value, and so on through the *n*th order statistic, which is the largest observed value. We then use the positioning-point formula

$$\frac{n+1}{2}$$

to find the place in the ordered array which corresponds to the median value.

RULE 1 If the number of observations in the sample is an *odd* number, the median is represented by the numerical value corresponding to the $(n + 1)/2$ ordered observation.

RULE 2 If the number of observations in the sample is an *even* number, the median is represented by the mean or average of the two middle values in the ordered array—the $n/2$ ordered observation and the $(n + 2)/2$ ordered observation.

For the comptroller's problem the raw data were

9.0 13.0 12.0 7.0 6.0 11.0 12.0

The ordered array becomes

ordered observation:	6.0	7.0	9.0	11.0	12.0	12.0	13.0
				↑ median ↓			
numerical order:	1	2	3	4	5	6	7

The median for this odd-sized sample of seven is 11.0 thousands of dollars, the value corresponding to the $(n + 1)/2 = (7 + 1)/2 =$ 4th ordered observation.

As can be observed, the median is unaffected by extreme observations. Regardless of whether the smallest value is 1.0, 2.0, or 6.0, the median is still 11.0 thousands of dollars.

To summarize, the calculation of the median value is affected by the number of observations, not by the magnitude of any extreme(s). Moreover, any observation selected at random is just as likely to exceed the median as it is to be exceeded by it.

EXAMPLE 3.1. Computing the median for an even-sized sample.

Given the following set of weights for a sample of $n = 6$ linebackers in the National Football League (NFL):

235 220 242 239 230 240

The ordered array is

ordered observation:	220	230	235	239	240	242
numerical order:	1	2	3	4	5	6

The median positioning point is the $(n + 1)/2 = (6 + 1)/2 = 3.5$th ordered observation. Thus, the median is the average of the third and the fourth ordered observations:

$$\text{median} = \frac{235 + 239}{2} = 237 \text{ pounds}$$

Example 3.2. Computing the median with ties in the data.

Given the following set of weights for a sample of $n = 9$ defensive backs in the NFL:

205 190 195 185 188 200 195 198 212

The ordered array becomes

ordered observation:	185	188	190	195	195	198	200	205	212
numerical order:	1	2	3	4	5	6	7	8	9

For this odd-sized sample the median positioning point is the $(n + 1)/2 = (9 + 1)/2 = 5$th ordered observation. Thus the median is 195 pounds, the middle value in the ordered sequence (even though the 4th ordered observation is also 195 pounds).

3.4.3 The Mode

Sometimes, when describing or summarizing a set of data, the mode is used as a measure of central tendency. The **mode** is the value in a set of data which appears most frequently. Unlike the arithmetic mean, the mode is not affected by the occurrence of any extreme values. However, the mode is not used for more than descriptive purposes because it lacks the stability of other measures of central tendency. That is, a change in a few observations could change the mode by quite a bit.

Using the ordered array for the comptroller's data

6.0 7.0 9.0 11.0 12.0 12.0 13.0

the mode is 12.0 thousands of dollars. That value was projected more frequently than any other. Notice that if one of the values of 12.0 were 12.5, and the value of 7.0 were 6.0, then the mode would change from 12.0 to 6.0. This is what we meant by the instability of the mode.

Example 3.3. Finding the mode.

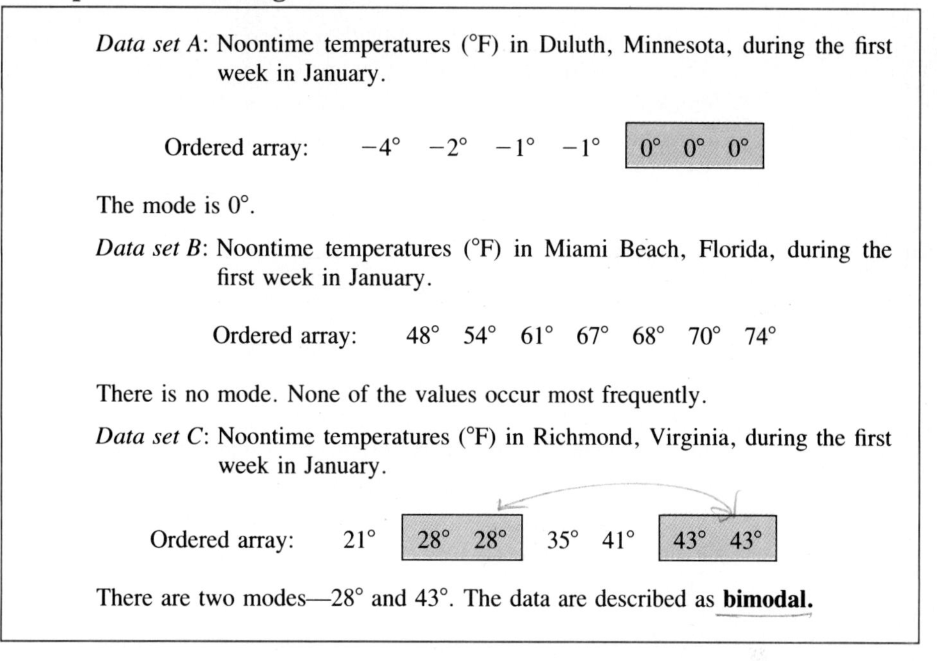

Data set A: Noontime temperatures (°F) in Duluth, Minnesota, during the first week in January.

Ordered array: −4° −2° −1° −1° 0° 0° 0°

The mode is 0°.

Data set B: Noontime temperatures (°F) in Miami Beach, Florida, during the first week in January.

Ordered array: 48° 54° 61° 67° 68° 70° 74°

There is no mode. None of the values occur most frequently.

Data set C: Noontime temperatures (°F) in Richmond, Virginia, during the first week in January.

Ordered array: 21° 28° 28° 35° 41° 43° 43°

There are two modes—28° and 43°. The data are described as **bimodal.**

3.4.4 The Midrange

The **midrange** is the average of the smallest and largest observations in a set of data. This can be written as

$$\text{midrange} = \frac{X_{\text{smallest}} + X_{\text{largest}}}{2} \qquad \textbf{(3.3)}$$

Another notation for X_{smallest} is $X_{(1)}$, which stands for the first order statistic; similarly, another notation for X_{largest} is $X_{(n)}$, which is the nth (largest) order statistic.

The midrange is a "quick and simple" measure of central tendency. It is easily obtained from an ordered array.

Using the ordered array from the comptroller's data

6.0 7.0 9.0 11.0 12.0 12.0 13.0

the midrange is computed from Equation (3.3) as

$$\text{midrange} = \frac{X_{\text{smallest}} + X_{\text{largest}}}{2} = \frac{6.0 + 13.0}{2}$$
$$= 9.5 \text{ thousands of dollars}$$

The midrange is most frequently used as a summary measure by financial analysts and by weather reporters. As examples, the ''average'' price per share of common stock in IBM over the twelve-month period January 1986–December 1986 is simply computed by averaging the yearly low value with the yearly high value. Moreover, our local TV weather reporter will tell us the ''average'' temperature today by simply averaging the lowest recorded reading with the highest recorded reading.

CAUTION

Despite its simplicity, the midrange must be used cautiously. Since it involves only the smallest and largest observations in a data set, it would become distorted as a summary measure of central tendency if an extreme value were present.[4] On the other hand, for the applications just discussed the midrange can give an adequate, ''quick and simple'' value to describe or summarize the entire data set—be it a series of daily closing stock prices over a whole year or a series of recorded hourly temperature readings over a whole day. In such situations, an outlying value is not likely to occur.

Problems

3.1 Which of the following statements are objective (analytic) and which are subjective (interpretive):

- **(a)** The average price of a home in Bergen County is $154,700.
- **(b)** Housing is expensive in Palo Alto.
- **(c)** The police in New York City are smarter, better educated, and more honest now than they were 30 years ago.
- **(d)** There were more burglaries per 1,000 homes reported in Chicago last year than in Des Moines.
- **(e)** The average score on an IQ test of students at the Wallace School of Science is 145.
- **(f)** Foreign product dumping is crippling our industry.

For each subjective statement, list some objective statements which might have been used to make the interpretation given. For each objective statement, tell what interpretation might be intended by the person who made the statement.

---

[4] For this reason we may prefer to use the midhinge. (See Section 5.4.)

3.2 Given the following set of data from a sample of size $n = 7$:

10 2 3 2 4 2 5

Compute the mean, median, mode, and midrange.

3.3 Given the following set of data from a sample of size $n = 7$:

20 12 13 12 14 12 15

(a) Compute the mean, median, mode, and midrange.
(b) Compare your results with those from Problem 3.2.

3.4 Given the following set of data from a sample of size $n = 7$:

−10 2 3 2 −4 2 5

Compute the mean, median, mode, and midrange.

3.5 Upon examining the monthly billing records of a mail-order book company, the auditor takes a sample of 20 of its unpaid accounts. The amounts (in $) owed the company were:

4, 18, 11, 7, 7, 10, 5, 33, 9, 12,
3, 11, 10, 6, 26, 37, 15, 18, 10, 21

(a) Compute the mean, median, mode, and midrange.
(b) If a total of 350 bills were still outstanding, use the mean to estimate the total amount owed to the company.

3.6 A track coach must decide on which one of two sprinters to select for the 100-meter dash at an upcoming meet. The coach will base the decision on the results of five races between the two athletes run over a one-hour period with 15-minute rest intervals between. The following times (in seconds) were recorded for the five races:

	Race				
Athlete	**1**	**2**	**3**	**4**	**5**
Sharyn	12.1	12.0	12.0	16.8	12.1
Tamara	12.3	12.4	12.4	12.5	12.4

(a) Based on the data above, which of the two sprinters should the coach select? Why?
(b) Should the selection be different if the coach knew that Sharyn had fallen at the start of the fourth race?
(c) Discuss the differences in the concepts of the mean and the median as measures of central tendency and how this relates to (a) and (b).

3.7 Suppose that, owing to an error, a data set of ages of a group of children was recorded as 3, 5, 4, 7, 3, 6, 5, 6, and 60, where the last value should have been 6 instead of 60. Show how much the mean, median, and midrange are affected by the error (that is, compute these statistics for the "bad" and "good" data sets, and compare the results of using different estimators of central tendency).

3.8 Suppose that the average family in the United States has 2.3 children, and that there are 40,000,000 families in the United States.

(a) Which formula should be used to calculate the total number of children in the United States?

(b) Use the formula to compute the total number of children.

(c) Suppose the median number of children per family is 2; does the median seem to be a more or a less reasonable measure of central tendency for the variable "number of children per family" than the mean? Why?

(d) If you knew only the median, and not the mean, could you calculate the total number of children in the United States?

3.9 A manufacturer of light bulbs took a sample of 13 bulbs from a day's production and burned them continuously until they failed. The numbers of hours they burned were:

342, 426, 317, 545, 264, 451, 1049,
631, 512, 266, 492, 562, 298

Compute the mean, median, and midrange. Looking at the distribution of times, which descriptive measures seem best, and which worst? (And why?)

• 3.10 Suppose that a city has 400 fifth-grade classrooms. A random sample of 8 classrooms contains 35, 29, 27, 32, 36, 29, 35, and 37 children.

(a) Compute the mean, median, and midrange.

(b) Estimate the total number of fifth-graders in the school system.

3.11 The physics laboratory at the University of Gorgon has discovered a new subatomic particle, which it has named the riton. It has measured the mass of the riton on 12 occasions, with the following results (all measurements are in 10^{-24} grams):

2.11, 2.13, 2.21, 2.19, 2.09, 2.14,
2.17, 2.12, 2.20, 2.16, 2.14, 2.15

Calculate the mean, median, and midrange masses.

3.12 One widely applied test of creativity is to have people list as many uses as they can for a brick. Suppose that the following represent the scores of 24 students on this test (by score, we mean the number of uses listed by a student):

3, 13, 8, 6, 18, 12, 10, 16,
48, 6, 13, 21, 15, 19, 7, 17,
11, 24, 12, 16, 29, 14, 19, 9

Compute the mean, median, and midrange. Do they agree closely? (If so, why do you think so? If not, why not?)

3.13 The death rate per 100,000 population in the early 1970s in several industrialized countries was:

Australia	866	Italy	996
Austria	1,339	Japan	680
Belgium	1,249	Netherlands	835
Canada	734	Norway	1,012
Denmark	930	Sweden	1,045
Finland	978	Switzerland	943
France	1,134	United Kingdom	1,187
West Germany	1,223	United States	932
Ireland	1,155		

SOURCE: *Social Indicators, 1976*, p. 254. Reprinted from United Nations, World Health Organization, *World Health Statistics Annual, 1973–1976*, Vol. I, *Vital Statistics and Cause of Death.*

Compute the mean, median, and midrange death rate.

3.14 To establish developmental standards, researchers recorded the age at which each of a sample of 18 normal children began speaking full sentences. The ages (in months) were

19, 12, 10, 13, 15, 12, 9, 7, 10,
14, 13, 19, 15, 12, 18, 9, 15, 12

What is the mean age at which the children began to speak in sentences? What is the median and midrange?

3.15 You are writing a newsletter about microcomputers, and in this issue you want to describe the available printers of a certain type, called dot-matrix printers. The dollar prices of the models available are

575, 259, 1049, 340, 499, 675,
599, 450, 649, 475, 520, 550,
398, 490, 560, 625, 875, 749

(a) What is the mean price? The median price? The midrange?
(b) If three new models came on the market, with prices of 345, 375, and 355, what would the new mean and median prices be? Which measure changed the most, and why?

3.16 The price of a single room on a weekday in several hotels in New York was

145, 210, 125, 110, 135, 90,
120, 105, 130, 124, 118, 122

(a) What is the median price of a hotel room in New York? The mean price?
(b) If an additional hotel were included in the survey, and by mistake the Presidential Suite was counted (at $3,000 a day) instead of an ordinary room, what would the mean and median be?

3.17 The batting averages of 15 major league baseball players at midseason were

.285, .299, .215, .314, .278,
.199, .267, .295, .268, .301,
.283, .258, .289, .263, .293

What are the mean, median, and midrange batting averages?

3.18 In order to detect unusual conditions, U.S. intelligence agencies routinely monitor radio transmissions on all frequencies coming from various countries of interest. Unusually high numbers of transmissions on a frequency indicate that something out of the ordinary is happening. Suppose that over the past few weeks measurements were made every day, yielding the following number of minutes (over a two-hour period) used for transmission in a particular frequency range:

23, 5, 33, 14, 31, 22, 18, 12, 19, 6,
29, 16, 26, 47, 28, 16, 24, 21, 17, 28

What are the mean and median number of minutes for transmissions per two-hour period?

3.19 The Department of Agriculture wanted to estimate the corn crop in the United States, so it measured the yield per acre in 13 locations in the main corn-growing states. These yields, in bushels per acre, were

126, 111, 129, 121, 102, 147,
104, 133, 93, 83, 118, 109, 103

(a) What are the median and mean yield per acre for these locations?
(b) If 20,000,000 acres are devoted to growing corn, how many bushels do you estimate will come to market this year?

3.20 A drug manufacturer is testing a new method of using genetic engineering to produce a drug. The manufacturer wants to make sure that the resulting drug attains a high enough level of purity. To do this, samples are tested to see what percentage is unwanted residual from the production process rather than active drug. The percentage impurity for 17 such samples is given below:

1.3, 3.1, 2.6, 6.0, 0.3, 0.1, 0.8, 1.6,
2.5, 0.9, 1.4, 0.8, 3.3, 2.8, 1.2, 0.7, 1.6

(a) What level of measurement is "percentage impurity"?
✗**(b)** Is this a continuous or discrete variable? If discrete, how could it be made continuous; if continuous, how could it be made discrete?
(c) What is the mean percentage impurity? The median?

3.21 For the last ten days in June, the "Shore Special" train was late arriving at its destination by the following times (in minutes; a negative number means that the train was early by that number of minutes): −3, 6, 4, 10, −4, 124, 2, −1, 4, 1.
(a) If you were hired by the railroad as a statistician to show that the railroad is providing good service, what are some of the summary measures you would use to accomplish this?
(b) If you were hired by a TV station which was producing a documentary to show that the railroad is providing bad service, what summary measures would you use?
✗**(c)** If you were trying to be "objective" and "unbiased" in assessing the railroad's performance, which summary measures would you use? (This is the hardest part, because you cannot answer it without making additional assumptions about the relative costs of being late by various amounts of time. The field of decision theory deals with what happens when these costs are explicitly accounted for.)

• 3.22 An airline is trying to characterize the amount of traffic from Boston to New York so that it can plan how many flights per day it needs. For simplicity, consider only one day of the week (since the number of people flying a route depends on day of the week). Suppose that on 13 consecutive Thursdays the number of passengers flying from Boston to New York on this airline has been

735, 587, 639, 867, 612, 538, 669,
751, 927, 698, 751, 712, 637

Compute the mean, median, and midrange number of passengers flying that route on Thursdays.

3.23 A nutritionist wanted to determine how much vitamin C is in a typical orange. He measured 19 oranges; the amount of vitamin C they contained (in milligrams) was

109, 88, 76, 136, 93, 101, 89, 97, 115, 92,
114, 106, 91, 109, 85, 94, 89, 117, 105

What are the mean, median, and midrange of the vitamin C counts in the oranges?

3.24 **(Project)** Buy a dozen bags of candy, all of which are of the same size (for example, jelly beans or honey-coated peanuts). Count the number of pieces of candy in each bag. Weigh each bag on a scale which records to at least $\frac{1}{10}$ of an ounce.

(a) What are the mean, median, and midrange number of pieces of candy in the bags?
(b) Repeat part (a) for the measurements of weight.

3.25 **(Project)** Have all class members take their temperature when they get up in the morning. Compute the mean, median, and midrange of the measurements.

3.26 In order to estimate how much water will be needed to supply the community of Falling Rock in the next decade, the town council asked the city manager to find out how much water a sample of families currently uses. The sample of 15 families used the following number of gallons (in thousands) in the past year:

11.2, 21.5, 16.4, 19.7, 14.6, 16.9, 32.2, 18.2,
13.1, 23.8, 18.3, 15.5, 18.8, 22.7, 14.0

(a) What is the mean amount of water used per family? The median? The midrange?
(b) Suppose that ten years from now the town council expects that there will be 45,000 families living in Falling Rock. How many gallons of water per year will be needed, if the rate of consumption per family stays the same?
(c) Why might the town council have used the data from a survey rather than just measuring the total consumption in the town? (Think about what types of users are not yet included in the estimation process.)

3.5 MEASURES OF DISPERSION

A second important property which describes a set of data is dispersion. **Dispersion** is the amount of variation or "spread" in the data. Two sets of data may differ in both central tendency and dispersion; or, as shown in Figure 3.2, two sets of data

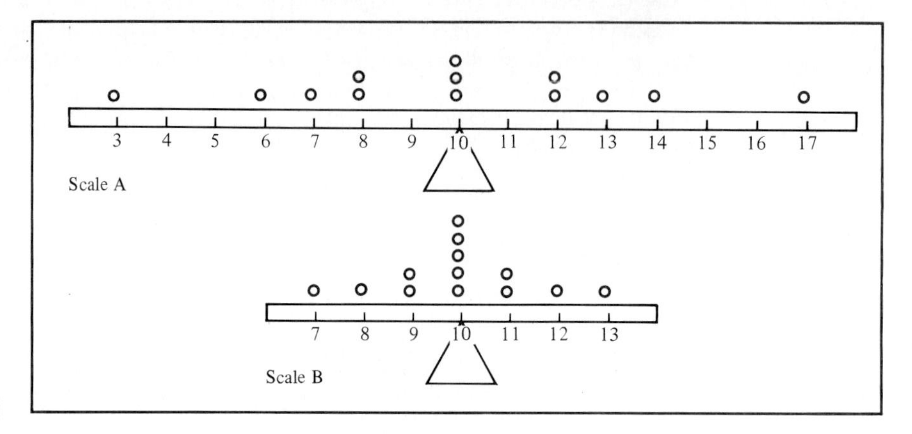

FIGURE 3.2 □ Two data sets differing in dispersion.

may have the same measures of central tendency but differ greatly in terms of dispersion. The data set depicted on scale B is much more concentrated about the measure of central tendency and therefore is less variable than that depicted by scale A.

In dealing with numerical data, then, it is insufficient to summarize those data by merely presenting some descriptive measures of central tendency. We must also characterize the data in terms of their dispersion or variability. Three such measures are the range, the variance, and the standard deviation.

3.5.1 The Range

The **range** is the difference between the largest and smallest observations in a set of data. That is,

$$\text{Range} = X_{\text{largest}} - X_{\text{smallest}} \quad \textbf{(3.4)}$$

The range measures the "total spread" in the set of data. It can be readily obtained from the ordered array.

Using the ordered array for our comptroller's data

6.0 7.0 9.0 11.0 12.0 12.0 13.0

the range is 13.0 − 6.0 = 7.0 thousands of dollars.

Although the range is a simple, easily calculated measure of dispersion, its distinct weakness is that it fails to take into account how the data are distributed *between* the smallest and largest values. This can be observed from Figure 3.3. As scale D

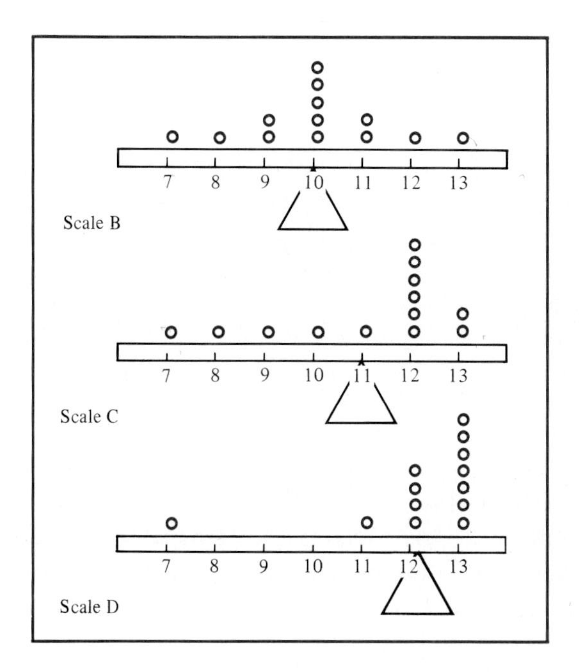

FIGURE 3.3 □ **Comparing three data sets with the same range.**

shows, it would be unwise to use the range as a measure of dispersion if either one or both of its components are extreme observations (outliers). This is the same problem we had with the midrange, and for the same reasons.

3.5.2 The Variance and the Standard Deviation

Two commonly used measures of dispersion which do take into account how *all* the values in the data are distributed are the variance and its square root, the standard deviation. These measures evaluate how the values fluctuate about the mean.

DEFINING THE SAMPLE VARIANCE

The **sample variance** is (almost) the average of the squared differences between each of the observations in a set of data and the mean.[5] Thus, for a sample containing n observations $X_1, X_2, \ldots, X_n$ the sample variance (given by the symbol S^2) can be written as

$$S^2 = \frac{(X_1 - \bar{X})^2 + (X_2 - \bar{X})^2 + \cdots + (X_n - \bar{X})^2}{n - 1}$$

[5] Had the denominator been n instead of $n - 1$, we would have exactly the average squared deviation from the mean. But $n - 1$ is used because it has a desirable statistical property—it provides an unbiased estimate of the true variance. We will make use of this property later in the book. If the sample size is large, division by either n or $n - 1$ will give approximately the same result.

Using our summation notation, the above formulation can be more simply expressed as

$$S^2 = \frac{\Sigma\,(X - \overline{X})^2}{n - 1} \tag{3.5}$$

where $\overline{X}$ = sample arithmetic mean
n = sample size
$\Sigma\,(X - \overline{X})^2$ = sum of all the squared differences between X and $\overline{X}$

Why is this formula sensible? For each observation, we subtract the mean. This tells how far the observation is from the mean. But the number may be either positive or negative, and we don't care whether the observation is above or below the mean, only how far it is from the mean. By squaring each difference, we get rid of any minus signs. Summing the values gives the total of the squared distances between the observations and the mean, and dividing by $n - 1$ gives (almost) an average of these values.

If most of the observations are close to the mean, then most of the differences will be small, and the variance will be small. (In the most extreme case, where all observations are equal, then each difference will be zero and the variance will be zero.) If many of the observations are far from the mean, then many of the differences between the observations and the mean will be large, and the variance will be large.

DEFINING THE SAMPLE STANDARD DEVIATION

The **sample standard deviation** (given by the symbol S) is simply the square root of the sample variance. That is

$$S = \sqrt{S^2} = \sqrt{\frac{\Sigma\,(X - \overline{X})^2}{n - 1}} \tag{3.6}$$

CALCULATING S^2 AND S

To compute the variance we

1. calculate the mean;
2. take the difference between each observation and the mean;
3. square it;
4. add the squared results together;
5. divide this sum by $n - 1$.

To obtain the standard deviation we merely take the square root of the variance.

For the comptroller's problem the raw data were

9.0 13.0 12.0 7.0 6.0 11.0 12.0

The sample variance is computed as follows:

Since $\bar{X} = 10.0$,

$$S^2 = \frac{\Sigma (X - \bar{X})^2}{n - 1}$$

$$= \frac{(9.0 - 10.0)^2 + (13.0 - 10.0)^2 + \cdots + (12.0 - 10.0)^2}{7 - 1}$$

$$= \frac{44}{6} = 7.33 \qquad \text{(in squared thousands of dollars)}$$

and the sample standard deviation is computed as

$$S = \sqrt{S^2} = \sqrt{7.33} = 2.7 \qquad \text{thousands of dollars}$$

Since, in the preceding computations, we are squaring the differences, *neither the variance nor the standard deviation can ever be negative*. S^2 and S could be zero only if there were no variation at all in the data—if each observation in the sample were exactly the same. In such an unusual case the range would also be zero. But data are inherently variable—not constant. Any random phenomenon of interest that we can think of usually takes on a variety of values. For example, people have differing IQs, incomes, weights, heights, ages, pulse rates, etc. It is because data inherently vary that it becomes so important to study not only measures of central tendency, which locate the center of the data, but also measures of dispersion, which reflect how the data are varying.

WHAT THE VARIANCE AND THE STANDARD DEVIATION SHOW

The variance and the standard deviation are measuring the "average" scatter around the mean—that is, how larger observations fluctuate above it and how smaller observations distribute below it.

While the variance possesses certain useful mathematical properties, its computation is measured in squared units—squared dollars, squared inches, squared kilograms, squared percentage points, squared seconds, etc. Thus, for practical work our primary measure of dispersion will be the standard deviation, whose measurement is in the original units of the data—dollars, inches, kilograms, percentage points, seconds, etc.

WHY WE SQUARE THE DEVIATIONS

The formulas for variance and standard deviation could not merely use $\Sigma (X - \overline{X})$ as a numerator, because the mean acts as a "balancing point" for observations larger and smaller than it. Therefore, the sum of the deviations about the mean is always zero,[6] —that is,

$$\Sigma (X - \overline{X}) = 0$$

To demonstrate this, let us refer to the comptroller's raw data:

9.0 13.0 12.0 7.0 6.0 11.0 12.0

If we subtract the mean of 10.0 from each observation, we obtain the following deviations from the mean:

(9.0 − 10.0) (13.0 − 10.0) (12.0 − 10.0) (7.0 − 10.0) (6.0 − 10.0) (11.0 − 10.0) (12.0 − 10.0)

−1.0 3.0 2.0 −3.0 −4.0 1.0 2.0

This is depicted in the dot scale diagram displayed in Figure 3.4.

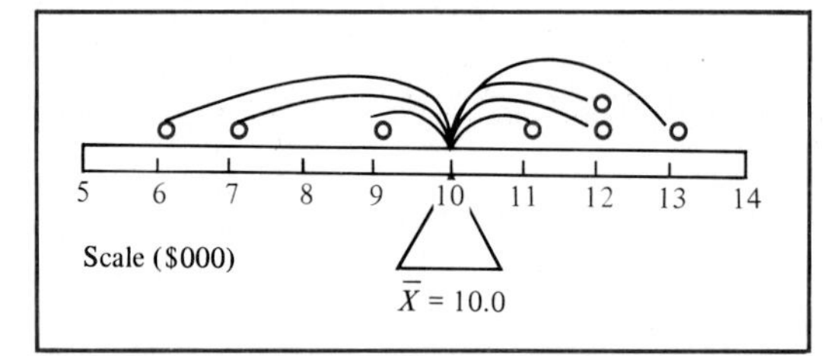

FIGURE 3.4 □ The mean as a "balancing point."

As already noted, three of the observations are smaller than the mean, while four are larger. Although the sum of the seven deviations (−1.0, 3.0, 2.0, −3.0, −4.0, 1.0, and 2.0) is zero, the sum of the squared deviations allows us to study the dispersion in the data. Hence we use $\Sigma (X - \overline{X})^2$ when computing the variance and standard deviation. In the squaring process, observations that are further from the mean get more weight than observations closer to the mean.

The respective squared deviations for the comptroller's data are

1.0 9.0 4.0 9.0 16.0 1.0 4.0

---

[6] Using the summation rules from Appendix A, we show the following proof:

$$\Sigma (X - \overline{X}) = \Sigma X - \Sigma \overline{X} = \Sigma X - n\overline{X} = \Sigma X - \Sigma X = 0$$

We note that the second observation ($X_2 = 13.0$) is 3.0 thousands of dollars higher than the mean, while the fourth observation ($X_4 = 7.0$) is 3.0 thousands of dollars lower. In the squaring process both these values contribute substantially more to the calculation of S^2 and S than do the first and sixth observations ($X_1 = 9.0$ and $X_6 = 11.0$), which are only 1.0 thousand dollars different from the mean.

Therefore we may generalize as follows:

1. The more "spread out" or dispersed the data are, the larger will be the range, the variance, and the standard deviation.
2. The more "concentrated" or homogeneous the data are, the smaller will be the range, the variance, and the standard deviation.
3. If the observations are all the same (so that there is no variation in the data), the range, variance, and standard deviation will all be zero.

CALCULATING S^2 AND S: "SHORTCUT" FORMULAS

The formulas for variance and standard deviation, Equations (3.5) and (3.6), are **definitional formulas,** but they are not often practical to use—even with an electronic calculator. For our comptroller's problem the mean is an integer—10.0. For more usual situations where the observations and the mean are unlikely to be integers, the following **"shortcut" formulas** for the variance and the standard deviation are given for practical use:

observation values

$$S^2 = \frac{\sum X^2 - \frac{(\sum X)^2}{n}}{n - 1} \tag{3.7}$$

$$S = \sqrt{\frac{\sum X^2 - \frac{(\sum X)^2}{n}}{n - 1}} \tag{3.8}$$

where $\sum X^2$ = sum of the squares of each observation

$(\sum X)^2$ = square of the sum of all observations

The shortcut formulas, Equations (3.7) and (3.8), are identical to the definitional formulas, Equations (3.5) and (3.6). The denominators are the same, and it is easy to show through expansion and the use of summation rules (see Appendix A) that

$$\sum (X - \overline{X})^2 = \sum X^2 - \frac{(\sum X)^2}{n}$$

Moreover, since $\Sigma (X - \overline{X})^2$ can never be negative, ΣX^2, the sum of squares, must always equal or exceed $(\Sigma X)^2/n$, the square of the sum divided by n.

Returning to the comptroller's data, the variance and standard deviation are recomputed using Equations (3.7) and (3.8) as follows:

$$S^2 = \frac{\Sigma X^2 - \dfrac{(\Sigma X)^2}{n}}{n - 1}$$

$$= \frac{(9.0^2 + 13.0^2 + \cdots + 12.0^2) - \dfrac{(9.0 + 13.0 + \cdots + 12.0)^2}{7}}{7 - 1}$$

$$= \frac{(81.0 + 169.0 + \cdots + 144.0) - \dfrac{(70.0)^2}{7}}{6}$$

$$= \frac{744.0 - 700.0}{6} = \frac{44.0}{6} = 7.33 \text{ (in squared thousands of dollars)}$$

and

$$S = \sqrt{7.33} = 2.7 \quad \text{thousands of dollars}$$

Problems

3.27 For the following, refer to the data in Problem 3.9 on page 46:

(a) Calculate the variance, standard deviation, and range.

✓**(b)** For many sets of data, the range is about six times the standard deviation. Is this true here? (If not, why do you think it is not?)

✓**(c)** Using the information above, what would you advise the manufacturer to do if he wanted to be able to say in advertisements that these bulbs "should last 400 hours"? (*Note*: There is no right answer to this question; the point is to consider how to make such a statement precise.)

• 3.28 For the data set in Problem 3.10 on page 46 compute the range, variance, and standard deviation.

3.29 For the data set in Problem 3.11 on page 46 compute the range, variance, and standard deviation of the masses.

3.30 For the data set in Problem 3.12 on page 46 compute the range, variance, and standard deviation.

3.31 Refer to the data in Problem 3.13 on page 46:

(a) Compute the range, variance, and standard deviation.

(b) If we define an "unusually high" death rate as one which is more than two standard deviations above the mean, which countries have unusually high death rates? Using a similar definition of "unusually low," which countries have unusually low death rates?

3.32 Answer the following using the data in Problem 3.14 on page 47:
(a) What are the range, variance, and standard deviation?
(b) At what age would a child be more than 2 standard deviations earlier than the mean?

3.33 For the data in Problem 3.15 on page 47, what is the range in prices? The standard deviation?

3.34 Refer to the data in Problem 3.16 on page 47:
(a) What are the mean and midrange? What are the range, variance, and standard deviation?
(b) If the data are changed as indicated in part (c) of the original problem, what is the ratio of the range to the standard deviation? How does this compare with the same ratio for the original data, and why?

3.35 For the data of Problem 3.17 on page 47, what are the range and standard deviation?

3.36 For the data of Problem 3.18 on page 48, what are the range and standard deviation of the number of minutes for transmissions?

3.37 For the data in Problem 3.19 on page 48, what are the range and standard deviation of the yields?

3.38 For the data in Problem 3.20 on page 48, what is the range of impurity? The standard deviation?

• 3.39 For the data in Problem 3.22 on page 49, compute the range and standard deviation.

3.40 For the data in Problem 3.23 on page 49, what proportion of the oranges have vitamin C contents within 2.5 standard deviations of the mean?

3.41 For the data you collected for the project in Problem 3.24 on page 49:
(a) What are the range and standard deviation of the number of pieces of candy per bag?
(b) What proportion of the bags are within 2 standard deviations of the mean?
(c) Repeat parts (a) and (b) for the measurements of weight.

3.42 For the data you gathered for the project in Problem 3.25 on page 49:
(a) Compute the range and standard deviation.
(b) Calculate the proportion of the observations within 2 standard deviations of the mean; within 3 standard deviations of the mean.

3.43 Using the data from Problem 3.2 on page 45, compute the range, variance, and standard deviation.

3.44 Using the data from Problem 3.3 on page 45:
(a) Compute the range, variance, and standard deviation.
(b) Compare your results with those from Problem 3.43. Discuss.

3.45 Using the data from Problem 3.4 on page 45, compute the range, variance, and standard deviation.

3.46 Using the data from Problem 3.5 on page 45, compute the range, variance, and standard deviation of the amount owed to the company.

3.47 Using the data from Problem 3.6 on page 45, compute the range, variance, and standard deviation for each of the two athletes.

3.6 SHAPE

A third important property of a set of data is its shape. In describing a set of numerical data it is not only necessary to summarize the data by presenting appropriate measures of central tendency and variability; it is also necessary to consider the shape of the data—the manner in which they are distributed. Either the distribution of the data is symmetrical or it is not. In visual terms, the distribution is symmetric if its lower half is a mirror image of its upper half. If the distribution of data is not symmetrical, it is called asymmetrical or **skewed.**

A simple way to describe the property of shape is to compare the mean and the median. If these two measures are equal, we may generally consider the data to be symmetrical (or **zero-skewed**). On the other hand, if the mean exceeds the median, the data may generally be described as **positive** or **right-skewed,** while if the mean is exceeded by the median it can generally be called **negative** or **left-skewed.** That is,

mean > median:	positive or right-skewness
mean = median:	symmetry or zero-skewness
mean < median:	negative or left-skewness

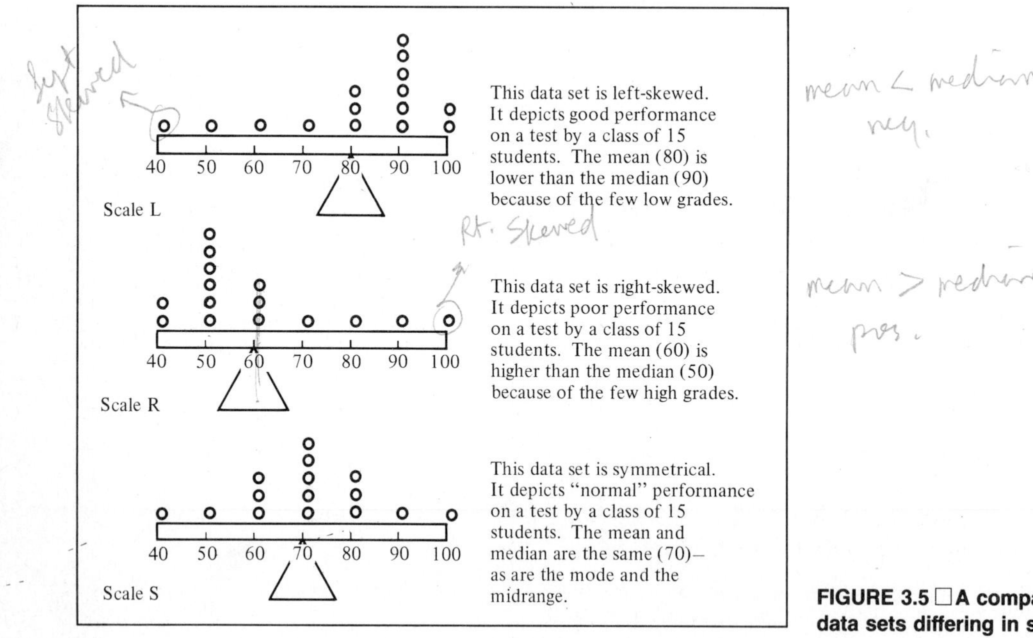

FIGURE 3.5 □ A comparison of three data sets differing in shape.

Positive skewness arises when the mean is increased by some unusually high values, while negative skewness occurs when the mean is reduced by some extremely low values. Data are symmetrical when there are no really extreme values in a particular direction so that low and high values balance each other out.

Figure 3.5 depicts the shapes of three data sets: the data on Scale L are negative or left-skewed (since the distortion to the left is caused by extremely small values); the data on Scale R are positive or right-skewed (since the distortion to the right is caused by extremely large values); and the data on Scale S are symmetrical (the low and high values on the scale balance and the mean equals the median).

For our comptroller's problem the data were displayed along the dot scale in Figure 3.1 (see page 39). The mean (10.0) is smaller than the median (11.0), and the shape appears to be slightly left-skewed.

Problems

3.48 Using the data from Problem 3.2 on page 45, describe the shape.

3.49 Using the data from Problem 3.3 on page 45:
- **(a)** Describe the shape.
- **(b)** Compare your results to those from Problem 3.48.

3.50 Using the data from Problem 3.4 on page 45, describe the shape.

3.51 Using the data from Problem 3.5 on page 45, describe the shape.

3.7 CALCULATING DESCRIPTIVE SUMMARY MEASURES FROM A POPULATION

In the previous sections we examined various measures which are used to describe or summarize a set of quantitative data. In particular, for the sales tax receipts projected from a sample of $n = 7$ stores we obtained nine different statistics. The mean, median, mode, and midrange described the property of central tendency, and the mean was also used to estimate a total amount; the range, variance, and standard deviation described the property of dispersion; and the relative position of the mean and the median was used to describe the property of shape.

Suppose, however, that we have access to the data for an entire population. Such would be the case here if the comptroller of Heldman Township were to obtain projected sales tax receipt statements from each of the listed ($N = 50$) small businesses in the township (i.e., the population). The resulting measures (i.e., **parameters**) computed from the data to summarize or describe the properties of central tendency, dispersion, and shape would then be used by the comptroller when writing the revised report for the governor's office. Almost all of the formulas for calculating

Store	Receipts*	Location†	Hours‡	Store	Receipts*	Location†	Hours‡
1	8.0	E	N	26	10.0	E	N
2	12.0	W	A	27	13.0	W	A
3	9.0	E	A	28	9.0	W	N
4	11.0	W	S	29	10.0	W	N
5	9.0	W	N	30	13.0	W	A
6	9.0	W	N	31	12.0	W	A
7	12.0	W	A	32	9.0	E	N
8	11.0	W	S	33	10.0	E	A
9	12.0	W	A	34	12.0	W	A
10	8.0	E	N	35	11.0	W	A
11	10.0	E	A	36	10.0	W	S
12	8.0	E	S	37	11.0	W	A
13	10.0	W	S	38	10.0	W	N
14	9.0	E	A	39	9.0	W	N
15	8.0	E	N	40	14.0	W	A
16	11.0	W	A	41	13.0	W	S
17	10.0	W	A	42	7.0	E	N
18	11.0	W	A	43	11.0	W	S
19	10.0	E	A	44	8.0	E	N
20	10.0	E	A	45	10.0	E	A
21	11.0	W	A	46	13.0	W	A
22	15.0	W	A	47	11.0	W	A
23	12.0	E	A	48	7.0	E	N
24	6.0	E	N	49	5.0	E	N
25	8.0	E	S	50	12.0	W	A

*Projected sales tax receipts (in $000).
† Location: E = East Central St., W = West Central St.
‡Sunday operations: A = Always, S = Sometimes, N = Never.

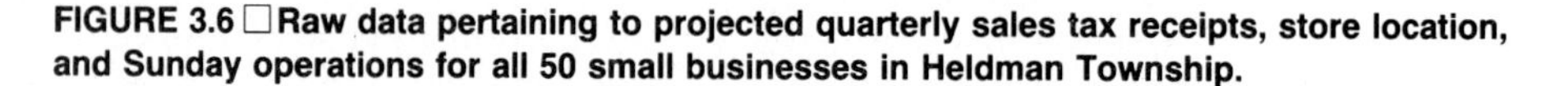

FIGURE 3.6 □ Raw data pertaining to projected quarterly sales tax receipts, store location, and Sunday operations for all 50 small businesses in Heldman Township.

these parameters have the same form as in the calculation of the corresponding statistics in a sample; the only minor difference is dividing by N instead of $N - 1$ for the variance and standard deviation.

Figure 3.6 lists the projected sales tax receipt statements for each of the small businesses in Heldman Township. In addition, some qualitative information is displayed. The location of each business (East Central Street or West Central Street) is specified by an "E" or a "W." Furthermore, "A," "S," or "N" indicate whether a business is always, sometimes, or never open on Sundays.[7]

Recall that when numerical data are listed in the random order in which they have been collected, the data are said to be in *raw form*. The projected sales tax receipts listed in Figure 3.6 provide an example of data in raw form. As seen from Figure 3.6, as the number of observations gets large, it becomes more difficult to focus on the major features of the data.

Although the mean, total amount, variance, and standard deviation can be computed from the raw data, all our other summary measures require the data to be placed into an *ordered array*. Figure 3.7 depicts the ordered array for the projected sales tax receipts data.

When the data are all organized into an ordered array, our evaluation is facilitated.

[7] The categorical responses to these qualitative questions will be examined in Section 4.5.

5.0	9.0	10.0	11.0	12.0
6.0	9.0	10.0	11.0	12.0
7.0	9.0	10.0	11.0	12.0
7.0	9.0	10.0	11.0	12.0
8.0	9.0	10.0	11.0	13.0
8.0	9.0	10.0	11.0	13.0
8.0	9.0	10.0	11.0	13.0
8.0	10.0	10.0	12.0	13.0
8.0	10.0	11.0	12.0	14.0
8.0	10.0	11.0	12.0	15.0

FIGURE 3.7 □ Ordered array of projected sales tax receipts.

It becomes easy to pick out extremes, typical values, and concentrations of values. On the other hand, as the number of observations gets large, the process of organizing the raw data into an ordered array becomes more cumbersome.

3.7.1 Measures of Central Tendency

THE POPULATION MEAN

The **true arithmetic mean** of a finite population is calculated by summing all the observations and then dividing this *total amount* by the number of observations involved (i.e., the size of the population). This is the same procedure as we used to calculate the sample mean. Thus, for a population containing N observations $X_1, X_2, \ldots, X_N$ the arithmetic mean (given by the symbol μ_X, where μ is the Greek lower-case letter "mu") can be written as

$$\mu_X = \frac{\Sigma X}{N} \tag{3.9}$$

where μ_X = population arithmetic mean
N = population size
ΣX = sum of all of the X values in the population

THE TOTAL AMOUNT

When appropriate to use as a summary measure, the **population total amount** (given by the symbol τ—the Greek lower-case letter "tau") is simply the sum of all the observations. That is,

$$\text{total amount:} \quad \tau = \Sigma X = N\mu_X \tag{3.10}$$

where τ = total amount in the population
ΣX = sum of all X values in the population
N = population size
μ_X = population arithmetic mean

THE MEDIAN

For a population of size N the **median** is the value pertaining to the $[(N + 1)/2]$th observation in the ordered array. If the population size is an odd number, the median is the value corresponding to the middle observation in the ordered array. On the other hand, if the population size is an even number, the median is simply the mean of the two middle observations in the ordered array.

THE MODE

The **mode** in the population (if it exists) is the observation which appears most frequently. Some populations have more than one mode; others have no mode (there are no duplicate values).

THE MIDRANGE

The **midrange** in the population is the mean of the smallest and largest observations in the ordered array.

$$\text{midrange} = \frac{X_{\text{smallest}} + X_{\text{largest}}}{2} \qquad \textbf{(3.3a)}$$

3.7.2 Measures of Dispersion

THE RANGE

The **range** in the population is the difference between the largest and smallest values.

$$\text{range} = X_{\text{largest}} - X_{\text{smallest}} \qquad \textbf{(3.4a)}$$

THE VARIANCE

The **variance** in the population can be defined as the average of the squared differences between each of the observations and the mean. Thus, for a population containing N observations $X_1, X_2, \ldots, X_N$ the population variance (given by the symbol σ_X^2, where σ is the Greek lower-case letter "sigma") can be written as:

$$\sigma_X^2 = \frac{\Sigma (X - \mu_X)^2}{N} \quad \textbf{(3.11)}$$

where μ_X = population arithmetic mean
N = population size
$\Sigma (X - \mu_X)^2$ = sum of squared differences between each observation and the mean

Note that we divide by N, not by $N - 1$, because we are not estimating the population value; in this case, we are computing the exact value in the population, because we have all of the observations. We are calculating exactly the average squared difference between the observations and the population mean.

THE STANDARD DEVIATION

The **population standard deviation** (given by the symbol σ_X) is the square root of the population variance. That is,

$$\sigma_X = \sqrt{\sigma_X^2} = \sqrt{\frac{\Sigma (X - \mu_X)^2}{N}} \quad \textbf{(3.12)}$$

The standard deviation is called the *root-mean-square* (or RMS) *deviation* in some contexts, because it is the (square) root of the mean (or average) squared difference between observations and the mean.

3.7.3 Shape

The **shape** of the population is obtained through a relative comparison of the mean and the median. The same principles apply as those discussed in Section 3.6 for sample data.

3.7.4 Results

Using the ordered array of the projected sales tax receipts (Figure 3.7) we compute the following:

Mean: $\mu_X = \frac{\Sigma X}{N} = \frac{510.0}{50} = 10.2$ thousands of dollars

Total: $\tau = \Sigma X = 510.0$ thousands of dollars for the quarter

Estimated annual total amount:

$$4\tau = 4 \sum X = 2.04 \quad \text{millions of dollars}$$

Median:

$$\text{positioning point} = \frac{N+1}{2} \quad \text{ordered observation}$$

$$= \frac{50+1}{2} = 25.5\text{th} \quad \text{ordered observation}$$

Thus, the median is the average of the 25th and 26th ordered observations.

$$\text{Median} = \frac{X_{(25)} + X_{(26)}}{2} = \frac{10.0 + 10.0}{2} = 10.0 \quad \text{thousands of dollars.}$$

(Remember from the discussion of order statistics that the notation $X_{(25)}$ means the 25th smallest observation.)

Mode: The most frequently projected sales tax receipt is 10.0 thousands of dollars.

Midrange:

$$\frac{X_{\text{smallest}} + X_{\text{largest}}}{2} = \frac{5.0 + 15.0}{2} = 10.0 \quad \text{thousands of dollars}$$

Range:

$$X_{\text{largest}} - X_{\text{smallest}} = 15.0 - 5.0 = 10.0 \quad \text{thousands of dollars}$$

Variance:

$$\sigma_X^2 = \frac{\sum (X - \mu_X)^2}{N} = \frac{(5.0 - 10.2)^2 + (6.0 - 10.2)^2 + \cdots + (15.0 - 10.2)^2}{50}$$

$$= 4.04 \quad \text{(in squared thousands of dollars)}$$

Standard deviation:

$$\sigma_X = \sqrt{\sigma_X^2} = \sqrt{4.04}$$

$$\cong 2.0 \quad \text{thousands of dollars}$$

Shape: In Heldman Township, the population of reported projected sales tax receipts can be considered as approximately symmetrical in *shape* because the mean (10.2) and the median (10.0) are sufficiently close—relative to the size of the standard deviation.

TABLE 3.1 □ Using descriptive measures on two data sets

Descriptive measure	Projected sales tax receipts (in $000)	
	Sample ($n = 7$)	Population ($N = 50$)
Mean	10.0	10.2
Total amount per quarter	500.0	510.0
Annual total amount	2.0 (millions)	2.04 (millions)
Median	11.0	10.0
Mode	12.0	10.0
Midrange	9.5	10.0
Range	7.0	10.0
Variance	7.33	4.04
Standard deviation	2.7	2.0
Shape	Slightly left-skewed	Approximately symmetrical

SUMMARIZING THE FINDINGS

Table 3.1 summarizes the results of utilizing the various descriptive measures we have investigated.

In most cases the computed statistics from the sample of size 7 appear to adequately represent the corresponding characteristics from the population of size 50. Note the following:

1. The estimated mean ($\bar{X} = 10.0$) is quite close to the true mean ($\mu_X = 10.2$), so that the total amount estimated from the sample (500.0) is within 2% of the quarterly total amount projected in the population (510.0). The revised report that the comptroller must send to the governor's office needs to be modified only slightly.
2. In comparing the measures of dispersion from the sample to the population, at first glance an anomoly appears to exist. The range has increased (from 7.0 to 10.0) while the variance and standard deviation have decreased. But this can be explained if we realize that by examining 43 more observations there was a strong likelihood that we would find new extremes (so that the range would increase); however, there is no expectation that the variance and standard deviation would change.
3. Although the data in the sample were slightly negatively skewed, the data in the population appear to be distributed (approximately) symmetrically. This is depicted on the dot scale diagram in Figure 3.8 on page 66, which nevertheless indicates that the sample data (color dots) were an adequate representation of the population data.

USING THE STANDARD DEVIATION: THE BIENAYME-CHEBYSHEV RULE

More than a century ago, the mathematicians Bienayme and Chebyshev (Reference 3) independently examined the property of data variability around the mean. They found that regardless of how a set of data is distributed, the percentage of observations

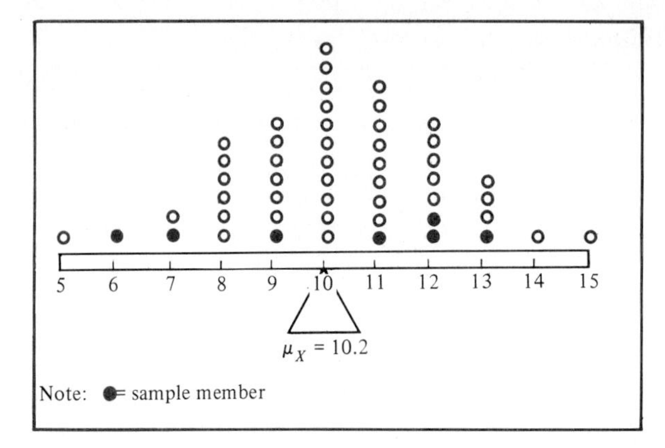

FIGURE 3.8 □ Dot scale showing projected sales tax receipts for a population of 50 stores.

SOURCE: Data are taken from Figure 3.6.

that are contained within distances of k standard deviations around the mean[8] must be at least

$$\left(1 - \frac{1}{k^2}\right) 100\%$$

Therefore, for data with any shape whatsoever,

1. at least $\left[1 - \frac{1}{2^2}\right]100\% = \left[1 - \frac{1}{4}\right]100\% = \left(\frac{3}{4}\right)100\% = 75.0\%$ of the observations must be contained within distances of ± 2 standard deviations around the mean;
2. at least $\left[1 - \frac{1}{3^2}\right]100\% = 88.89\%$ of the observations must be contained within distances of ± 3 standard deviations around the mean;
3. and at least $\left[1 - \frac{1}{4^2}\right]100\% = 93.75\%$ of all the observations must be included within distances of ± 4 standard deviations about the mean.

While the Bienayme-Chebyshev rule is general in nature and applies to any kind of distribution of data, we will see in Chapter 7 that if the data form the "bell-shaped" normal distribution, 68.26% of all the observations will be contained within a distance of 1 standard deviation from the mean, while 95.44%, 99.73%, and 99.99% of the observations will be within 2, 3, and 4 standard deviations (respectively) of the mean. These results (among others) are summarized in Table 3.2.

Specifically, if we knew that a particular random phenomenon followed the pattern of the normal distribution, we would then know (as will be shown in Chapter 7) *exactly* how likely it is that any particular observation was close to or far from its mean. Generally, however, for any kind of distribution, the Bienayme-Chebyshev rule tells us *at least* how likely it must be that any particular observation falls

[8] The Bienayme-Chebyshev rule can apply only to distances beyond 1 standard deviation about the mean.

TABLE 3.2 ☐ How data vary around the mean

Number of standard deviation units k	Percentage of observations contained between the mean and k standard deviations from the mean based on: Bienayme-Chebyshev rule for any distribution	Normal distribution	Heldman Township data
1	Not calculable	68.26%	68.0%
2	At least 75.00%	95.44%	94.0%
3	At least 88.89%	99.73%	100.0%
4	At least 93.75%	99.99%	100.0%

within a given distance about the mean. Such information is quite valuable in data analysis.

From the ordered array (Figure 3.7 on page 61) we note that 34 out of 50 stores (68.0%) have projected sales tax receipts within one standard deviation of the mean (10.2 ± 2.0). Moreover, we see that 47 out of 50 stores (94.0%) have projected sales tax receipts in the range $\mu_X \pm 2\sigma_X$ (that is, 6.2 to 14.2 thousands of dollars). Finally, we note that all 50 stores (100%) project receipts to be in the range $\mu_X \pm 3\sigma_X$ (that is, 4.2 to 16.2 thousands of dollars). Thus the projected sales tax data appear to be approximately normally distributed.

Case B—The Heldman Township Comptroller's Problem Revisited

Suppose that at the end of the quarterly period the comptroller obtains the *reported* sales tax receipts data for each of the 50 small businesses in Heldman Township. The data are presented in raw form in Figure 3.9, and the ordered array is listed in Figure 3.10. From the latter, the comptroller calculates the various descriptive measures and prepares a *final* report for the governor's office. Table 3.3 presents a summary of these results.

TABLE 3.3 ☐ Obtaining descriptive measures

Descriptive measure	Reported sales tax receipts (in $000): Population $N = 50$
Mean	10.28
Total amount for quarter	514.0
Estimated annual total amount	2.056 (millions)
Median	10.30
Mode	11.1 and 12.5
Midrange	10.20
Range	9.80
Variance	4.0984
Standard deviation	2.02
Shape	Approximately symmetrical

SOURCE: Computed from Figures 3.9 and 3.10.

Store	Receipts*	Location†	Hours‡	Store	Receipts*	Location†	Hours‡
1	8.0	E	N	26	10.0	E	N
2	11.8	W	A	27	12.9	W	A
3	8.7	E	A	28	9.2	W	N
4	10.6	W	S	29	10.0	W	N
5	9.5	W	N	30	12.8	W	A
6	9.3	W	N	31	12.5	W	A
7	11.5	W	A	32	9.3	E	N
8	10.7	W	S	33	10.4	E	A
9	11.6	W	A	34	12.7	W	A
10	7.8	E	N	35	10.5	W	A
11	10.5	E	A	36	10.3	W	S
12	7.6	E	S	37	11.1	W	A
13	10.1	W	S	38	9.6	W	N
14	8.9	E	A	39	9.0	W	N
15	8.6	E	N	40	14.5	W	A
16	11.1	W	A	41	13.0	W	S
17	10.2	W	A	42	6.7	E	N
18	11.1	W	A	43	11.0	W	S
19	9.9	E	A	44	8.4	E	N
20	9.8	E	A	45	10.3	E	A
21	11.6	W	A	46	13.0	W	A
22	15.1	W	A	47	11.2	W	A
23	12.5	E	A	48	7.3	E	N
24	6.5	E	N	49	5.3	E	N
25	7.5	E	S	50	12.5	W	A

*Reported sales tax receipts (in $000).
†Location: E = East Central St., W = West Central St.
‡Sunday operations: A = Always, S = Sometimes, N = Never.

FIGURE 3.9 □ Raw data pertaining to reported quarterly sales tax receipts, store location, and Sunday operations for all 50 small businesses in Heldman Township.

5.3	8.7	10.0	10.7	12.5
6.5	8.9	10.0	11.0	12.5
6.7	9.0	10.1	11.1	12.5
7.3	9.2	10.2	11.1	12.7
7.5	9.3	10.3	11.1	12.8
7.6	9.3	10.3	11.2	12.9
7.8	9.5	10.4	11.5	13.0
8.0	9.6	10.5	11.6	13.0
8.4	9.8	10.5	11.6	14.5
8.6	9.9	10.6	11.8	15.1

FIGURE 3.10 □ Ordered array of reported sales tax receipts.

SOURCE: Data are taken from Figure 3.9.

Problems

3.52 Using the ordered array in Figure 3.10, verify the calculations displayed in Table 3.3.

3.53 Using the raw data listed in Figures 3.6 and 3.9, compare the differences between the projected sales tax receipts and the reported sales tax receipts for each store. Comment on these discrepancies. (*Hint*: Look at the last digit of each value.)

3.54 Consider the 24 observations in Problem 3.12 on page 46 to be a population of values. Take 10 random samples each of size 8 from this population. Calculate the mean and variance of each sample. How close do these estimates seem to be to the true value for all 24 observations?

3.55 For the data set in Problem 3.11 on page 46, and for which you calculated the standard deviation in Problem 3.29 on page 56, what proportion of the observations are within 2 standard deviations of the mean? Is this consistent with the Bienayme-Chebyshev rule?

• 3.56 For the data set in Problem 3.12 on page 46, and for which you computed the standard deviation in Problem 3.30 on page 56:

- **(a)** What proportion of the observations are less than 2 standard deviations from the mean?
- **(b)** What proportion are less than 3 standard deviations from the mean?
- **(c)** How do the results in parts (a) and (b) compare with what the Bienyame-Chebyshev inequality says should happen?

3.57 The following questions concern the data of Problem 3.18 on page 48, for which you calculated the standard deviation in Problem 3.36 on page 57.

- **(a)** According to the Bienayme-Chebyshev rule, what is the range within which 99% of the observations should occur?
- **(b)** Does this range seem unusual in any way?

3.58 For the data in Problem 3.23 on page 49, and for which you calculated the standard deviation in Problem 3.40 on page 57, what proportion of the sample should be within 2.5 standard deviations of the mean according to the Bienayme-Chebyshev rule?

3.59 Compare the results obtained in Problem 3.41(b) on page 57 with the percentage that would be predicted by the Bienayme-Chebyshev rule.

Supplementary Problems

3.60 Suppose you take your temperature once a week for 7 weeks; you find it to be 98.2, 99.0, 98.5, 99.1, 98.6, 97.9, and 98.2.

- **(a)** Find the mean, median, and midrange.
- **(b)** Do they all give about the same answer? (If so, why do you think this is? If not, why not?)
- **(c)** Find the range, variance, and standard deviation.
- **(d)** What percentage of the observations are within 2 standard deviations of the mean temperature? Is this consistent with what the Bienayme-Chebyshev rule says should be the case?

3.61 Tell why the mode is not a useful statistic for a continuous variable.

3.62 To give the administration an estimate of the total revenue which will be produced by income taxes, the IRS Commissioner took a random sample of 15 tax returns, in which the taxes paid were (in thousands of dollars):

34, 2, 12, 39, 16, 7, 9, 72, 23, 15, 0, 19, 6, 12, 43

(a) Compute the mean, median, and midrange.
(b) Estimate the total revenue to the government, if there are 50 million taxpayers.
(c) Compute the range, variance, and standard deviation.
(d) Tell why the mean differs by so much from the median and midrange.
(e) For many sets of data, the range is about six times greater than the standard deviation. Is this true here? (If not, can you tell why?)

3.63 Given the following set of data for a population of size $N = 10$:

7 5 11 8 3 6 2 1 9 8

(a) Compute the mean, median, mode, and midrange.
(b) Compute the range, variance, and standard deviation.
(c) Are these data skewed? If so, how?

3.64 Given the following set of data for a population of size $N = 10$:

7 5 6 6 6 4 8 6 9 3

(a) Compute the mean, median, mode, and midrange.
(b) Compute the range, variance, and standard deviation.
(c) Are these data skewed? If so, how?
(d) Compare the measures of central tendency to those of Problem 3.63(a). Discuss.
(e) Compare the measures of dispersion to those of Problem 3.63(b). Discuss.

3.65 In your own words, explain the difference between raw data and an ordered array.

3.66 In your own words, explain the difference between a statistic and a parameter.

3.67 A batch of numerical data has three major properties. In your own words, define these properties and give examples of each.

3.68 The following data are the monthly rental prices for a sample of 10 unfurnished studio apartments in Manhattan and a sample of 10 unfurnished studio apartments in Brooklyn Heights:

Manhattan	$655,	$700,	$685,	$680,	$640,	$675,	$665,	$699,	$947,	$819
Brooklyn Hts.	$450,	$475,	$425,	$405,	$394,	$425,	$390,	$445,	$275,	$500

(a) For each set of data compute the mean, median, range, and standard deviation.
(b) What can be said about unfurnished studio apartments renting in Brooklyn Heights versus those renting in Manhattan?

• 3.69 The following data represent a random sample of the on-base averages (hits, walks, and hit by pitches divided by total plate appearances) for a random sample of 15 starting players during a recent National League baseball season:

.393, .379, .285, .331, .331,
.297, .295, .374, .362, .293,
.362, .349, .285, .307, .349

(a) Compute the:

(1) mean
(2) median
(3) mode
(4) midrange
(5) range
(6) variance
(7) standard deviation

(b) Are these data skewed? If so, how?

3.70 The following data represent net income as a percent of sales for a random sample of 15 of the "500 largest industrial companies":

Company	Net income to sales (%)
Meredith	9.0
American Home Products	15.3
SCM	1.9
Bemis	2.9
Rockwell International	5.3
Pennzoil	5.1
Triangle Industries	2.2
Scott Paper	6.6
VF	9.4
American Cyanimid	3.5
International Minerals and Chem.	7.4
Sealed Power	4.8
SCI Systems	2.6
Dataproducts	5.9
Tracor	2.4

SOURCE: "Fortune's Directory of the 500 Largest Industrial Corporations," *Fortune*, April 28, 1986.

(a) Compute the:

(1) mean
(2) median
(3) mode
(4) midrange
(5) range
(6) variance
(7) standard deviation

(b) Are these data skewed? If so, how?

3.71 The following data represent the seven-day yield percentage on a specific date for a random sample of 25 money market mutual funds with assets of at least $100 million:

5.92, 5.71, 5.77, 5.91, 5.21,
5.62, 6.19, 4.47, 5.80, 5.92,
6.06, 6.07, 5.83, 6.02, 5.57,
6.09, 6.11, 6.06, 6.13, 5.90,
5.82, 6.22, 5.64, 6.01, 6.08

(a) Compute the:

(1) mean
(2) median
(3) mode
(4) midrange
(5) range
(6) variance
(7) standard deviation

(b) Are these data skewed? If so, how?

3.72 The following data represent the price (in cents) per 8 ounces for a sample of 20 brands of sparkling spring and mineral water sold in New York City supermarkets:

31	40	28	30	63
35	38	33	42	22
36	68	31	32	38
34	46	34	34	28

(a) Compute the:
(1) mean
(2) median
(3) mode
(4) midrange
(5) range
(6) variance
(7) standard deviation

(b) Are these data skewed? If so, how?

• **3.73** The following data are the price-to-earnings ratios (recorded to the last full value) for a random sample of 50 stocks from the New York Stock Exchange:

7	9	8	6	12	6	9	15	9	16
8	5	14	8	7	6	10	8	11	4
10	6	16	5	10	12	7	10	15	7
10	8	8	10	18	8	10	11	7	10
7	8	15	23	13	9	8	9	9	13

(a) Compute the:
(1) mean
(2) median
(3) mode
(4) midrange
(5) range
(6) variance
(7) standard deviation

(b) Are these data skewed? If so, how?

(c) Based on the Bienayme-Chebyshev rule, between what two values would we estimate that at least 75% of the data are contained?

(d) What percentage of the data are actually contained within ±2 standard deviations of the mean? Compare the results with those from (c).

3.74 The following data represent the projections of percent changes in real GNP for a recent year by a sample of 27 economists:

5.0, 4.5, 5.6, 4.6, 4.5, 4.1, 4.6, 4.7, 4.3,
4.5, 4.2, 3.9, 4.0, 4.3, 4.3, 3.6, 4.6, 3.5,
4.6, 4.9, 3.7, 5.1, 4.3, 4.4, 4.4, 4.0, 4.3

(a) Compute the:
(1) mean
(2) median
(3) mode
(4) midrange
(5) range
(6) variance
(7) standard deviation

(b) Are these data skewed? If so, how?

3.75 The following data represent the amount of time (in seconds) to get from 0 to 60 mph during a road test for a sample of 17 German-made automobile models and a sample of 18 Japanese-made automobile models:

German-made cars				Japanese-made cars			
11.2	8.5	10.9	13.0	11.5	11.5	9.2	11.8
11.6	6.2	7.0	10.6	9.7	14.2	12.8	8.9
8.2	13.2	8.3		7.9	14.3	15.0	10.5
10.3	10.4	8.1		15.1	9.7	11.3	
8.3	10.1	11.7		13.0	13.2	14.8	

(a) For each set of data compute the mean, median, range, and standard deviation.
(b) Summarize your findings. Discuss.

3.76 The following data represent the tuition charged (in $000) at a sample of 15 private colleges in the northeast and at a sample of 15 private colleges in the midwest during the academic year 1986–7:

Northeast colleges			Midwest colleges		
9.5	7.9	8.6	6.9	9.6	7.4
9.1	8.3	8.1	7.2	9.1	8.2
9.0	8.7	10.2	8.1	7.5	9.7
10.0	9.4	9.5	8.3	6.5	8.5
8.8	9.0	8.9	7.8	8.3	8.8

(a) For each set of data compute the mean, median, range, and standard deviation.
(b) Summarize your findings. Discuss.

3.77 The following data are the ages of a sample of 25 automobile salesmen in the United States and a sample of 25 automobile salesmen in Western Europe.

United States					Western Europe				
23	63	25	22	32	43	26	30	27	40
56	30	34	56	30	35	48	36	47	41
25	48	44	27	26	34	45	30	38	33
38	26	30	39	30	35	44	24	33	40
36	32	36	38	33	31	23	29	37	28

(a) For each set of data compute the mean, median, range, and standard deviation.
(b) Summarize your findings. Discuss.

3.78 Using the annual reports from a *population* of 19 companies associated with leisure-time and recreational activities, the following data represent the earnings-per-share during the year 1985:

Company	Earnings-per-share	Company	Earnings-per-share	Company	Earnings-per-share
American Greetings	2.35	Fleetwood Enterprises	2.09	Mattel	1.00
Bally Mfg.	0.95	Gibson Greetings	1.81	Outboard Marine	1.43
Brunswick	2.34	Hasbro	3.55	Polaroid	1.19
Caesars World	1.15	Lorimar	2.60	Ramada Inns	0.37
Coleco Ind.	3.87	MCA	2.02	UA Communications	0.68
Disney	5.24	MGM/UA Entertainment	−2.91	Warner Communications	2.87
Eastman Kodak	1.46				

(a) Compute the:
- **(1)** mean
- **(2)** median
- **(3)** mode
- **(4)** midrange
- **(5)** range
- **(6)** variance
- **(7)** standard deviation

(b) Are these data skewed? If so, how?

3.79 The following data represent the fringe benefits as a percentage of salaries paid at the 34 institutions of higher learning in the state of Colorado during the 1983–1984 academic year:

Institution	Ben. as % of salary	Institution	Ben. as % of salary
Adams State College	22	Northeastern Jr. College	18
Arapahoe Cmty. College	17	Otero Junior College	18
Blair J.C.	13	Parks College	16
Colorado College	23	Pikes Peak Cmty. College	17
Colorado Mountain C.	20	Pueblo Vocational C.C.	17
Colorado Northwestern C.C.	17	Regis College	17
Colorado School of Mines	21	Saint Thomas Seminary	22
Colorado State University	17	Trinidad State Jr. College	17
Colo. Technical College	14	U. of Colorado at Boulder	15
C.C. of Denver Auraria Cam.	20	U. of Colo. Colo. Springs	15
C.C. of Denver North Campus	20	U. of Colo. at Denver	15
C.C. Denver Red Rocks Cam.	18	U. of Colo. Hlth. Sci. Center	20
Denver Cons. Bapt. Sem.	29	University of Denver	16
Lamar Community College	17	U. of Northern Colorado	19
Mesa College	18	U. of Southern Colorado	16
Metropolitan St. College	21	Western Bible College	14
Nazarene Bible College	31	Western St. College Colo.	21

SOURCE: Data are taken from *Academe—Bulletin of the AAUP*, vol. 70, no. 2, July–August 1984, p. 24.

(a) Compute the:
- **(1)** mean
- **(2)** median
- **(3)** mode
- **(4)** midrange
- **(5)** range
- **(6)** variance
- **(7)** standard deviation

(b) Are these data skewed? If so, how?

3.80 The following data represent the annual per capita alcohol consumption (in gallons) in all 50 states plus the District of Columbia in a recent year:

State	Per capita consumption	State	Per capita consumption
District of Columbia	5.34	Louisiana	2.63
Nevada	5.19	Michigan	2.60
New Hampshire	4.91	Maine	2.57
Alaska	3.86	North Dakota	2.55
Wisconsin	3.19	Virginia	2.55
California	3.19	South Carolina	2.50
Delaware	3.17	Georgia	2.48
Vermont	3.12	Idaho	2.43
Florida	3.12	Nebraska	2.41
Colorado	3.09	South Dakota	2.33
Arizona	3.08	Missouri	2.27
Massachusetts	3.04	Ohio	2.26
Hawaii	2.97	Pennsylvania	2.25
Montana	2.95	Indiana	2.19
Rhode Island	2.92	North Carolina	2.13
Wyoming	2.86	Iowa	2.09
Maryland	2.84	Mississippi	2.06
New Jersey	2.83	Kansas	1.95
Texas	2.82	Tennessee	1.95
Connecticut	2.80	Oklahoma	1.91
Illinois	2.77	Kentucky	1.85
New Mexico	2.75	Alabama	1.80
Washington	2.71	Arkansas	1.78
Minnesota	2.68	West Virginia	1.68
New York	2.67	Utah	1.53
Oregon	2.63		

(a) Compute the:
- **(1)** mean
- **(2)** median
- **(3)** mode
- **(4)** midrange
- **(5)** range
- **(6)** variance
- **(7)** standard deviation

(b) Are these data skewed? If so, how?

Database Exercises

The following problems refer to the sample data obtained from the questionnaire of Figure 2.5 and presented in Figure 2.10. They should be solved (to the greatest extent possible) with the aid of a computer package.

□ **3.81** From the responses to question 1 of the alumni association survey obtain the
- **(a)** mean,
- **(b)** median,
- **(c)** mode,
- **(d)** midrange,
- **(e)** range,
- **(f)** standard deviation

for the number of years since graduation.

Note: Box □ indicates those problems that refer specifically to alumni survey (Case Study A).

□ 3.82 From the responses to question 7 of the alumni association survey obtain the
(a) mean, **(d)** midrange,
(b) median, **(e)** range,
(c) mode, **(f)** standard deviation
for the current salaries of alumni association members.

□ 3.83 From the responses to question 10 of the alumni association survey obtain the
(a) mean, **(d)** midrange,
(b) median, **(e)** range,
(c) mode, **(f)** standard deviation
for the number of years working in the current job area.

□ 3.84 From the responses to question 12 of the alumni association survey obtain the
(a) mean, **(d)** midrange,
(b) median, **(e)** range,
(c) mode, **(f)** standard deviation
for the number of job areas worked in since graduation.

□ 3.85 From the responses to question 13 of the alumni association survey obtain the
(a) mean, **(d)** midrange,
(b) median, **(e)** range,
(c) mode, **(f)** standard deviation
for the number of full-time jobs held over the past ten years.

□ 3.86 From the responses to question 15 of the alumni association survey obtain the
(a) mean, **(d)** midrange,
(b) median, **(e)** range,
(c) mode, **(f)** standard deviation
for the grade-point indexes of alumni association members.

□ 3.87 From the responses to question 19 of the alumni association survey obtain the
(a) mean, **(d)** midrange,
(b) median, **(e)** range,
(c) mode, **(f)** standard deviation
for the heights of alumni association members.

□ 3.88 From the responses to question 20 of the alumni association survey obtain the
(a) mean, **(d)** midrange,
(b) median, **(e)** range,
(c) mode, **(f)** standard deviation
for the ages of alumni association members.

□ 3.89 From the responses to questions 1 and 16 of the alumni association survey obtain the
(a) mean, **(d)** midrange,
(b) median, **(e)** range,
(c) mode, **(f)** standard deviation
for the number of years since graduation based on whether or not the alumnus would have attended now.

□ 3.90 From the responses to questions 7 and 2 of the alumni association survey obtain the
(a) mean, **(d)** midrange,
(b) median, **(e)** range,
(c) mode, **(f)** standard deviation
for the current salaries of alumni association members based on whether or not their undergraduate major was in business.

□**3.91** From the responses to questions 7 and 5 of the alumni association survey obtain the
(a) mean, **(d)** midrange,
(b) median, **(e)** range,
(c) mode, **(f)** standard deviation
for the current salaries of alumni association members based on whether or not they attended graduate school.

□**3.92** From the responses to questions 15 and 2 of the alumni association survey obtain the
(a) mean, **(d)** midrange,
(b) median, **(e)** range,
(c) mode, **(f)** standard deviation
for the grade-point indexes of alumni association members based on whether or not their undergraduate major was in business.

□**3.93** From the responses to questions 15 and 5 of the alumni association survey obtain the
(a) mean, **(d)** midrange,
(b) median, **(e)** range,
(c) mode, **(f)** standard deviation
for the grade-point indexes of alumni association members based on whether or not they attended graduate school.

□**3.94** From the responses to questions 15 and 18 of the alumni association survey obtain the
(a) mean, **(d)** midrange,
(b) median, **(e)** range,
(c) mode, **(f)** standard deviation
for the grade-point indexes of alumni association members based on gender.

□**3.95** From the responses to questions 19 and 18 of the alumni association survey obtain the
(a) mean, **(d)** midrange,
(b) median, **(e)** range,
(c) mode, **(f)** standard deviation
for the heights of alumni association members based on gender.

References

1. ARKIN, H., AND R. COLTON, *Statistical Methods*, 5th ed. (New York: Barnes & Noble College Outline Series, 1970).
2. CROXTON, F., D. COWDEN, AND S. KLEIN, *Applied General Statistics*, 3d ed. (Englewood Cliffs, N.J.: Prentice-Hall, 1967).
3. KENDALL, M. G., AND A. STUART, *The Advanced Theory of Statistics*, Vol. I (London: Charles W. Griffin, 1958).

Data Presentation: Tables and Charts

4.1 INTRODUCTION: WHAT'S AHEAD

As pointed out in Section 2.5, we usually deal with sample information rather than data from the entire population. Nevertheless, regardless of whether we are dealing with a sample or a population, as a rule of thumb, whenever a set of data contains more than a few observations, a useful way to examine such "mass data" is to present it in summary form by constructing appropriate tables and charts. From these we can then extract the important features of the data.

The goal of this chapter is to demonstrate how data sets can be organized and effectively presented in terms of tables and charts as an aid to data analysis and interpretation.

4.2 TABULATING QUANTITATIVE DATA: THE FREQUENCY DISTRIBUTION

Using the raw data and ordered array of the reported sales tax receipts displayed in Figures 3.9 and 3.10 (page 68), the comptroller wishes to construct the appropriate tables and charts that will enhance the final report which must be prepared for the governor's office. Even though an ordered array has already been developed for *organizing* data, as the number of observations gets large it becomes necessary to further condense or summarize the data into appropriate summary tables. Thus we may wish to arrange the data into **class groupings** or **categories** according to conveniently established divisions of the range of the observations. Such an arrangement of data in tabular form is called a **frequency distribution.**

> **A frequency distribution is a summary table in which the data are arranged into conveniently established numerically ordered class groupings or categories.**

When the data are "grouped" or condensed into frequency-distribution tables, the process of data analysis and interpretation is made much more manageable and meaningful. In such summary form the major data characteristics are very easily approximated, thus compensating for the fact that when the data are so grouped, the initial information pertaining to individual observations that was previously available is lost through the grouping or condensing process.

In constructing the frequency-distribution table, attention must be given to:

1. Selecting the appropriate number of class groupings for the table.
2. Obtaining a suitable *class interval* or "width" of each class grouping.
3. Establishing the *boundaries* of each class grouping to avoid overlapping.

4.2.1 Selecting the Number of Classes

The number of class groupings to be used depends primarily upon the number of observations in the data. That is, the larger the number of observations, the larger the number of class groups, and vice versa. In general, however, the frequency distribution should have at least five class groupings, but no more than 15. If there are not enough class groupings, there is too much concentration of data; if there are too many groupings, there is too little concentration of data. In either case little information would be obtained.

As an example, a frequency distribution having but one class grouping could be formed for the reported sales tax receipts data as follows:

Reported sales tax receipts (in \$000)	Number of stores
0–20	50
Total	50

However, no additional information is obtained from such a summary table that was not already known from scanning either the raw data or the ordered array. A table with too much data concentration is not meaningful. The same would be true at the other extreme—if a table had too many class groupings, there would be an underconcentration of data, and very little would be learned.

4.2.2 Obtaining the Class Intervals

When developing the frequency distribution table, it is desirable to have each class grouping of equal width. To determine the width of each class, the range of the data is divided by the number of class groupings desired:

$$\text{width of interval} = \frac{\text{range}}{\text{number of class groupings}}$$

Since there are only 50 observations in our reported sales tax receipts data, we decide that approximately seven class groupings would be appropriate. From the ordered array in Figure 3.10 (page 68), the range is computed as $15.1 - 5.3 = 9.8$, and the width of the class interval is approximated by

$$\text{width of interval} = \frac{9.8}{7} = 1.4$$

For convenience and ease of reading, the selected interval or width of each class grouping is rounded to 1.5.

4.2.3 Establishing the Boundaries of the Classes

To construct the frequency-distribution table, we must establish clearly defined class boundaries for each class grouping so that the observations either in the raw form (Figure 3.9) or in an ordered array (Figure 3.10) can be properly tallied. Overlapping of classes must be avoided.

Since the width of each class interval for the sales tax data has been set at 1.5, the boundaries of the various class groupings must be established so as to include the entire range of observations. Whenever possible these boundaries should be chosen to facilitate the reading and interpreting of data. Thus the first class interval is established from 5.0 to under 6.5, the second from 6.5 to under 8.0, and so on. The data in their raw form (Figure 3.9) or from the ordered array (Figure 3.10) are then tallied into each class as shown:

Reported sales tax receipts (in $000)	Frequency tallies	
5.0 but less than 6.5	/	1
6.5 but less than 8.0	𝍸 /	6
8.0 but less than 9.5	𝍸 ////	9
9.5 but less than 11.0	𝍸 𝍸 𝍸	15
11.0 but less than 12.5	𝍸 ////	9
12.5 but less than 14.0	𝍸 ///	8
14.0 but less than 15.5	//	2
Total		50

By establishing the boundaries of each class as above, we have tallied all 50 observations into seven classes, each having an interval width of 1.5, without overlapping. From this "worksheet" the frequency distribution is presented in Table 4.1.

The main advantage of using such a summary table is that the major data characteristics become immediately clear to the reader. For example, we see from Table 4.1 that the *approximate* range of the 50 reported sales tax receipts is from 5.0 to 15.5 thousands of dollars. Most observations tend to cluster between 9.5 and 11.0 thousands of dollars. We can also see that this interval must contain the median observation, since fewer than 50% of the observations (i.e., 16 out of 50) are less than 9.5, but more than 50% of the observations (i.e., 31 out of 50) are less than 11.0.

TABLE 4.1 □ Frequency distribution of reported sales tax receipts from 50 stores in Heldman Township

Reported sales tax receipts (in $000)	Number of stores
5.0 but less than 6.5	1
6.5 but less than 8.0	6
8.0 but less than 9.5	9
9.5 but less than 11.0	15
11.0 but less than 12.5	9
12.5 but less than 14.0	8
14.0 but less than 15.5	2
Total	50

SOURCE: Data taken from Figure 3.9 on page 68.

On the other hand, the major disadvantage of such a summary table is that we cannot know how the individual values are distributed within a particular class interval without access to the original data. Thus for the two stores which report sales tax receipts between 14.0 and 15.5 thousands of dollars, it is not clear from Table 4.1 whether the values are distributed throughout the interval, are close to 14.0, or are close to 15.5. The **class midpoint,** however, is the value used to represent all the data summarized into a particular interval.

> **The class midpoint is the point halfway between the boundaries of each class and is representative of the data within that class.**

The class midpoint for the interval ''14.0 but less than 15.5'' is 14.75. (The other class midpoints are, respectively, 5.75, 7.25, 8.75, 10.25, 11.75, 13.25.)

4.2.4 Subjectivity in Selecting Class Boundaries

The selection of class boundaries for frequency-distribution tables is highly subjective. Hence for data sets which do not contain many observations, the choice of particular class boundaries might yield an entirely different picture than would other choices. For example, for the reported sales tax receipts data, using a class-interval width of 2.0 thousands of dollars instead of 1.5 (as was used in Table 4.1) may cause shifts in the way in which the set of data is distributed among the classes. This is particularly true if the number of observations in the set is not very large.

However, such shifts in data concentration do not occur only because the width of the class interval is altered. We may keep the interval width at 2.0 thousands of dollars but choose different lower and upper class boundaries. Such manipulation may also cause shifts in the way in which the data are distributed—especially if the size of the data set is not very large. Fortunately, as the number of observations increases, alterations in the selection of class boundaries affect the concentration of data less and less.

Problems

4.1 Using the ordered array in Figure 3.10 on page 68 set up the frequency distributions with the following class boundaries:
- **(a)** 4.0 but less than 5.5, 5.5 but less than 7.0, and so on.
- **(b)** 4.0 but less than 6.0, 6.0 but less than 8.0, and so on.
- **(c)** 5.0 but less than 7.0, 7.0 but less than 9.0, and so on.
- **(d)** 5.0 but less than 7.5, 7.5 but less than 10.0, and so on.
- **(e)** 3.0 but less than 6.0, 6.0 but less than 9.0, and so on.

(f) Compare and contrast the different results and state whether you believe any of these distributions should be preferred to the one selected in Table 4.1. [*Hint*: To help you decide, compare the *approximate* mode, midrange, and range for each frequency distribution to the *actual* mode, midrange, and range in the (ungrouped) data.]

4.2 Construct a frequency distribution for the data on unpaid accounts from Problem 3.5 on page 45. (Consider both the range of values and number of data values in deciding the number of class intervals.)

4.3 Construct a frequency distribution for the creativity test data in Problem 3.12 on page 46.

4.4 Construct a frequency distribution for the death-rate data in Problem 3.13 on page 46 using **(a)** 5 classes and **(b)** 8 classes. Does one seem to be better? Why?

4.5 Construct a frequency distribution for the water-price data in Problem 3.72 on page 72.

4.6 Construct a frequency distribution for the stock price-to-earnings ratio data in Problem 3.73 on page 72. Now pick a different class interval and construct another frequency distribution. Does one seem more informative than the other? Why?

• **4.7** Construct a frequency distribution for the forecasts of GNP changes in Problem 3.74 on page 72 using a class interval of **(a)** 0.5 and **(b)** 1.0. (In each case, make the lower limit of the first class boundary start at 3.0.) Is one more easily interpreted than the other?

4.8 Construct frequency distributions of the acceleration data on automobiles in Problem 3.75 on page 73. Put the German-made cars and Japanese-made cars in separate distributions, but next to each other so that they can be compared. Describe the distributions, and see if looking at the distributions leads to any different conclusions than you had in answering part (b) of Problem 3.75.

4.9 Construct frequency distributions of the tuition data in Problem 3.76 on page 73. Put the distribution of the colleges in the Northeast next to the distribution of the colleges in the Midwest. Is there anything you see in the distributions which would change the opinion expressed in your answer to part (b) of Problem 3.76?

4.10 Construct frequency distributions of the data on ages of automobile sales representatives presented in Problem 3.77 on page 73. Do sales representatives seem older in the United States or in Western Europe? Do the spreads seem the same in both locations?

4.11 The following data represent the noise level generated by several different brands of chainsaws (in decibels):

103, 103, 105, 106, 108, 108, 106, 107, 106, 105,
104, 106, 107, 97, 93, 94, 93, 91, 95, 92, 94, 95

(a) Construct a frequency distribution of these data, and write a description of what you see.

(b) There are two kinds of chainsaws: gas-powered and electric-powered. From what you know (or can find out) about how much noise gas and electric engines make, guess why the frequency distribution shows two "clumps."

• **4.12** The following data are the calories in a 4-oz. serving of cottage cheese from various producers:

113, 98, 123, 105, 95, 107, 106, 100, 105, 108,
110, 100, 109, 107, 97, 106, 103, 118, 122, 116,
119, 113, 107, 105,
99, 88, 76, 83, 92, 89, 81, 110, 89, 79,
75, 81, 95, 82, 79, 87, 82, 85, 75, 93,
83, 99, 76, 79, 90, 73, 84, 85, 107, 91

SOURCE: Copyright 1986 by Consumers Union of United States, Inc., Mount Vernon, NY 10553. Excerpted by permission from *Consumer Reports*, March 1986.

(a) Construct a frequency distribution of this data set, and write a description of what you see.
(b) Construct another frequency distribution of this data set using different class intervals, and tell whether anything looks different.
(c) Construct two frequency distributions next to each other, with the first 24 observations (i.e., rows 1 through 3) in the first distribution and the last 30 observations in the second distribution. Do they appear to be similar?
(d) Suppose that there are two kinds of cottage cheese: one kind has 4% milkfat, the other kind 2% milkfat or less. Now what do you think is represented by the data, and what do you think the two sections of observations mentioned in part (c) are?

4.13 The high temperature (in degrees Fahrenheit) for a recent day in several U.S. cities was:

40, 59, 48, 51, 47, 45, 47, 50, 65, 59,
45, 81, 73, 56, 62, 72, 78, 38, 54, 62,
74, 48, 72, 50, 58, 50

(a) Construct a frequency distribution of this data set.
(b) Guess what time of year it was when the readings were made, and give reasons for your guess. If there is more than one possibility, with varying likelihoods of occurrence, describe them.

4.14 A random sample of 50 executive vice-presidents was selected from among the various public relations firms in the United States; among other information their salaries were obtained. The salaries ranged from $52,000 to $137,000. Set up the class boundaries for a frequency distribution if:
(a) 5 class intervals are desired.
(b) 6 class intervals are desired.
(c) 7 class intervals are desired.
(d) 8 class intervals are desired.

4.15 If the annual salaries of city employees varied from $12,700 to $56,200:
(a) Indicate the class boundaries of 10 classes into which these values can be grouped.
(b) What class-interval width did you choose?
(c) What are the 10 class midpoints?

4.16 If the asking price of one-bedroom cooperative and condominium apartments in New York City varied from $23,000 to $245,000:
(a) Indicate the class boundaries of 10 classes into which these values can be grouped.

(b) What class-interval width did you choose, and why?
(c) What are the 10 class midpoints?

4.17 The raw data displayed below are the electric and gas utility charges during the month of July 1987 for a random sample of 50 three-bedroom apartments in Manhattan:

Raw data on utility charges ($)

96	171	202	178	147	102	153	197	127	82
157	185	90	116	172	111	148	213	130	165
141	149	206	175	123	128	144	168	109	167
95	163	150	154	130	143	187	166	139	149
108	119	183	151	114	135	191	137	129	158

(a) Form a frequency distribution:
(1) having 5 class intervals.
(2) having 6 class intervals.
(3) having 7 class intervals.
(b) Form a frequency distribution having 7 class intervals with the following class boundaries: $80 but less than $100, $100 but less than $120, and so on.

4.18 Given the ordered arrays in the accompanying table dealing with the lengths of life (in hours) of a sample of 40 100-watt light bulbs produced by Manufacturer A and a sample of 40 100-watt light bulbs produced by Manufacturer B:

Ordered arrays of length of life of two brands of 100-watt light bulbs (in hours)

Manufacturer A				
684	697	720	773	821
831	835	848	852	852
859	860	868	870	876
893	899	905	909	911
922	924	926	926	938
939	943	946	954	971
972	977	984	1,005	1,014
1,016	1,041	1,052	1,080	1,093
Manufacturer B				
819	836	888	897	903
907	912	918	942	943
952	959	962	986	992
994	1,004	1,005	1,007	1,015
1,016	1,018	1,020	1,022	1,034
1,038	1,072	1,077	1,077	1,082
1,096	1,100	1,113	1,113	1,116
1,153	1,154	1,174	1,188	1,230

(a) Form the frequency distribution for each brand. [*Hint*: For purposes of comparison, choose class interval widths of $100 for each distribution.]

(b) Form the frequency distribution for each brand according to the following schema [if you had not already done so in part (a) of this problem]:

Manufacturer A: 650 but less than 750, 750 but less than 850, and so on
Manufacturer B: 750 but less than 850, 850 but less than 950, and so on

4.19 The raw data displayed in the accompanying table are the starting salaries for a random sample of 100 computer science or computer systems majors who earned their baccalaureate degrees during a recent year:

Starting salaries ($000)									
24.2	29.9	23.4	23.0	25.5	22.0	33.9	20.4	26.6	24.0
28.9	22.5	18.7	32.6	26.1	26.2	26.7	20.4	22.2	24.7
18.6	18.5	19.6	24.4	24.8	27.8	27.6	27.2	20.8	22.1
19.7	25.3	28.2	34.2	32.5	30.8	26.8	20.6	21.2	20.7
25.2	25.7	32.2	28.8	24.7	18.7	20.5	25.5	19.1	25.5
22.1	27.5	25.8	25.2	25.6	25.2	25.2	27.9	18.9	37.3
29.9	23.2	19.8	20.8	29.5	27.6	21.2	38.7	21.3	24.8
32.3	20.1	26.8	25.4	26.3	21.2	19.5	22.8	21.7	25.3
32.3	28.1	27.5	25.3	19.3	27.4	26.4	20.9	34.5	25.9
31.4	27.4	27.3	20.6	31.8	25.8	25.2	21.9	26.8	26.5

(a) Place the raw data in an ordered array.
(b) Form the frequency distribution.

4.20 The ordered arrays displayed in the accompanying table are the batting averages from a sample of 40 players in the American League and a sample of 40 players in the National League. Form the frequency distributions for each league.

Batting averages*			
American League		National League	
.184	.255	.201	.263
.204	.268	.205	.263
.207	.271	.214	.265
.209	.272	.220	.274
.210	.278	.227	.275
.219	.279	.233	.279
.220	.280	.234	.287
.220	.282	.242	.290
.225	.291	.242	.300
.227	.292	.245	.306
.227	.294	.247	.308
.235	.298	.248	.310
.240	.299	.250	.319
.241	.301	.253	.320
.247	.302	.253	.320
.248	.313	.253	.338
.252	.316	.253	.342
.253	.321	.261	.356
.255	.327	.261	.359
.255	.348	.261	.375

* For individuals with at least 75 "at bats."

4.3 TABULATING QUANTITATIVE DATA: THE RELATIVE FREQUENCY DISTRIBUTION AND PERCENTAGE DISTRIBUTION

The frequency distribution is a summary table into which the original data are condensed or grouped to facilitate data analysis. To facilitate the analysis further, however, it is often desirable to form either the **relative frequency distribution** or the **percentage distribution,** depending on whether we prefer proportions or percentages. These two equivalent distributions are shown in Tables 4.2 and 4.3, respectively.

The relative frequency distribution depicted in Table 4.2 is formed by dividing the frequencies in each class of the frequency distribution (Table 4.1) by the total number of observations. A percentage distribution (Table 4.3) may then be formed by multiplying each relative frequency or proportion by 100.0. Thus from Table

TABLE 4.2 □ Relative frequency distribution of reported sales tax receipts from 50 stores in Heldman Township

Reported sales tax receipts (in $000)	Proportion of stores
5.0 but less than 6.5	.02
6.5 but less than 8.0	.12
8.0 but less than 9.5	.18
9.5 but less than 11.0	.30
11.0 but less than 12.5	.18
12.5 but less than 14.0	.16
14.0 but less than 15.5	.04
Total	1.00

SOURCE: Data are taken from Table 4.1.

TABLE 4.3 □ Percentage distribution of reported sales tax receipts from 50 stores in Heldman Township

Reported sales tax receipts (in $000)	Percentage of stores
5.0 but less than 6.5	2.0
6.5 but less than 8.0	12.0
8.0 but less than 9.5	18.0
9.5 but less than 11.0	30.0
11.0 but less than 12.5	18.0
12.5 but less than 14.0	16.0
14.0 but less than 15.5	4.0
Total	100.0

SOURCE: Data are taken from Table 4.1.

4.2 it is clear that the proportion of stores reporting sales tax receipts from 5.0 to under 6.5 thousands of dollars is .02, while from Table 4.3 we see that 2.0% of the stores report such sales tax receipts.

Working with a base of 1 for proportions or of 100.0 for percentages is usually more meaningful than using the frequencies themselves. Indeed, the use of the relative frequency distribution or percentage distribution becomes essential whenever one set of data is being compared to other sets of data, especially if the numbers of observations in each set differ.

As a case in point, let us suppose that at the county level a researcher wishes to compare the reported sales tax receipts from the 50 stores in Heldman Township with those reported from the 35 stores in Safran Township (the only other township in the county). Figure 4.1 presents the raw data for the 35 reported sales tax receipts from Safran Township.

8.0	10.5	10.4	8.1
7.7	11.8	7.9	9.6
6.4	8.7	7.5	6.6
9.4	10.4	9.3	9.7
7.9	9.0	9.4	7.8
9.7	8.9	8.8	
9.2	10.8	6.2	
12.3	7.0	10.6	
8.6	11.5	8.3	
9.8	10.3	5.5	

FIGURE 4.1 □ Raw data pertaining to reported quarterly sales tax receipts for all 35 small businesses in Safran Township.

To compare the reported sales tax receipts from the 50 stores in Heldman Township with those from the 35 stores in Safran Township we will develop a percentage distribution for the latter group. This new table will be compared with Table 4.3.

Table 4.4 depicts both the frequency distribution and the percentage distribution of the reported sales tax receipts from the 35 stores in Safran Township. Constructing one such table, in lieu of two separate tables, to save space is both permissible and desirable. This is also true when comparing two or more percentage distributions. Table 4.5 shows how this is done.

TABLE 4.4 □ Frequency distribution and percentage distribution of reported sales tax receipts from 35 stores in Safran Township

Reported sales tax receipts (in $000)	Number of stores	Percentage of stores
5.0 but less than 6.5	3	8.6
6.5 but less than 8.0	7	20.0
8.0 but less than 9.5	12	34.3
9.5 but less than 11.0	10	28.6
11.0 but less than 12.5	3	8.6
Totals	35	100.1*

* Error due to rounding.
SOURCE: Data are taken from Figure 4.1.

TABLE 4.5 ☐ Percentage distribution of reported sales tax receipts from stores in Heldman and Safran Townships

Reported sales tax receipts (in $000)	Percentage of stores Heldman ($n = 50$)	Safran ($n = 35$)
5.0 but less than 6.5	2.0	8.6
6.5 but less than 8.0	12.0	20.0
8.0 but less than 9.5	18.0	34.3
9.5 but less than 11.0	30.0	28.6
11.0 but less than 12.5	18.0	8.6
12.5 but less than 14.0	16.0	0.0
14.0 but less than 15.5	4.0	0.0
Total	100.0	100.1*

* Error due to rounding.
SOURCE: Data are taken from Tables 4.3 and 4.4.

Note that the class groupings selected in Table 4.4 for the stores in Safran Township match, where possible, those selected in Tables 4.1 through 4.3 for the stores in Heldman Township. The boundaries of the classes should match or be multiples of each other in order to facilitate comparisons.

4.3.1 Grouped Data Comparisons: Central Tendency and Dispersion

Using the percentage distributions of Tables 4.3 and 4.4, it is now meaningful to compare the two townships in terms of reported quarterly sales tax receipts. For example, from the latter table it is clear that most typically the reports from Safran Township are clustering between 8.0 and 9.5 thousands of dollars. The most frequently observed class grouping is called the **modal group.** This group accounts for 34.3% of the reports from Safran Township. On the other hand, using Table 4.3, we observe that in Heldman Township the modal group (accounting for 30.0% of the reports) is between 9.5 and 11.0 thousands of dollars.[1] This may be an indication of more affluence in Heldman Township and/or better shopping facilities (i.e., access, parking, location, etc.), higher-priced merchandise, or larger sales volume per store.

What other kinds of comparisons can be made from the data in Tables 4.3 and 4.4? The **midrange** can be approximated by averaging the possible extremes—the upper boundary of the last class grouping and the lower boundary of the first class grouping—while the **range** can be approximated by simply taking the difference between these two values.

[1] Since the numbers of observations differ (50 stores versus 35 stores), it would obviously not be appropriate to compare the actual numbers (that is, frequency tallies) of responses in these class groupings without first adjusting for differences in the two totals.

In Heldman Township the midrange is approximately 10.25 thousands of dollars (the average of 5.0 and 15.5), while the range in reported quarterly sales tax receipts is approximately 10.5 thousands of dollars (15.5 − 5.0). On the other hand, in Safran Township the midrange is approximately 8.75 thousands of dollars (the average of 5.0 and 12.5), while the range in reported quarterly sales tax receipts is approximately 7.5 thousands of dollars (12.5 − 5.0).

From this we can observe that the difference in the two midranges (that is, measures of central tendency) has given support to our previous notions regarding possibly more affluence and/or better shopping facilities, higher-priced merchandise, or larger sales volume per store in Heldman Township. On the other hand, the difference in the two ranges (that is, measures of dispersion) adds to our insight. The reported quarterly sales tax receipts from Safran Township are much more homogeneous (that is, heavily concentrated about the measures of central tendency) than those from Heldman Township. Is there, perhaps, a wider variety of shops and stores in the latter? Or are the socioeconomic status characteristics of the residents in Safran Township more homogeneous than those from Heldman Township? (We leave it to the reader to list other potential reasons as well.) What can be gleaned from the data, however, is that while only 8.6% of the stores in Safran Township report sales tax receipts of *at least* 11.0 thousands of dollars, 38.0% of the stores do so in Heldman Township. Moreover, at the other end of the distributions we observe that 62.9% of stores in Safran Township report sales tax receipts under 9.5 thousands of dollars, as compared with only 32.0% of the stores in Heldman Township.[2]

4.3.2 Grouped Data Comparisons: Shape

A final comparison between the data from the two townships is based on shape. As will be indicated in Figure 4.4 on page 93, the shapes appear to be similar—a heavy concentration of reported sales tax receipts in the middle with a light concentration at the extremes. Hence both data sets tend to "balance out" and may be considered as approximately symmetrical, "bell-shaped normal" distributions.

Problems

4.21 Construct a percentage distribution for the frequency distribution you constructed in Problem 4.2 on page 83 of data on unpaid accounts.

[2] Formulas exist for approximating the mean, median, variance, and standard deviation when data have been grouped into a frequency distribution. (See Reference 1.) In this text, however, we take the position that such approximations should be made only if the original raw data are unavailable. If the individual observations $X_1, X_2, \ldots, X_n$ are available, we may use an electronic calculator to obtain the actual results. For large data sets, we may enter the data into a computer and employ statistical software packages to obtain the results. (See Chapter 15.)

4.22 Construct a percentage distribution for the frequency distribution you constructed in Problem 4.3 on page 83 of data from the creativity test.

4.23 Construct a percentage distribution for the frequency distribution you constructed in Problem 4.4 on page 83 of data on death rates. (Use the 8-class distribution.)

4.24 Construct a percentage distribution for the frequency distribution you constructed in Problem 4.5 on page 83 of water-price data.

4.25 Construct a percentage distribution for the frequency distribution you constructed in Problem 4.6 on page 83 for the stock price-to-earnings-ratio data.

• 4.26 Construct a percentage distribution for the frequency distribution you constructed in Problem 4.7 on page 83 for the GNP forecasts.

4.27 Construct a percentage distribution for the frequency distribution in Problem 4.13 on page 84 of the data on temperature in U.S. cities.

4.28 Construct a percentage distribution for the frequency distribution in Problem 4.11 on page 83 of the data on noise level of chainsaws.

4.29 Construct a table with the percentage distributions of the acceleration of the German-made and Japanese-made cars, for which you constructed frequency distributions in Problem 4.8 on page 83. Compare the central tendency, dispersion, and shape of the two distributions.

4.30 For the data on tuition for which you constructed frequency distributions in Problem 4.9 on page 83, construct percentage distributions. Comment on the location, dispersion, and shape of the distribution of tuition in the colleges of the Northeast versus those in the Midwest.

4.31 Construct percentage distributions for the ages of automobile salesmen in the United States and in Western Europe, for which you constructed frequency distributions in Problem 4.10 on page 83. Do the two distributions have similar shapes? Do they have about the same central tendency? Dispersion?

• 4.32 Construct percentage distributions for the 4% milkfat and 2% milkfat cottage cheeses for which you constructed frequency distributions in Problem 4.12 on page 83. How similar are the two distributions in central tendency? Dispersion? Shape?

4.33 Construct percentage distributions for the light-bulb data in Problem 4.18 on page 85. (Construct separate distributions for the data from each manufacturer.) Compare the central tendency, dispersion, and shape of the two distributions.

4.34 Construct a percentage distribution for the salary data in Problem 4.19 on page 86.

4.35 Construct percentage distributions for the batting averages of the American and National League players, which you encountered in Problem 4.20 on page 86. Compare the two distributions.

4.4 GRAPHING QUANTITATIVE DATA: THE HISTOGRAM AND POLYGON

It is often said that "one picture is worth a thousand words." Indeed, statisticians have employed graphic techniques to describe sets of data more vividly than numbers alone can do. In particular, **histograms** and **polygons** are used to describe quantitative data which have been grouped into frequency, relative frequency, or percentage distributions.

4.4.1 Histograms

Histograms are vertical bar charts in which the rectangular bars are constructed at the boundaries of each class.

When plotting histograms, the random variable or phenomenon of interest is plotted along the horizontal axis; the vertical axis represents the number, proportion, or percentage of observations per class interval—depending on whether or not the particular histogram is, respectively, a frequency histogram, a relative frequency histogram, or a percentage histogram.

Vertical axis label	Type of chart
Number of observations	*Frequency* histogram or polygon
Proportion of observations	*Relative frequency* histogram or polygon
Percentage of observations	*Percentage* histogram or polygon

A percentage histogram is depicted in Figure 4.2 for the reported sales tax receipts from the 50 stores in Heldman Township.

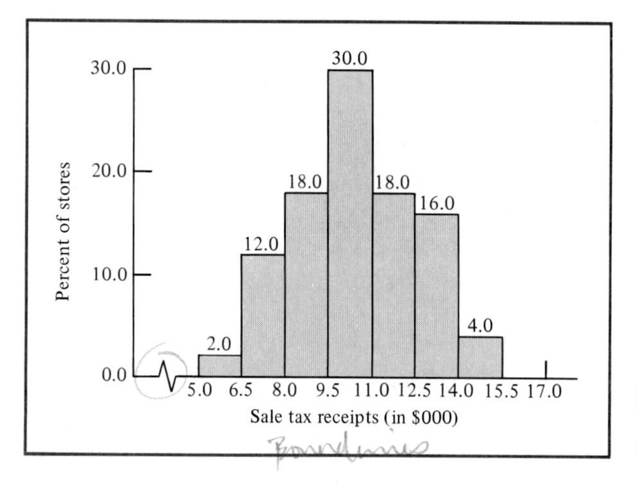

FIGURE 4.2 □ Percentage histogram of reported sales tax receipts from 50 stores in Heldman Township.

SOURCE: Data are taken from Table 4.3.

When comparing two or more sets of data, however, the various histograms cannot be constructed on the same graph, because superimposing the vertical bars of one on another would cause difficulty in interpretation. For such cases it is necessary to construct relative frequency or percentage polygons.

4.4.2 Polygons

As with histograms, when plotting polygons the phenomenon of interest is plotted along the horizontal axis, while the vertical axis represents the number, proportion, or percentage of observations per class interval.

The polygon is formed by letting the midpoint of each class represent the data in that class and then connecting the sequence of midpoints.

Figure 4.3 shows the percentage polygon for the reported sales tax receipts from the 50 stores in Heldman Township, and Figure 4.4 compares the percentage polygons of the 50 stores in Heldman Township versus the 35 stores in Safran Township in terms of their reported quarterly sales tax receipts. The similarities and differences between these two townships, previously discussed when examining Table 4.5, are clearly indicated here.

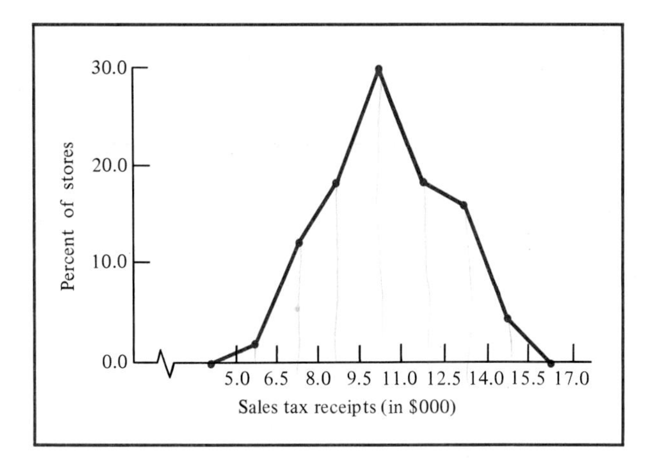

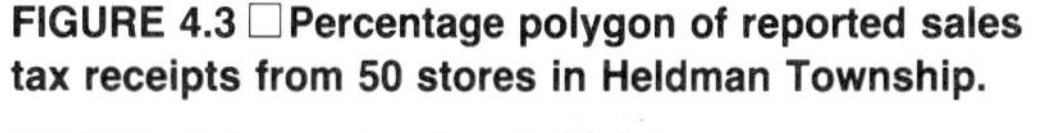

FIGURE 4.3 □ Percentage polygon of reported sales tax receipts from 50 stores in Heldman Township.

SOURCE: Data are taken from Table 4.3.

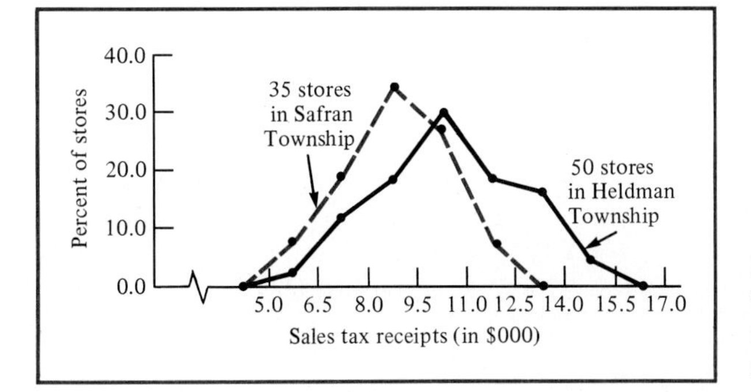

FIGURE 4.4 □ Percentage polygons of reported sales tax receipts from 50 stores in Heldman Township and 35 stores in Safran Township.

SOURCE: Data are taken from Tables 4.3 and 4.4.

POLYGON CONSTRUCTION

The polygon is a representation of the shape of the particular distribution. To complete the polygon, we connect the first and last midpoints with the horizontal axis so as to enclose the area of the observed distribution. In Figure 4.3 this is accomplished by connecting the first observed midpoint with the midpoint of a "fictitious preceding" class (4.25) having 0.0% observations and by connecting the last observed midpoint with the midpoint of a "fictitious succeeding" class (16.25) having 0.0% observations.

Note that when polygons (Figure 4.3) or histograms (Figure 4.2) are constructed, the vertical axis must show the true zero or ''origin'' so as not to distort or otherwise misrepresent the character of the data. The horizontal axis, however, does not need to specify the zero point for the phenomenon of interest. For aesthetic reasons, the range of the random variable should constitute the major portion of the chart. When zero is not included, ''breaks'' (—/\—) in the axis are appropriate.

USING POLYGONS FOR COMPARING GROUPED DATA SETS

Polygons provide us with a useful visual aid for comparing two or more sets of data. Expanding on the dot diagram scales displayed in Chapter 3, the properties of central tendency, dispersion, and shape can be depicted by comparing the polygons of particular distributions of data.

Figure 4.5 depicts a perfectly symmetrical bell-shaped distribution called the **normal distribution,** which will be discussed in greater detail in Chapter 7. The mean, median, mode, and midrange are (theoretically) identical.

Figure 4.6 displays two identical normal distributions. Polygons A and B are superimposed on one another.

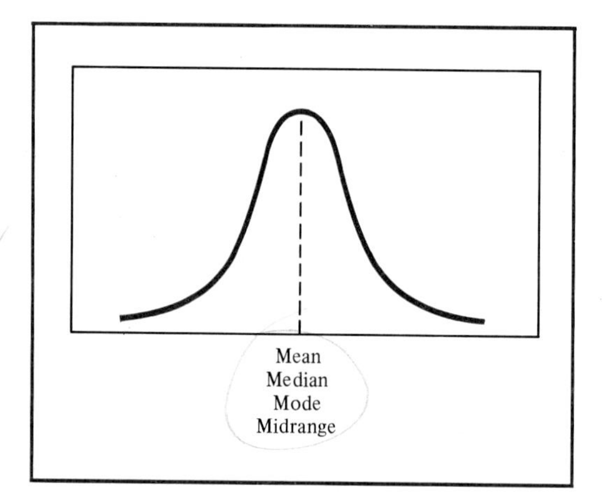

FIGURE 4.5 □ Bell-shaped curve.

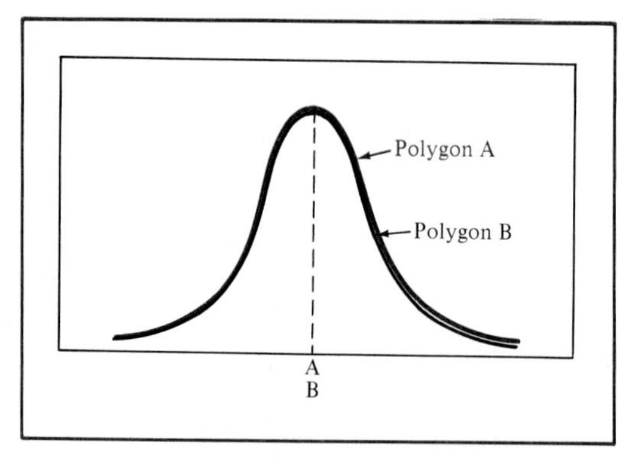

FIGURE 4.6 □ Two identical symmetrical bell-shaped normal distributions.

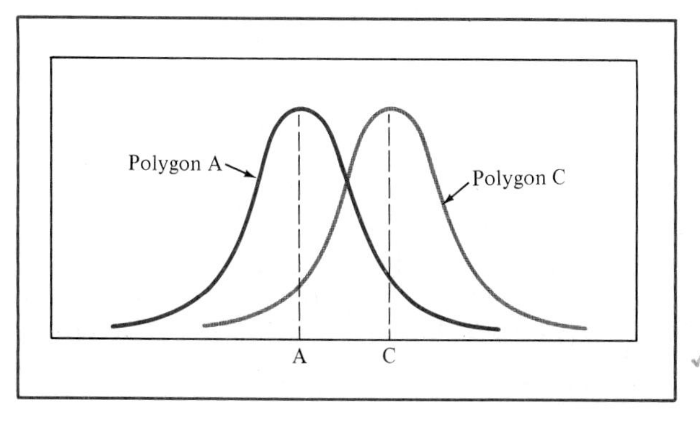

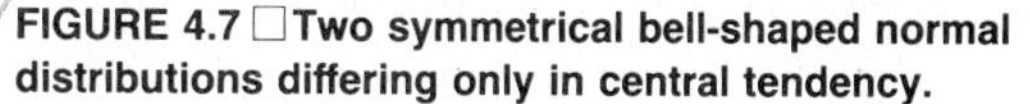
FIGURE 4.7 □ Two symmetrical bell-shaped normal distributions differing only in central tendency.

Figure 4.7 presents two normal distributions which differ only in central tendency. The mean, median, mode, and midrange in polygon C exceed (are to the right of) those for polygon A.

Figure 4.8 demonstrates two normal distributions which differ only in dispersion. The range, variance, and standard deviation in polygon D are smaller than those for polygon A. Polygon D is more *homogeneous*, polygon A is more *heterogeneous*.

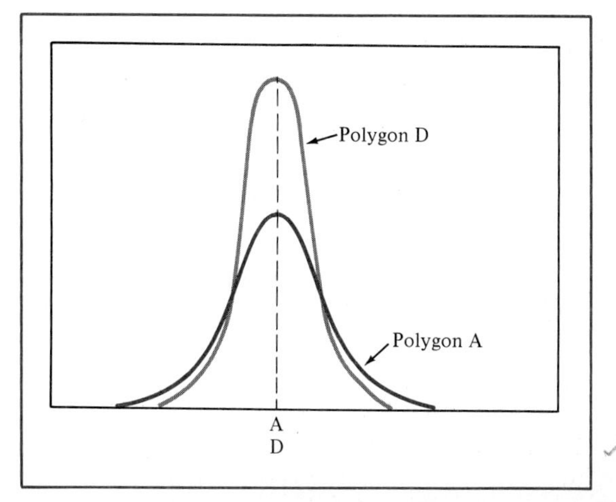

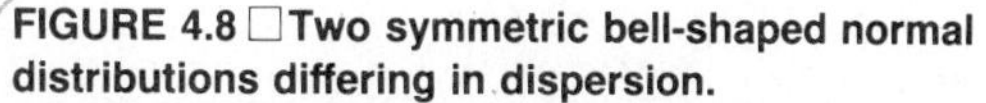
FIGURE 4.8 □ Two symmetric bell-shaped normal distributions differing in dispersion.

Figure 4.9 on page 96 depicts three hypothetical polygons: polygon A is a symmetrical bell-shaped normal distribution; polygon L is negative or left-skewed (since the distortion to the left is caused by extremely small values); and polygon R is positive or right-skewed (since the distortion to the right is caused by extremely large values).

The *relative positions* of the various measures of central tendency (the mean, median, mode, and midrange) in skewed distributions can best be examined from Figures 4.10 and 4.11 on page 96.

In left-skewed distributions (Figure 4.10) the few extremely small observations distort the midrange and mean toward the left tail. Hence we would expect the mode to be the largest value and the midrange to be the smallest. That is

$$\text{midrange} < \text{mean} < \text{median} < \text{mode}$$

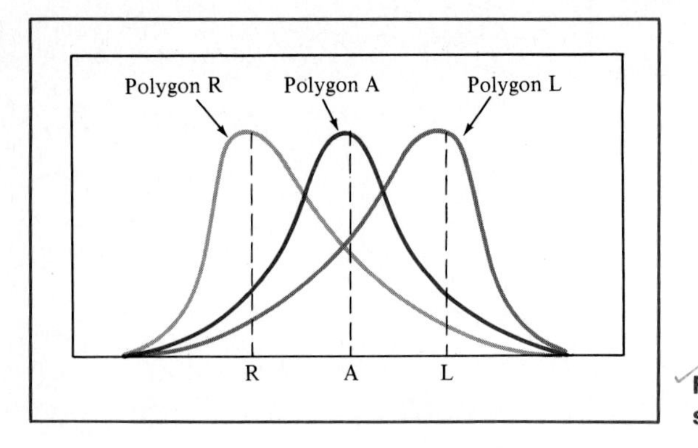

FIGURE 4.9 □ Three distributions differing primarily in shape.

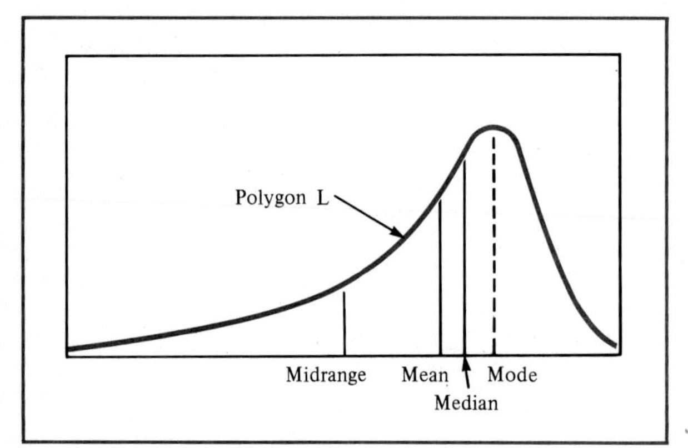

FIGURE 4.10 □ Left-skewed distribution.

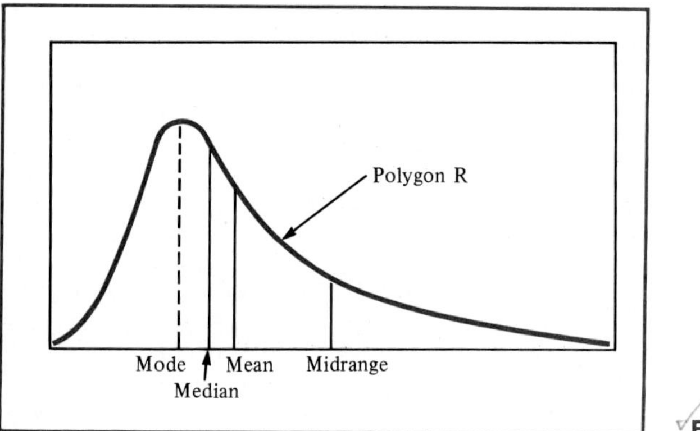

FIGURE 4.11 □ Right-skewed distribution.

However, in right-skewed distributions (Figure 4.11) the reverse is true. The few extremely large observations distort the midrange and the mean toward the right tail. Hence, we would expect the midrange to exceed (that is, be to the right of) all other measures. That is,

$$\text{mode} < \text{median} < \text{mean} < \text{midrange}$$

On the other hand, in perfectly symmetric distributions the mean, median, and midrange will all be identical. As displayed in Figures 4.12 through 4.14, the shape of the curve to the left side of these measures of central tendency is the mirror image of the shape of the curve to their right.

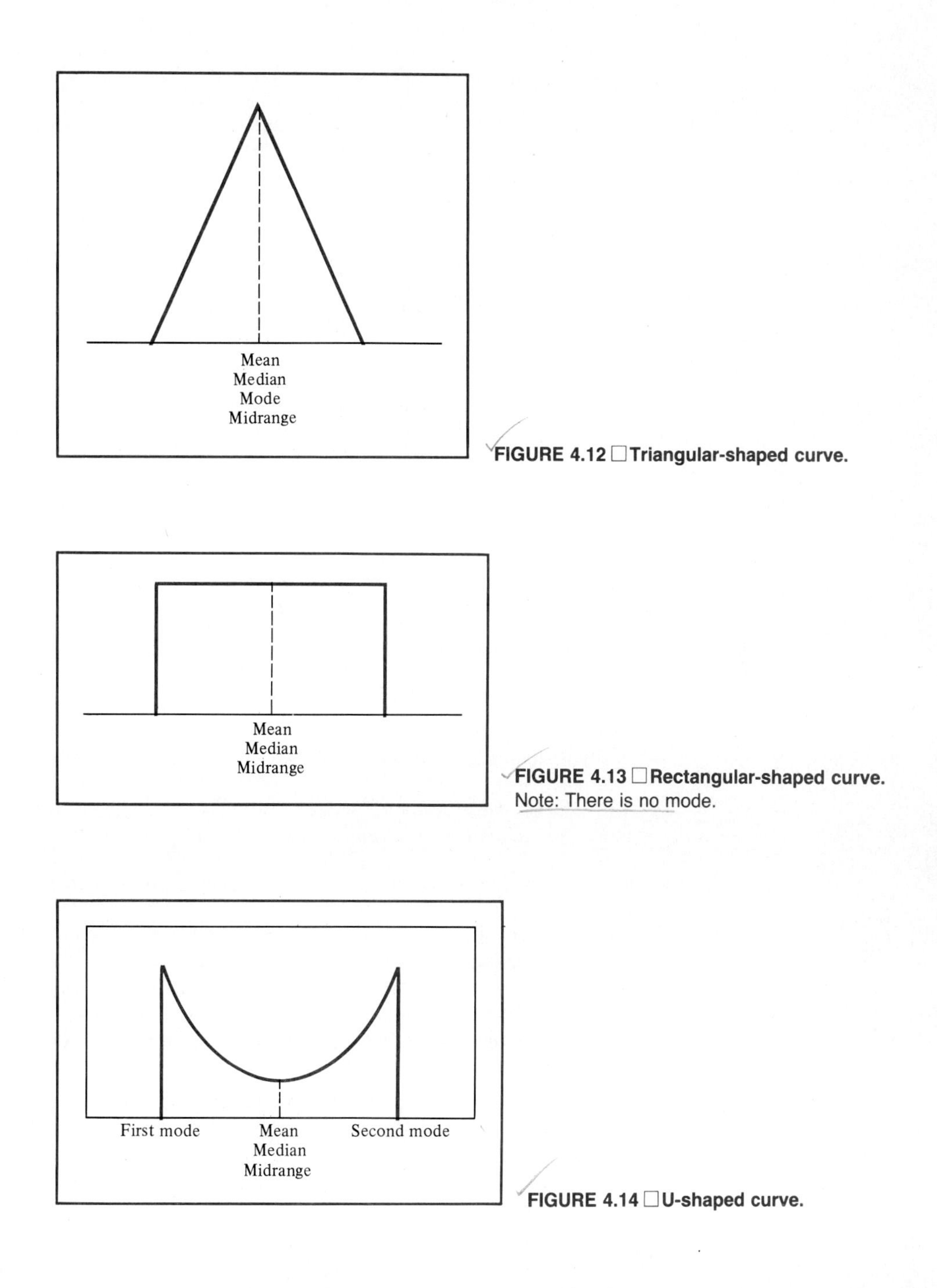

FIGURE 4.12 □ Triangular-shaped curve.

FIGURE 4.13 □ Rectangular-shaped curve.
Note: There is no mode.

FIGURE 4.14 □ U-shaped curve.

Problems

4.36 From the percentage distribution of the data on unpaid accounts which you constructed in Problem 4.21 on page 90:
- **(a)** Plot the percentage histogram.
- **(b)** Plot the percentage polygon.
- **(c)** Tell whether the distribution looks skewed; is the ordering of the mean, median, mode, and midrange consistent with this? (See Problem 3.5 on page 45.)

4.37 From the percentage distribution which you constructed of the data on creativity from Problem 4.22 on page 91:
- **(a)** Plot the percentage histogram.
- **(b)** Plot the percentage polygon.
- **(c)** Tell whether the distribution looks skewed; is the ordering of the mean, median, mode, and midrange consistent with this? (See Problem 3.12 on page 46.)

4.38 From the percentage distribution you constructed in Problem 4.25 (page 91) on stock price-to-earnings ratios:
- **(a)** Plot the percentage histogram.
- **(b)** Plot the percentage polygon.
- **(c)** Tell whether the distribution looks skewed; is the ordering of the mean, median, mode, and midrange consistent with this? (See Problem 3.73 on page 72.)

• 4.39 From the percentage distribution you constructed in Problem 4.26 (page 91) for the data on GNP forecasts:
- **(a)** Plot the percentage histogram.
- **(b)** Plot the percentage polygon.
- **(c)** Tell whether the distribution looks skewed; is the ordering of the mean, median, mode, and midrange consistent with this? (See Problem 3.74 on page 72.)

4.40 From the percentage distribution you constructed in Problem 4.27 on page 91 of the temperature in various U.S. cities:
- **(a)** Plot the percentage histogram.
- **(b)** Plot the percentage polygon.
- **(c)** Compute the mean, median, mode, and midrange from the data in Problem 4.13 on page 84.
- **(d)** Tell whether the distribution looks skewed; is the ordering of the mean, median, mode, and midrange consistent with this?

4.41 From the chainsaw data's percentage distribution which you constructed in Problem 4.28 on page 91:
- **(a)** Plot the percentage histogram.
- **(b)** Plot the percentage polygon.
- **(c)** Compute the mean, median, mode, and midrange from the data in Problem 4.11 on page 83.
- **(d)** Tell whether the distribution looks skewed; is the ordering of the mean, median, mode, and midrange consistent with this?
- **(e)** What is the most notable characteristic in these plots?

4.42 Using the percentage distributions of acceleration of cars (Problem 4.29 on page 91), construct percentage polygons on the same graph for the Japanese and German automobiles. Describe each distribution verbally, and describe similarities and differences between them.

4.43 Construct percentage polygons from the percentage distributions of tuition data which you constructed in Problem 4.30 on page 91. Describe each distribution verbally, and describe similarities and differences between the distributions.

• **4.44** Construct percentage polygons from the percentage distributions of data on caloric content of cottage cheese which you constructed in Problem 4.32 on page 91. Describe each distribution verbally, and describe similarities and differences between the distributions.

4.45 Using the distributions of the lifetime of light bulbs from two manufacturers (Problem 4.33 on page 91), construct the percentage polygons for the data from each of the two manufacturers. Describe each distribution verbally, and describe similarities and differences between the distributions.

4.46 From the percentage distribution you constructed in Problem 4.34 on page 91 on salaries:
(a) Plot the percentage histogram.
(b) Plot the percentage polygon.

4.47 Construct percentage polygons for the batting averages of American and National League players using the distributions you constructed in Problem 4.35 on page 91. Describe each distribution verbally, and describe similarities and differences between the distributions. Are you surprised by these results?

4.5 QUALITATIVE DATA PRESENTATION: SUMMARY TABLES

The responses to categorical variables may be tallied into **summary tables** and then graphically displayed as either **bar charts** or **pie diagrams.** For example, from Figure 3.6 on page 60 we see that of the 50 small stores in Heldman Township, 25 are *always* open on Sundays, 10 are *sometimes* open on Sundays, while 15 are *never* open on Sundays. This information is presented in the accompanying frequency and percentage summary table (Table 4.6).

To obtain the percentages, each of the aforementioned tallies (i.e., "frequencies") is divided by the total number of stores, and the result is multiplied by 100.

From Table 4.6 we may conclude that small stores in Heldman Township typically are open on Sundays; on some occasions (most likely during holiday periods), as many as 70.0% will be open on Sundays.

TABLE 4.6 ☐ Frequency and percentage summary table pertaining to Sunday operations for 50 stores in Heldman Township

Sunday operations	Number of stores	Percentage of stores
Always	25	50.0
Sometimes	10	20.0
Never	15	30.0
Totals	50	100.0

SOURCE: Data are taken from Figure 3.6.

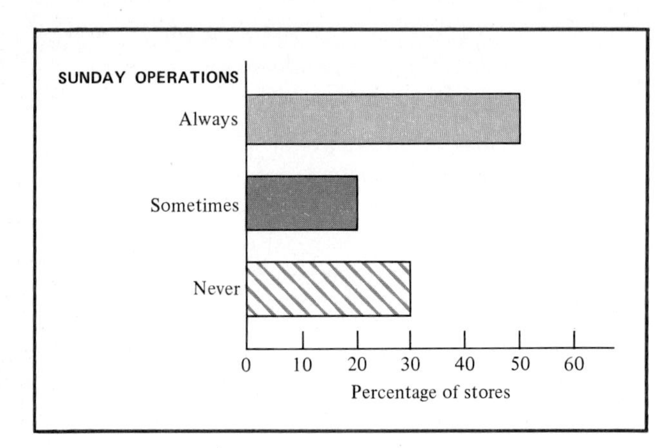

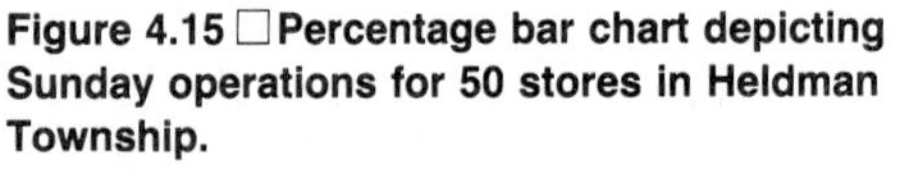

Figure 4.15 □ Percentage bar chart depicting Sunday operations for 50 stores in Heldman Township.

SOURCE: Data are taken from Table 4.6.

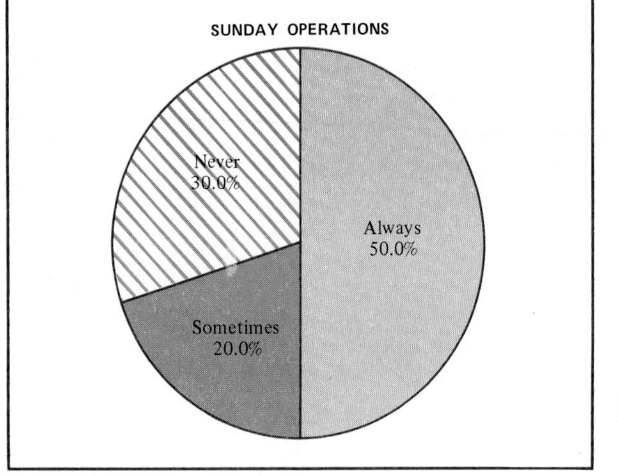

FIGURE 4.16 □ Percentage pie diagram depicting Sunday operations for 50 stores in Heldman Township.

SOURCE: Data are taken from Table 4.6.

4.6 QUALITATIVE DATA PRESENTATION: BAR CHARTS AND PIE DIAGRAMS

To express this information graphically, either a percentage bar chart (Figure 4.15) or a percentage pie diagram (Figure 4.16) can be displayed.

The following suggestions are made for constructing bar charts:

1. For categorical responses that are qualitative the bars should be constructed horizontally as in Figure 4.15; for categorical responses that are numerical the bars should be constructed vertically.
2. All bars should have the same width (as in Figure 4.15) so as not to mislead the reader. Only the lengths may differ.
3. Spaces between bars should range from one-half the width of a bar to the width of a bar.
4. Scales and guidelines are useful aids in reading a chart and should be included.
5. The axes of the chart should be clearly labeled.

6. Any "keys" to interpreting the chart may be included within the body of the chart or below the body of the chart.
7. The title of the chart appears below the body.
8. Footnotes or source notes, when appropriate, are given below the title of the chart.

To construct pie diagrams we may use both the compass and the protractor—the former to draw the circle, the latter to measure off the appropriate pie sectors. Since the circle has 360 degrees, the protractor may be used to divide up the pie based on the percentage "slices" desired. As an example, in Figure 4.16, 30.0% of the stores are never open on Sundays. Thus, we would multiply .300 by 360, mark off the resulting 108 degrees with the protractor, and then connect the appropriate points to the center of the pie, forming a slice comprising 30.0% of the area. In such a manner the entire pie diagram can be constructed.

360° in circle   percentage × 360° = Area of degrees

Problems

• 4.48 A well-known newspaper conducted a telephone poll of New Yorkers' attitudes toward New York City. A total of 419 people were selected in a simple random sample. The following data reflect the responses to a question regarding the adequacy of police and fire protection:

Is the police and fire protection in your neighborhood adequate?	
Yes	293
No	80
Don't know or refused to answer	46
Total	419

(a) Convert the data to percentages and construct:
(1) a bar chart.
(2) a pie chart.

(b) Which of these charts do you prefer to use here? Why?

4.49 The board of directors of a large housing cooperative wished to investigate the possibility of hiring a supervisor for an outdoor playground. All 616 households in the cooperative were polled, with each household having one vote, regardless of its size. The following data were collected:

Should the co-op hire a supervisor?	
Yes	146
No	91
Not sure	58
No response	321
Total	616

(a) Convert the data to percentages and construct:
(1) a bar chart.
(2) a pie chart.
(b) Which of these charts do you prefer to use here, and why?
(c) Eliminating the "no response" group, convert the 295 remaining responses to percentages and construct:
(1) a bar chart.
(2) a pie chart.
(d) Based on your findings in (a) and (c), what would you recommend that the board of directors do?

4.50 Based on sales during a recent year, the following data represent the market shares (in percent) held by the leading producers of soft drinks sold in the United States:

Soft drink producers	Market share (%)
Coca-Cola	39.6
Pepsi-Cola	29.4
7-Up	6.0
Dr. Pepper	6.1
Royal Crown	4.5
Crush	1.4
A&W	0.9
All others	12.1
Total	100.0

(a) Construct a bar chart of these data.
(b) Construct a pie chart of these data.
(c) Which of these charts do you prefer to use here, and why?
(d) **(Class Project)** Let each student in the class respond to the question: "Which soft drink do you most prefer?" The teacher should tally the results into a summary table.
(1) Convert the data to percentages and construct a bar chart and a pie chart.
(2) Compare and contrast the findings from the class with those obtained nationally based on market shares. What do you conclude?

4.51 The number of nuclear energy plants operating in the world's developed countries is:

Britain	35
Canada	12
France	36
West Germany	12
Japan	27
Soviet Union	37
United States	83
Others	38

(a) Construct the percentage summary table. Then reorganize this table so that the countries are listed in order from most to fewest atomic plants. Does this seem easier to read than the original table?
(b) Construct a pie chart and a bar chart; tell which seems more informative, and why.

4.52 States get money from a variety of sources. The table below shows the tax revenue (in billions of dollars) which state governments in the 50 states received from different sources in a recent year:

Sales and gross receipts	72
Licenses	9
Individual income	41
Corporation income	14
Property	3
Death and gift	2
Severance	6
Documentary and stock transfer	1

(a) Construct a percentage summary table.
(b) Construct a bar chart and a pie chart of these data; tell which seems more informative, and why.
(c) If you worked for a state government which needed to raise more money for state programs, how would you use this information to decide how to raise that money?

4.53 As of 1980, the number of people living in each region of the United States was:

Northeast	49,135,000
Midwest	58,866,000
South	75,372,000
West	43,172,000

(a) Does anything about this distribution surprise you? What additional information might help put this distribution in perspective?
(b) Construct a pie chart of this distribution.

4.54 As of 1982, the number of people in the United States who belonged to various religious congregations was:

Buddhist	100,000
Eastern	3,860,000
Jewish	5,725,000
Old Catholic, Polish Catholic, Armenian	924,000
Roman Catholic	52,089,000
Protestant	73,479,000
Others	161,000

(a) Construct a percentage summary table.
(b) Construct a pie chart and a bar chart.

4.55 The recent death rates per 100,000 population in the United States from several sources of death were:

Cardiovascular	436
Malignancies	184
Accidents	47
Pulmonary diseases	25
Pneumonia, influenza	24
Diabetes	15
Liver disease	14
Suicide	12
Homicide	11
Others	110

(a) Construct a percentage summary table.
(b) Construct a bar chart and a pie chart.
(c) Tell how government officials and other interested people might use these data to plan programs which would increase the lifespan of U.S. citizens.

• 4.56 The amount of money that public elementary and secondary schools received from various sources in a recent year was (in billions of dollars):

Federal	8
State	62
Local	57

(a) Construct a percentage summary table.
(b) Construct a pie chart.
(c) Does anything surprise you about these numbers?

4.7 QUALITATIVE DATA PRESENTATION: CROSS-CLASSIFICATION TABLES

With survey data, it is often useful to examine the categorical responses to two qualitative variables simultaneously. For example, we may be interested in examining whether or not there is any "pattern" or relationship between the location of a store in Heldman Township and its policy toward operating on Sundays. Table 4.7 depicts this information for all 50 stores in the township. Such two-way tables of

TABLE 4.7 □ Cross-classification table of store location and policy toward Sunday operations for 50 stores in Heldman Township

	Sunday operations			
Location	**Always open**	**Sometimes open**	**Never open**	**Totals**
East Central St.	8	2	10	20
West Central St.	17	8	5	30
Totals	25	10	15	50

two way = two totals

cross classification, known as **contingency tables,** will be studied again in Chapters 6 and 11.

To construct Table 4.7, for example, the joint responses for each of the 50 stores with respect to location and Sunday operations are tallied into one of the six possible "cells" of the table. Thus from Figure 3.6 (on page 60), the first store is located on East Central Street (code E) and is never open on Sundays (code N). These responses were tallied into the cell composed of the first row and third column. The second store is located on West Central Street and is always open on Sundays. These responses were tallied into the cell composed of the second row and first column. The remaining 48 joint responses were recorded in a similar manner.

In order to explore any possible pattern or relationship between store location and policy toward Sunday operations it is useful to first convert these results into percentages based on

1. The overall total (i.e., 50 stores in Heldman Township).
2. The row totals (i.e., East Central Street versus West Central Street).
3. The column totals (i.e., always, sometimes, or never open on Sundays).

This is accomplished in Tables 4.8, 4.9, and 4.10, respectively.

From Table 4.8 we note that:

1. 60% of the stores are located on West Central Street.
2. 50% of the stores are always open on Sundays.
3. 34% of the stores are located on West Central Street and always open on Sundays.

TABLE 4.8 Cross-classification table of store location and policy toward Sunday operations (Percentages based on overall total—50 stores)

	Sunday operations			
Location	**Always open**	**Sometimes open**	**Never open**	**Totals**
East Central St.	16.0	4.0	20.0	40.0
West Central St.	34.0	16.0	10.0	60.0
Totals	50.0	20.0	30.0	100.0

TABLE 4.9 Cross-classification table of store location and policy toward Sunday operations (Percentages based on row totals)

	Sunday operations			
Location	**Always open**	**Sometimes open**	**Never open**	**Totals**
East Central St.	40.0	10.0	50.0	100.0
West Central St.	56.7	26.7	16.7	100.1*
Totals	50.0	20.0	30.0	100.0

* Error due to rounding.

TABLE 4.10 □ Cross-classification table of store location and policy toward Sunday operations (Percentages based on column totals)

Location	Sunday operations			
	Always open	Sometimes open	Never open	Totals
East Central St.	32.0	20.0	66.7	40.0
West Central St.	68.0	80.0	33.3	60.0
Totals	100.0	100.0	100.0	100.0

From Table 4.9 we note that:

1. The majority of West Central Street stores (56.7%) are always open on Sundays, while the majority of East Central Street stores (50.0%) are never open on Sundays.
2. 56.7% of the West Central Street stores are always open on Sundays, while only 40.0% of the East Central Street stores are.
3. 50.0% of the East Central Street stores are never open on Sundays, while only 16.7% of the West Central Street stores are.

From Table 4.10 we note that:

1. 66.7% of the stores that are never open on Sundays are located on East Central Street.
2. 68.0% of the stores that are always open on Sundays and 80.0% of the stores that are sometimes open on Sundays are located on West Central Street.

From the above a pattern seems to emerge: shoppers in Heldman Township are more apt to find stores open on Sundays on West Central Street.

Problems

4.57 People between the ages of 65 and 70 were studied to see if those who exercised more were healthier. One measure of health was blood pressure. The number of people at each level of blood pressure and amount of exercise are:

Blood pressure	Amount of Exercise		
	None	Moderate	High
Too high	54	23	7
Normal	86	76	32

(a) Convert this table into either row or column proportions, depending on which you think will be more informative.

(b) Describe the results.

✗**(c)** Some people might conclude from such a study that if older people exercised more, they would have lower blood pressure. What is wrong with this reasoning, and what is a plausible alternative reason for the relationship in the table?

• 4.58 Researchers were looking at the relationship between the type of college which was attended and the level of job which people who graduated in 1960 held at the time of the study. The researchers only examined graduates who went into industry. The cross-tabulation of the data is presented below:

	Type of college		
Management level	**Ivy league**	**Other private**	**Public**
High (Sr. V-P or above)	45	62	75
Middle	231	563	962
Low	254	341	732

(a) Construct a table with either row or column percentages, depending on which you think is more informative.
(b) Interpret the results of the study.
(c) What other variable or variables might you want to know before advising someone to attend an Ivy League or other private school if they want to get to the top in business?

4.59 When inoculation to prevent disease was first being introduced, studies were done to see if the process was successful. A typical study of a test of smallpox vaccination in a town might have yielded results such as the following:

	Outcome	
Vaccinated?	**Live**	**Die**
Yes	832	243
No	1643	968

(a) Find the row or column proportions, depending on which you think are more informative.
(b) Interpret the results, supposing that there is no relationship between general health level and who gets vaccinated.
(c) If healthier people actually got the vaccine (perhaps because rich people were healthier, and the rich could afford doctors), how would it change your interpretation?

4.60 People returning from vacations in different countries were asked how they enjoyed their vacation. The number of people who went to each country and gave each response is as follows:

	Response to country			
Country	**Yuck**	**So-So**	**Good**	**Great**
England	5	32	65	45
Italy	3	12	32	43
France	8	23	28	25
Guatemala	9	12	6	2

(a) Construct a table of row proportions.
(b) What would you conclude from this study?
(c) What assumptions might you make about the personality of people who would go to each country, and how might that affect the results?

Supplementary Problems

4.61 In the original experiment to test penicillin, eight rats were infected with a dangerous variety of bacteria, and then four of the rats were given penicillin. The results of the experiment were:

	Outcome	
Treatment	**Live**	**Die**
Penicillin	4	0
None	0	4

(a) Construct a table with the row or column percentages.
(b) Did the penicillin appear to be effective in this study?
✗**(c)** Does the small number of subjects in the study bother you in any way? If so, why; if not, why not?

✗4.62 In a recent year, the crime rate per 100,000 people in the United States for violent crimes was:

Crime	Large cities	Small cities	Rural areas
Murder, manslaughter	9	5	6
Forcible rape	39	21	15
Robbery	273	49	17
Aggravated assault	306	240	124

(a) Tell what these data show you in their current form.
(b) Tell why none of the techniques in this chapter is needed to clarify the data.

4.63 Using the ordered array (Figure 3.10) on page 68 for the reported receipts:
(a) Set up the various frequency distributions with the following class boundaries:
(1) $3.0 but less than $5.5, $5.5 but less than $8.0, and so on.
(2) $4.5 but less than $6.5, $6.5 but less than $8.5, and so on.
(3) $4.0 but less than $6.5, $6.5 but less than $9.0, and so on.
(b) Compare and contrast the different results and state whether you believe any of these distributions should be preferred to the one selected in Table 4.1 on page 81 for the reported receipts. [*Hint*: To help you decide, compare the descriptive measures obtained from each of the frequency distributions to the actual values computed from the (ungrouped) data and recorded in Table 3.3 on page 67.]

4.64 Explain the differences between frequency distributions, relative frequency distributions, and percentage distributions.

4.65 When comparing two or more sets of data with different sample sizes, why is it necessary to compare their respective relative frequency or percentage distributions?

4.66 Use the data on price-to-earnings ratios (Problem 3.73 on page 72):
 (a) Form the frequency and percentage distributions in the same table.
 (b) Plot the frequency polygon.
 (c) Approximate:
  (1) the mode.
  (2) the midrange.
  (3) the range.
 (d) How would you express the shape of the data? Discuss.
 (e) Compare and contrast your approximations with the actual results computed from the ungrouped data in part (a) of Problem 3.73.

4.67 Use the data on projections of percentage changes in real GNP (Problem 3.74 on page 72):
 (a) Form the frequency and percentage distributions in the same table.
 (b) Plot the frequency polygon.
 (c) Approximate:
  (1) the mode.
  (2) the midrange.
  (3) the range.
 (d) How would you express the shape of the data? Discuss.
 (e) Compare and contrast your approximations with the actual results computed from the ungrouped data in part (a) of Problem 3.74.

4.68 Refer to Problem 3.77 on page 73 concerning the ages of automobile salesmen in the United States and in Western Europe. For each set of data:
 (a) Form the frequency and percentage distributions in the same table.
 (b) Plot the frequency polygon.
 (c) Approximate the mode, range, and midrange.
 (d) Compare and contrast your approximations with the actual results computed from the ungrouped data set in part (a) of Problem 3.77.

4.69 Use the data on earnings per share for all 19 companies in the leisure time and recreational activity industry (Problem 3.78 on page 73):
 (a) Form the frequency and percentage distributions in the same table.
 (b) Plot the frequency polygon.
 (c) Approximate the mode, range, and midrange.
 (d) Compare and contrast your approximations with the actual results computed from the ungrouped data set in part (a) of Problem 3.78.

4.70 Use the data on fringe benefits as a percentage of salaries paid at the 34 institutions of higher learning in the state of Colorado (Problem 3.79 on page 74):
 (a) Form the frequency and percentage distributions in the same table.
 (b) Plot the frequency polygon.
 (c) Approximate the mode, range, and midrange.
 (d) Compare and contrast your approximations with the actual results computed from the ungrouped data set in part (a) of Problem 3.79.

• **4.71** Use the data on annual per capita alcohol consumption in all 50 states plus the District of Columbia. (Problem 3.80 on page 75):
(a) Form the frequency and percentage distributions in the same table.
(b) Plot the frequency polygon.
(c) Approximate the mode, range, and midrange.
(d) Compare and contrast your approximations with the actual results computed from the ungrouped data set in part (a) of Problem 3.80.

4.72 A wholesale appliance distributing firm wished to study its accounts receivable for two successive months. Two independent samples of 50 accounts were selected for each of the two months. The results have been summarized in the following table:

Amount	March frequency	April frequency
$0–under $2,000	6	10
$2,000–under $4,000	13	14
$4,000–under $6,000	17	13
$6,000–under $8,000	10	10
$8,000–under $10,000	4	0
$10,000–under $12,000	0	3
Totals	50	50

(a) Plot the frequency histogram for each month.
(b) On one graph, plot the percentage polygon for each month.
(c) Write a brief report comparing and contrasting the accounts receivable of the two months.

4.73 A home-heating oil delivery firm wished to compare how fast the oil bills were paid in two different Long Island suburbs. A random sample of 50 vouchers from Suburb A and 100 vouchers from Suburb B were selected, and the number of days between delivery and payment were recorded as shown in the following table:

Number of days	Suburb A frequency	Suburb B frequency
0–4	4	6
5–9	14	21
10–14	16	24
15–19	10	30
20–24	5	7
25–29	1	6
30–34	0	6
Totals	50	100

(a) Plot the percentage histogram for each suburb.
(b) On one graph, plot the percentage polygon for each suburb.
(c) Write a brief report comparing and contrasting the number of days between delivery and payment in the two suburbs.

4.74 The accompanying table contains the **cumulative distributions** and **cumulative percentage distributions** of steering turn revolutions (lock-to-lock) for a sample of 45 U.S.-manufactured compact/small-car automobile models and 92 foreign-manufactured compact/small-car automobile models. (We read a cumulative distribution as follows: No U.S. models have less than 2.0 steering turn revolutions; 1 U.S. model has less than 2.5 steering turn revolutions; 5 U.S. models have less than 3.0 revolutions; . . . ; and all 45 U.S. models have less than 6.5 revolutions.) Based on these data, answer the following questions:

(a) How many models of U.S.-made automobiles have 4.5 or more revolutions?

(b) What is the percentage of U.S.-made automobiles with less than 3.5 revolutions?

(c) Which group of car models—U.S.-made or foreign-made—have the wider range in steering turn revolutions?

(d) How many foreign-made automobile models have steering turn revolutions between 3.50 and 3.99 (inclusive)?

(e) Write a brief report comparing and contrasting the steering turn revolutions information for the two groups of car models.

Cumulative frequency and percentage distributions for the number of steering turn revolutions for U.S.-manufactured and foreign-made compact/small-car automobile models

Steering turn revolutions	U.S.-made automobile models less than indicated values		Foreign-made automobile models less than indicated values	
	Number	Percentage	Number	Percentage
2.0	0	0.0	0	0.0
2.5	1	2.2	3	3.3
3.0	5	11.1	15	16.3
3.5	16	35.6	44	47.8
4.0	33	73.3	85	92.4
4.5	43	95.6	91	98.9
5.0	44	97.8	92	100.0
5.5	44	97.8	92	100.0
6.0	44	97.8	92	100.0
6.5	45	100.0	92	100.0

SOURCE: Data are extracted from *Road & Track*, vol. 34, no. 7 (March 1983), pp. 70–73.

4.75 Explain the differences between histograms and polygons.

4.76 Explain the differences between histograms and bar charts.

4.77 Explain the differences between bar charts and pie charts.

• 4.78 Based on sales during a recent year, the following data represent the market shares (in percent) held by the leading types of instant decaffeinated coffees sold in the United States:

Instant decaffeinated coffee	Market share (%)
Sanka	33.9
Tasters' Choice	20.8
High Point	17.4
Brim	10.4
Nescafe Decaf	7.9
Sanka Freeze Dried	7.0
All others	2.6
Total	100.0

Construct a bar chart and pie chart. Which of these charts do you prefer to use here? Why?

4.79 Based on sales during a recent year, the following data represent the market share for heavy-weight motorcycles in the United States:

Manufacturer	Market share (in %)
Honda	42
Harley-Davidson	17
Yamaha	16
Kawasaki	14
Suzuki	8
Other	3

(a) Construct a bar chart.
(b) Construct a pie chart.
(c) Which do you prefer here? Why?

4.80 Based on sales during two different years (separated by a ten-year interval), the following data represent market shares (in percent) held by competing liquors sold in the United States:

	Market share (in %)	
Type of liquor	Last year	Ten years ago
Bourbon	11.8	17.0
Canadian Whiskey	12.8	12.0
Scotch	10.1	13.6
Vodka	21.5	17.5
Gin	9.0	9.8
Cordials	12.2	6.4
Rum	7.9	3.9
Brandy	4.6	3.8
Other	10.1	16.0

(a) Construct bar charts for each year.
(b) Construct pie charts for each year.
(c) Which of these charts do you prefer to use here? Why?
(d) Describe the sales results.

4.81 Based on sales during a recent year, the following data represent the market shares (in percent) held by the leading types of nonprescription pain relievers sold in the United States:

Pain reliever	Market share (in %)
Tylenol	30
Anacin	15
Aspirin	10
Bufferin	8
Excedrin	8
All others	29
Total	100

(a) Construct a bar chart.
(b) Construct a pie chart.
(c) Which of these charts do you prefer to use here, and why?
(d) **(Class Project)** Let each student in the class respond to the question "Which pain reliever do you most prefer?" so that the teacher may tally the results into a summary table on the blackboard.
(1) Convert the data to percentages; construct a bar chart and a pie chart.
(2) Compare and contrast the findings from the class with those obtained nationally based on market shares. List possible reasons for differences between the two.

Database Exercises

The following problems refer to the sample data obtained from the questionnaire of Figure 2.5 and presented in Figure 2.10. They should be solved (to the greatest extent possible) with the aid of a computer package.

□ 4.82 From the responses to question 2 of the alumni association survey regarding undergraduate major:
(a) Construct a percentage summary table and bar chart.
(b) Construct a pie chart.
(c) Discuss your findings.

□ 4.83 From the responses to question 3 of the alumni association survey regarding the primary reason for attending college:
(a) Construct a percentage summary table and bar chart.
(b) Construct a pie chart.
(c) Discuss your findings.

□ 4.84 From the responses to question 4 of the alumni association survey regarding current philosophical ideology:
(a) Construct a percentage summary table and bar chart.
(b) Construct a pie chart.
(c) Discuss your findings.

□ 4.85 From the responses to question 5 of the alumni association survey regarding graduate school attendance:
(a) Construct a percentage summary table and bar chart.
(b) Construct a pie chart.
(c) Discuss your findings.

□ 4.86 From the responses to question 6 of the alumni association survey regarding the perceived relationship between present job and undergraduate major:
(a) Construct a percentage summary table and bar chart.
(b) Construct a pie chart.
(c) Discuss your findings.

□ 4.87 From the responses to question 8 of the alumni association survey regarding the type of organization at which members are employed:
(a) Construct a percentage summary table and bar chart.
(b) Construct a pie chart.
(c) Discuss your findings.

□ **4.88** From the responses to question 9 of the alumni association survey regarding current job activity:
(a) Construct a percentage summary table and bar chart.
(b) Construct a pie chart.
(c) Discuss your findings.

□ **4.89** From the responses to question 11 of the alumni association survey regarding the first full-time job held after graduation from college:
(a) Construct a percentage summary table and bar chart.
(b) Construct a pie chart.
(c) Discuss your findings.

□ **4.90** From the responses to question 14A of the alumni association survey regarding the most valuable course taken with respect to career:
(a) Construct a percentage summary table and bar chart.
(b) Construct a pie chart.
(c) Discuss your findings.

□ **4.91** From the responses to question 14B of the alumni association survey regarding the most valuable course taken with respect to daily (nonworking) activities:
(a) Construct a percentage summary table and bar chart.
(b) Construct a pie chart.
(c) Discuss your findings.

□ **4.92** From the responses to question 16 of the alumni association survey regarding the choice of college the alumni would have made now:
(a) Construct a percentage summary table and bar chart.
(b) Construct a pie chart.
(c) Discuss your findings.

□ **4.93** From the responses to question 17 of the alumni association survey regarding the advice the alumni would give:
(a) Construct a percentage summary table and bar chart.
(b) Construct a pie chart.
(c) Discuss your findings.

□ **4.94** From the responses to question 18 of the alumni association survey regarding gender:
(a) Construct a percentage summary table and bar chart.
(b) Construct a pie chart.
(c) Discuss your findings.

□ **4.95** From the responses to the alumni association survey:
(a) Cross-classify undergraduate major (question 2) with primary reason for attending the college (question 3).
(b) Analyze your findings.

□ **4.96** From the responses to the alumni association survey:
(a) Cross-classify undergraduate major (question 2) with graduate school attendance (Question 5).
(b) Analyze your findings.

□ **4.97** From the responses to the alumni association survey:
(a) Cross-classify undergraduate major (question 2) with gender (question 18).
(b) Analyze your findings.

□ **4.98** From the responses to the alumni association survey:
- **(a)** Cross-classify primary reason for attending the college (question 3) with choice that would be made now (question 16).
- **(b)** Analyze your findings.

□ **4.99** From the responses to the alumni association survey:
- **(a)** Cross-classify philosophical ideology (question 4) with gender (question 18).
- **(b)** Analyze your findings.

□ **4.100** From the responses to the alumni association survey:
- **(a)** Cross-classify type of organization employed at (question 8) with gender (question 18).
- **(b)** Analyze your findings.

□ **4.101** From the responses to the alumni association survey:
- **(a)** Cross-classify the choice that would be made now (question 16) with advice given to others (question 17).
- **(b)** Analyze your findings.

□ **4.102** From the responses to the alumni association survey:
- **(a)** Cross-classify advice given to others (question 17) with gender (question 18).
- **(b)** Analyze your findings.

References

1. BERENSON, M. L., AND D. M. LEVINE, *Basic Business Statistics: Concepts and Applications*, 3rd ed. (Englewood Cliffs, N.J.: Prentice-Hall, 1986).
2. CROXTON, F., D. COWDEN, AND S. KLEIN, *Applied General Statistics*, 3rd ed. (Englewood Cliffs, N.J.: Prentice-Hall, 1967).
3. TUFTE, E. R., *The Visual Display of Quantitative Information* (Cheshire, Conn.: Graphics Press, 1983).
4. WAINER, H., "How to Display Data Badly," *The American Statistician*, vol. 38 (May 1984), pp. 137–147.

An Introduction to Exploratory Data Analysis Techniques

5.1 INTRODUCTION: WHAT'S AHEAD

In addition to the more traditional descriptive statistical techniques for analyzing quantitative data characteristics, recently developed **exploratory data analysis** methods (References 1 and 2) may also be used.

The goal of this chapter is to introduce exploratory data analysis. We will see how these useful techniques can enhance a descriptive statistical analysis and assist us to a better understanding of all of the information in the data.

5.2 THE STEM-AND-LEAF DISPLAY

We previously noted that when dealing with large amounts of raw data (see, for example, Figure 3.9 on page 68) it is useful to place the data into an ordered array before developing tables and charts or computing descriptive summary measures. We also noted, however, that the greater the number of observations present in a data set, the more difficult or cumbersome it is to form the ordered array. In such situations it becomes particularly useful to organize the data into **stem-and-leaf displays** in order to study its characteristics.

Figure 5.1 depicts the stem-and-leaf display for the reported quarterly sales tax receipts for all 50 stores in Heldman Township. In developing this display, note that the "stems" are listed in a column to the left of the vertical line, while in each row the "leaves" branch out to the right of the vertical line. Since the reported sales tax receipts are recorded in thousands of dollars, we see that the stems consist of *all* sales tax receipts in *thousands of dollars*, while the leaves give the *particular* values in *fractions of thousands of dollars*.

5.2.1 Constructing the Stem-and-Leaf Display

Using the data from Figure 3.9, we can easily construct the stem-and-leaf display. Note that the first store reported quarterly sales tax receipts of 8.0 thousands of dollars. Therefore the "rightmost" digit of 0 is listed as the first leaf value next to the stem value of 8. The second store reported quarterly sales tax receipts of 11.8 thousands of dollars. Here the "rightmost" digit of 8 is listed as the first leaf value next to the stem value of 11. Continuing, the third store reported quarterly sales tax receipts of 8.7 thousands of dollars, so that the "rightmost" digit of 7 is listed as the second leaf value next to the stem value of 8.

At this point in its construction, our stem-and-leaf display would appear as follows:

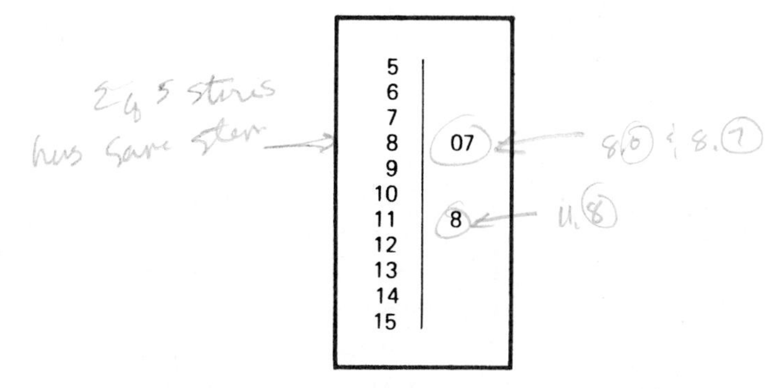

Stem	Leaves
5	
6	
7	
8	07
9	
10	
11	8
12	
13	
14	
15	

Note that two of the three stores have the same stem. As more and more stores are included, those possessing the same stems and, perhaps, even the same leaves within stems (that is, the same reported quarterly sales tax receipts) will be observed. Such leaf values will be recorded adjacent to the previously recorded leaves, opposite the appropriate stem—resulting in Figure 5.1 on page 118.

To assist us in further examining the data, we may wish to rearrange the leaves within each of the stems by placing the digits in ascending order, row-by-row. The **revised stem-and-leaf display** is presented in Figure 5.2 on page 118.

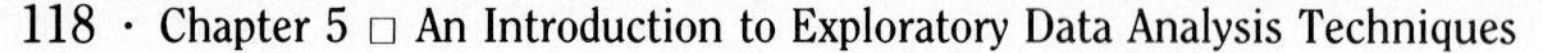

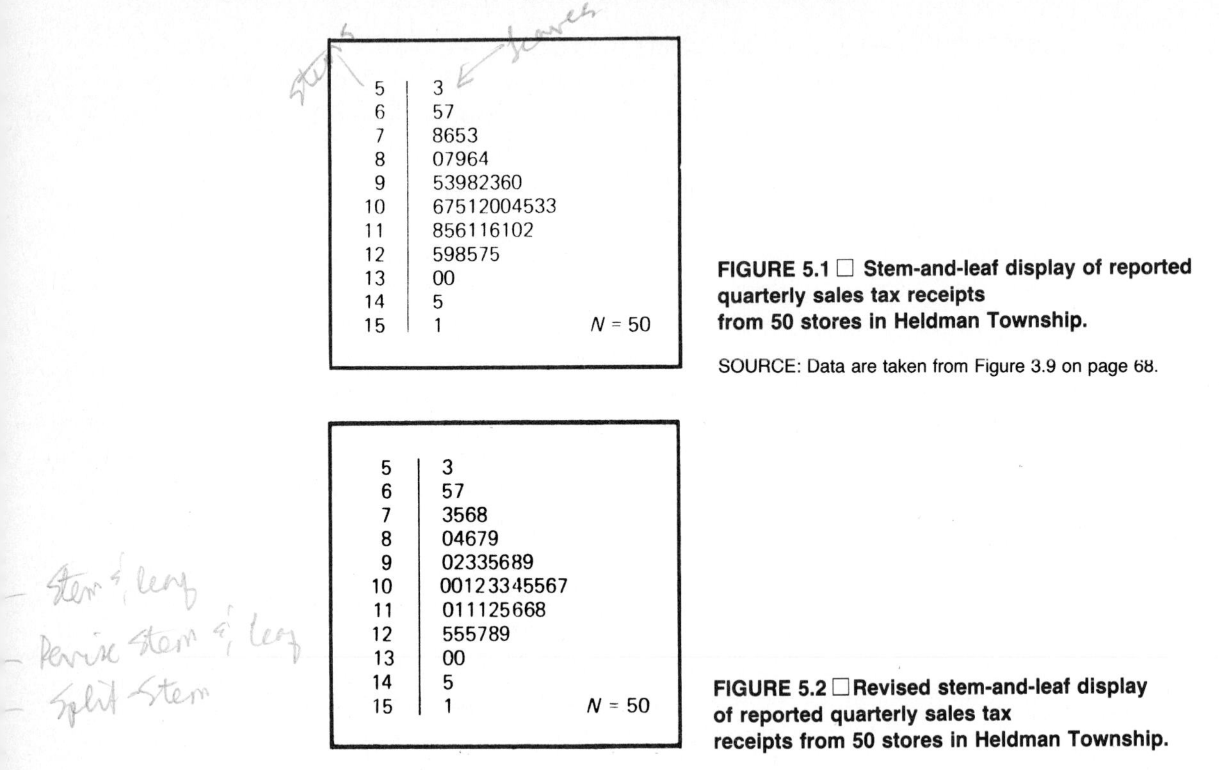

Stem	Leaves
5	3
6	57
7	8653
8	07964
9	53982360
10	67512004533
11	856116102
12	598575
13	00
14	5
15	1

$N = 50$

FIGURE 5.1 □ Stem-and-leaf display of reported quarterly sales tax receipts from 50 stores in Heldman Township.

SOURCE: Data are taken from Figure 3.9 on page 68.

Stem	Leaves
5	3
6	57
7	3568
8	04679
9	02335689
10	00123345567
11	011125668
12	555789
13	00
14	5
15	1

$N = 50$

FIGURE 5.2 □ Revised stem-and-leaf display of reported quarterly sales tax receipts from 50 stores in Heldman Township.

Another type of rearrangement, the **split-stem,** is also useful. Suppose we want to increase the number of stems so that we can attain a lighter concentration of leaves on the remaining stems. One way of doing this is illustrated in Figure 5.3. Each stem from Figure 5.1 has been split into two new stems, one for the Low unit digits 0, 1, 2, 3, or 4; and the other for the High unit digits 5, 6, 7, 8, or 9. These are represented by the L and H, respectively, as indicated in the stem listings of Figure 5.3 on page 119.

5.2.2 Importance of the Stem-and-Leaf Display

The (revised) stem-and-leaf display is, perhaps, the most versatile technique in descriptive statistics. It serves the function of data organization needed for further descriptive analyses and also serves the function of data presentation in both tabular and chart form. Thus, the (revised) stem-and-leaf display is, essentially, an ordered array (Section 3.4.2), a frequency distribution (Section 4.2), and a frequency histogram (Section 4.4.1)—all in one—without sacrificing the original information pertaining to the individual observations themselves.

To demonstrate these features of the stem-and-leaf display let us turn to Figure 5.2. If we were to attach the leaves from one stem to another, consecutively, we would obtain the ordered array for the 50 reported quarterly sales tax receipts. Moreover, if we were to merely tally the leaves within each stem, a frequency

Stem	Leaves
5L	3
5H	
6L	
6H	57
7L	3
7H	865
8L	04
8H	796
9L	3230
9H	5986
10L	1200433
10H	6755
11L	11102
11H	8566
12L	
12H	598575
13L	00
13H	
14L	
14H	5
15L	1

FIGURE 5.3 □ Stem-and-leaf display with split stems; same data as in Figure 5.1.

distribution would be constructed. Finally, if we were to rotate the stem-and-leaf display 90 degrees (that is, hold our book ''sideways''), a frequency histogram would be depicted in such manner that the vertical bars would be represented by individual leaves on each stem.

Problems

5.1 What are the advantages of using a stem-and-leaf display rather than a frequency distribution?

• 5.2 Do a stem-and-leaf display of the creativity test data in Problem 3.12 on page 46. Repeat, using a split stem (as in Figure 5.3). Tell why the second display is better. Construct a revised stem-and-leaf display using the original stems, putting the leaves in order.

5.3 Do a stem-and-leaf display of the death rate data in Problem 3.13 on page 46. [*Hint*: Make the stem the hundreds, and make the leaf either the number of 10s, or use the whole remainder as the leaf. For example 1339 would have 13 as the stem, and either 4—the rounded tens digit—or 39 as the leaf.] Construct a revised stem-and-leaf display, putting the leaves in order.

5.4 Construct a stem-and-leaf display of the price-earnings ratio data in Problem 3.73 on page 72. [*Hint*: The usual display—with the tens digit as the stems and units as the leaves—is not very informative here. Splitting the stems in two helps, but try splitting the stems into five parts—0, 1; 2, 3; 4, 5; 6, 7; and 8, 9—to get the best feel for this distribution.] Construct a revised stem-and-leaf display, putting the leaves in order.

5.5 Construct stem-and-leaf displays for the data on college tuition from Problem 3.76 on page 73. Put the data for the two regions side by side so that you can compare them. Construct a revised side-by-side stem-and-leaf display, putting the leaves in order. Discuss the shape of each distribution separately, and compare them.

5.6 Construct stem-and-leaf displays for the data on fringe benefits in Problem 3.79 on page 74, first using the tens digit as the stem, and then splitting the stem into two parts. The best display here comes from splitting the stems into five parts (see the *Hint* for Problem 5.4), so do the display that way, and tell what you see that was hard to see in the first two displays. Construct a revised stem-and-leaf display, putting the leaves in order.

5.7 Construct a stem-and-leaf display for the alcohol consumption data from Problem 3.80 on page 75. Construct a revised stem-and-leaf display, putting the leaves in order.

5.8 Construct a stem-and-leaf display for the data on chainsaw noise levels in Problem 4.11 on page 83. What choice of stems (the usual; split in two parts; split in five parts) gives the most informative display? [*Hint*: To answer this, see which makes the distinction between gasoline- and electric-powered models most obvious, as well as giving the best impression of the distribution for each type of saw.] Construct a revised stem-and-leaf display, putting the leaves in order.

• 5.9 Construct a stem-and-leaf display for the data on the caloric content of cottage cheeses presented in Problem 4.12 on page 83. Construct one display using the usual stems and another using split stems. Which one best shows the distinction between the 4% and 2% fat content brands? (Notice that the distinction is not so clear as with the chainsaw data in Problem 5.8.) Construct a revised stem-and-leaf display, putting the leaves in order.

5.10 Construct a stem-and-leaf display for the data on the temperatures in U.S. cities in Problem 4.13 on page 84. Notice that both the usual stems and split stems correspond to ways which people actually talk about the weather—some people might say that the temperature was in the 70s, others might say whether the high or low 70s. Construct a revised stem-and-leaf display, putting the leaves in order.

5.11 Construct a stem-and-leaf display for the data on batting averages from Problem 4.20 on page 86. [*Hint*: Put the data for the two leagues side by side on the same display so that you can compare them more easily.] As in Problems 5.4 and 5.6, splitting the stems into either two or five parts may be useful. Construct a revised stem-and-leaf display, putting the leaves in order.

5.3 OBTAINING AND STUDYING THE QUARTILES

Aside from the measures of central tendency, dispersion, and shape, some useful measures of "noncentral" location are often employed when summarizing or describing the properties of data. The measures are called **quantiles.** The most familiar quantiles are the **quartiles** and the **percentiles.** The latter are frequently used in such fields as educational research and industrial psychology to compare the relative performance of subjects on tests. Our concern here will be with the quartiles.

Whereas the median is a value that splits the ordered array in half (50.0% of the observations are smaller and 50.0% are larger), the quartiles are descriptive measures that split the ordered data into four quarters.

The first quartile, Q_1, is the value such that 25.0% of the observations are smaller and 75.0% are larger than Q_1.

The second quartile, Q_2, is the median—50.0% of the observations are smaller and 50.0% are larger than Q_2.

The third quartile, Q_3, is the value such that 75.0% of the observations are smaller and 25.0% are larger than Q_3.

To approximate the quartiles from a population containing N observations, the following **positioning-point formulas**[1] are used:

$$Q_1 = \text{value corresponding to the } \frac{N+1}{4} \text{ ordered observation}$$

$$Q_2 = \text{median, the value corresponding to the } \frac{2(N+1)}{4} = \frac{N+1}{2} \text{ ordered observation}$$

$$Q_3 = \text{value corresponding to the } \frac{3(N+1)}{4} \text{ ordered observation}$$

The following set of rules is used for obtaining the quartile values:

1. If the resulting positioning point is an integer, the particular numerical observation corresponding to that positioning point is chosen for the quartile.
2. If the resulting positioning point is halfway between two integer positioning points, the average of their corresponding values is selected.
3. If the resulting positioning point is neither an integer nor a value halfway between two other positioning points, a simple rule of thumb used to approximate the particular quartile is to merely round off to the nearest integer positioning point and select the numerical value of the corresponding observation.

Thus, for example, from the (revised) stem-and-leaf display (Figure 5.2) pertaining to the reported quarterly sales tax receipts for the 50 stores in Heldman Township we have:

$$Q_1 = \frac{N+1}{4} \text{ ordered observation}$$

$$= \frac{50+1}{4} = 12.75 \rightarrow 13\text{th ordered observation (when rounded)}$$

Therefore $Q_1 = 9.0$ thousands of dollars.

$$Q_2 = \text{median} = \frac{N+1}{2} \text{ ordered observation}$$

$$= \frac{50+1}{2} = 25.5\text{th ordered observation}$$

Therefore $Q_2 = 10.3$ thousands of dollars.

[1] When dealing with a sample, we simply replace N by n in the positioning-point formulas.

$$Q_3 = \frac{3(N+1)}{4} \text{ ordered observation}$$

$$= \frac{3(50+1)}{4} = 38.25 \rightarrow 38\text{th ordered observation (when rounded)}$$

Therefore $Q_3 = 11.6$ thousands of dollars.

These quartiles, along with the smallest and largest observations, are highlighted in the accompanying stem-and-leaf display (Figure 5.4).

1. To obtain Q_1 we simply count (left to right, row by row) to the 13th ordered observation. In our data, the "leaf" is 0 so that the result is 9.0 thousands of dollars.
2. To obtain the median we must take the mean of the 25th and 26th ordered observations. In our data, both the observations are 10.3 so that the median is 10.3 thousands of dollars.
3. To obtain Q_3 we count *down* (left to right, row by row) to the 38th ordered observation or count *up* (right to left, row by row) to the 13th ordered observation. In our data, the "leaf" is 6, so that the result is 11.6 thousands of dollars.

	Stem	Leaf
$X_{smallest} = 5.3$	5	3
	6	57
	7	3568
	8	04679
$Q_1 = 9.0$	9	02335689
Median = 10.3	10	00123345567
$Q_3 = 11.6$	11	011125668
	12	555789
	13	00
	14	5
$X_{largest} = 15.1$	15	1

FIGURE 5.4 □ Stem-and-leaf display and summary measures.

Problems

- • 5.12 Find the quartiles for the creativity test data from the revised stem-and-leaf display you constructed in Problem 5.2 on page 119.
- 5.13 Find the quartiles for the data on death rates from the revised stem-and-leaf display you constructed in Problem 5.3 on page 119.
- 5.14 Find the quartiles for the data on price-earnings ratios from the revised stem-and-leaf display you constructed in Problem 5.4 on page 119.
- 5.15 Find the quartiles for the data on fringe benefits from the revised stem-and-leaf display you constructed in Problem 5.6 on page 120.
- 5.16 Find the quartiles for the data on alcohol consumption from the revised stem-and-leaf display you constructed in Problem 5.7 on page 120.

• 5.17 Find the quartiles for the data on cottage-cheese calories from the revised stem-and-leaf display you constructed in Problem 5.9 on page 120. (Do not separate the 4% from the 2% brands when constructing the quartiles.)

5.18 Find the quartiles for the data on U.S. temperatures from the revised stem-and-leaf display which you constructed in Problem 5.10 on page 120.

5.4 THE MIDHINGE—A MEASURE OF CENTRAL TENDENCY

The **midhinge** is the mean of the first and third quartiles in a set of data. That is,

$$\text{midhinge} = \frac{Q_1 + Q_3}{2} \quad \textbf{(5.1)}$$

One use of the midhinge is to overcome potential problems introduced by extreme values in the data. We pointed out earlier that the midrange can be changed greatly by one extreme value; the midhinge, however, is not affected unless a large proportion of values are unusual.

For the reported quarterly sales tax data we have

$$\text{midhinge} = \frac{9.0 + 11.6}{2} = 10.3 \text{ thousands of dollars}$$

5.5 THE INTERQUARTILE RANGE—A MEASURE OF DISPERSION

The **interquartile range** (also called "midspread") is the difference between the third and first quartiles in a set of data. That is,

$$\text{interquartile range} = Q_3 - Q_1 \quad \textbf{(5.2)}$$

This simple measure considers the spread in the middle 50% of the data and, like the midhinge, is not influenced by a few extreme values.

For the reported quarterly sales tax data we have

$$\text{interquartile range} = 11.6 - 9.0 = 2.6 \text{ thousands of dollars}$$

This is the range in reported sales tax receipts for the "middle group of stores."

Problems

• 5.19 Find the midhinge and interquartile range for the data on creativity test scores for which you found the quartiles in Problem 5.12 on page 122.

5.20 Find the midhinge and interquartile range for the data on death rates for which you found the quartiles in Problem 5.13 on page 122.

5.21 Find the midhinge and interquartile range for the data on price-earnings ratios for which you found the quartiles in Problem 5.14 on page 122.

5.22 Find the midhinge and interquartile range for the data on fringe benefits for which you found the quartiles in Problem 5.15 on page 122.

5.23 Find the midhinge and interquartile range for the data on alcohol consumption for which you found the quartiles in Problem 5.16 on page 122.

• 5.24 Find the midhinge and interquartile range for the data on the caloric content of cottage cheeses for which you found the quartiles in Problem 5.17 on page 123.

5.25 Find the midhinge and interquartile range for the data on temperatures in U.S. cities for which you found the quartiles in Problem 5.18 on page 123.

5.6 FIVE-NUMBER SUMMARIES

To identify and describe the major features of the data, the ''exploratory data analysis'' approach utilizes measures of central tendency and dispersion which have the property of **resistance**—that is, statistics which are relatively insensitive to extreme values or to changes in some of the data. The median, the midhinge, and the interquartile range are three widely used resistant statistics. If we combine these resistant measures with information regarding the occurrence of extreme values, we have a better idea as to the shape of the distribution. A **five-number summary** consists of

$$X_{\text{smallest}} \quad Q_1 \quad \text{median} \quad Q_3 \quad X_{\text{largest}}$$

The five-number summary may be used to study the shape of the distribution. If the data were perfectly symmetric, the following would be true:

1. The distance from Q_1 to the median would equal the distance from the median to Q_3.
2. The distance from X_{smallest} to Q_1 would equal the distance from Q_3 to X_{largest}.
3. The median, the midhinge, and the midrange would all be equal. (Moreover, these measures would also equal the mean in the data.)

Using the reported quarterly sales tax data (as displayed in Figure 5.4), only minor differences from these three rules are observed. Hence we may conclude that the distribution is approximately symmetrical.

On the other hand, for nonsymmetrical distributions the following would be true:

1. In right-skewed distributions the distance from Q_3 to $X_{largest}$ greatly exceeds the distance from $X_{smallest}$ to Q_1.
2. In left-skewed distributions the distance from $X_{smallest}$ to Q_1 greatly exceeds the distance from Q_3 to $X_{largest}$.

Problems

• 5.26 List the five-number summary for the data on creativity test scores from which you found the quartiles in Problem 5.12 on page 122. Use this summary to describe the shape of the distribution.

5.27 List the five-number summary for the data on death rates from which you found the quartiles in Problem 5.13 on page 122. Use this summary to describe the shape of the distribution.

5.28 List the five-number summary for the data on price-earnings ratios of stocks from which you found the quartiles in Problem 5.14 on page 122. Use this summary to describe the shape of the distribution.

5.29 List the five-number summary for the data on fringe benefits from which you found the quartiles in Problem 5.15 on page 122. Use this summary to describe the shape of the distribution.

5.30 List the five-number summary for the data on alcohol consumption from which you found the quartiles in Problem 5.16 on page 122. Use this summary to describe the shape of the distribution.

• 5.31 List the five-number summary for the data on the caloric content of cottage cheeses from which you found the quartiles in Problem 5.17 on page 123. Use this summary to describe the shape of the distribution.

5.32 List the five-number summary for the data on temperatures in U.S. cities from which you found the quartiles in Problem 5.18 on page 123. Use this summary to describe the shape of the distribution.

5.7 THE BOX-AND-WHISKER PLOT

In its simplest form a **box-and-whisker plot** provides a graphical representation of the data through its five-number summary. Such a plot is depicted in Figure 5.5 for the reported quarterly sales tax receipts data.

The vertical line drawn within the box represents the location of the median value in the data. Further, the vertical line at the left side of the box represents the location of Q_1, while the vertical line at the right side of the box represents the location of Q_3. Therefore, the box contains the middle 50% of the observations in the distribution. The lower 25% of the data are represented by a dashed line (that is, a *whisker*) connecting the left side of the box to the location of the smallest value, $X_{smallest}$. Similarly, the upper 25% of the data are represented by a dashed line connecting the right side of the box to $X_{largest}$.

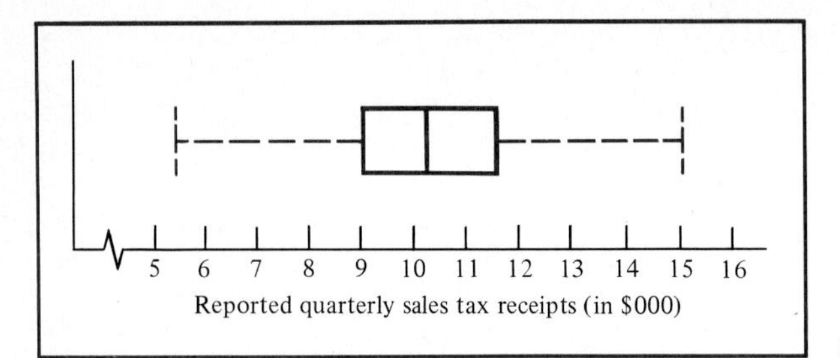

FIGURE 5.5 □ Box-and-whisker plot of reported quarterly sales tax receipts from 50 stores.

SOURCE: Data are taken from Figure 5.1.

The visual representation of the reported quarterly sales tax receipts data depicted in Figure 5.5 indicates again the approximate symmetry in the shape of that distribution. First, observe that the vertical median line appears in the center of the box. (In a perfectly symmetrical distribution the quartiles are equidistant from the median.) Moreover, note the approximate equivalence in the lengths of the whiskers—the dashed lines representing the lower and upper 25% of the data.

In order to illustrate how the shape of a distribution affects the box-and-whisker plot, Figure 5.6 depicts six such plots—one for each of the distributions represented in Figures 4.5 and 4.10 to 4.14, respectively. (See pages 94 to 97.)

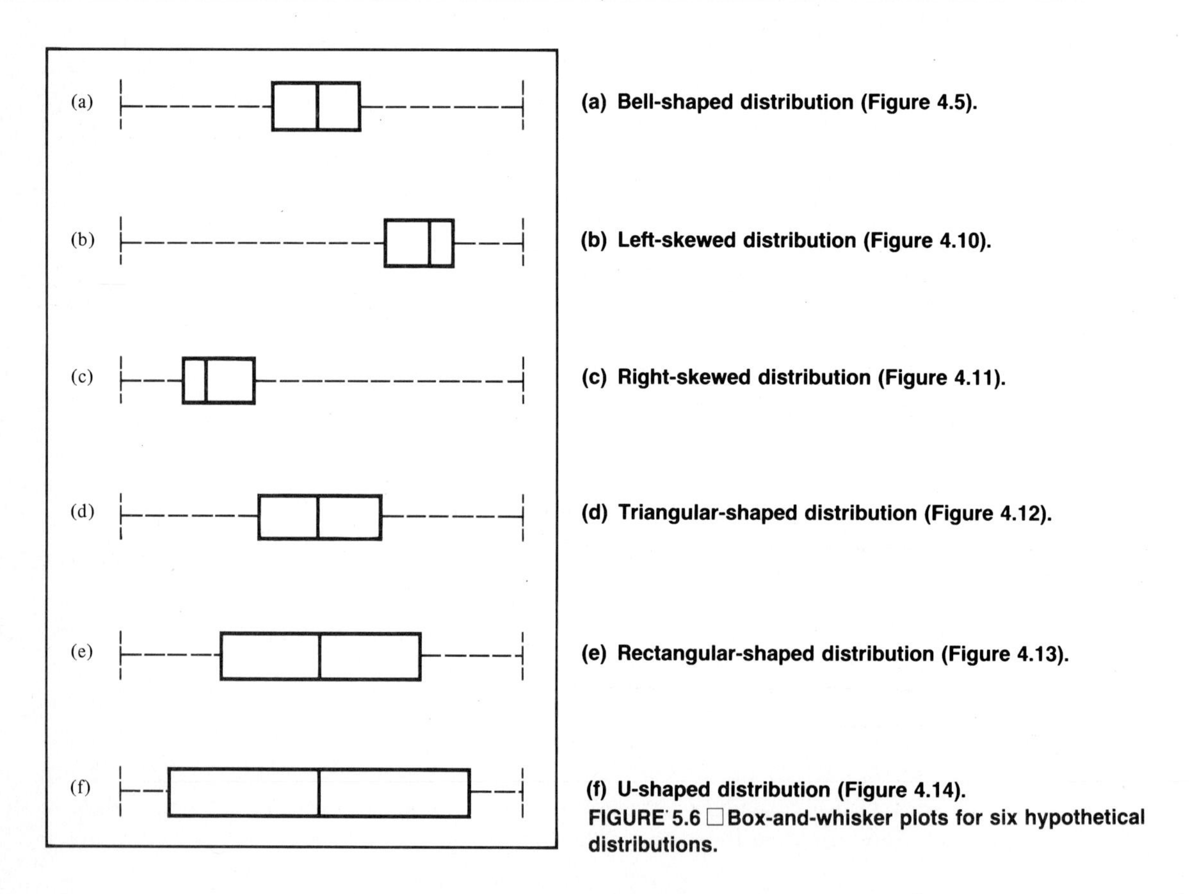

(a) Bell-shaped distribution (Figure 4.5).

(b) Left-skewed distribution (Figure 4.10).

(c) Right-skewed distribution (Figure 4.11).

(d) Triangular-shaped distribution (Figure 4.12).

(e) Rectangular-shaped distribution (Figure 4.13).

(f) U-shaped distribution (Figure 4.14).

FIGURE 5.6 □ Box-and-whisker plots for six hypothetical distributions.

Problems

• 5.33 Construct a box-and-whisker plot of the data on creativity test scores from the five-number summary you created in Problem 5.26 on page 125.

5.34 Construct a box-and-whisker plot of the data on death rates from the five-number summary you created in Problem 5.27 on page 125.

5.35 Construct a box-and-whisker plot of the data on price-earnings ratios from the five-number summary you created in Problem 5.28 on page 125.

5.36 Construct a box-and-whisker plot of the data on fringe benefits from the five-number summary you created in Problem 5.29 on page 125.

5.37 Construct a box-and-whisker plot of the data on alcohol consumption from the five-number summary you created in Problem 5.30 on page 125.

• 5.38 Construct a box-and-whisker plot of the data on the caloric content of cottage cheese using the five-number summary you created in Problem 5.31 on page 125.

5.39 Construct a box-and-whisker plot of the data on temperatures in U.S. cities from the five-number summary you created in Problem 5.32 on page 125.

5.8 DESCRIPTIVE STATISTICS: AN OVERVIEW

In Chapters 3 through 5 we have studied how collected data are prepared, presented, and characterized. Qualitative data were tallied into summary tables and graphically displayed by bar charts and pie diagrams. Quantitative data were characterized through the calculation of various descriptive summary measures representing the properties of central tendency, dispersion, and shape. Moreover, when quantitative data were obtained, various methods of data preparation and presentation (ordered arrays, frequency and percentage distributions, histograms and polygons) were described. Finally, in an effort to further enhance our analysis of quantitative data, exploratory data analysis techniques (using stem-and-leaf displays, five-number summaries, and box-and-whisker plots) were also considered.

Supplementary Problems

5.40 For the data on the seven-day yield percentages of money-market mutual funds in Problem 3.71 on page 71:

(a) Construct a stem-and-leaf display.
(b) Find the quartiles.
(c) Find the midhinge and interquartile range.
(d) List the five-number summary.
(e) Draw a box-and-whisker plot.
(f) Write a description of the data set, using your answers to (a) through (e).

5.41 For the data on the price of spring and mineral water in Problem 3.72 on page 72:
- **(a)** Construct a stem-and-leaf display.
- **(b)** Find the quartiles.
- **(c)** Find the midhinge and interquartile range.
- **(d)** List the five-number summary.
- **(e)** Draw a box-and-whisker plot.
- **(f)** Write a description of the data set, using your answers to (a) through (e).

5.42 For the data on projected changes in the Gross National Product in Problem 3.74 on page 72:
- **(a)** Construct a stem-and-leaf display.
- **(b)** Find the quartiles.
- **(c)** Find the midhinge and interquartile range.
- **(d)** List the five-number summary.
- **(e)** Draw a box-and-whisker plot.
- **(f)** Write a description of the data set, using your answers to (a) through (e).

• 5.43 For the data on automobile acceleration in Problem 3.75 on page 73, for each country's automobiles:
- **(a)** Construct a stem-and-leaf display.
- **(b)** Find the quartiles.
- **(c)** Find the midhinge and interquartile range.
- **(d)** List the five-number summary.
- **(e)** Draw a box-and-whisker plot.
- **(f)** Write a description of the data set, using your answers to (a) through (e). Include a comparison of the German- and Japanese-made automobile models.

5.44 For the data on earnings per share of companies in the leisure time and recreational activities industry in Problem 3.78 on page 73:
- **(a)** Construct a stem-and-leaf display.
- **(b)** Find the quartiles.
- **(c)** Find the midhinge and interquartile range.
- **(d)** List the five-number summary.
- **(e)** Draw a box-and-whisker plot.
- **(f)** Write a description of the data set, using your answers to (a) through (e).

5.45 For the data on utility charges in Problem 4.17 on page 85:
- **(a)** Construct a stem-and-leaf display.
- **(b)** Find the quartiles.
- **(c)** Find the midhinge and interquartile range.
- **(d)** List the five-number summary.
- **(e)** Draw a box-and-whisker plot.
- **(f)** Write a description of the data set, using your answers to (a) through (e).

5.46 For the data from each manufacturer on the length of life of light bulbs, shown in Problem 4.18 on page 85:
- **(a)** Construct a stem-and-leaf display.
- **(b)** Find the quartiles.
- **(c)** Find the midhinge and interquartile range.
- **(d)** List the five-number summary.

(e) Draw a box-and-whisker plot.
(f) Write a description of the data set, using your answers to (a) through (e). Include a comparison of the data from the two manufacturers.

5.47 For the data on starting salaries in the computer industry, shown in Problem 4.19 on page 86:
(a) Construct a stem-and-leaf display.
(b) Find the quartiles.
(c) Find the midhinge and interquartile range.
(d) List the five-number summary.
(e) Draw a box-and-whisker plot.
(f) Write a description of the data set, using your answers to (a) through (e).

Database Exercises

The following problems refer to the sample data obtained from the questionnaire of Figure 2.5 and presented in Figure 2.10. They should be solved (to the greatest extent possible) with the aid of a computer package.

□ 5.48 From the responses to question 1 of the alumni association survey, regarding number of years since graduation:
(a) Form the stem-and-leaf display.
(b) Obtain a five-number summary.
(c) Construct a box-and-whisker plot.
(d) Obtain the midhinge and interquartile range.
(e) Analyze your findings.

□ 5.49 From the responses to question 7 of the alumni association survey, regarding current salary:
(a) Form the stem-and-leaf display.
(b) Obtain a five-number summary.
(c) Construct a box-and-whisker plot.
(d) Obtain the midhinge and interquartile range.
(e) Analyze your findings.

□ 5.50 From the responses to question 10 of the alumni association survey, regarding number of years working in the current job area:
(a) Form the stem-and-leaf display.
(b) Obtain a five-number summary.
(c) Construct a box-and-whisker plot.
(d) Obtain the midhinge and interquartile range.
(e) Analyze your findings.

□ 5.51 From the responses to question 12 of the alumni association survey, regarding number of job areas worked in since graduation:
(a) Form the stem-and-leaf display.
(b) Obtain a five-number summary.
(c) Construct a box-and-whisker plot.
(d) Obtain the midhinge and interquartile range.
(e) Analyze your findings.

□ 5.52 From the responses to question 13 of the alumni association survey, regarding number of full-time jobs held over the past ten years:
- **(a)** Form the stem-and-leaf display.
- **(b)** Obtain a five-number summary.
- **(c)** Construct a box-and-whisker plot.
- **(d)** Obtain the midhinge and interquartile range.
- **(e)** Analyze your findings.

□ 5.53 From the responses to question 15 of the alumni association survey, regarding grade-point indexes:
- **(a)** Form the stem-and-leaf display.
- **(b)** Obtain a five-number summary.
- **(c)** Construct a box-and-whisker plot.
- **(d)** Obtain the midhinge and interquartile range.
- **(e)** Analyze your findings.

□ 5.54 From the responses to question 19 of the alumni association survey, regarding heights:
- **(a)** Form the stem-and-leaf display.
- **(b)** Obtain a five-number summary.
- **(c)** Construct a box-and-whisker plot.
- **(d)** Obtain the midhinge and interquartile range.
- **(e)** Analyze your findings.

□ 5.55 From the responses to question 20 of the alumni association survey, regarding ages:
- **(a)** Form the stem-and-leaf display.
- **(b)** Obtain a five-number summary.
- **(c)** Construct a box-and-whisker plot.
- **(d)** Obtain the midhinge and interquartile range.
- **(e)** Analyze your findings.

□ 5.56 From the responses to questions 7 and 5 of the alumni association survey, regarding current salary based on graduate school attendance:
- **(a)** Form the stem-and-leaf displays.
- **(b)** Obtain the five-number summaries.
- **(c)** Construct box-and-whisker plots.
- **(d)** Obtain the midhinges and interquartile ranges.
- **(e)** Compare and contrast the results over the two groups.

□ 5.57 From the responses to questions 15 and 5 of the alumni association survey, regarding grade-point indexes based on graduate-school attendance:
- **(a)** Form the stem-and-leaf displays.
- **(b)** Obtain the five-number summaries.
- **(c)** Construct box-and-whisker plots.
- **(d)** Obtain the midhinges and interquartile ranges.
- **(e)** Compare and contrast the results over the two groups.

□ 5.58 From the responses to questions 15 and 18 of the alumni association survey, regarding grade-point indexes based on gender:
- **(a)** Form the stem-and-leaf displays.
- **(b)** Obtain the five-number summaries.
- **(c)** Construct box-and-whisker plots.
- **(d)** Obtain the midhinges and interquartile ranges.
- **(e)** Compare and contrast the results over the two groups.

□ 5.59 From the responses to questions 19 and 18 of the alumni association survey, regarding height based on gender:

(a) Form the stem-and-leaf displays.
(b) Obtain the five-number summaries.
(c) Construct box-and-whisker plots.
(d) Obtain the midhinges and interquartile ranges.
(e) Compare and contrast the results over the two groups.

References

1. TUKEY, J., *Exploratory Data Analysis* (Reading, Mass.: Addison-Wesley, 1977).
2. VELLEMAN, P. F., AND D. C. HOAGLIN, *Applications, Basics, and Computing of Exploratory Data Analysis* (Boston: Duxbury Press, 1981).

Basic Probability

6.1 INTRODUCTION: WHAT'S AHEAD

In the past four chapters we have examined methods of collecting, presenting, and describing a set of data. In this chapter we begin to study various rules of basic probability that can be used to evaluate the chance of occurrence of different phenomena. These rules will be expanded so that we can use probability theory in making inferences about population parameters based only on sample statistics.

Case C—The Medical Researcher's Study of Common Cold Remedies

A medical researcher in a small city wishes to study the various ways in which individuals pursue the treatment of the common cold (without resorting to prescription medicines). In order to simplify the study, the researcher divides common-cold treatments into two categories: (1) primarily using natural remedies, including vitamin C, soup, and fruit juice, (2) primarily using nonprescription cold remedies, such as decongestants, aspirin, and aspirin substitutes. The researcher also wishes to study other characteristics that might be related to self-treatment of the common cold. Among those that might be considered are: (1) whether the individual considers himself or herself as a "heavy junk food" eater, (2) whether the individual smokes, and (3) the sex of the individual. A random sample of 200 adults (aged 18 and above) was selected from the community. Assuming that the sample was representative, the medical researcher would like to draw conclusions about the probability that an individual:

1. Primarily uses "natural remedies" against the common cold.
2. Primarily uses nonprescription cold remedies.
3. Classifies himself as a "heavy junk food" eater.
4. Primarily uses "natural remedies" *and* is not a "heavy junk food" eater.
5. Prefers nonprescription remedies *or* is a "heavy junk food" eater.
6. Prefers natural remedies *given that* he or she is a "heavy junk food" eater.

The goal of this chapter is to learn how the laws of probability may be used to answer these and similar questions.

6.2 OBJECTIVE AND SUBJECTIVE PROBABILITY

In the case study of concern to the medical researcher, we indicated an interest in determining the probability of occurrence of various types of events. What did we mean by the word "probability"? We can think of **probability** as the chance or likelihood that a particular event will occur. If each possible outcome is equally likely, this chance of occurrence of the event may be defined as follows:

$$\text{probability of occurrence} = \frac{X}{T} \qquad \textbf{(6.1)}$$

where X = the number of outcomes for which the event we are looking for occurs
T = the total number of possible outcomes

The determination of the probability of occurrence of an event is not always easy since it depends on the type of application being studied. For instance, how would we determine the probability that our favorite professional football team (let us say, the New York Jets) will win the Super Bowl this year? This type of probability, called **subjective probability,** would depend on who is assigning the probability. If a Jets fan were determining his subjective probability, then a high value might be assigned to the Jets' winning the Super Bowl. On the other hand, if a national sportscaster were determining her probability, we could be reasonably certain that a different probability would be assigned.

As a second example of probability we might wish to determine the probability of winning a state lottery in which six numbers are to be selected out of a total of 44 different numbers. This type of probability, called **a priori classical probability,** is based upon prior knowledge of the process involved. Such probabilities will be discussed in Chapter 7. In this chapter, however, we shall focus on the problem of interest to the medical researcher. Since it involves obtaining probabilities based on an actual sample of data, this approach is referred to as **empirical classical probability.**

6.3 BASIC PROBABILITY CONCEPTS

The most obvious rule for probabilities is that they range in value from 0 to 1. An impossible event has probability 0 of occurring, while a certain event has probability 1 of occurring. To develop further rules, we return to the problem of the medical researcher.

We have mentioned that the medical researcher was evaluating different types of treatments for the common cold along with other factors that might relate to the treatment preferred. In the remainder of this chapter we shall focus on two variables: (1) the treatment preferred: natural remedies or nonprescription cold remedies, and (2) whether or not the individual considers himself or herself a ''junk food'' eater. Thus if we wish to present the results relating to these two variables, we can cross-classify them in a **contingency table** (see Section 4.7). Table 6.1 represents the contingency table for the sample of 200 adults.

In the contingency table each possible outcome is referred to as an event. A **simple event** is one that can be described by a single characteristic. In Table 6.1, the event ''prefers natural remedies'' would be considered a simple event, as would the event ''does not eat junk food.'' The collection of all possible events is called the **sample space.** The sample space for the medical researcher's problem consists of the responses from the 200 individuals who have been interviewed. In Table 6.1 the sample space has been subdivided according to two different kinds of simple

TABLE 6.1 □ Contingency table for treatment preferred and eating junk food

	Eats junk food		
Treatment preferred	**Yes**	**No**	**Total**
Natural cold remedies	45	50	95
Nonprescription cold remedies	85	20	105
Total	130	70	200

events: cold treatment preferred and eating junk food. Since we have a two-way table, we may also define a **joint event** as an event that has two or more characteristics. For example, the event "eats junk food *and* prefers nonprescription cold remedies" would be a joint event, since it involves two characteristics.

It is also useful to define the complement of an event as follows: The **complement** of event *A* includes all events that are not part of event *A*. Thus the complement of the event "prefers natural remedies" is "prefers nonprescription cold remedies," while the complement of the event "eats junk food" is, of course, "does not eat junk food."

6.4 SIMPLE OR MARGINAL PROBABILITY

Thus far in our discussion of probability we have focused on the several types of probability and on defining different events. We shall now begin to answer some of the researcher's questions about the respondents in the study by developing rules for different types of probability.

Simple probability refers to the probability of occurrence of a simple event (an event described by a single characteristic) such as the probability of using natural cold remedies. The notation $P(A)$ represents the probability that event A occurs.

How would we find the probability of selecting an individual who prefers natural cold remedies? We would find the number of respondents who prefer natural cold remedies *and* eat junk food, and the number of respondents who prefer natural cold remedies *and* do not eat junk food. Then we would have:

$$P\begin{pmatrix}\text{preferring natural}\\ \text{cold remedies}\end{pmatrix} = \frac{\text{number preferring natural cold remedies}}{\text{total number of respondents}}$$

$$= \frac{\begin{pmatrix}\text{number preferring}\\ \text{natural cold remedies}\\ \textit{and}\\ \text{eating junk food}\end{pmatrix} + \begin{pmatrix}\text{number preferring}\\ \text{natural cold remedies}\\ \textit{and}\\ \text{not eating junk food}\end{pmatrix}}{\text{total number of respondents}}$$

$$= \frac{45 + 50}{200} = \frac{95}{200} = .475$$

We may observe that the simple probability is also called **marginal probability,** since the total number of successes (preferring natural cold remedies) can be obtained from the appropriate margin of the contingency table (see Table 6.1). Thus if we also wanted to determine the probability that a respondent eats junk food, we would have:

$$P(\text{eating junk food}) = \frac{\text{number preferring junk food}}{\text{total number of respondents}}$$

$$P(\text{eating junk food}) = \frac{130}{200} = .65$$

Problems

6.1 In your own words, tell what the following are, and give an example of each:

(a) Subjective probability.

(b) A priori classical probability.

(c) Empirical classical probability.

6.2 Dr. Frank N. Stein, a physicist at the University of Transylvania, fires a particle beam at a magnetic field. The beam is a mixture of two types of particles: one-third are triminos and two-thirds are caryons. The triminos are deflected up by the magnetic field one-sixth of the time and down five-sixths of the time. The caryons are deflected up three-fourths of the time and down one-fourth of the time.

(a) If 720 particle traces are recorded from this process, how many should fall in each cell of the following contingency table?

	Direction deflected	
Particle	**Up**	**Down**
Triminos		
Caryons		

(b) What is the simple (marginal) probability that a particle will be deflected up? Down?

(c) What is the complement of the event "deflected up"?

(d) Is the event "caryon particle" a simple or joint event?

(e) The table you prepared in part (a) is what is expected on the basis of a theory. Would you classify the probabilities from this table as a priori classical, empirical classical, or subjective?

(f) If a similar table were prepared of results from an actual experiment, would probabilities from that table be a priori classical, empirical classical, or subjective?

6.3 A psychologist devises a test to diagnose schizophrenia. Her test is a good one in the sense that if a person is schizophrenic, the probability is .9 that the test will say that the person is schizophrenic (and the probability is .1 that it will say the person is not schizophrenic). On the other hand, if the person is not schizophrenic, there is a probability of .85 that the test will say that the person is not schizophrenic. About 1% of the population is schizophrenic.

(a) If the test were given to a random sample of 2,000 people, what would the expected outcome be?

	Test says	
Person really is	**Schizophrenic**	**Normal**
Schizophrenic		
Normal		

(b) What is the probability of the simple event "test says the person is normal"?

• **6.4** A researcher counts the number of people who appear to have blonde, brunette, red, and black hair:

Gender	Apparent hair color			
	Blonde	**Brunette**	**Red**	**Black**
Female	68	43	7	23
Male	12	24	3	14

Compute the marginal probabilities for each simple event.

6.5 A drug company is testing a new formula for its nonprescription pain reliever, Cureall-X. It decides to give the new formula, the old formula, and placebos to different groups of people after dental surgery to compare the pain-relieving ability of the new and old drugs to each other and to the placebo. (A placebo is a pill which contains no active ingredients; in spite of this, placebos sometimes work as well as supposedly active ingredients.) The results of the company's study are:

Formula	Degree of pain relief		
	None	**Some**	**Total**
New	4	16	30
Old	3	27	9
Placebo	12	21	6

(a) Compute the marginal probabilities for each pain-relief event (none, some, and total relief from pain).

(b) What would the comparable marginal probabilities for type of drug indicate, and why is it of little importance in this type of study? (Remember that the researchers decide who gets each formula.) If a study were done in which people were asked which brand of pain reliever they used, these marginals would be more meaningful. Why?

6.6 Use the data on blood pressure in people between 65 and 70 years old (Problem 4.57 on page 106) to answer the following:

(a) What are the simple events for this data set?

(b) What are the marginal probabilities for the simple events?

(c) What information would you want to know about how people were chosen to be in the study before you tried to interpret these marginal probabilities? Discuss this issue, giving examples of how marginals could be misleading.

6.7 For the data on type of college attended and level of job (Problem 4.58 on page 107):

(a) What are the simple events?

(b) What are the marginal probabilities for the simple events?

(c) What information would you want to know about how people were chosen to be in the study before you tried to interpret these marginal probabilities? Discuss this issue, giving examples of how marginals could be misleading.

• **6.8** For the data on smallpox vaccination (Problem 4.59 on page 107):

(a) What are the simple events?

(b) What are the marginal probabilities for the simple events?

6.5 JOINT PROBABILITY

Marginal probability referred to the occurrence of simple events, such as the probability of preferring natural cold remedies. **Joint probability** refers to phenomena containing two or more events, such as the probability of eating junk food *and* preferring nonprescription cold remedies. This joint event "*A and B*" means that event *A* and event *B* must both occur. The event "eats junk food *and* prefers nonprescription cold remedies" consists only of those respondents who satisfy both these criteria. Referring to Table 6.1, we may observe that respondents satisfying both these conditions can be found in the single cell "eats junk food *and* prefers nonprescription cold remedies." Since there are 85 respondents in this cell, the joint probability is

$$P\begin{pmatrix}\text{eats junk food } \textit{and} \text{ prefers} \\ \text{nonprescription cold remedies}\end{pmatrix} = \frac{85}{200} = .425$$

Now that we have discussed the concept of joint probability, the marginal probability of a particular event can be viewed in an alternative manner. We have already shown that the marginal probability contains a set of joint probabilities. For example, the event "eats junk food" contains all respondents who "eat junk food *and* prefer natural cold remedies" along with those who "eat junk food *and* prefer nonprescription cold remedies." Thus in general we can state that

$$P(A) = P(A \textit{ and } B_1) + P(A \textit{ and } B_2) + \cdots + P(A \textit{ and } B_k) \qquad \textbf{(6.2)}$$

where $B_1, B_2, \ldots, B_k$ are mutually exclusive and collectively exhaustive events. That is, none of the B_i can occur at the same time **(mutually exclusive)** and at least one of the B_i must occur **(collectively exhaustive).** For example, being male and being female are mutually exclusive and collectively exhaustive events. No one is both, and everyone is one or the other.

Therefore, the probability of eating junk food can be expressed as:

$$\begin{aligned} P(\text{junk food}) &= P(\text{junk food } \textit{and} \text{ natural remedies}) \\ &\quad + P(\text{junk food } \textit{and} \text{ nonprescription cold remedies}) \\ &= \frac{45}{200} + \frac{85}{200} \\ &= \frac{130}{200} \end{aligned}$$

This is, of course, the same result that we would obtain if we added up the frequencies for outcomes that made up the simple event "eating junk food."

Problems

6.9 Using the data on deflection of elementary particles from the physics experiment (Problem 6.2 on page 136):
- **(a)** Construct a table with the joint probabilities of each joint event in the table.
- **(b)** Calculate the marginal probability of each simple event by adding the appropriate joint probabilities. (Compare with your answers to Problem 6.2; they should be the same.)

6.10 Using the data on the test for schizophrenia in Problem 6.3 on page 136:
- **(a)** Construct a table with the joint probabilities of each joint event in the table.
- **(b)** Calculate the marginal probability of each simple event by adding the appropriate joint probabilities. (Compare with your answers to Problem 6.3; they should be the same.)

• **6.11** Using the data on sex and hair color (Problem 6.4 on page 137):
- **(a)** Construct a table with the joint probabilities of each joint event in the table.
- **(b)** Calculate the marginal probability of each simple event by adding the appropriate joint probabilities. (Compare with your answers to Problem 6.4; they should be the same.)

6.12 Using the data on pain relievers (Problem 6.5 on page 137):
- **(a)** Construct a table with the joint probabilities of each joint event in the table.
- **(b)** Calculate the marginal probability of each simple event by adding the appropriate joint probabilities. (Compare with your answers to Problem 6.5; they should be the same.)

6.13 Using the data on blood pressure and exercise (Problem 6.6 on page 137, from Problem 4.57 on page 106):
- **(a)** Construct a table with the joint probabilities of each joint event in the table.
- **(b)** Calculate the marginal probability of each simple event by adding the appropriate joint probabilities. (Compare with your answers to Problem 6.6; they should be the same.)

6.14 Using the data on type of college attended and job level held (Problem 6.7 on page 137, from Problem 4.58 on page 107):
- **(a)** Construct a table with the joint probabilities of each joint event in the table.
- **(b)** Calculate the marginal probability of each simple event by adding the appropriate joint probabilities. (Compare with your answers to Problem 6.7; they should be the same.)

• **6.15** Using the data on smallpox (Problem 6.8 on page 137, from Problem 4.59 on page 107):
- **(a)** Construct a table with the joint probabilities of each joint event in the table.
- **(b)** Calculate the marginal probability of each simple event by adding the appropriate joint probabilities. (Compare with your answers to Problem 6.8; they should be the same.)

6.6 ADDITION RULE

Now that we have developed a means of finding $P(A)$, the probability of the event A, and $P(A \text{ and } B)$, the probability of the event A *and* B, we would like to find $P(A \text{ or } B)$, the probability of the event A *or* B. The rule involved, called the **addition rule,** refers to the occurrence of either event A, or event B, or both A and B. For example, the event "prefers nonprescription cold remedies *or* eats junk food" would include all respondents who use nonprescription cold remedies, or eat junk food, or had both these characteristics.

In order to determine the number of respondents who are included in this event, each cell of the contingency table (Table 6.1) can be examined. From Table 6.1, the cell "prefers natural cold remedies *and* eats junk food" is part of the event, since it includes respondents who eat junk food. The cell "prefers nonprescription cold remedies *and* does not eat junk food" is also part of the event, because it includes respondents who prefer nonprescription cold remedies. Finally, the cell "prefers nonprescription cold remedies *and* eats junk food" is included in the event because it contains respondents who prefer nonprescription cold remedies and also eat junk food. Therefore, given this information, the desired probability can be obtained as follows:

$$
\begin{aligned}
P\begin{pmatrix}\text{prefers nonprescription} \\ \text{cold remedies } \textit{or} \text{ eats} \\ \text{junk food}\end{pmatrix} &= P\begin{pmatrix}\text{prefers natural remedies} \\ \textit{and} \text{ eats junk food}\end{pmatrix} \\
&+ P\begin{pmatrix}\text{prefers nonprescription cold remedies} \\ \textit{and} \text{ does not eat junk food}\end{pmatrix} \\
&+ P\begin{pmatrix}\text{prefers nonprescription cold remedies} \\ \textit{and} \text{ eats junk food}\end{pmatrix} \\
&= \frac{45}{200} + \frac{20}{200} + \frac{85}{200} = \frac{150}{200}
\end{aligned}
$$

More readily, however, the computation of $P(A \text{ or } B)$, the probability of the event A *or* B, can be expressed by the **general addition rule:**

$$P(A \text{ or } B) = P(A) + P(B) - P(A \text{ and } B) \tag{6.3}$$

Applying this addition rule to the previous example, we obtain the following result:

$$P\begin{pmatrix}\text{nonprescription cold}\\ \text{remedies } \textit{or} \text{ eats junk food}\end{pmatrix} = P\,(\text{nonprescription cold remedies}) + P\,(\text{eats junk food}) - P\begin{pmatrix}\text{nonprescription cold remedies } \textit{and}\\ \text{eats junk food}\end{pmatrix}$$

$$= \frac{105}{200} + \frac{130}{200} - \frac{85}{200} = \frac{150}{200}$$

The addition rule takes the probability of A and adds it to the probability of B. However, the joint probability of A *and* B must be subtracted from this total, since it has already been included twice—once in the probability of A and also in the probability of B. Therefore, since this joint probability has been double-counted, it must be subtracted to provide the correct result.

MUTUALLY EXCLUSIVE EVENTS

In certain circumstances, however, the joint probability need not be subtracted because it is equal to zero. Such situations occur when no outcomes exist for a particular event. For example, suppose that we wanted to know the probability of picking from a standard deck of 52 playing cards a single card that was either a queen or a king. Using the addition rule, we have the following:

$$P(\text{queen } \textit{or} \text{ king}) = P(\text{queen}) + P(\text{king}) - P(\text{queen } \textit{and} \text{ king})$$

$$= \frac{4}{52} + \frac{4}{52} - \frac{0}{52} = \frac{8}{52}$$

Certainly we realize that the probability that a card will be both a queen and a king simultaneously is zero, since in a standard deck each card may take on only one particular value. As mentioned previously, whenever the joint probability does not contain any outcomes, the events involved are considered to be *mutually exclusive*. This refers to the fact that the occurrence of one event (a queen) means that the other event (a king) cannot occur. Thus, the addition rule for mutually exclusive events reduces to:

$$P(A \text{ or } B) = P(A) + P(B) \qquad \textbf{(6.4)}$$

COLLECTIVELY EXHAUSTIVE EVENTS

Now consider what the probability would be of selecting a card that was red or black. Since red and black are mutually exclusive events, using Equation (6.4) we would have

$$P(\text{red } or \text{ black}) = P(\text{red}) + P(\text{black})$$

$$= \frac{26}{52} + \frac{26}{52} = \frac{52}{52} = 1.0$$

The probability of red *or* black adds up to 1.0. This means that the card selected must be red or black, since they are the only colors in a standard deck. Since one of these events must occur, they are considered to be *collectively exhaustive* events.

Problems

6.16 For each of the following, tell whether the events which are created are (i) mutually exclusive (ii) collectively exhaustive. If they are not, either reword the categories to make them mutually exclusive and exhaustive, or tell why this would not be useful. [*Hint*: Some are measuring a single concept, others more than one concept, and some mix categories from different possible levels of analysis of a concept.]

- **(a)** People are classified on religion into the categories Protestant, Catholic, Christian, Jewish, none.
- **(b)** Registered voters were asked whether they registered as Republican or Democrat.
- **(c)** People are asked, "When you get a cold, do you treat it using (i) natural remedies, (ii) nonprescription cold remedies?" (They can answer yes or no to each; note the difference between this and the example used in this chapter.)
- **(d)** Letters of the alphabet (capital letters, in plain block style) are classified in terms of whether they (i) have any straight lines in them, (ii) have any curved lines in them.
- **(e)** Words are classified as being either nouns, verbs, adverbs, or adjectives.
- **(f)** Living organisms are classified as being plants, animals, mammals, or invertebrates.

6.17 Suppose that people are asked whether they usually use natural remedies when they get a cold, and also whether they usually use nonprescription cold remedies when they get a cold. (Note the difference between this and the example used in this chapter.) The data from this survey are

	Nonprescription cold remedy	
Natural remedy	**Yes**	**No**
Yes	45	75
No	125	25

- **(a)** Compute the probabilities of each of the joint events in the table above.
- **(b)** Compute the probabilities of each of the marginal events.
- **(c)** Use the addition rule to compute the probability that a person uses either natural *or* nonprescription remedies (or both) for colds.

6.18 The probability of each of the following events is zero. For each, tell why. Tell what common characteristic of these events makes their probability zero.

(a) A person is a Jewish atheist.
(b) An animal has feathers *and* is a horse.
(c) A person is male *and* has had a hysterectomy.

6.19 What is the probability that:

(a) A person is either male *or* female.
(b) A body of water is either fresh water *or* salt water.
(c) A student either passes his final exam *or* fails his final exam.

6.20 Referring to Problem 6.2 on page 136, what is the probability that a particle chosen at random:

(a) Is a trimino *or* is deflected up?
(b) Is a caryon *or* is deflected down?
(c) Is a trimino *or* is a caryon?

6.21 Referring to Problem 6.3 on page 136, what is the probability that a randomly chosen person:

(a) Is schizophrenic *or* the test says is schizophrenic?
(b) Is normal *or* the test says is normal?
(c) Is schizophrenic *or* is normal?

• 6.22 Referring to Problem 6.4 on page 137, what is the probability that a randomly chosen person:

(a) Is a male *or* appears to have red hair?
(b) Is a female *or* appears to have black hair?
(c) Appears to have blonde hair *or* brunette hair *or* red hair *or* black hair? Are these events mutually exclusive? Why? Are they also collectively exhaustive? Why?

6.23 Referring to Problem 6.5 on page 137, what is the probability that a randomly chosen person:

(a) Received the new formula *or* achieved total pain relief?
(b) Received the placebo *or* achieved no pain relief?
(c) Achieved some pain relief *or* total pain relief? Are these events mutually exclusive? Why? Are they also collectively exhaustive? Why?

6.24 Referring to Problem 4.57 on page 106, what is the probability that a randomly selected person:

(a) Had blood pressure that was too high *or* had no exercise?
(b) Had normal blood pressure *or* a moderate amount of exercise?
(c) Had a moderate amount of exercise *or* a high amount of exercise? Are these events mutually exclusive? Why? Are they also collectively exhaustive? Why?

6.25 Referring to Problem 4.58 on page 107, what is the probability that a randomly selected person:

(a) Was in a high management level *or* attended an Ivy League college?
(b) Was in a middle management level *or* attended a public college?
(c) Attended an Ivy League *or* other private college? Are these events mutually exclusive? Why? Are they also collectively exhaustive? Why?

6.7 CONDITIONAL PROBABILITY

Each example that we have studied thus far in the chapter has involved the probability of a particular event selected from the complete sample space. However, we might like to know how these probabilities would be affected if certain information about the events were known in advance. For example, what if we knew that the person selected was a "junk food eater"? What then would be the probability that he or she preferred natural cold remedies?

In situations where we are computing the probability of a particular event (A) given information about the occurrence of another event (B), the probability involved is referred to as **conditional probability** ($A | B$). This conditional probability can be defined as:

$$P(A | B) = \frac{P(A \text{ and } B)}{P(B)} \qquad \textbf{(6.5)}$$

where $P(A \text{ and } B)$ is the joint probability of A and B
$P(B)$ is the marginal probability of B

Thus if we wished to determine the probability that a respondent prefers natural remedies given that he or she is a junk food eater, using Equation (6.5) we would have:

$$P(\text{natural remedies} \mid \text{junk food}) = \frac{P(\text{natural remedies } and \text{ junk food})}{P(\text{junk food})}$$

$$= \frac{45/200}{130/200}$$

$$= \frac{45}{130} = .346$$

We may observe that this probability is not the same as the marginal probability of any individual preferring natural remedies ($95/200 = .475$). This result reveals some important information: the prior knowledge of the individual's eating habits has affected our prediction of his preferred cold treatment. Thus from a statistical perspective we can state that these two events can be considered to be associated—i.e., not independent. The events A and B are said to be **statistically independent** if $P(A | B) = P(A)$. That is, A and B are independent if the probability of A's occurring does not depend on whether or not the event B occurs.

Problems

6.26 In Problem 6.2 on page 136, which dealt with an experiment to measure the deflection of atomic particles:

(a) Is the statement that the beam consists of 1/3 triminos a statement about a marginal probability or a conditional probability?

(b) Is the statement that 3/4 of the caryons are deflected up a statement about a marginal probability or a conditional probability?

(c) What is the conditional probability that a particle is a trimino, given that the particle was deflected down?

(d) What is the conditional probability that a particle is a caryon, given that the particle was deflected down? [Compare with your answer to part (c).]

(e) Does it appear that the events "particle is a caryon" *and* "particle is deflected down" are independent? Why or why not?

6.27 For the data on a test to diagnose schizophrenia (Problem 6.3 on page 136):

(a) What is the probability of being schizophrenic, given that the test is positive? Why does this seem strange?

(b) Suppose now that the marginal probability of the simple event "person is schizophrenic" is .25—that is, that schizophrenia is very common. Redo the calculations of part (a). Discuss the importance of this marginal probability (called the *base rate* or *prevalence* of the disease) in detecting the disease.

• 6.28 For the data in Problem 6.4 on page 137:

(a) What is the conditional probability of being blonde for males? For females?

(b) How do these conditional probabilities compare with the probability of the simple event "being blonde"? What does this say about the independence of the events "male" *and* "blonde"? Of the events "female" *and* "blonde"?

(c) What is the conditional probability of being male, given that the person is blonde?

6.29 For the data on pain relievers in Problem 6.5 on page 137:

(a) What is the conditional probability of total pain relief, given that the new formula was taken?

(b) How does this compare to the conditional probabilities, given the old formula and given the placebo?

(c) Does the event "total relief from pain" appear to be independent of the event "new formula was taken"?

6.30 For the data on type of college attended and job level attained (Problem 6.7 on page 137; originally from Problem 4.58 on page 107):

(a) What is the conditional probability of being in high-level management, given that the person graduated from an Ivy League school?

(b) What is the conditional probability of being in middle-level management, given graduation from an Ivy League school?

(c) Is being a high-level manager independent of graduating from an Ivy League school?

6.31 Suppose that the following results were obtained in a random sample of males between the ages of 16 and 25 who were in the labor force (i.e., were working or looking for work) in a large urban area:

	Employment status	
Race	**Employed**	**Unemployed**
Black	46	40
White	220	75

(a) What is the probability of being unemployed?
(b) What is the conditional probability of being unemployed, given that the person is black?
(c) What is the conditional probability of being unemployed, given that the person is white?
(d) What is the probability that an unemployed male is black? Does this seem strange, given the answers to parts (b) and (c)?

6.32 Referring to Problem 4.57 on page 106:
(a) Given that the person selected had high blood pressure, what is probability that he or she did no exercise?
(b) Given that the person did no exercise, what is the probability that he or she has blood pressure that is too high?
(c) Explain the difference in the results in parts (a) and (b).
(d) Are blood-pressure level *and* amount of exercise statistically independent? Explain.

6.8 MULTIPLICATION RULE

The formula for conditional probability can be manipulated algebraically so that the joint probability can be determined from the conditional probability of an event. From Equation (6.5) we have:

$$P(A \mid B) = \frac{P(A \text{ and } B)}{P(B)}$$

Solving for the joint probability (A *and* B), we have the **general multiplication rule:**

$$P(A \text{ and } B) = P(A \mid B)P(B) \qquad \textbf{(6.6)}$$

The use of this multiplication rule can be illustrated by an example concerning a deck of playing cards. Suppose that instead of selecting only one card, we wished to select two cards from a complete 52-card deck. What would be the probability that both cards selected are red? In this example the multiplication rule would be used in the following way:

$$P(A \text{ and } B) = P(A \mid B)P(B)$$

Therefore if

$$A_R = \text{second card selected is red}$$
$$B_R = \text{first card selected is red}$$

we need to find

$$P(A_R \text{ and } B_R) = P(A_R \mid B_R)P(B_R)$$

$P(B_R)$, the probability that the first card selected is red, is 26/52 = 1/2, since there are 26 red cards in a 52-card deck. However, the probability that the second card is red depends on the result of the first selection. If the first card is not returned to the deck after its color is determined (sampling without replacement—see Section 2.7), then there will be 51 cards remaining from which to select. If the first card selected is red, the probability that the second is red is 25/51, since there would be 25 red cards remaining in the deck. Therefore, using Equation (6.6), we have the following:

$$P(A_R \text{ and } B_R) = \left(\frac{25}{51}\right)\left(\frac{26}{52}\right)$$
$$= \frac{25}{102} = .245$$

However, what if the first card is returned to the deck after its color is determined? Then the probability of picking a red card on the second selection is the same as for the first selection (sampling with replacement—Section 2.7), since there still would be 26 red cards from which to select. Therefore, we have the following:

$$P(A_R \text{ and } B_R) = P(A_R \mid B_R)P(B_R)$$
$$= \left(\frac{26}{52}\right)\left(\frac{26}{52}\right) = \left(\frac{1}{2}\right)\left(\frac{1}{2}\right)$$
$$= \frac{1}{4} = .25$$

This example of sampling with replacement illustrates that the second selection is independent of the first, since the second probability was not at all influenced by the first selection. Therefore, the **multiplication rule for independent events** can be expressed as follows [by substituting $P(A)$ for $P(A \mid B)$]:

$$P(A \text{ and } B) = P(A)P(B) \qquad \textbf{(6.7)}$$

If the rule holds for two events A and B, then A and B are statistically independent. Therefore, we have two ways to determine statistical independence:

1. Events A and B are statistically independent if and only if $P(A \mid B) = P(A)$.
2. Events A and B are statistically independent if and only if $P(A \text{ and } B) = P(A)P(B)$.

We may note that in a 2×2 contingency table, if this is true for one joint event, it will be true for all joint events. That is, in a 2×2 table, if A and B are independent, then so are A and not-B, not-A and B, and not-A and not-B. For example, if the probability that a person has blue eyes is independent of that person's being a male, then the probability of having blue eyes is independent of being female, the probability of not having blue eyes is independent of being male, and the probability of not having blue eyes is independent of being female.

Problems

6.33 Prove that, in a 2×2 table, if A and B are independent, then so are A and not-B, not-A and B, and not-A and not-B. [*Hint*: Construct a table and put in the letters a, b, c, and d to stand for the observed probabilities in the four cells of the table. Then write the equation which represents events A and B being independent in terms of the product of the marginal probabilities. From this, develop similar formulas for the remaining combinations of events making up the table.]

6.34 Suppose that, according to genetic theory, a species of plant would have a 1-in-4 chance of being yellow rather than green and a 1-in-4 chance of having smooth rather than wrinkled leaves. If these traits are independent, fill in the following table with the probabilities of each of the joint events involving leaf color and texture:

	Leaf texture	
Leaf color	**Wrinkled**	**Smooth**
Green		
Yellow		

Suppose that 1,600 plants were grown in a study to investigate the genetic theory described above. Using the above probabilities, fill in the following table showing how many plants out of 1,600 should have each combination of characteristics:

	Leaf texture	
Leaf color	**Wrinkled**	**Smooth**
Green		
Yellow		

6.35 Suppose that you believe that the probability that you will get an A in statistics is .7, and the probability that you will get an A in English is .6. If these events are independent, what is the probability that you will get an A in both statistics and English? Give some plausible reasons why these events may not be independent, even though the teachers in these two subjects may not communicate about your work.

• 6.36 You enter two sweepstakes, each of which has a 1-in-100 chance that you will win some prize. If winning each is independent of the other, then what is the probability that you will win nothing? [*Hint*: Winning nothing means you lose the first sweepstakes and you lose the second sweepstakes.]

6.37 In the Wallace County highway system there are 10 billboards behind which traffic officers can hide. If there are two officers, each of whom hides independently (and could, therefore, be behind the same billboard), and you take a trip which goes past 4 billboards, what is the probability that you will get caught if you speed the whole way? [*Hint*: First decide the probability that you would not get caught by Officer A, then do the same for Officer B. To calculate the probability that you will not get caught, use one minus the probability that you will get caught.]

6.38 Referring to Problem 6.2 on page 136, use Equation (6.7) to determine whether the type of particle is statistically independent from the direction deflected.

6.39 Referring to Problem 6.3 on page 136, use Equation (6.7) to determine whether the person's actual condition is statistically independent from the condition diagnosed by the test.

• 6.40 Referring to Problem 6.4 on page 137, use Equation (6.7) to determine whether the gender of the person is statistically independent from the apparent hair color.

6.41 Referring to problem 6.5 on page 137, use Equation (6.7) to determine whether the pain-reliever formula used is statistically independent from the degree of pain relief achieved.

6.42 Referring to Problem 4.58 on page 107, use Equation (6.7) to determine whether the management level achieved is statistically independent from the type of college attended.

6.9 COUNTING RULES

Each rule of probability that we have discussed has involved the counting of the number of favorable outcomes and the total number of outcomes. In many instances, however, because of the large number of possibilities, it is not feasible to list each of the outcomes. In these circumstances, rules for counting have been developed. Five different counting rules will be discussed here.

Suppose that we are tossing a coin 4 times. How can we determine the number of possible sequences of heads and tails?

COUNTING RULE 1 If any one of k different mutually exclusive and collectively exhaustive events can occur on each of n trials, the number of possible outcomes is equal to

$$k^n \qquad \textbf{(6.8)}$$

Therefore, there would be $2^4 = 16$ different outcomes for the coin that was tossed four times.

The second counting rule is a more general version of the first. To illustrate this rule, suppose that the number of possible events is different on some of the trials. For example, New York State currently uses three letters followed by three digits for motor vehicle license-plate identification. The fact that there are letters (each having 26 outcomes) and numerical digits (each having 10 outcomes) leads to the second counting rule.

COUNTING RULE 2 If there are k_1 events on the first trial, k_2 events on the second trial, . . . , and k_n events on the nth trial, the number of possible outcomes is

$$k_1 k_2 \cdots k_n \tag{6.9}$$

Thus the total number of possible license plates (having three letters followed by three digits) is

$$(26)(26)(26)(10)(10)(10) = 17{,}576{,}000$$

The third counting rule involves the number of ways that a set of objects can be arranged in order. For example, the Eastern Division of the American Football Conference includes the Buffalo Bills, Indianapolis Colts, Miami Dolphins, New England Patriots, and New York Jets. How can we determine the possible number of orders of finish for these five teams? We may begin by realizing that any one of the five teams could finish first. Once the first position is filled, four teams remain to be assigned to the second position. This assignment procedure is continued until all the positions are occupied. The total number of assignments would be equal to (5) (4) (3) (2) (1) = 120. This situation may be generalized as

COUNTING RULE 3 The number of ways n objects may be arranged in order is

$$n! = n(n-1)\cdots(1) \tag{6.10}$$

where $n!$ is called n "factorial" and $0!$ is defined as 1.

The number of ways that the five teams could be arranged in order is

$$n! = 5(4)(3)(2)(1) = 120$$

In many instances we need to know the number of ways in which a subset of the entire group of objects can be arranged in order. Each possible arrangement is called a **permutation.** For example, a teacher may want to know the number of ways that a class president, vice-president, and student council representative can be selected from a class of ten students. Since only three students of the ten are to be selected for the different positions, we need the following rule:

COUNTING RULE 4: PERMUTATIONS — The number of ways of arranging X objects selected from n objects in order is

$$\frac{n!}{(n-X)!} \quad \textbf{(6.11)}$$

Therefore the number of ways of selecting the students for the three different jobs is equal to

$$\frac{n!}{(n-X)!} = \frac{10!}{(10-3)!} = \frac{(10)(9)(8)(7)(6)(5)(4)(3)(2)(1)}{(7)(6)(5)(4)(3)(2)(1)} = 720$$

Finally, what if the teacher were designating only three students for the job of student council representative (rather than president, vice-president, and a single student council representative). In this case, the order of selection would be ignored, since the three students are being selected for the same job. This rule is called the rule of **combinations.**

COUNTING RULE 5: COMBINATIONS — The number of ways of selecting X objects out of n objects irrespective of order is equal to

$$\frac{n!}{X!(n-X)!} \quad \textbf{(6.12)}$$

This expression may also be denoted by the symbol

$$\binom{n}{X}$$

Comparing this rule to the previous one, we see that it differs only in the inclusion of a term $X!$ in the denominator. This is because when we counted permutations, all of the arrangements of the X objects were distinguishable; with combinations, the $X!$ possible arrangements of objects are irrelevant, as in the example above of selecting members of a council. For example, selecting Sharyn, Jessica, and Ryan is the same as selecting Jessica, Sharyn, and Ryan, and so on. The division by $X!$ corrects for this overcounting.

Therefore the number of combinations of three students selected from ten students is equal to

$$\frac{n!}{X!(n-X)!} = \frac{10!}{3!(10-3)!} = \frac{10!}{3!7!} = \frac{(10)(9)(8)(7)(6)(5)(4)(3)(2)(1)}{(3)(2)(1)(7)(6)(5)(4)(3)(2)(1)} = 120$$

Problems

6.43 Dr. Brainglow was attempting to order a computer system for her department. She had a choice of 10 computers, 5 printers, and 12 word processing programs. She wanted to make sure that each item she bought would work with the others, so she went to her computer dealer to try all possible systems. How many systems (computer plus printer plus program) did she have to try? Given your answer, do you see why some companies (and individuals) stick with one manufacturer when selecting equipment?

6.44 The operators of Gunilla's Restaurant want to advertise that they can serve over 10,000 different meals. They have 5 kinds of soup, 3 kinds of salad, 12 entrees, 6 vegetables, and 7 desserts.

- **(a)** How many different meals can they serve, if a meal consists of one item from each category?
- **(b)** If people might skip any part of the meal except the entree, how many kinds of meals can they serve? [*Very slight hint*: This is just as easy as part (a) if you think about it the right way.]

6.45 Wacko Wayne's record store has 50 different heavy metal albums, 30 punk albums, and 3 classical albums in stock. If you buy one album of each kind, in how many different ways could you make your selection?

• 6.46 Bruce's Ice Cream Parlour has 11 flavors to choose from. If you have a double-dip cone:

- **(a)** How many different cones can you have, if you can use the same flavor for both dips?
- **(b)** How many different cones can you have, if you use each flavor at most once in a cone?
- **(c)** How do your answers to (a) and (b) change if you count flavor A on top and B on the bottom as different than B on top and A on the bottom? [If you already assumed that in doing parts (a) and (b), now do the reverse.]

6.47 To find the best way to display his collection of rocks, Jonathan decides to try all different ways and pick the one that looks best. How many arrangements must he try if he has 7 rocks and has

- **(a)** 7 slots in his display case.
- **(b)** 11 slots in his display case.
- **(c)** 4 slots in his display case (so that not all of the rocks can be displayed at once).

• 6.48 An art director for a magazine has 12 pictures to choose from for 5 positions in her magazine.
- **(a)** How many different sets of 5 pictures could she choose from the 12 available pictures?
- **(b)** Once she has picked her 5 pictures, in how many ways can she arrange them in the magazine?
- **(c)** How many permutations are there of 12 objects taken 5 at a time? [Show how this answer is related to your answers to (a) and (b).]

6.49 The programming director for television station KORN has to pick a situation comedy, a drama, and a nonfiction series for his Wednesday night prime time slots at 8, 9, and 10 P.M. If there are 4 situation comedies, 5 dramas, and 2 nonfiction series to choose from:
- **(a)** How many sets of three programs can he choose?
- **(b)** Once he has picked a set of three programs, in how many ways can he arrange them in the three time slots?
- **(c)** Find out how many different lineups he can choose by multiplying your answers to (a) and (b).

Supplementary Problems

6.50 Explain the difference between a simple event and a joint event. Provide an example of each.

6.51 A manager of a women's store wishes to determine the relationship between the type of customer and the type of payment. She has collected the following data:

	Payment	
Customer	**Credit**	**Cash**
Regular	70	50
Nonregular	40	40

- **(a)** Give an example of a simple event.
- **(b)** Give an example of a joint event.
- **(c)** What is the complement of a cash payment?
- **(d)** Why is "regular customer who makes a cash payment" a joint event?

• 6.52 The statistics association at a large state university would like to determine whether there is a relationship between a student's interest in statistics and his or her ability in mathematics. A random sample of 200 students is selected, and they are asked whether their ability in mathematics and interest in statistics is low, average, or high. The results are as follows:

	Ability in mathematics			
Interest in statistics	**Low**	**Average**	**High**	**Totals**
Low	60	15	15	90
Average	15	45	10	70
High	5	10	25	40
Totals	80	70	50	200

(a) Give an example of a simple event.
(b) Give an example of a joint event.
(c) Why is "high interest in statistics" *and* "high ability in mathematics" a joint event?
(d) If a student is selected at random, what is the probability that he or she:
(1) Has a high ability in mathematics?
(2) Has an average interest in statistics?
(3) Has a low ability in mathematics?
(4) Has a high interest in statistics?
(e) What is the probability that a student chosen at random:
(1) Has a low ability in mathematics *and* a low interest in statistics?
(2) Has a high ability in mathematics *and* an average interest in statistics?
(3) Has a high ability in mathematics *and* a high interest in statistics?

6.53 A standard deck of cards is being used to play a game. There are four suits (hearts, diamonds, clubs, and spades), each having 13 cards (ace, 2, 3, 4, 5, 6, 7, 8, 9, 10, jack, queen, and king), making a total of 52 cards. This complete deck is thoroughly mixed, and you will receive the first two cards from the deck *without* replacement.
(a) What is the probability that both cards are queens?
(b) What is the probability that the first card is a 10 *and* the second card is a 5 *or* 6?
(c) If we were sampling *with* replacement, what would be the answer in (a)?
(d) In the game of blackjack the picture cards (jack, queen, king) count as 10 points while the ace counts as either 1 or 11 points. All other cards are counted at their face value. Blackjack is achieved if your two cards sum to 21 points. What is the probability of getting blackjack in this problem?

• 6.54 A bin contains two defective tubes and five good tubes. Two tubes are randomly selected from the bin *without* replacement.
(a) What is the probability that both tubes are defective?
(b) What is the probability that the first tube selected is defective *and* the second tube is good?

6.55 A lock on a bank vault consists of three dials, each with 30 positions. In order for the vault to open when closed, each of the three dials must be in the correct position.
(a) How many different possible "dial combinations" are there for this lock?
(b) What is the probability that if you randomly select a position on each dial, you will be able to open the bank vault?
(c) Explain why "dial combinations" are not mathematical combinations expressed by Equation (6.12)?

6.56 **(a)** If a coin is tossed seven times, how many different outcomes are possible?
(b) If a die is tossed seven times, how many different outcomes are possible?
(c) Discuss the differences in your answers to (a) and (b).

6.57 A particular brand of women's jeans can be ordered in seven different sizes, three different colors, and three different styles. How many different jeans would have to be ordered if a store wanted to have one pair of each type?

• 6.58 **(a)** The Daily Double at the local racetrack consists of picking the winners of the first two races. If there are 10 horses in the first race and 13 horses in the second race, how many Daily Double possibilities are there?
(b) The Quinella at the local racetrack consists of picking the horses that will place

first and second in a race irrespective of order. If eight horses are entered in a race, how many Quinella combinations are there?

6.59 When rolling a die once, what is the probability that:

(a) The face of the die is odd?
(b) The face is even *or* odd?
(c) The face is even *or* a one?
(d) The face is odd *or* a one?
(e) The face is both even *and* a one?
(f) Given the face is odd, it is a one?

6.60 Suppose that a survey has been undertaken to determine if there is a relationship between place of residence and ownership of a foreign-made automobile. A random sample of 200 car owners from large cities, 150 from suburbs, and 150 from rural areas was selected with the following results:

	Type of area			
Car ownership	**Large city**	**Suburb**	**Rural**	**Totals**
Own foreign car	90	60	25	175
Do not own foreign car	110	90	125	325
Totals	200	150	150	500

(a) If a car owner is selected at random, what is the probability that he or she:
(1) Owns a foreign car?
(2) Lives in a suburb?
(3) Owns a foreign car *or* lives in a large city?
(4) Lives in a large city *or* a suburb?
(5) Lives in a large city *and* owns a foreign car?
(6) Lives in a rural area *or* does not own a foreign car?
(b) Assume we know that the person selected lives in a suburb. What is the probability that he or she owns a foreign car?
(c) Is area of residence statistically independent of whether the person owns a foreign car? Explain.

Database Exercises

The following problems refer to the sample data obtained from the questionnaire of Figure 2.5 and presented in Figure 2.10. They should be solved with the aid of a computer package.

□ 6.61 Do a cross-tabulation of the respondents' area of undergraduate major (question 2) with gender (question 18). From the resulting table, answer the following questions:

(a) What is the probability that a randomly selected person in the sample is female? Is male?
(b) What is the probability that a randomly selected person in the sample majored in business? In humanities and social science? In either business *or* science and mathematics? In neither business nor humanities and social science?
(c) What is the probability that a randomly selected person in the sample was female

and majored in business? Are these two events (at least approximately) statistically independent?

(d) What is the conditional probability of being a business major, given that the student is female?

□ **6.62** Do a cross-tabulation of current philosophical ideology (question 4) with the type of organization in which the respondent is currently employed (question 8). Use this information to answer the following questions:

(a) What is the probability that a randomly selected person in the sample is working for a small (less than 100 employees) corporation? A large corporation? A corporation of any size, small or large? What is the probability that a randomly selected person in the sample does not work for a corporation?

(b) What is the probability that a randomly selected person in the sample works in the area of education? Is self-employed?

(c) What is the probability that a randomly selected person in the sample is more conservative now than as an undergraduate? Is about the same? Is more liberal?

(d) What is the probability that a randomly selected person in the sample who works for a corporation is more conservative? What is the probability that a person works for a corporation *and* is more conservative? Are these two outcomes independent?

□ **6.63** Do a cross-tabulation of whether the respondent would attend the college if he or she had to make the choice again (question 16), with the area in which he or she majored (question 2). Use this table to answer the following questions:

(a) What is the probability that a randomly selected person in the sample would attend the college again? Would not attend? Is not sure?

(b) What is the probability that a randomly selected person in the sample would attend the college again, given that he or she majored in business? Humanities and social science? Science and mathematics?

(c) Does it appear that whether he or she would attend again is independent of whether he or she majored in business?

□ **6.64** Referring to Problem 4.95 on page 114, what is the probability that a randomly selected respondent:

(a) Majored in business?

(b) Majored in business *or* humanities and social science?

(c) Majored in science and mathematics *or* attended primarily due to reputation?

(d) Majored in business *and* attended primarily due to cost?

(e) Suppose you knew that the person majored in business. What then would be the probability that he or she attended primarily due to cost?

(f) Are undergraduate major *and* primary reason for attending statistically independent?

□ **6.65** Referring to Problem 4.98 on page 115, what is the probability that a randomly selected respondent:

(a) Primarily attended due to convenient location?

(b) Would attend the college now?

(c) Primarily attended due to reputation *or* would attend the college now?

(d) Primarily attended due to cost *and* would not attend now?

(e) Suppose we know that the respondent attended due to cost. What then is the probability that he or she would not attend now?

(f) Are primary reason for attending *and* the choice that would be made now statistically independent?

□ **6.66** Referring to Problem 4.99 on page 115, what is the probability that a randomly selected respondent:

(a) Is more conservative now?
(b) Is more conservative now *or* about the same?
(c) Is more conservative now *and* is a male?
(d) Is about the same now *or* is a female?
(e) Suppose that a male respondent is selected. What then is the probability that he is more conservative now?
(f) Is philosophical ideology statistically independent of gender?

□ **6.67** Referring to Problem 4.96 on page 114, what is the probability that a randomly selected respondent:
(a) Attended graduate school?
(b) Majored in business *and* did not attend graduate school?
(c) Majored in business *or* attended graduate school?
(d) Suppose that a respondent who majored in business is selected. What then is the probability that he or she did not attend graduate school?
(e) Are graduate-school attendance *and* major statistically independent?

□ **6.68** Referring to problem 4.100 on page 115, what is the probability that a randomly selected respondent:
(a) Is employed by a corporation/company with over 100 employees?
(b) Is employed in education *or* by a nonprofit or government agency?
(c) Is a male *and* is self-employed?
(d) Is a female *or* is employed by a corporation/company with under 100 employees?
(e) Suppose a male respondent is selected. What then is the probability that he is self-employed?
(f) Are type of organization *and* gender statistically independent?

□ **6.69** Referring to Problem 4.101 on page 115, what is the probability that a randomly selected respondent:
(a) Would attend the college again?
(b) Would recommend *or* highly recommend the college?
(c) Would attend the college again *or* highly recommend the college?
(d) Would not attend the college *and* would recommend the college?
(e) Suppose the respondent selected would choose to attend the college again. What then is the probability that he or she would recommend the college?
(f) Are the choice that would be made now *and* the advice given to others statistically independent?

□ **6.70** Referring to Problem 4.102 on page 115, what is the probability that a randomly selected respondent:
(a) Would recommend *or* highly recommend the college?
(b) Would not recommend the college?
(c) Is a male *or* would not recommend the college?
(d) Is a female *and* is undecided about recommending the college?
(e) Suppose that a male respondent is selected. What then is the probability that he highly recommends the college?
(f) Are advice given to others *and* gender statistically independent?

References and Other Sources

1. HAYS, W. L., *Statistics*, 3d ed. (New York: Holt, Rinehart and Winston, 1981).

2. Kirk, R. E., ed., *Statistical Issues: A Reader for the Behavioral Sciences* (Belmont, Calif.: Wadsworth, 1972).
3. Mosteller, F., R. Rourke, and G. Thomas, *Probability with Statistical Applications*, 2d ed. (Reading, Mass.: Addison-Wesley, 1970).

Basic Probability Distributions

7.1 INTRODUCTION: WHAT'S AHEAD

In the previous chapter we established various rules of probability and examined some counting techniques. In this chapter we will utilize this information to develop the concept of mathematical expectation and to explore various probability models which represent phenomena of interest.

The goals of this chapter are:

1. To develop an understanding of the concept of mathematical expectation and its applications in decision making.
2. To acquire the ability to use the binomial probability distribution in decision making.
3. To demonstrate how the normal probability distribution can be used: (a) to represent certain types of continuous phenomena and (b) to approximate the binomial model under specific conditions.

In order to introduce the important aspects of this chapter let us consider an illustrative example.

Case D—The Governor's Football Lottery Program

To obtain needed revenue to support the funding of public educational programs in the state, the Governor proposes the establishment of a State Football Lottery Program. The program is to be conducted on a weekly basis for each of the 16 weeks which constitute the regularly scheduled season-of-play by the 28 teams in the National Football League. The rules for participation are:

1. People may purchase, for $1, a card listing the 14 scheduled games.
2. When filling out the card the purchaser has the option of "selecting the potential winners" in as few as four games or in as many as ten games.

- If four games are selected and all are correctly picked, $12 is won.
- If five games are selected and all are correctly picked, $25 is won.
- If six games are selected and all are correctly picked, $50 is won.
- If seven games are selected and all are correctly picked, $100 is won.
- If eight games are selected and all are correctly picked, $225 is won.
- If nine games are selected and all are correctly picked, $450 is won.
- If ten games are selected and all are correctly picked, $1,000 is won.
- No matter how many games are selected, if all are correctly picked, the amount wagered (i.e., $1) is also returned to the purchaser.
- No matter how many games are selected, if any is incorrectly picked, the dollar is lost.

For simplicity, we will assume that all of the teams are evenly matched, so that the probability of selecting the winner is one-half. In reality, weaker teams are given a handicap (point spread) to make the bets even.[1]

A typical weekly football card is displayed in Figure 7.1.

The following questions arise for the potential purchaser of such a football card:

1. If we bet on a certain number of games, what proportion of the time will none of the teams we bet on win? What proportion of the time will one team win? (Etc.)
2. Realizing that in order to win the lottery we must be correct in all the selections we make, is it a better strategy to make selections in
   (a) Only four games?

[1] To be "picked as a winner" in a real bet, the team selected must cover the point spread established by professional oddsmakers in Las Vegas. The purpose of the point spread is to make the game "even," so that the selecting process becomes similar to choosing the outcome (heads or tails) which results from tossing a fair coin. That is, in theory the process of "picking with the point spread" is akin to guessing the results when tossing a coin. As an example, suppose Team A is favored to defeat Team B by 3 points. If Team A is selected, the purchaser has picked the winner, provided that A wins the actual game by 4 or more points; if Team B is selected, the purchaser has picked the winner provided that B loses by only 1 or 2 points, or ties, or actually wins the game. If Team A wins by exactly 3 points (i.e., the point spread) the purchaser loses no matter which team had been selected.

Circle the team you think will win in at least four of the following games:

Game Number	Team 1	Team 2
1	A	B
2	C	D
3	E	F
4	G	H
5	I	J
6	K	L
7	M	N
8	O	P
9	Q	R
10	S	T
11	U	V
12	W	X
13	Y	Z
14	AA	BB

FIGURE 7.1 ☐ Example of proposed football lottery card.

(b) Five games? (See Problem 7.15.)
(c) As many as ten games? (See Problem 7.16.)
3. How much money would we expect to win (or lose) in making each possible bet?

To answer such questions we must develop the concept of mathematical expectation.

7.2 MATHEMATICAL EXPECTATION FOR DISCRETE RANDOM VARIABLES

As discussed in Section 2.3, a quantitative random variable is some phenomenon of interest whose responses or outcomes may be expressed numerically. Such a random variable may also be classified as discrete or continuous—the former arising from a counting process and the latter from a measuring process.

7.2.1 The Probability Distribution for a Random Variable

We may define the probability distribution for a discrete random variable as follows:

> **A probability distribution for a discrete random variable is a mutually exclusive listing of all possible numerical outcomes for that random variable such that a particular probability of occurrence is associated with each outcome.**

Assuming that each game is "even" (i.e., a fair one), Table 7.1 on page 163 represents the probability distribution corresponding to the selection of four games on the football lottery card.[2]

[2] This type of discrete probability distribution is called the *binomial probability distribution*. It is developed in Section 7.3.

When selecting four games, five possible outcomes may occur. Either none of the picks will be correct, one will be correct, two will be correct, three will be correct, or all four will be correct. Moreover, there are a total of 16 possible events which give rise to these five possible outcomes. These are depicted in the accompanying display. Note that a "thumbs-up" sign is used to represent a correct selection, a "thumbs-down" sign an incorrect selection.

Possible Event	Selections Game 1	Game 2	Game 3	Game 4	Results of Selections	Results of Lottery
1	👎	👎	👎	👎	0 Correct	☹ Lose $1
2	👎	👎	👎	👍	1 Correct	☹ Lose $1
3	👎	👎	👍	👎	1 Correct	☹ Lose $1
4	👎	👍	👎	👎	1 Correct	☹ Lose $1
5	👍	👎	👎	👎	1 Correct	☹ Lose $1
6	👎	👎	👍	👍	2 Correct	☹ Lose $1
7	👎	👍	👎	👍	2 Correct	☹ Lose $1
8	👍	👎	👎	👍	2 Correct	☹ Lose $1
9	👎	👍	👍	👎	2 Correct	☹ Lose $1
10	👍	👎	👍	👎	2 Correct	☹ Lose $1
11	👍	👍	👎	👎	2 Correct	☹ Lose $1
12	👍	👍	👍	👎	3 Correct	☹ Lose $1
13	👍	👍	👎	👍	3 Correct	☹ Lose $1
14	👍	👎	👍	👍	3 Correct	☹ Lose $1
15	👎	👍	👍	👍	3 Correct	☹ Lose $1
16	👍	👍	👍	👍	4 Correct	☺ Win $12

The probabilities associated with the number of correct selections in Table 7.1 are obtained from the display. (Because the teams are evenly matched, the probability of each outcome listed is the same.) We note that there is but 1 chance in 16 that none of the selections will be correct. There is a 25% chance (i.e., 4 out of 16) that one selection will be correct, and so on. Since all possible outcomes are included, the listing in Table 7.1 is complete (i.e., collectively exhaustive) and thus the probabilities must sum to 1.

TABLE 7.1 □ Theoretical probability distribution of the results of making four selections from a football lottery card

No. of correct selections in 4 games	Probability
0	1/16
1	4/16
2	6/16
3	4/16
4	1/16

In order to describe the major features of a probability distribution, we must compute its mean and standard deviation. For example, as prospective purchasers of the football lottery card we should want to know:

1. On average, how many games can we be expected to correctly select if our strategy is to make selections from only four games on a weekly basis?
2. How much variability (or stability) is there in the data? That is, can we expect our weekly results to cluster "close to" the mean, or can we expect a reasonable number of occurrences "far from" the mean (so that we may win the lottery)?

The mean of a probability distribution (μ_X) is the expected value of its random variable. Thus, the answer to the first question is obtained by computing the expected value of the discrete random variable "number of correct selections."

EXPECTED VALUE OF A DISCRETE RANDOM VARIABLE

The expected value of a discrete random variable may be considered as its weighted average over all possible outcomes.

This summary measure can be obtained by multiplying each possible outcome x by its corresponding probability $P(x)$ and then summing up the resulting products. Thus, the expected value of the discrete random variable X, symbolized as $E(X)$, may be expressed as follows:

$$E(X) = \mu_X = \Sigma\, x\, P(x) \tag{7.1}$$

where x = a particular value of the discrete random variable of interest
$P(x)$ = the probability of obtaining the particular value x
Σ = sum over all values of the random variable X

For the theoretical probability distribution of Table 7.1, the expected value of the strategy is

x	P(x)	xP(x)
0	1/16	0
1	4/16	4/16
2	6/16	12/16
3	4/16	12/16
4	1/16	4/16
	1	32/16 = 2.0 = $E(X) = \mu_X = \Sigma\, x\, P(x)$

That is, on average over the long run, we should expect to correctly select two games out of the four every time we purchase a card and choose to utilize this (four-game) selection strategy. Hence, on average, we should not be expected to win the football lottery on a weekly basis.

To address question 2 on page 163 we must obtain the variance (σ_X^2) and the standard deviation (σ_X) of the probability distribution.

VARIANCE AND STANDARD DEVIATION OF THE PROBABILITY DISTRIBUTION OF A DISCRETE RANDOM VARIABLE

The variance of the probability distribution (σ_X^2) of a discrete random variable may be defined as the weighted average of the squared discrepancies between each possible value and its mean—the weights being the probabilities of each of the respective values.

This summary measure can be obtained by multiplying each possible squared discrepancy $(x - \mu_X)^2$ by its corresponding probability $P(x)$ and then summing up the resulting products. Hence, $\sigma_X^2 = \text{Var}(X)$, the variance of the discrete random variable X, may be expressed as follows:

$$\sigma_X^2 = \text{Var}(X) = \Sigma\,(x - \mu_X)^2 P(x) \qquad \textbf{(7.2)}$$

where x = a particular value of the discrete random variable of interest
$P(x)$ = the probability of observing the particular value x
Σ = sum over all values of the random variable X

Moreover, the **standard deviation of the probability distribution (σ_X) of a discrete random variable** is given by

$$\sigma_X = \sqrt{\text{Var}(X)} = \sqrt{\Sigma\,(x - \mu_X)^2 P(x)} \qquad \textbf{(7.3)}$$

For the theoretical probability distribution of Table 7.1, the variance and the standard deviation may be computed by

x	$(x - \mu_X)$	$(x - \mu_X)^2$	$P(x)$	$(x - \mu_X)^2 P(x)$
0	0 − 2.0	4.0	1/16	4/16
1	1 − 2.0	1.0	4/16	4/16
2	2 − 2.0	0.0	6/16	0
3	3 − 2.0	1.0	4/16	4/16
4	4 − 2.0	4.0	1/16	4/16
			1	$16/16 = 1.0 = \sigma_X^2 = \text{Var}(X)$

and

$$\sigma_X = \sqrt{\text{Var}(X)} = \sqrt{1.0} = 1.0$$

In terms of our selection strategy, the Bienayme-Chebyshev rule (Section 3.7) tells us that the majority of the weekly results would be expected to be within $\sqrt{2} = 1.414$ standard deviations of the mean (that is, $\mu_X \pm 1.414\ \sigma_X$). Most likely, then, on a weekly basis we would be expected to make between one and three correct selections [that is, these are the integer outcomes between the values $2.0 \pm 1.414(1.0)$] and not win the lottery. (Of course, zero correct selections would not win either.)

7.2.2 Expected Value of a Function of a Random Variable: Expected Monetary Value of the Lottery

As prospective purchasers of the football lottery card, the most important question that we must address is whether or not it is *profitable* for us to select this particular strategy. From the above discussion we realize that with this strategy we are not expected to win on a regular basis. But is it financially worthwhile to use it at all?

To answer this question we must look at the problem from another perspective. The random variable of interest from the gambling viewpoint is not really X, the *number of games* we correctly select (as in Table 7.1), but rather V, the *dollar value* of the strategy. Thus the random variable V takes on only two possible results—win \$12 or lose \$1. Using the display on page 162, our new probability distribution is presented in Table 7.2.

To understand how Table 7.2 is constructed, we note from the display on page

TABLE 7.2 □ Theoretical probability distribution representing the dollar value in making four selections from a football lottery

Result	Dollar value	Probability
Does not correctly select 4 games ☹	− 1	15/16
Correctly selects 4 games ☺	+12	1/16
		1

162 (as well as from Table 7.1) that of the 16 possible events, the only way we can win with our strategy is to make correct selections in all four games. If so, we make a $12 profit (i.e., we get back $13 from the Lottery Commission). On the other hand, if we do not correctly select all four games, we lose the $1 we spent when purchasing the card. Unfortunately, the chance that this happens is 15/16—the complement of correctly selecting all four games.

To determine whether or not it is financially profitable to purchase the card and use this strategy we compute the expected value of the random variable V as follows:

$$E(V) = \sum vP(v) = (-1)\left(\frac{15}{16}\right) + (12)\left(\frac{1}{16}\right) = -\left(\frac{3}{16}\right) = -.1875$$

Thus our long-run expected payoff is negative. On average, we'd be losing $18\frac{3}{4}$ cents each time we purchased a one-dollar card and used the four-game selection strategy.[3]

This is an example of a **zero-sum game**—that is, it is a game in which what one player loses, the other player wins. (The sum of the players' gains and losses is zero.) If this gaming situation were "fair," neither we nor the Lottery Commission would have an advantage. Here, however, the game is "unfair." The Lottery Commission has the advantage—it expects to win on average $18\frac{3}{4}$ cents per one-dollar card purchased—provided that the four-game selection strategy is employed.

EXAMPLE 7.1

Demonstrate that from the Lottery Commission's standpoint, offering the public the opportunity to use a four-game selection strategy is profitable.

Solution:

For the Lottery Commission we have:

Dollar value	Probability
−12	1/16
+ 1	15/16
Total	1

The expected value for the Lottery Commission is

$$E(V) = (-12)\left(\frac{1}{16}\right) + (+1)\left(\frac{15}{16}\right) = \left(+\frac{3}{16}\right) = +.1875$$

[3] Over the 16-week season we'd be expected to win $12 once and lose the $1 the other 15 occasions. Thus, at the end of the season we'd be expected to lose $3 on the $16 "wagered"—an average of $18\frac{3}{4}$ cents per dollar "wagered."

Thus, if 100,000 such cards were purchased per week, the Lottery Commission would be expected to collect, on average, $18,750 in weekly revenues just from the four-game selection strategy. Add to this the weekly revenues obtained from the other possible selection strategies and we realize that even after deducting for operating expenses the Lottery Commission would be expected to turn over to the state a large sum of money to be used for educational programs.

Nevertheless, now that we know what is expected to happen if we purchase the card, we may decide not to play. That is our option. Since we as individuals can expect to lose from such gambling, on the average, $18\frac{3}{4}$ cents per card over time, unless we derive some intrinsic satisfaction (or, as in Examples 7.5 and 7.6 on pages 176–177, we feel we are so special or possess such rare insight so as to alter the probabilities), we should refrain from participating in such a game.

This situation, however, is not unique. In any lottery, casino, or carnival-type game the expected long-run payoff to the participant is designed to be negative; otherwise the "house" would not be in business (References 3 and 4). Such games as Craps, Keno, Bingo, Lotto, Chuck-a-Luck, Under-or-Over Seven, or Roulette attract large numbers of participants, and, in each case, the expected return over time favors the house. Since a decision maker who acts rationally would want to maximize gains or minimize losses, rational persons would freely partake in such transactions only if something other than expected long-run monetary value were the ultimate criterion. For such rational persons, the criterion used for gambling is the **expected utility of money**—that is, the "action," the excitement, the added interest, the thrill of winning, etc. It is this criterion which rational participants are considering, implicitly or explicitly, when they partake in such games. On the other hand, however, it should be realized that it is the house which uses the expected-monetary-value criterion when it participates in such games.

7.2.3 Mathematical Models of Discrete Random Variables: The Probability Distribution Function

A probability distribution for a discrete random variable may generally be thought of as a *theoretical* listing of outcomes and probabilities [as in Table 7.1] which often can be obtained from a mathematical model representing the phenomenon of interest.

A model is considered to be a miniature representation of some underlying phenomenon. In particular, a mathematical model is a mathematical expression representing some underlying phenomenon. For discrete random variables, this mathematical expression is known as a probability distribution function.

When such mathematical expressions are available, the exact probability of occurrence of any particular outcome of the random variable can be computed. In such cases, then, the entire probability distribution can be obtained and listed. For example, the probability distribution represented in Table 7.1 is one in which the discrete random variable of interest is said to follow the **binomial probability distribution function.** This type of mathematical model will be discussed in Section 7.3.

Problems

7.1 In Problem 7.15 we shall verify that for a five-game selection strategy the expected number of correct selections is 2.5 and the standard deviation is 1.1. Moreover, the probability of winning the lottery (with a profit of $25) by correctly selecting all five games is 1/32. For now, assume that these are true, and use them to respond to the following:
- **(a)** Comment on the meaning of μ_X and σ_X for this strategy.
- **(b)** Determine the expected (monetary) value for using this strategy.
- **(c)** Compare and contrast your results with the four-game selection strategy.

7.2 In Problem 7.16 we shall verify that for a ten-game selection strategy the expected number of correct selections is 5.0 and the standard deviation is 1.6. Moreover, the probability of winning the lottery (with a profit of $1,000) by correctly selecting all ten games is 1/1,024. For now, assume that these are true, and use them to respond to the following:
- **(a)** Comment on the meaning of μ_X and σ_X for this strategy.
- **(b)** Determine the expected (monetary) value for using this strategy.
- **(c)** Compare and contrast your results with the four- and five-game selection strategies. (See the text and Problem 7.1.)

✓7.3 Why does the term *expected value* have that name, even though in many cases (such as the football lottery) you will never see the expected value as the result of any single experiment? (That is, in what sense is a value which never occurs expected?)

• 7.4 Given the following probability distributions:

Distribution A		Distribution B	
X	P(X)	X	P(X)
0	.50	0	.05
1	.20	1	.10
2	.15	2	.15
3	.10	3	.20
4	.05	4	.50

- **(a)** Compute the mean for each distribution.
- **(b)** Compute the standard deviation for each distribution.
- **(c)** Compare and contrast the results in parts (a) and (b).

7.5 Given the following probability distributions:

Distribution C		Distribution D	
X	$P(X)$	X	$P(X)$
0	.20	0	.10
1	.20	1	.20
2	.20	2	.40
3	.20	3	.20
4	.20	4	.10

(a) Compute the mean for each distribution.
(b) Compute the standard deviation for each distribution.
(c) Compare and contrast the results in parts (a) and (b).

7.6 Using the company records for the past 500 working days, the manager of Johnson & Merke Motors, a suburban automobile dealership, has summarized the number of cars sold per day into the following table:

Number of cars sold per day	Frequency of occurrence
0	40
1	100
2	142
3	66
4	36
5	30
6	26
7	20
8	16
9	14
10	8
11	2
Total	500

(a) Form the empirical probability distribution (that is, relative frequency distribution) for the discrete random variable X, the number of cars sold per day.
(b) Compute the mean or expected number of cars sold per day.
(c) Compute the standard deviation.
(d) What is the probability that on any given day:
(1) Fewer than 4 cars will be sold?
(2) At most 4 cars will be sold?
(3) At least 4 cars will be sold?
(4) Exactly 4 cars will be sold?
(5) More than 4 cars will be sold?

• 7.7 An employee for a vending concession at a ball park must choose between working behind the "hot dog counter" and receiving a fixed sum of $50 for the evening versus walking around the stands selling beer on a commission basis. If the latter is chosen, the employee can make $90 on a warm night, $70 on a moderate night, $45 on a cool night, and $15 on a cold night. At this time of year the probabilities of a warm, moderate, cool, or cold night are, respectively, 0.1, 0.3, 0.4, and 0.2.

(a) Determine the mean or expected value to be earned by selling beer that evening.
(b) Compute the standard deviation.
(c) Which product should the employee sell? Why?

7.8 A state lottery is to be conducted in which 10,000 tickets are to be sold for $1 each. Six winning tickets are to be randomly selected: one grand-prize winner of $5,000, one second-prize winner of $2,000, one third-prize winner of $1,000, and three other winners of $500 each.

(a) Compute the expected value of playing this game.
(b) Should the game be played? Why?

7.9 Let us consider rolling a pair of six-sided dice. The random variable of interest represents the sum of the two numbers that are "facing up" when the pair of fair dice are rolled. The probability distribution is given below:

X	P(X)
2	1/36
3	2/36
4	3/36
5	4/36
6	5/36
7	6/36
8	5/36
9	4/36
10	3/36
11	2/36
12	1/36
	1

(a) Determine the mean or expected sum from rolling a pair of fair dice.
(b) Compute the variance and the standard deviation.

7.3 BINOMIAL DISTRIBUTION

Many different types of mathematical models have been developed to represent various discrete phenomena which occur in the social and natural sciences, in medical research, and in business. In particular, one discrete probability distribution which is extremely useful for describing many phenomena is the binomial distribution. The binomial model also plays an important role as an approximation to the hypergeometric distribution (which we shall consider in Section 7.3.5).

The binomial distribution possesses four essential properties:

1. The possible observations may be obtained by two different sampling methods. Either each observation may be considered as having been selected from an **infinite population without replacement** or from a **finite population with replacement.**

2. Each observation may be classified into one of two mutually exclusive and collectively exhaustive categories, usually called *success* and *failure*. (This terminology is inappropriate in some contexts, such as medical research, where the occurrence of death or disease is being counted.)
3. The probability of an observation's being classified as success, p, is constant from observation to observation. Thus, the probability of an observation's being classified as failure, $1 - p$, is constant over all observations.
4. The outcome (that is, success or failure) of any observation is independent of the outcome of any other observation.

The discrete random variable or phenomenon of interest which follows the binomial distribution is the number of successes obtained in a sample of n observations. Thus the binomial distribution has enjoyed numerous applications:

- *In games of chance*:
  What is the probability that red will come up 15 or more times in 19 spins of the roulette wheel?
- *In product quality control*:
  What is the probability that in a sample of 20 tires of the same type none will be defective if 8% of all such tires produced at a particular plant are defective?
- *In education*:
  What is the probability that a student can pass a ten-question multiple-choice exam (each containing four choices) if the student guesses on each question? (Passing is defined as getting 60% of the items correct—that is, getting at least six out of ten items correct.)
- *In finance*:
  What is the probability that a particular stock will show an increase in its closing price on a daily basis over the next ten (consecutive) trading sessions if stock market price changes really are random?

In each of these examples the four properties of the binomial distribution are clearly satisfied. For the roulette example, a particular set of spins may be construed as the sample taken from an infinite population of spins without replacement. When spinning the roulette wheel, each observation is categorized as red (success) or not red (failure). The probability of spinning red, p, on an American roulette wheel is 18/38 and is assumed to remain stable over all observations. Thus the probability of failure (spinning black or green), $1 - p$, is 20/38 each and every time the roulette wheel spins. Moreover, the roulette wheel has no memory—the outcome of any one spin is independent of preceding or following spins, so that, for example, the probability of obtaining red on the 32nd spin, given that the previous 31 spins were all red, remains equal to p, 18/38, if the roulette wheel is a fair one (See Figure 7.2).[4]

[4] It has been reported (Reference 5) that one time at Monte Carlo red came up on 32 consecutive spins. The probability of such an occurrence is indeed a very small number—$(18/37)^{32}$ = .0000000000969. The Monte Carlo roulette wheel, like other European wheels, has 37 equal-sized sectors—18 red, 18 black, and one green.

FIGURE 7.2 □ Roulette wheel.

In the product quality-control example, the sample of tires is also selected without replacement from an ongoing production process, an infinite population of manufactured tires.[5] As each tire in the sample is inspected, it is categorized as defective or nondefective.[6] Over the entire sample of tires the probability of any particular tire's being classified as defective, p, is .08, so that the probability of any tire being categorized as nondefective, $1 - p$, is .92. The production process is assumed to be stable. This is the case if:

1. The machinery producing the tires does not wear down.
2. The raw materials used are uniform.
3. The labor is consistent.

Moreover, for such a production process, the probability of one tire's being classified as defective or nondefective is independent of the classification for any other tire.

Similar statements pertaining to the binomial distribution's properties (characteristics) in the education example and in the finance example can be made as well. This is left to the reader. (See Problems 7.10 and 7.11.)

---

[5] As an example of a binomial random variable arising from sampling with replacement from a finite population, consider the probability of obtaining two clubs in five draws from a randomly shuffled deck of cards, where the selected card is replaced and the deck well shuffled after each draw.

[6] Note that when we are looking for defective tires, the discovery of such an event is deemed a success. This is one of the instances referred to above where, for statistical purposes, the term "success" may refer to business failures, deaths due to a particular illness, and other phenomena that, in nonstatistical terminology, would be deemed unsuccessful.

The four examples of binomial probability models described above are distinguished by the parameters n and p. Each time a set of parameters—the number of observations in the sample, n, and the probability of success, p—is specified, a particular binomial probability distribution can be generated.

- For the roulette example, $n = 19$ and $p = 18/38$.
- For the quality-control example, $n = 20$ and $p = .08$.
- For the education example, $n = 10$ and $p = 1/4$.
- For the finance example, $n = 10$ and $p = 1/2$.

Problems

7.10 Describe how the four properties of the binomial distribution could be satisfied in the education example.

7.11 Describe how the four properties of the binomial distribution could be satisfied in the finance example.

7.3.1 Mathematical Model

The following mathematical model represents the binomial probability distribution for obtaining the number of successes (X), given a knowledge of the parameters n and p:

$$P(X = x|n, p) = \frac{n!}{x!(n-x)!}p^x(1-p)^{n-x} \tag{7.4}$$

where
$P(X = x|n, p)$ = the probability that $X = x$, given a knowledge of n and p
n = sample size
p = probability of success
$1 - p$ = probability of failure
X = number of successes in the sample ($X = 0, 1, 2, \ldots, n$)

The binomial random variable X can have any integer value from 0 through n. In Equation (7.4) the product

$$p^x(1-p)^{n-x}$$

tells us the probability of obtaining exactly x successes out of n observations *in a particular sequence*, while the term

$$\frac{n!}{x!(n-x)!}$$

tells us *how many sequences* or arrangements (i.e., *combinations*—see Section 6.9) of the x successes out of n observations are possible. Hence, given the number of observations n and the probability of success p, we may determine the probability of x successes:

$$\begin{aligned} P(X = x|n, p) &= (\text{number of possible sequences}) \\ &\quad \times (\text{probability of a particular sequence}) \\ &= \frac{n!}{x!(n-x)!} p^x (1-p)^{n-x} \end{aligned}$$

by substituting the desired values for n, p, and x and computing the result.

Returning to our football lottery situation, the binomial probabilities listed in Table 7.1 can now be calculated. The parameters for our four-game selection strategy are $n = 4$ (the number of games selected) and $p = 1/2$ (the probability of correctly guessing the outcome of any given game). The random variable X, the number of correct selections, ranges from 0 through $n = 4$. Therefore we compute the desired probabilities as follows:

The probability of making no correct selections is:

$$\begin{aligned} P(X = 0|n = 4, p = .5) &= \frac{4!}{0!(4-0)!}(.5)^0(.5)^{4-0} \\ &= \frac{4!}{0!4!}(.5)^0(.5)^4 \\ &= (1)(1)(.5)(.5)(.5)(.5) = .0625 = \frac{1}{16} \end{aligned}$$

The probability of making one correct selection is:

$$\begin{aligned} P(X = 1|n = 4, p = .5) &= \frac{4!}{1!(4-1)!}(.5)^1(.5)^{4-1} \\ &= \frac{4!}{1!3!}(.5)^1(.5)^3 \\ &= (4)(.5)(.5)(.5)(.5) = .2500 = \frac{4}{16} \end{aligned}$$

The probability of making two correct selections is:

$$P(X = 2|n = 4, p = .5) = \frac{4!}{2!(4-2)!}(.5)^2(.5)^{4-2}$$

$$= \frac{4!}{2!2!}(.5)^2(.5)^2$$

$$= (6)(.5)(.5)(.5)(.5) = .3750 = \frac{6}{16}$$

The probability of making three correct selections is:

$$P(X = 3|n = 4, p = .5) = \frac{4!}{3!(4-3)!}(.5)^3(.5)^{4-3}$$

$$= \frac{4!}{3!1!}(.5)^3(.5)^1$$

$$= (4)(.5)(.5)(.5)(.5) = .2500 = \frac{4}{16}$$

The probability of making all four correct selections is:

$$P(X = 4|n = 4, p = .5) = \frac{4!}{4!(4-4)!}(.5)^4(.5)^{4-4}$$

$$= \frac{4!}{4!0!}(.5)^4(.5)^0$$

$$= (1)(.5)(.5)(.5)(.5)(1) = .0625 = \frac{1}{16}$$

These are the probabilities displayed in Table 7.1.

7.3.2 Using the Binomial Probability Distribution in Data Analysis

To demonstrate various applications of the binomial probability distribution we turn to some examples.

EXAMPLE 7.2

> What is the probability of making *at most* two correct selections if a four-game selection strategy is used?

***Solution*:**

From Table 7.1 we have

$$P(X \le 2|n = 4, p = .5) = \frac{1}{16} + \frac{4}{16} + \frac{6}{16} = \frac{11}{16}$$

EXAMPLE 7.3

What is the probability of making *at least* three correct selections if a four-game selection strategy is used:

***Solution*:**

From Table 7.1 we have

$$P(X \ge 3|n = 4, p = .5) = \frac{4}{16} + \frac{1}{16} = \frac{5}{16}$$

Also, from Example 7.2 we have

$$\begin{aligned} P(X \ge 3|n = 4, p = .5) &= 1 - P(X \le 2|n = 4, p = .5) \\ &= 1 - \frac{11}{16} \\ &= \frac{5}{16} \end{aligned}$$

EXAMPLE 7.4

What is the probability of making either no correct selections or all correct selections if a four-game selection strategy is used?

***Solution*:**

From Table 7.1 we have

$$P(X = 0|n = 4, p = .5) + P(X = 4|n = 4, p = .5) = \frac{1}{16} + \frac{1}{16} = \frac{2}{16}$$

EXAMPLE 7.5

Suppose that Lloyd believes he has such a thorough understanding of professional football—including the strengths and weaknesses of the NFL teams—that he can correctly predict the winner 70% of the time. If Lloyd is correct about his picking ability, what is the probability that he would win the football lottery if he were to purchase a card and use the four-game selection strategy on a particular week?

Solution:

$$P(X = 4|n = 4, p = .7) = \frac{4!}{4!(4-4)!}(.7)^4(.3)^{4-4}$$
$$= \frac{4!}{4!0!}(.7)^4(.3)^0$$
$$= (1)(.7)(.7)(.7)(.7)(1) = .2401$$

Thus, if Lloyd's assessment is correct, he has a .2401 chance of winning the football lottery each time he purchases a card and uses the four-game selection strategy.

EXAMPLE 7.6

What is Lloyd's expected monetary value for participating in such a manner in the State Football Lottery Program?

Solution:

If Lloyd believes he has a .2401 chance of winning the football lottery each time he uses the four-game selection strategy, then he also feels he has a .7599 chance that he will lose. Thus:

Dollar value	Probability
− 1	.7599
12	.2401
Total	1.0000

Lloyd's expected monetary value for this strategy is:

$$(-1)(.7599) + (12)(.2401) = \$2.12$$

That is, in the long run Lloyd expects to win, on average, \$2.12 for each dollar invested when he uses this strategy.

7.3.3 Using the Binomial Tables

Binomial probability computations using Equation (7.4) may be quite tedious, especially as n gets large. However, we may obtain the probabilities directly from Table C.6 and thereby avoid any computational drudgery. Table C.6 provides, for various

TABLE 7.3 □ Obtaining a binomial probability

		p															
n	*X*	0.37	0.38	0.39	0.40	0.41	0.42	0.43	0.44	0.45	0.46	0.47	0.48	0.49	0.50	*X*	*n*
2	0	0.3969	0.3844	0.3721	0.3600	0.3481	0.3364	0.3249	0.3136	0.3025	0.2916	0.2809	0.2704	0.2601	0.2500	2	
	1	0.4662	0.4712	0.4758	0.4800	0.4838	0.4872	0.4902	0.4928	0.4950	0.4968	0.4982	0.4992	0.4998	0.5000	1	
	2	0.1369	0.1444	0.1521	0.1600	0.1681	0.1764	0.1849	0.1936	0.2025	0.2116	0.2209	0.2304	0.2401	0.2500	0	2
3	0	0.2500	0.2383	0.2270	0.2160	0.2054	0.1951	0.1852	0.1756	0.1664	0.1575	0.1489	0.1406	0.1327	0.1250	3	
	1	0.4406	0.4382	0.4354	0.4320	0.4282	0.4239	0.4191	0.4140	0.4084	0.4024	0.3961	0.3894	0.3823	0.3750	2	
	2	0.2587	0.2686	0.2783	0.2880	0.2975	0.3069	0.3162	0.3252	0.3341	0.3428	0.3512	0.3594	0.3674	0.3750	1	
	3	0.0507	0.0549	0.0593	0.0640	0.0689	0.0741	0.0795	0.0852	0.0911	0.0973	0.1038	0.1106	0.1176	0.1250	0	3
4	0	0.1575	0.1478	0.1385	0.1296	0.1212	0.1132	0.1056	0.0983	0.0915	0.0850	0.0789	0.0731	0.0677	0.0625	4	
	1	0.3701	0.3623	0.3541	0.3456	0.3368	0.3278	0.3185	0.3091	0.2995	0.2897	0.2799	0.2700	0.2600	0.2500	3	
	2	0.3260	0.3330	0.3396	0.3456	0.3511	0.3560	0.3604	0.3643	0.3675	0.3702	0.3723	0.3738	0.3747	0.3750	2	
	3	0.1276	0.1361	0.1447	0.1536	0.1627	0.1719	0.1813	0.1908	0.2005	0.2102	0.2201	0.2300	0.2400	0.2500	1	
	4	0.0187	0.0209	0.0231	0.0256	0.0283	0.0311	0.0342	0.0375	0.0410	0.0448	0.0488	0.0531	0.0576	0.0625	0	4
5	0	0.0992	0.0916	0.0845	0.0778	0.0715	0.0656	0.0602	0.0551	0.0503	0.0459	0.0418	0.0380	0.0345	0.0312	5	
	1	0.2914	0.2808	0.2700	0.2592	0.2484	0.2376	0.2270	0.2164	0.2059	0.1956	0.1854	0.1755	0.1657	0.1562	4	
	2	0.3423	0.3441	0.3452	0.3456	0.3452	0.3442	0.3424	0.3400	0.3369	0.3332	0.3289	0.3240	0.3185	0.3125	3	
	3	0.2010	0.2109	0.2207	0.2304	0.2399	0.2492	0.2583	0.2671	0.2757	0.2838	0.2916	0.2990	0.3060	0.3125	2	
	4	0.0590	0.0646	0.0706	0.0768	0.0834	0.0902	0.0974	0.1049	0.1128	0.1209	0.1293	0.1380	0.1470	0.1562	1	
	5	0.0069	0.0079	0.0090	0.0102	0.0116	0.0131	0.0147	0.0165	0.0185	0.0206	0.0229	0.0255	0.0282	0.0312	0	5

SOURCE: Extracted from Table C.6.

selected combinations of the parameters n and p, the probabilities that the binomial random variable takes on values of $X = 0, 1, 2, \ldots, n$.[7]

As a case in point we refer to our football lottery card. To determine the probability of correctly selecting all four games when using the four-game selection strategy, we first find in Table C.6 the combination of $n = 4$ with $p = .50$. To obtain the probability of four successes, we read the probability corresponding to the row $X = 4$; the result is .0625 (that is, 1/16). This is demonstrated in Table 7.3. (Note that for $p > .5$, we read across the bottom and up the right-hand side of the table.)

To demonstrate the use of the binomial table for various applications we turn to some examples.

EXAMPLE 7.7 ROULETTE EXAMPLE:

What is the probability red will come up 15 or more times in 19 spins of the roulette wheel?

Solution:

The parameters are $n = 19$ and $p = 18/38 \cong .47$. Using Table C.6:

[7] However, the reader should be cautioned that the values for p in Table C.6 are taken to only two decimal places. Thus, if your value of p must be rounded off, the probabilities will be only approximations to the true result.

n	X	p = .47
19	15	.0037
	16	.0008
	17	.0001
	18	.0000
	19	.0000

Therefore,

$$P(X \geq 15 \mid n = 19, p \cong .47) = .0037 + .0008 + .0001 + .0000 + .0000 = .0046$$

EXAMPLE 7.8 QUALITY-CONTROL EXAMPLE:

What is the probability that, in a sample of 20 tires of the same type, all will be satisfactory, if 92% of all such tires produced at a particular plant are satisfactory?

Solution:

The parameters are $n = 20$ and $p = .92$. Using Table C.6, since $p > .5$, we read p from the bottom row of the page and X from the right-hand margin.

	n	X
.1887	20	20
p = .92		

Therefore, $P(X = 20 \mid n = 20, p = .92) = .1887$.

EXAMPLE 7.9 EDUCATION EXAMPLE:

What is the probability that a student can pass (that is, score at least 60% on) a 10-question multiple-choice exam (each containing 4 possible answers) if the student guesses on each question?

Solution:

The parameters are $n = 10$ and $p = .25$. To pass the test, the student must get at least 6 questions out of 10 right. Using Table C.6, we set up a table:

n	X	p = .25
10	6	.0162
	7	.0031
	8	.0004
	9	.0000
	10	.0000

Therefore,

$$P(X \geq 6 \mid n = 10, p = .25) = .0162 + .0031 + .0004 + .0000 + .0000$$
$$= .0197$$

EXAMPLE 7.10 FINANCE EXAMPLE:

What is the probability that a particular stock will show an increase in its closing price on a daily basis over the next ten (consecutive) trading sessions if stock market price changes are a random phenomenon, rising half the time and falling half the time?

Solution:

The parameters are $n = 10$ and $p = .50$. Using Table C.6:

n	X	p = .50
10	10	.0010

Therefore, $P(X = 10 \mid n = 10, p = .50) = .0010$.

7.3.4 Characteristics of the Binomial Distribution

By examining Table C.6 for various combinations of n and p, we have seen that each time a set of parameters—n and p—is specified, a particular binomial probability distribution can be generated.

SHAPE

We note that a binomial distribution may be symmetric or skewed. Whenever $p = .5$, the binomial distribution will be symmetric regardless of how large or small the value of n. However, when $p \neq .5$, the distribution will be skewed. The closer p is to .5 and the larger the number of observations, n, the less skewed the distribution will be.

THE MEAN

The mean of the binomial distribution can be readily obtained as the product of its two parameters, n and p. That is, instead of using Equation (7.1), which holds for all discrete probability distributions, for data that are binomially distributed we simply compute

$$\mu_X = E(X) = np \qquad \textbf{(7.5)}$$

Intuitively, this makes sense. For example, if we spin the roulette wheel 19 times, how frequently should we "expect" the color red to come up? On the average, over the long run, we would theoretically expect

$$\mu_X = E(X) = np = (19)\left(\frac{18}{38}\right) = 9$$

occurrences of red in 19 spins. In a similar vein, if we refer to our football lottery, the expected number of correct selections when using the four-game selection strategy is simply

$$\mu_X = E(X) = np = (4)\left(\frac{1}{2}\right) = 2$$

—the same result obtained from the more general expression shown in Equation (7.1).

THE STANDARD DEVIATION

The standard deviation of the binomial distribution is easily calculated using the formula

$$\sigma_X = \sqrt{\text{Var}(X)} = \sqrt{np(1-p)} \qquad \textbf{(7.6)}$$

Referring to our football lottery, we simply compute

$$\sigma_X = \sqrt{\text{Var}(X)} = \sqrt{(4)\left(\frac{1}{2}\right)\left(\frac{1}{2}\right)} = \sqrt{1} = 1.0$$

—the same result obtained from the more general expression shown in Equation (7.2).

7.3.5 Using the Binomial Distribution to Approximate the Hypergeometric Distribution

In this section we have developed the binomial model as a useful discrete probability distribution in its own right. The binomial distribution, however, also plays an important role when it is used to approximate another discrete probability distribution called the **hypergeometric probability distribution.**

The major distinction between the binomial distribution and the hypergeometric distribution (and the mathematical expressions derived therefrom) is the manner in which the data are obtained. For the hypergeometric distribution, the sample data are drawn without replacement from a finite population. For example, in a town of 700 Democrats and 300 Republicans, a panel of 10 people is selected at random to form an advisory committee to the mayor. What is the probability that at least 7 committee members will be Democrats?

Even though the selection is without replacement from a finite population, as a rule of thumb if the sample size is less than 5% of the population size (that is, $n < .05N$), we may use the binomial probability distribution to approximate the hypergeometric distribution. The results will be quite similar, and the binomial model is much less cumbersome. The importance of this application of the binomial distribution should not be overlooked, since much sample survey work (such as that undertaken by the statistician in developing the alumni questionnaire presented in Chapter 2) deals with hypergeometric type data. Estimation and hypothesis testing for the proportion of successes will be studied in Chapters 9 and 11.

Problems

✓7.12 William and Mary have decided to have 5 children. If the probability of having a boy is .5, use the formula given in Equation (7.4) to construct a probability distribution and answer the following:

- **(a)** What is the probability that they will have all boys? All girls? (Why is it so easy in this case to answer the part about girls, once you found the probability of all boys?)
- **(b)** What is the probability of 3 or more boys?
- **(c)** What is the probability of 2 or fewer girls?
- **(d)** What is the probability of 2 or 3 boys?
- **(e)** Use Table C.6 to check your calculations of the probability distribution.
- **(f)** If the actual probability of having a boy is .52, use Table C.6 to obtain the results for parts (a) through (d).
- **(g)** Discuss the difference in the results of using $p = .5$ versus $p = .52$.

7.13 Abe Lincoln said that "you can't please all the people all the time." Suppose that you can please each individual 9 times out of 10 and that there are 8 people you want to please. Use Equation (7.4) to calculate:

- **(a)** The probability that you'll please all of them.
- **(b)** The probability that you'll please at least 6 of them.
- **(c)** The probability that you'll please 4 or fewer.
- **(d)** Use Table C.6 to check your calculations of the probability distribution.

(e) What is the expected number of people you will please? How likely is it that you will please exactly that number?

(f) What is the standard deviation of the number of people you will please? From this and the expected value, find out approximately how many people you will please at least three-fourths of the time.

• **7.14** Suppose that on a very long arithmetic test, Jim would get 70% of the items right. For a 10-item quiz, use Equation (7.4) to calculate

(a) The probability that Jim will get at least 7 items right.

(b) The probability that Jim will get less than 6 items right (and therefore fail the quiz).

(c) The probability that Jim will get 9 or 10 items right (and get an A on the quiz).

(d) Use Table C.6 to check your calculations of the probability distribution.

(e) What is the expected number of items that Jim will get right? What proportion of the time will he get that number right?

(f) What is the standard deviation of the number of items that Jim will get right? Compare the proportion of time that Jim will be within 2 standard deviations according to the distribution you just calculated with the same probability calculated from the Bienayme-Cheybshev inequality.

7.15 Verify that for a 5-game selection strategy in the football lottery example the expected number of correct selections is 2.5 and the standard deviation is approximately 1.1. Also use Equation (7.4) to find the probability of winning all 5 games.

7.16 Verify that for a 10-game selection strategy in the football lottery example the expected number of correct selections is 5 and the standard deviation is approximately 1.6. Also use Equation (7.4) to find the probability of winning all 10 games.

• **7.17** It is known that 30% of the defective parts produced in a certain manufacturing process can be made satisfactory by rework.

(a) What is the probability that in a batch of 6 such defective parts at least 3 can be satisfactorily reworked?

(b) What is the probability that none of them can be reworked?

(c) What is the probability that all of them can be reworked?

7.18 Based on past experience, the main printer in a university computer center is operating properly 90% of the time. If a random sample of 10 inspections are made:

(a) What is the probability that the main printer is operating properly:

(1) Exactly 9 times?

(2) At least 9 times?

(3) At most 9 times?

(4) More than 9 times?

(5) Fewer than 9 times?

(b) How many times can the main printer be expected to operate properly?

7.19 The probability that a patient fails to recover from a particular operation is .1.

(a) What is the probability that exactly 2 of the next 8 patients having this operation will not recover?

(b) What is the probability that at most 1 patient of the 8 will not recover?

(c) Using the probabilities extracted from Table C.6, plot the histogram for this binomial distribution.

7.4 MATHEMATICAL MODELS OF CONTINUOUS RANDOM VARIABLES: THE PROBABILITY DENSITY FUNCTION

When we are dealing with a continuous random variable, the mathematical model representing some underlying phenomenon is known as a **probability density function.** When such a mathematical expression is available, the probability that various values of the random variable occur within certain ranges or intervals may be calculated. However, the *exact* probability of a *particular value* is zero.

As an example, the probability distribution represented in Table 7.4 is obtained by categorizing a distribution in which the continuous random phenomenon of interest is said to follow the bell-shaped **normal probability density function.** If the nonoverlapping (mutually exclusive) listing contains all possible class intervals (is collectively exhaustive), the probabilities will again sum to 1. This is demonstrated in Table 7.4. Such a probability distribution may be considered as a relative frequency distribution as described in Section 4.3, where, except for the two open-ended classes, the midpoint of every other class interval represents the data in that interval.

Continuous probability density functions are those that arise owing to some measuring process on various phenomena of interest. Continuous models have important applications in engineering and the physical sciences as well as in business and the social sciences. Examples of continuous random phenomena are: *time* between arrivals of telephone calls into a switchboard; *heights* of basketball players; *weights* of newborn chicks; and *distances* between cars on a highway. Unfortunately, obtaining probabilities or computing expected values and the standard deviations for continuous phenomena involves mathematical expressions which require a knowledge of integral

TABLE 7.4 □ Verbal SAT scores of 1,000 seniors from Entes County high schools

SAT score	Relative frequency or probability
Under 225	3/1,000 = .003
225 < 275	9/1,000 = .009
275 < 325	28/1,000 = .028
325 < 375	65/1,000 = .065
375 < 425	124/1,000 = .124
425 < 475	172/1,000 = .172
475 < 525	198/1,000 = .198
525 < 575	172/1,000 = .172
575 < 625	124/1,000 = .124
625 < 675	65/1,000 = .065
675 < 725	28/1,000 = .028
725 < 775	9/1,000 = .009
775 or Above	3/1,000 = .003
Total	1.000

calculus and are beyond the scope of this book. Nevertheless, one continuous probability density function which we shall focus upon has been deemed so important for applications that special probability tables (such as Table C.2) have been devised in order to eliminate the need for what otherwise would require laborious mathematical computations. This particular continuous probability density function is known as the **Gaussian** or **normal distribution.**

7.5 THE NORMAL DISTRIBUTION

7.5.1 Importance of the Normal Distribution

The normal distribution is vitally important in statistics for three main reasons:

1. Numerous continuous phenomena seem to follow it or can be approximated by it.
2. We can use it to approximate various discrete probability distributions and thereby avoid much computational drudgery (Section 7.6).
3. It provides the basis for *classical statistical inference* because of its relationship to the *central limit theorem* (Chapter 8).[8]

7.5.2 Properties of the Normal Distribution

The normal distribution has several interesting theoretical properties. Among these are:

1. It is "bell-shaped" and symmetrical in its appearance.
2. Its measures of central tendency (mean, median, mode, midrange, and midhinge) are all identical.
3. Its "middle spread" is equal to 1.33 standard deviations.[9] That is, the interquartile range is contained within an interval of two-thirds of a standard deviation below the mean to two-thirds of a standard deviation above the mean.
4. Its associated random variable has an infinite range ($-\infty < X < +\infty$).

In actual practice some of the variables we observe may only approximate these theoretical properties. This occurs for two reasons: (1) the underlying population distribution may be only approximately normal; and (2) any actual sample may deviate from the theoretically expected characteristics. For some phenomenon which may be approximated by the normal distribution model:

[8] This third (and most important) point will be developed in Chapter 8 and emphasized in Chapters 9 through 12.

[9] Thus, for normally distributed data, if the interquartile range equals 1.33 standard deviations, then the standard deviation equals three-fourths of the interquartile range.

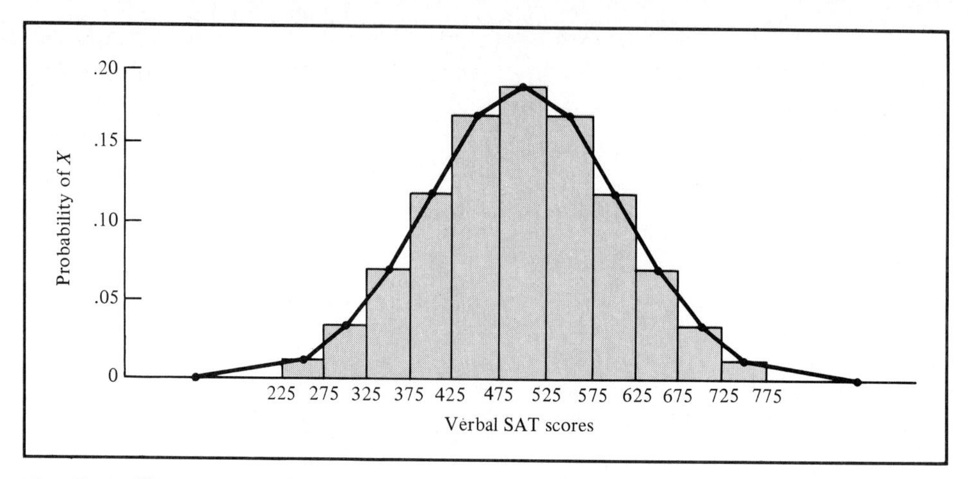

FIGURE 7.3 □ Relative frequency histogram and polygon of the verbal SAT scores of 1,000 seniors from Entes County high schools.

1. Its polygon may be only approximately "bell-shaped" and symmetrical in appearance.
2. Its measures of central tendency may differ slightly from each other.
3. The value of its interquartile range may differ slightly from 1.33 standard deviations.
4. Its *practical range* will not be infinite but will generally lie within 3 standard deviations above and below the mean. (That is, range ≈ 6 standard deviations.)[10]

As a case in point, let us refer to Figure 7.3, which depicts the relative frequency histogram and polygon for the verbal SAT scores of 1,000 seniors from Entes County high schools presented in Table 7.4. For these data, the first three theoretical properties of the normal distribution seem to have been satisfied; however, the fourth does not hold. Based on the Educational Testing Service scoring and scaling system, the random variable of interest (verbal SAT score) cannot possibly take on values below 200 or above 800. Hence the practical range of the verbal SAT scores is 600. Since it can be shown that the standard deviation σ_X for these data is 100, the practical range is equivalent to 6 standard deviation units, which satisfies the fourth "practical" property.

Caution: What about other data sets? Not all continuous random variables are normally distributed. That is, the reader must be cautioned that often the continuous random phenomenon we may be interested in studying neither will follow the normal distribution nor can be adequately approximated by it. In such a case the four "practical" properties described on page 185 will not hold. While some methods for studying such continuous phenomena are outside the scope of this text (see References 1 and 2), other (distribution-free) techniques which do not depend on the particular form of the underlying random variable will be discussed in Chapter 14. Nevertheless, Problem 7.21 (on page 193) will give the reader the opportunity

[10] Thus, for normally distributed data, if the range equals 6 standard deviations, then the standard deviation equals one-sixth the range.

to decide whether a given data set seems to approximate the normal distribution sufficiently to permit it to be examined using the methodology of this section.

7.5.3 The Mathematical Model

The mathematical model or expression representing a probability density function is denoted by the symbol $f(X)$. For the normal distribution, the model used to obtain the desired probabilities is

$$f(X) = \frac{1}{\sqrt{2\pi}\,\sigma_X} e^{-(1/2)[(X-\mu_X)/\sigma_X]^2} \qquad \textbf{(7.7)}$$

where e is the mathematical constant approximated by 2.71828
π is the mathematical constant approximated by 3.14159
μ_X is the population mean
σ_X is the population standard deviation
X is any value of the continuous random variable, where $-\infty < X < +\infty$

Let us examine the components of the normal probability density function in Equation (7.7). Since e and π are mathematical constants, the probabilities of the random variable X are only dependent upon the two parameters of the normal distribution—the population mean μ_X and the population standard deviation σ_X. Every time we specify a *particular combination* of μ_X and σ_X, a *different* normal probability distribution will be generated—each having its own set of probabilities. We illustrate this in Figure 7.4, where three different normal distributions are depicted. Distributions A and B have the same mean (μ_X) but have different standard deviations. On the other hand, distributions A and C have the same standard deviation (σ_X) but have different means. Furthermore, distributions B and C depict two normal probability density functions which differ with respect to both μ_X and σ_X.

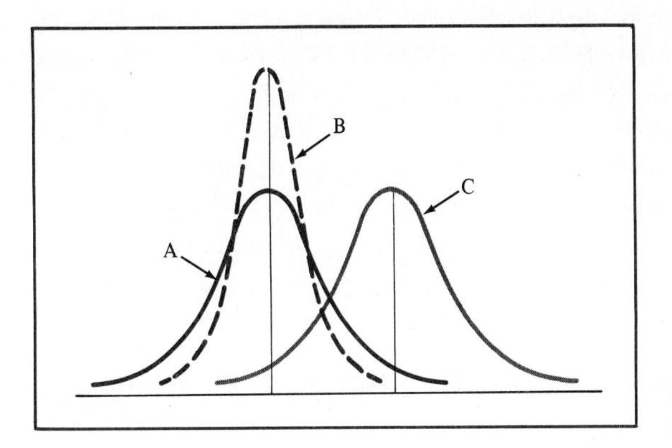

FIGURE 7.4 □ Three normal distributions having differing parameters μ_X and σ_X.

Unfortunately, the mathematical expression in Equation (7.7) is computationally tedious. To avoid such computations, it would be useful to have a set of tables that would provide the desired probabilities. However, since an infinite number of combinations of the parameters μ_X and σ_X exist, an infinite number of tables would be required.

7.5.4 Standardizing the Normal Distribution

Fortunately, by *standardizing* the data, we will only need one table. (See Table C.2.) By using the **transformation formula:**

$$Z = \frac{X - \mu_X}{\sigma_X} \tag{7.8}$$

any normal random variable X is converted to a standardized normal random variable Z. While the original data for the random variable X had mean μ_X and standard deviation σ_X, the standardized random variable Z will always have mean $\mu_Z = 0$ and standard deviation $\sigma_Z = 1$.

A standardized normal distribution is one whose random variable Z always has a mean $\mu_Z = 0$ and a standard deviation $\sigma_Z = 1$.

Substituting in Equation (7.7), we see that the probability density function of a standard normal variable Z is

$$f(Z) = \frac{1}{\sqrt{2\pi}} e^{-(1/2)Z^2} \tag{7.7a}$$

Thus, we can always convert any set of normally distributed data to its standardized form and then determine its desired probabilities from the table of the standardized normal distribution.

To see how the transformation formula (7.8) may be applied and how we may then use the results to read probabilities from the table of the standardized normal distribution (Table C.2), let us consider the data on verbal SAT scores displayed in Table 7.4 on page 184. For this high-school population the variable of interest (verbal SAT scores) is approximately normally distributed with a mean μ_X of 500 points and a standard deviation σ_X of 100 points.[11]

[11] As we noted in Section 7.5.2, observed data will often have properties that closely approximate the theoretical properties of the normal distribution. In the examples that follow, we will apply these theoretical properties to the Entes County high schools data, even though we will be making small errors in doing so.

TRANSFORMING THE DATA

Every measurement (X) from a normal distribution has a corresponding standardized measurement (Z) obtained from the transformation formula. For example, a SAT score of 600 points (X) is equivalent to a standardized Z value obtained as follows:

$$Z = \frac{X - \mu_X}{\sigma_X}$$

$$= \frac{600 - 500}{100} = +1.00$$

That is, an SAT score of 600 is exactly 1.00 standard deviation above the mean. Thus, the standard deviation has become the unit of measurement. In other words, a score of 600 is 100 points (i.e., 1.00 standard deviation) higher than the average verbal SAT score of 500 points. This is depicted in Figure 7.5.

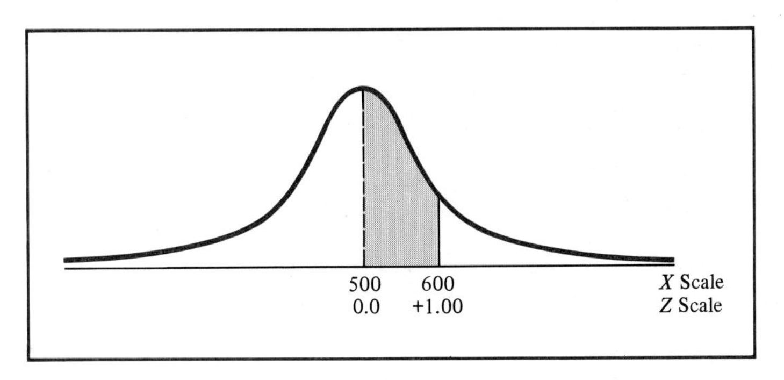

FIGURE 7.5 □ **Transformation of scales.**

In a similar manner, a score of 300 points is 2.00 standard deviations below the mean, since

$$Z = \frac{X - \mu_X}{\sigma_X}$$

$$= \frac{300 - 500}{100} = -2.00$$

A score of 648 is 1.48 standard deviations above the mean because

$$Z = \frac{X - \mu_X}{\sigma_X}$$

$$= \frac{648 - 500}{100} = +1.48$$

These results are displayed in Figures 7.6 and 7.7, respectively, on page 190.

Figure 7.8 depicts a variety of SAT scores along with their corresponding standardized values (measurements). We leave it to the reader to use Equation (7.8) to verify these Z values.

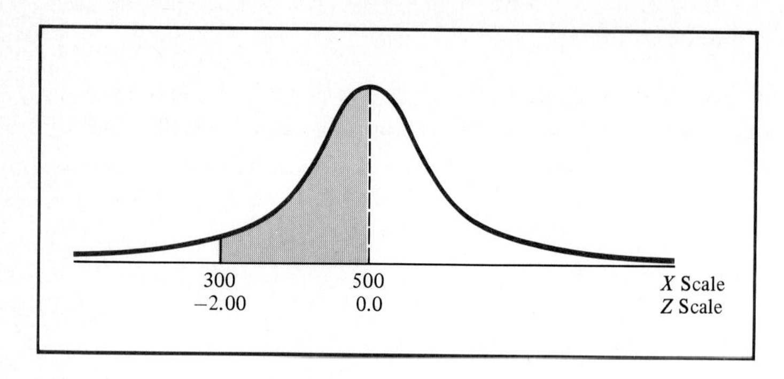

FIGURE 7.6 □ Transformation of scales.

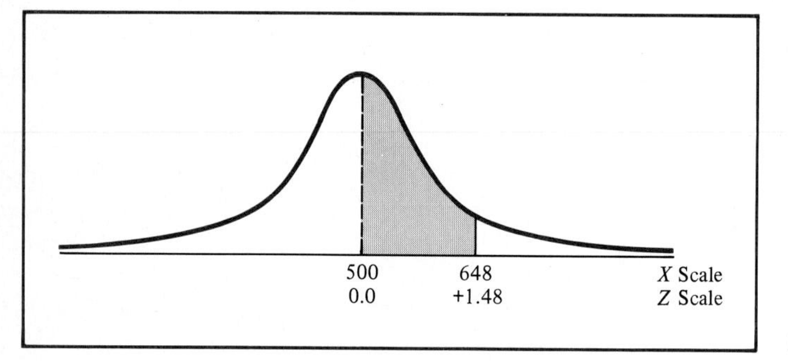

FIGURE 7.7 □ Transformation of scales.

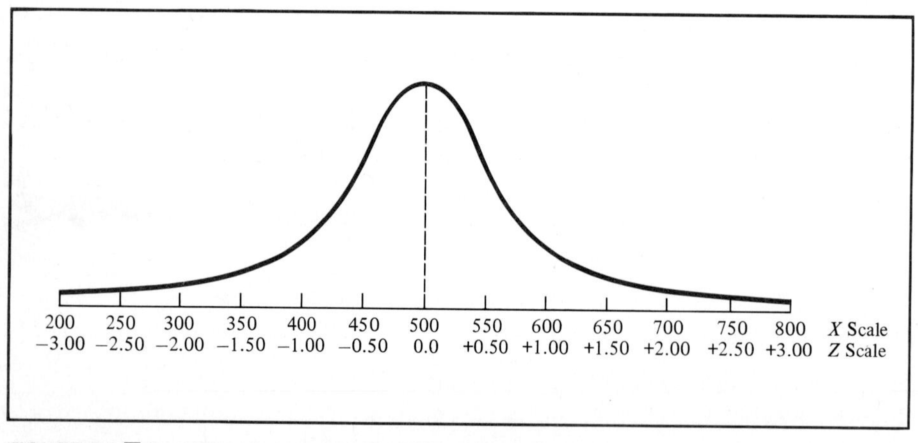

FIGURE 7.8 □ A variety of corresponding *X* and *Z* values.

EXAMPLE 7.11

Given the following normal distribution:

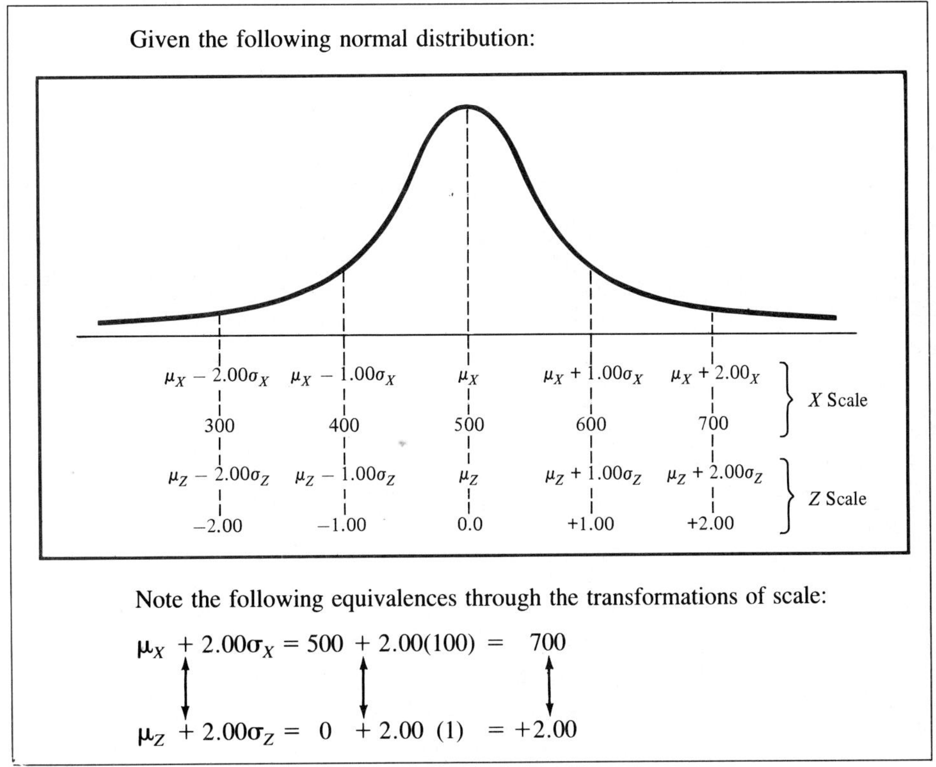

Note the following equivalences through the transformations of scale:

$$\mu_X + 2.00\sigma_X = 500 + 2.00(100) = 700$$

$$\updownarrow \qquad \updownarrow \qquad \updownarrow$$

$$\mu_Z + 2.00\sigma_Z = 0 + 2.00\ (1) = +2.00$$

USING THE TABLE OF THE STANDARDIZED NORMAL DISTRIBUTION

Figure 7.8 depicts the relative frequency polygon for the normal distribution representing verbal SAT scores for all 1,000 seniors at Entes County high schools. Since the obtained data constitute the entire senior classes at these high schools (i.e., the population), the probabilities or area under the curve must add up to 1. Clearly, the area under the curve between any two SAT scores represents but a portion of the total area possible.

Suppose now we wish to find, using Figure 7.7, how likely it is that a particular student selected at random from the senior class will obtain a verbal SAT score between 500 and 648 points. Using Table C.2 we may easily obtain this answer. From Figure 7.7 we note that the probability of obtaining a verbal SAT score between 500 and 648 points is equivalent to the probability of obtaining a standardized measurement between the mean ($\mu_Z = 0.0$) and +1.48 standard deviations above the mean.[12] Table 7.5 on page 192 is a replica of Table C.2.

[12] In mathematical terms

$$P(500 < X < 648) = P\ (0.0 < Z < +1.48)$$

TABLE 7.5 □ Obtaining an area under the normal curve

Z	.00	.01	.02	.03	.04	.05	.06	.07	.08	.09
0.0	.0000	.0040	.0080	.0120	.0160	.0199	.0239	.0279	.0319	.0359
0.1	.0398	.0438	.0478	.0517	.0557	.0596	.0636	.0675	.0714	.0753
0.2	.0793	.0832	.0871	.0910	.0948	.0987	.1026	.1064	.1103	.1141
0.3	.1179	.1217	.1255	.1293	.1331	.1368	.1406	.1443	.1480	.1517
0.4	.1554	.1591	.1628	.1664	.1700	.1736	.1772	.1808	.1844	.1879
0.5	.1915	.1950	.1985	.2019	.2054	.2088	.2123	.2157	.2190	.2224
0.6	.2257	.2291	.2324	.2357	.2389	.2422	.2454	.2486	.2518	.2549
0.7	.2580	.2612	.2642	.2673	.2704	.2734	.2764	.2794	.2823	.2852
0.8	.2881	.2910	.2939	.2967	.2995	.3023	.3051	.3078	.3106	.3133
0.9	.3159	.3186	.3212	.3238	.3264	.3289	.3315	.3340	.3365	.3389
1.0	.3413	.3438	.3461	.3485	.3508	.3531	.3554	.3577	.3599	.3621
1.1	.3643	.3665	.3686	.3708	.3729	.3749	.3770	.3790	.3810	.3830
1.2	.3849	.3869	.3888	.3907	.3925	.3944	.3962	.3980	.3997	.4015
1.3	.4032	.4049	.4066	.4082	.4099	.4115	.4131	.4147	.4162	.4177
1.4	.4192	.4207	.4222	.4236	.4251	.4265	.4279	.4292	.4306	.4319
1.5	.4332	.4345	.4357	.4370	.4382	.4394	.4406	.4418	.4429	.4441

SOURCE: Extracted from Table C.2.

We note from Table C.2 that only positive entries for Z are listed. It is clear that for such a symmetrical distribution with zero mean, the area from the mean to $+Z$ must be identical to the area from the mean to $-Z$.

To use Table C.2, we note that all Z values must first be recorded to two decimal places. Thus our particular Z value of interest is recorded as $+1.48$. To read the probability or area under the curve from the mean to $Z = +1.48$, we scan down the Z column until we locate the Z value of interest (in tenths). Hence we stop in the row $Z = 1.4$. Next we read across this row until we intersect the column that contains the hundredths place of the Z value—that is, .08. Therefore, in the body of the table the tabulated probability for $Z = 1.48$ corresponds to the intersection of the row $Z = 1.4$ with the column $Z = .08$ as shown in Table 7.5. This probability is .4306. As depicted in Figure 7.9, there is a 43.06% chance that a senior selected at random from the Entes County high schools will achieve a verbal SAT score between 500 and 648 points.

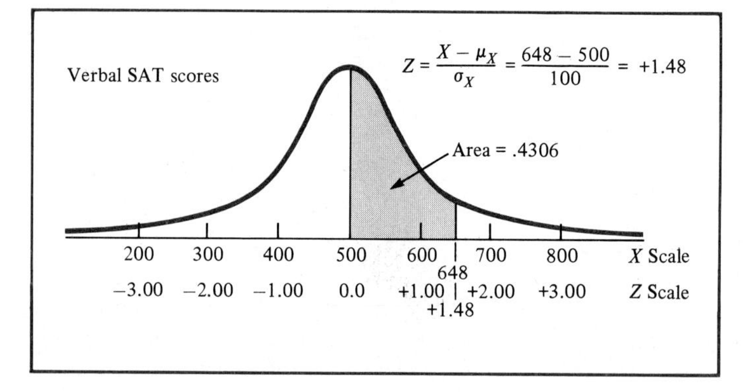

FIGURE 7.9 □ Determining the area between the mean and Z in a standardized normal distribution.

We may note further that there are several additional ways to express these results:

1. The *proportion* of seniors at Entes County high schools who achieve a verbal SAT score between 500 and 648 points is .4306.
2. The *percentage* of seniors at Entes County high schools who achieve a verbal SAT score between 500 and 648 points is 43.06%.
3. The *number* of seniors at Entes County high schools who are expected to achieve a verbal SAT score between 500 and 648 points is about 431.

Problems

7.20 Using Table C.2 with the results displayed in Figures 7.5 and 7.6, verify:
- **(a)** That there is a .3413 probability that a senior selected at random from Entes County high schools will achieve a verbal SAT score between 500 and 600 points.
- **(b)** That there is a 47.72% chance that a senior selected at random from Entes County high schools will obtain a verbal SAT score between 300 and 500 points.

7.21 Using the following data sets decide whether or not the random variable of interest appears to be approximately normally distributed:
- **(a)** Unpaid accounts (Problem 3.5, page 45, and Problem 4.2, page 83).
- **(b)** Price-to-earnings ratios (Problem 3.73, page 72, and Problem 4.6, page 83).
- **(c)** Automobile acceleration of German-made cars (Problem 3.75, page 73, and Problem 4.8, page 83).
- **(d)** Chainsaw noise levels (Problem 4.11, page 83).
- **(e)** Utility charges (Problem 4.17, page 85).
- **(f)** Life of light bulbs from Manufacturer A (Problem 4.18, page 85).
- **(g)** Starting salaries (Problem 4.19, page 86).
- **(h)** National League batting averages (Problem 4.20, page 86).

PROBABILITY APPLICATIONS

Now that we have learned to use Table C.2 in conjunction with Equation (7.8), we can resolve many different types of probability questions pertaining to the normal distribution. To illustrate this, let us refer once again to Case D—The Governor's Football Lottery Program.

Case D—The Governor's Football Lottery Program Revisited

Suppose the Governor projects that on a weekly basis the State Football Lottery Program is expected to average 10.0 million dollars in profits (to be turned over to the state for educational programs) with a standard deviation of 2.5 million dollars. Suppose further that the weekly profits data are assumed to be (approximately) normally distributed. The following questions may be raised (or anticipated at the Governor's next press conference):

1. What is the probability that on any given week profits will be between 10.0 and 12.5 million dollars?
2. What is the probability that on any given week profits will be between 7.5 and 10.0 million dollars?
3. What is the probability that on any given week profits will be between 7.5 and 12.5 million dollars?
4. What is the probability that on any given week profits will be at least 7.5 million dollars?
5. What is the probability that on any given week profits will be under 7.5 million dollars?
6. What is the probability that on any given week profits will be between 12.5 and 14.3 million dollars?
7. Fifty percent of the time weekly profits (in millions of dollars) are expected to be above what value?
8. Ninety percent of the time weekly profits (in millions of dollars) are expected to be above what value?
9. What is the interquartile range in weekly profits expected from the State Football Lottery Program?

QUESTION 1: FINDING $P(10.0 \leq X \leq 12.5)$

Using the transformation formula (7.8), we see that a weekly profit of 12.5 million dollars (X) is equivalent to +1.00 standard deviation above the mean. That is,

$$Z = \frac{X - \mu_X}{\sigma_X}$$

$$= \frac{12.5 - 10.0}{2.5} = +1.00$$

From Table C.2 we see that the area under the normal curve from the mean to +1.00 standard deviation above the mean is .3413. Thus, as depicted in Figure 7.10, there is a .3413 probability that on any given week profits will be between 10.0 and 12.5 million dollars.

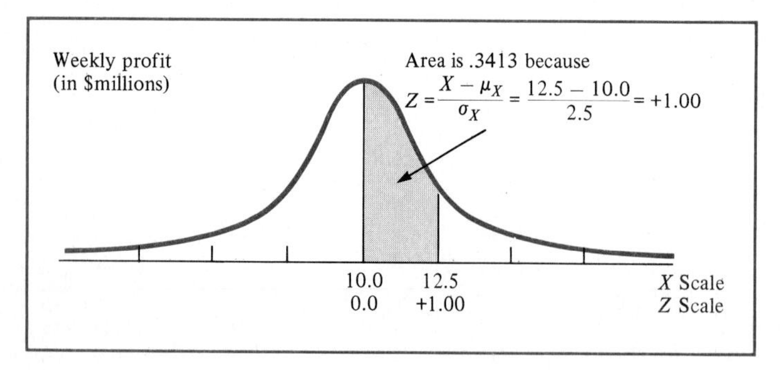

FIGURE 7.10 □ Finding $P(10.0 \leq X \leq 12.5)$.

QUESTION 2: FINDING $P(7.5 \leq X \leq 10.0)$

Using the transformation formula, we observe that a weekly profit of 7.5 million dollars (X) is equivalent to -1.00 standard deviation below the mean. That is,

$$Z = \frac{X - \mu_X}{\sigma_X}$$

$$= \frac{7.5 - 10.0}{2.5} = -1.00$$

Since Table C.2 shows only positive entries for Z (because of symmetry), it is clear that the area from the mean to $Z = -1.00$ must be identical to the area from the mean to $Z = +1.00$. (We already found this value to answer question 1, but for completeness we include it again.) Discarding the negative sign, we look up (in Table C.2) the value $Z = 1.00$ and find the probability to be .3413. Hence, as displayed in Figure 7.11, there is a .3413 probability that on any given week profits will be between 7.5 and 10.0 million dollars.

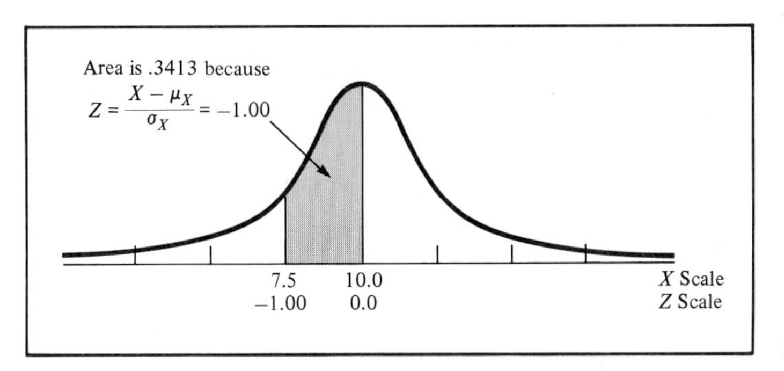

FIGURE 7.11 □ Finding $P(7.5 \leq X \leq 10.0)$.

QUESTION 3: FINDING $P(7.5 \leq X \leq 12.5)$

To determine the probability that on any given week profits will be between 7.5 and 12.5 million dollars, we note that one of the values of interest is above the mean while the other is below it. Since our transformation formula (7.8) permits us to find probabilities only from *a particular value of interest to the mean*, we can obtain our desired probability in three steps:

1. Determine the probability from the mean to 12.5 million dollars.
2. Determine the probability from the mean to 7.5 million dollars.
3. Sum up the two mutually exclusive results.

For this example, we have already completed steps 1 and 2 in answering questions 1 and 2. Hence, from step 3, the probability that on any given week profits will

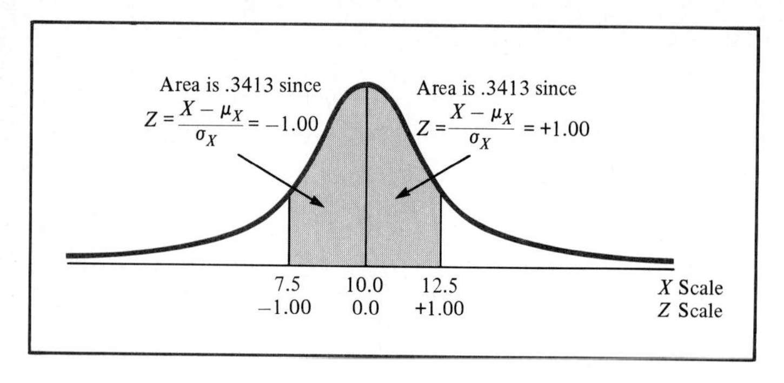

FIGURE 7.12 □ Finding $P(7.5 \leq X \leq 12.5)$.

This is a rather important finding—for, if we generalize, we can see that for any normal distribution there is a 68.26% chance that a randomly selected observation will fall within 1 standard deviation above or below the mean.[13]

EXAMPLE 7.12

Verify from Table C.2 that there is a 95.44% chance that any randomly selected normally distributed observation will fall within 2 standard deviations above or below the mean and that there is a 99.73% chance that the observation will fall between 3 standard deviations above or below the mean.

Solution:

Using the football lottery data, these results are depicted in panels A and B, respectively.

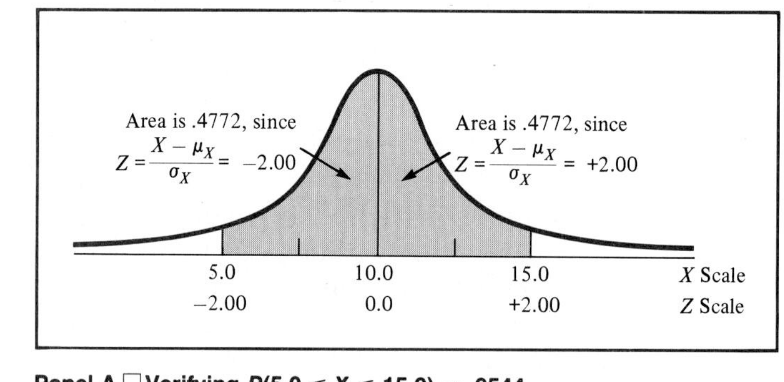

Panel A □ Verifying $P(5.0 \leq X \leq 15.0) = .9544$

[13] We may recall from Chapter 3 that when we defined the standard deviation as a measure of dispersion relative to the mean, we stated that for most data sets the majority of observations will cluster together within 1 standard deviation around the mean. For normally distributed data, more than two-thirds of the observations fall within 1 standard deviation around the mean.

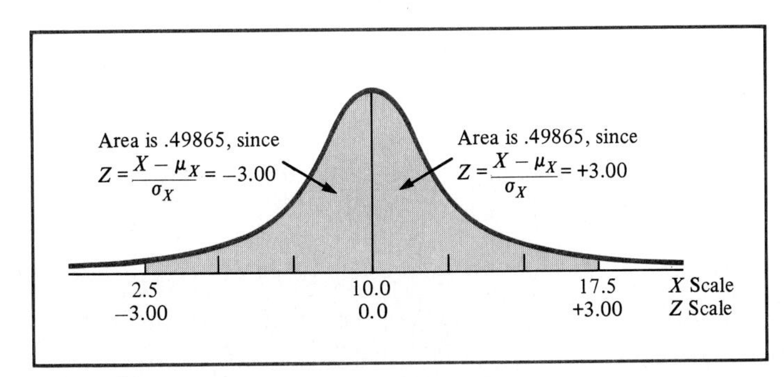

Panel B ☐ Verifying $P(2.5 \leq X \leq 17.5) = .9973$

Thus, for the State Football Lottery Program 95.44% of the time the weekly profits can be expected to range between 5.0 and 15.0 million dollars, while 99.73% of the time (i.e., "practically always") profits can be expected to fall between 2.5 and 17.5 million dollars. From this latter result it is clear why $6\sigma_X$ (that is, 3 standard deviations above the mean to 3 standard deviations below the mean) is often used as a practical approximation of the range for normally distributed data.

QUESTION 4: FINDING $P(X \geq 7.5)$

To obtain the probability that on any given week profits will be at least 7.5 million dollars we should examine the shaded region of Figure 7.13.

Our desired probability can be split into two parts—the probability that weekly profits will fall between 7.5 and 10.0 million dollars [that is, $P(7.5 \leq X \leq 10.0)$] and the probability that weekly profits will exceed 10.0 million dollars [that is, $P(X > 10.0)$].

For this example the first probability (.3413) was already obtained in answering question 2. The second probability is clearly .5000, since in a perfectly symmetrical distribution the mean and median are identical and, by definition, 50% of the observations exceed the median. Hence the desired result is .8413. That is, there is an 84.13% chance that in any given week profits will be at least 7.5 million dollars.

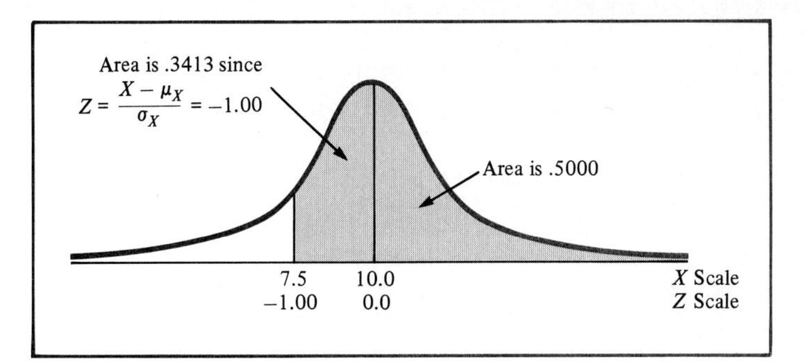

FIGURE 7.13 ☐ Finding $P(X \geq 7.5)$.

QUESTION 5: FINDING $P(X < 7.5)$

Determining the probability that in any given week profits will be less than 7.5 million dollars is an easy task now that we have answered question 4. Since all the probabilities (i.e., the entire area) under the curve (i.e., normal distribution) must add up to 1, we note that our desired probability is simply the complement of our previous result. That is,

$$\begin{aligned} P(X < 7.5) &= 1 - P(X \geq 7.5) \\ &= 1 - .8413 \\ &= .1587 \end{aligned}$$

In words, the probability that weekly profits are less than 7.5 million dollars is equal to one minus the probability that weekly profits are at least 7.5 million dollars.

Suppose, however, that we had not previously answered question 4. How then would we determine the probability that weekly profits are under 7.5 million dollars? Equation (7.8) permits us to find areas under the standardized normal distribution only from the mean to Z, not from Z to $-\infty$. Thus, if $Z < 0$, we must find the probability from the mean to Z and subtract this result from .5000 to obtain the desired answer. Since we know (from answering question 2) that the area or portion of the curve from the mean to $Z = -1.00$ is .3413, the area from $Z = -1.00$ to $Z = -\infty$ must be $.5000 - .3413 = .1587$.[14] This is indicated in Figure 7.14.

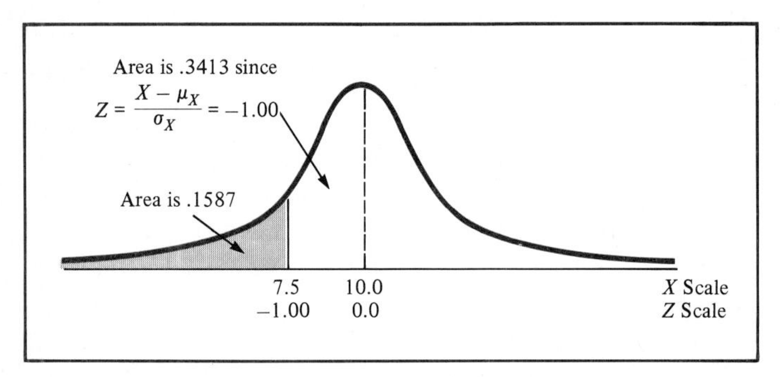

FIGURE 7.14 □ Finding $P(X < 7.5)$.

QUESTION 6: FINDING $P(12.5 \leq X \leq 14.3)$

As a final illustration of determining probabilities from the standardized normal distribution, suppose we wish to find how likely it is that on any given week profits

[14] Unlike the case of discrete random variables, where the wording of the problem is so essential, we note that for continuous random variables there is much more flexibility in the wording. Hence there are two ways to state our result: we can say that there is a 15.87% chance that weekly profits are under 7.5 million dollars, or we can say that there is a 15.87% chance that weekly profits are 7.5 million dollars or less. Semantics are unimportant, because with continuous random variables the probability that weekly profits are exactly 7.5 million dollars (or any other exactly specified amount) is 0.

will be between 12.5 and 14.3 million dollars. Since both values of interest are above the mean, it is clear from Figure 7.15 that the desired probability (or area under the curve between the two values) is less than .5000. Since our transformation formula (7.8) permits us to find probabilities only from a particular value of interest to the mean, we can obtain our desired probability in three steps:

1. Determine the probability (area under the curve) from the mean to 12.5 million dollars.
2. Determine the probability (area under the curve) from the mean to 14.3 million dollars.
3. Subtract the smaller area (probability) from the larger to avoid double counting.

For this example, we have already completed step 1 in answering question 1. To obtain the probability that weekly profits will be between 10.0 and 14.3 million dollars we have

$$Z = \frac{X - \mu_X}{\sigma_X}$$

$$= \frac{14.3 - 10.0}{2.5}$$

$$= +1.72$$

From Table C.2 we see that the resulting probability or area under the normal curve from the mean to +1.72 standard deviations above the mean is .4573. Hence, by subtracting the smaller area (.3413) from the larger one, we determine that there is only a .1160 probability that on any given week profits will be between 12.5 and 14.3 million dollars. That is,

$$P(12.5 \le X \le 14.3) = P(10.0 \le X \le 14.3) - P(10.0 \le X \le 12.5)$$

$$= .4573 - .3413$$

$$= .1160$$

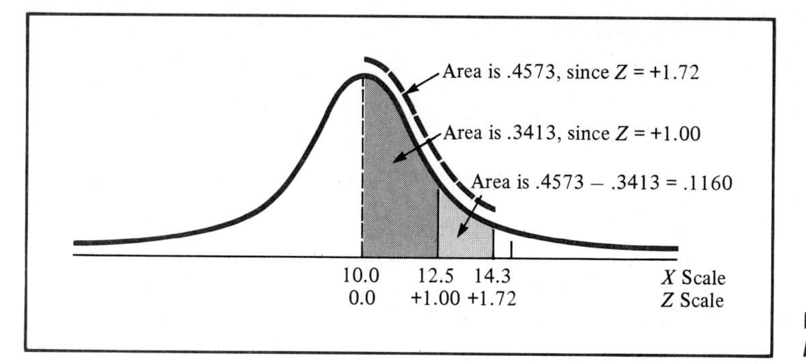

FIGURE 7.15 □ Finding $P(12.5 \le X \le 14.3)$.

Problem

• 7.22 Given a standardized normal distribution (with mean of 0 and standard deviation of 1), determine the following probabilities:

(a) $P(Z > +1.34)$
(b) $P(Z \leq +1.17)$
(c) $P(0 \leq Z \leq +1.17)$
(d) $P(Z < -1.17)$
(e) $P(-1.17 \leq Z \leq +1.34)$
(f) $P(-1.17 \leq Z \leq -0.50)$

FINDING THE VALUES CORRESPONDING TO KNOWN PROBABILITIES

In our previous applications regarding normally distributed data we have sought to determine the probabilities associated with various measured values. Now, however, suppose we wish to determine *particular values* of the variable of interest which correspond to known probabilities. As examples, let us respond to questions 7 through 9 of our case study.

QUESTION 7

Fifty percent of the time the Lottery Commission expects weekly profits to exceed a particular level. It is simple to determine that level. From Figure 7.16 we see that 50% of the time the profits are greater than the median value of 10.0 million dollars. Since the median and mean are equal in symmetric distributions, the answer to our question is 10.0 million dollars.

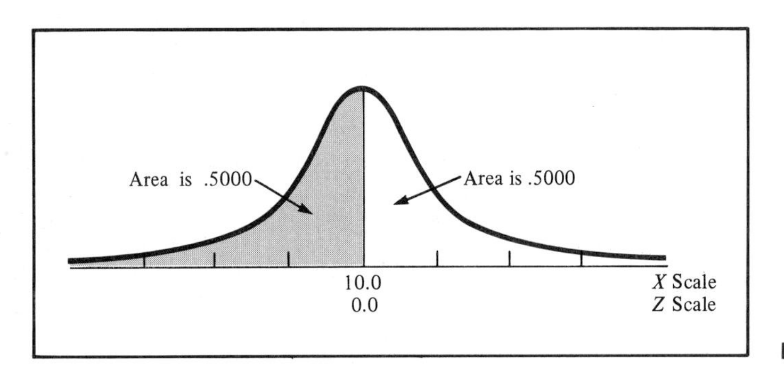

FIGURE 7.16 □ Finding X.

QUESTION 8

To determine the value of weekly profits that the Lottery Commission expects to exceed 90% of the time, we need to study Figure 7.17.

From Figure 7.17 it is clear that the 90% area can be split into two parts—

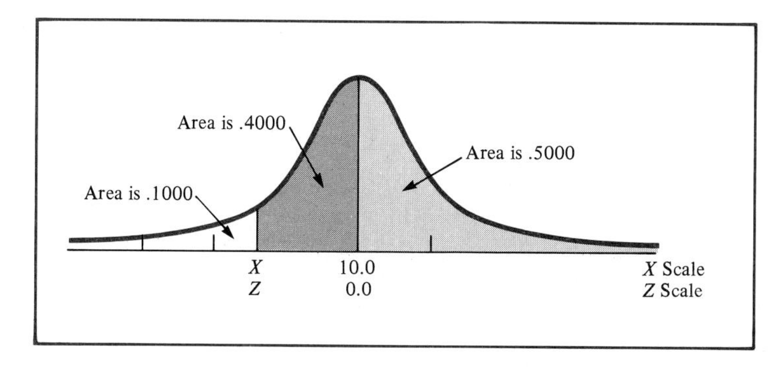

FIGURE 7.17 □ Finding *Z* to determine *X*.

weekly profits above the mean (area is 50%) and weekly profits between the mean and the desired value X (area is 40%). While we do not know X, we can easily determine the corresponding standardized value Z, since the area under the normal curve from the standardized mean 0 to this Z must be .4000. Using the body of Table C.2, we search for the area (or probability) .4000. As shown in Table 7.6, we observe that this area is located in the Z row of 1.2 and between the Z columns of .08 and .09. Because the value for the .08 column is closer to .4000 than that of the .09 column, we will use the value of 1.28 for Z.[15]

TABLE 7.6 □ Obtaining a *Z* value corresponding to a particular area under the normal curve

Z	.00	.01	.02	.03	.04	.05	.06	.07	.08	.09
0.0	.0000	.0040	.0080	.0120	.0160	.0199	.0239	.0279	.0319	.0359
0.1	.0398	.0438	.0478	.0517	.0557	.0596	.0636	.0675	.0714	.0753
0.2	.0793	.0832	.0871	.0910	.0948	.0987	.1026	.1064	.1103	.1141
0.3	.1179	.1217	.1255	.1293	.1331	.1368	.1406	.1443	.1480	.1517
0.4	.1554	.1591	.1628	.1664	.1700	.1736	.1772	.1808	.1844	.1879
0.5	.1915	.1950	.1985	.2019	.2054	.2088	.2123	.2157	.2190	.2224
0.6	.2257	.2291	.2324	.2357	.2389	.2422	.2454	.2486	.2518	.2549
0.7	.2580	.2612	.2642	.2673	.2704	.2734	.2764	.2794	.2823	.2852
0.8	.2881	.2910	.2939	.2967	.2995	.3023	.3051	.3078	.3106	.3133
0.9	.3159	.3186	.3212	.3238	.3264	.3289	.3315	.3340	.3365	.3389
1.0	.3413	.3438	.3461	.3485	.3508	.3531	.3554	.3577	.3599	.3621
1.1	.3643	.3665	.3686	.3708	.3729	.3749	.3770	.3790	.3810	.3830
1.2	.3849	.3869	.3888	.3907	.3925	.3944	.3962	.3980	.3997	.4015
1.3	.4032	.4049	.4066	.4082	.4099	.4115	.4131	.4147	.4162	.4177

SOURCE: Extracted from Table C.2.

However, from Figure 7.17 the Z value must be recorded as a negative (that is, $Z = -1.28$), since it is below (that is, lies to the left of) the standardized mean of 0.

Once Z is obtained, we can now use the transformation formula (7.8) to determine the value of interest, X. Since

[15] To be more accurate, we could interpolate between the values in the .08 and .09 columns.

$$Z = \frac{X - \mu_X}{\sigma_X}$$

then

$$X = \mu_X + Z\sigma_X \quad \textbf{(7.9)}$$

Substituting, we compute

$$\begin{aligned} X &= 10.0 + (-1.28)(2.5) \\ &= 10.0 - 3.20 \\ &= 6.8 \text{ million dollars} \end{aligned}$$

Thus the Lottery Commission can expect that 90% of the time weekly profits will exceed 6.8 million dollars.

As a review, to find a *particular value* associated with a known probability we take the following steps:

1. Sketch the normal curve and then place the values for the means (μ_X and μ_Z) on the respective X and Z scales.
2. Split the appropriate half of the normal curve into two parts—the portion from the desired X to the mean and the portion from the desired X to the tail.
3. Shade the area of interest.
4. Using Table C.2, determine the appropriate Z value corresponding to the area under the normal curve from the desired X to the mean μ_X.
5. Using Equation (7.9), solve for X; that is,

$$X = \mu_X + Z\sigma_X$$

EXAMPLE 7.13

97.5% of the time weekly profits will exceed how many millions of dollars?

Solution:

(Steps 1, 2, 3)

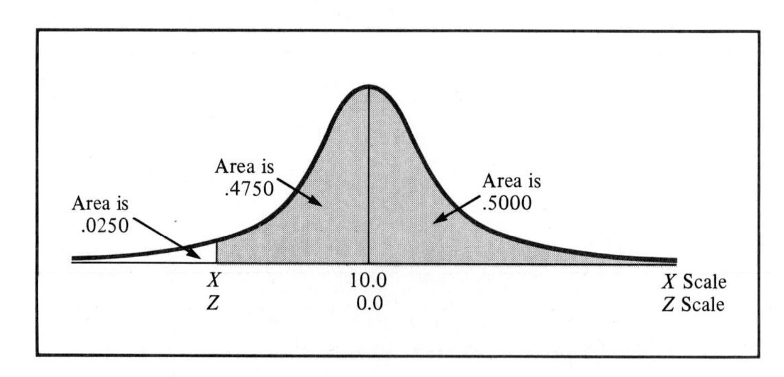

(Step 4) $Z = -1.96$ from Table C.2
(Step 5) $X = \mu_X + Z\sigma_X$
$= 10.0 + (-1.96)(2.5)$
$= 10.0 - 4.9$
$= 5.1$ million dollars

EXAMPLE 7.14

Determine the level of weekly profits (in millions of dollars) that the Lottery Commission expects to exceed only 2.5% of the time.

Solution:

(Steps 1, 2, 3)

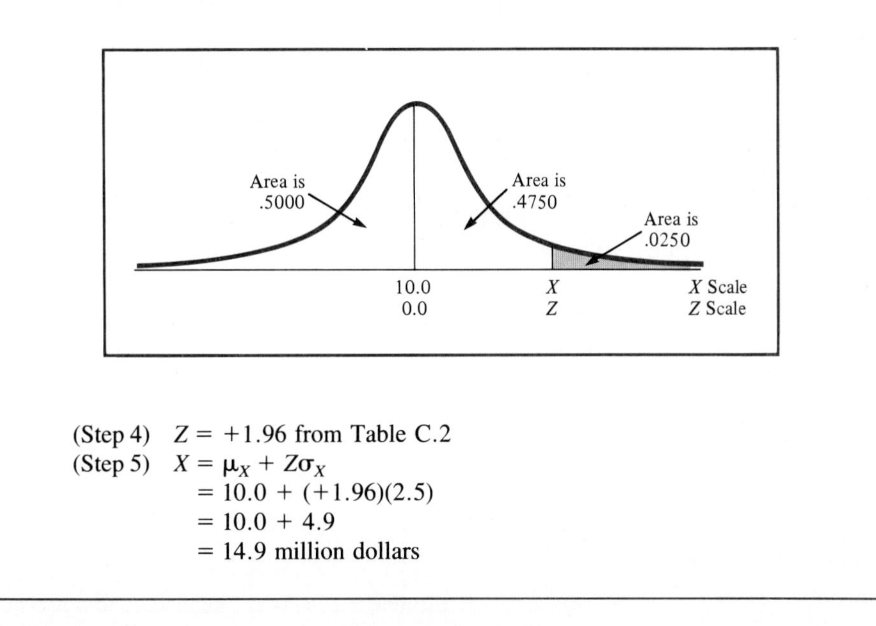

(Step 4) $Z = +1.96$ from Table C.2
(Step 5) $X = \mu_X + Z\sigma_X$
$= 10.0 + (+1.96)(2.5)$
$= 10.0 + 4.9$
$= 14.9$ million dollars

QUESTION 9

To obtain the interquartile range in weekly profits we must first find the value for Q_1 and the value for Q_3; then we must subtract the former from the latter.

To find the first quartile value, we must determine the level of weekly profits that the Lottery Commission expects to exceed 75% of the time. This is depicted in Figure 7.18 on page 204.

Although we do not know Q_1, we can easily obtain the corresponding standardized value Z, since the area under the normal curve from the standardized mean 0 to this Z must be .2500. Using the body of Table C.2, we search for the area or probability .2500. The closest result is .2486 as shown in Table 7.7 on the bottom of page 204 (which is a replica of Table C.2).

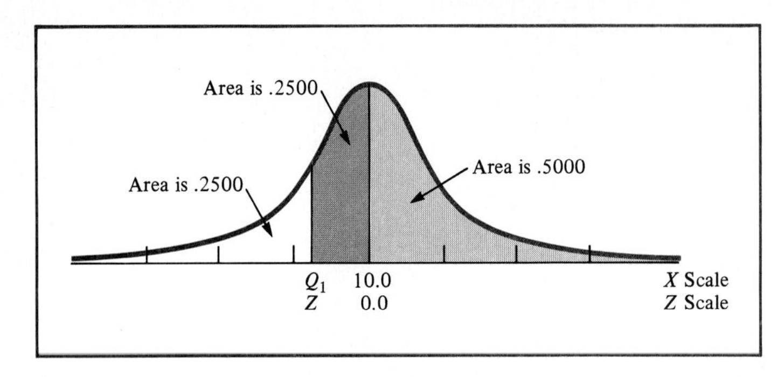

FIGURE 7.18 □ Finding Q_1.

Working from this area to the margins of the table, we see that the Z value corresponding to the particular Z row (0.6) and Z column (.07) is 0.67. However, from Figure 7.18 the Z value must be recorded as a negative (that is, $Z = -0.67$), since it lies to the left of the standardized mean of 0.

Once Z is obtained, the final step is to use Equation (7.9). Hence,

$$\begin{aligned} Q_1 = X &= \mu_X + Z\sigma_X \\ &= 10.0 + (-0.67)(2.5) \\ &\cong 10.0 - 1.7 \\ &= 8.3 \text{ million dollars} \end{aligned}$$

To find the third quartile, we must determine the level of weekly profits that the Lottery Commission expects to exceed 25% of the time. This is displayed in Figure 7.19.

From the symmetry of the normal distribution, our desired Z value must be $+0.67$ (since Z lies to the right of the standardized mean of 0). Therefore,

$$\begin{aligned} Q_3 = X &= \mu_X + Z\sigma_X \\ &= 10.0 + (+0.67)(2.5) \\ &\cong 10.0 + 1.7 \\ &= 11.7 \text{ million dollars} \end{aligned}$$

TABLE 7.7 □ Obtaining a Z value corresponding to a particular area under the normal curve

Z	.00	.01	.02	.03	.04	.05	.06	.07	.08	.09
0.0	.0000	.0040	.0080	.0120	.0160	.0199	.0239	.0279	.0319	.0359
0.1	.0398	.0438	.0478	.0517	.0557	.0596	0.636	.0675	.0714	.0753
0.2	.0793	.0832	.0871	.0910	.0948	.0987	.1026	.1064	.1103	.1141
0.3	.1179	.1217	.1255	.1293	.1331	.1368	.1406	.1443	.1480	.1517
0.4	.1554	.1591	.1628	.1664	.1700	.1736	.1772	.1808	.1844	.1879
0.5	.1915	.1950	.1985	.2019	.2054	.2088	.2123	.2157	.2190	.2224
0.6	.2257	.2291	.2324	.2357	.2389	.2422	.2454	.2486	.2518	.2549
0.7	.2580	.2612	.2642	.2673	.2704	.2734	.2764	.2794	.2823	.2852
0.8	.2881	.2910	.2939	.2967	.2995	.3023	.3051	.3078	.3106	.3133
0.9	.3159	.3186	.3212	.3238	.3264	.3289	.3315	.3340	.3365	.3389

SOURCE: Extracted from Table C.2.

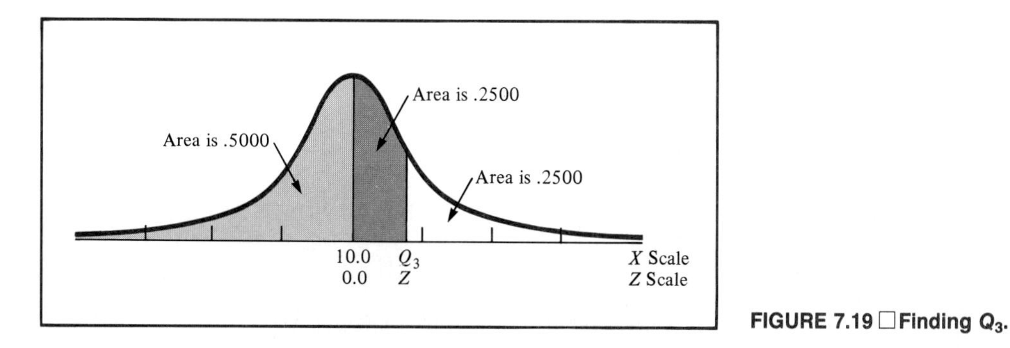

FIGURE 7.19 □ Finding Q_3.

The interquartile range or ''middle spread'' of the distribution is

$$\begin{aligned}\text{interquartile range} &= Q_3 - Q_1 \\ &\cong 11.7 - 8.3 \\ &= 3.4 \text{ million dollars}\end{aligned}$$

Alternatively, we should note that we did not need to calculate Q_1 and Q_3 directly if our interest was in computing the ''middle spread'' of the data. That is, once we determine that the distance from the mean to .67 standard deviations above the mean contains approximately 25% of the area under the normal curve, then, by symmetry, we know that approximately 25% of the area under the normal curve is also contained in the distance from the mean to .67 standard deviations below the mean. Since the standard deviation is 2.5 million dollars, this distance is $\pm Z\sigma_X = 0.67(2.5) \cong 1.7$ million dollars above the mean and 1.7 million dollars below the mean—that is, (a total distance of) approximately 3.4 million dollars, the interquartile range.

EXAMPLE 7.15

Show that for normally distributed data the interquartile range is approximately equal to 1.33 standard deviations.

***Solution*:**

The interquartile range is the interval from Q_1 to Q_3—that is, from $\mu_X - 0.67\sigma_X$ to $\mu_X + 0.67\sigma_X$. Thus,

$$\begin{aligned}\text{interquartile range} &= Q_3 - Q_1 \\ &= (\mu_X + 0.67\ \sigma_X) - (\mu_X - 0.67\ \sigma_X) \\ &= \mu_X + 0.67\ \sigma_X - \mu_X + 0.67\ \sigma_X \\ &\cong 1.33\sigma_X \quad \text{(owing to rounding)}\end{aligned}$$

EXAMPLE 7.16

Show that for normally distributed data the standard deviation can be approximated as .75 times the interquartile range.

Solution:

If

$$\text{interquartile range} \cong 1.33\sigma_X = \frac{4}{3}\sigma_X$$

then

$$\frac{3}{4}\begin{pmatrix}\text{interquartile}\\ \text{range}\end{pmatrix} = .75\begin{pmatrix}\text{interquartile}\\ \text{range}\end{pmatrix} \cong \sigma_X$$

Problems

7.23 Determine the "practical range" for the State Football Lottery Program data.

• **7.24** Suppose that an intelligence test is constructed to have a normal distribution with a mean of 100 and a standard deviation of 15.

- **(a)** Draw a picture of a normal curve with a mean of 100, and mark the abscissa (horizontal axis) where scores of 70, 85, 100, 115, and 130 should be.
- **(b)** What proportion of people have IQ scores less than 100? Less than 115? Greater than 130? Greater than 145?
- **(c)** What proportion of people have IQ scores between 85 and 115? Between 70 and 130?
- **(d)** If IQ scores were reported instead on a scale with a mean of 500 and a standard deviation of 100 (like SAT scores), what would scores of 400, 500, and 600 have been on the original IQ scale?

7.25 Suppose that the length of time it takes one variety of seeds of a plant to germinate is normally distributed with a mean of 15 days and a standard deviation of 4 days.

- **(a)** What proportion of the seeds should germinate within 19 days? Within 23 days?
- **(b)** By what day should three-fourths of the seeds have germinated?
- **(c)** Suppose that by the 15th day only 60% of the seeds have germinated; would you worry about having a bad bunch of seeds?

7.26 Suppose that the amount of time it takes the Internal Revenue Service (IRS) to send refunds to taxpayers is normally distributed with a mean of 12 weeks and a variance of 9.

- **(a)** What proportion of taxpayers should get a refund within 6 weeks? Within 9 weeks?
- **(b)** What proportion of refunds will be sent more than 15 weeks after the IRS receives the tax return?
- **(c)** How long will it take before 90% of taxpayers get their refunds?

7.27 Examine the frequency distributions, stem-and-leaf displays, and box-and-whisker plots you constructed in answering problems in Chapters 4 and 5. Find at least two which appear to be approximately normally distributed and at least two which appear not to be approximately normally distributed. For each, tell why you think it is or is not approximately normal. [*Hint*: Don't look at data sets with only a few values; it's harder to tell whether these really could be from a normal distribution.] For each distribution, calculate the proportion of observations which are in the intervals formed by cutting the distribution at the mean and at 1 and 2 standard deviations above and below the mean (i.e., you will cut the distribution into six parts). Compare these to the proportions from a standard normal distribution to see how closely the observed come to the theoretical proportions.

7.28 A set of final examination grades in an introductory statistics course was found to be normally distributed with a mean of 73 and a standard deviation of 8.

(a) What is the probability of getting at most a grade of 91 on this exam?
(b) What percentage of students scored between 65 and 89?
(c) What percentage of students scored between 81 and 89?
(d) What is the final exam grade if only 5% of the students taking the test scored higher?
(e) If the professor "curves" (gives A's to the top 10% of the class regardless of the score), are you better off with a grade of 81 on this exam or a grade of 68 on a different exam where the mean is 62 and the standard deviation is 3? Show statistically and explain.

7.29 At Tobias College the grade-point indexes of its 1,000 students are approximately normally distributed with mean $\mu_X = 2.83$ and standard deviation $\sigma_X = .38$.

(a) What is the probability that a randomly selected student has a grade-point index between 2.00 and 3.00?
(b) What percentage of the student body are on probation—that is, have grade-point indexes below 2.00?
(c) How many students at this college are expected to be on the dean's list—that is, have grade-point indexes equal to or exceeding 3.20?
(d) What grade-point index will be exceeded by only 15% of the student body?

7.6 THE NORMAL DISTRIBUTION AS AN APPROXIMATION TO VARIOUS DISCRETE PROBABILITY DISTRIBUTIONS

In the previous section we demonstrated the importance of the normal probability density function because of the numerous phenomena which seem to follow it or whose distributions can be approximated by it. In this section we shall demonstrate another important aspect of the normal distribution—how it may be used to approximate various important discrete probability distribution functions such as the binomial.

7.6.1 Approximating the Binomial Distribution

In Section 7.3.4 we stated that the binomial distribution will be symmetric (like the normal distribution) whenever $p = .5$. When $p \neq .5$ the binomial distribution will not be symmetric. However, the closer p is to .5 and the larger the number of sample observations n, the more symmetric the distribution becomes.

On the other hand, the larger the number of observations in the sample, the more tedious it is to compute the exact probabilities of success by use of Equation (7.4). Fortunately, though, whenever the sample size is large, the normal distribution can be used to approximate the exact probabilities of success that otherwise would have to be obtained through laborious computations.

As one rule of thumb this normal approximation can be used whenever both of the following two conditions are met:

1. The product of the two parameters n and p equals or exceeds 5.
2. The product $n(1 - p)$ equals or exceeds 5.

We recall from Section 7.3.4 that the mean of the binomial distribution is given by

$$\mu_X = np$$

while the standard deviation of the binomial distribution is obtained from

$$\sigma_X = \sqrt{np(1 - p)}$$

Substituting into the transformation formula (7.8),

$$Z = \frac{X - \mu_X}{\sigma_X}$$

we have

$$Z = \frac{X - np}{\sqrt{np(1 - p)}} \qquad \textbf{(7.10)}$$

where $\mu_X = np$, the mean of the binomial distribution
$\sigma_X = \sqrt{np(1 - p)}$, the standard deviation of the binomial distribution
X = number of successes in the sample ($X = 0, 1, 2, \ldots, n$)

and the approximate probabilities of success are obtained from Table C.2, the table of the standardized normal distribution.

To illustrate this, suppose, in the product quality-control example (described on page 171), that a sample of $n = 1{,}600$ tires of the same type are obtained at random from an ongoing production process in which 8% of all such tires produced are defective. What is the probability that in such a sample not more than 150 tires will be defective?

Since both $np = (1{,}600)(.08) = 128$ and $n(1 - p) = (1{,}600)(.92) = 1{,}472$ exceed 5, we may use the normal distribution to approximate the binomial:

$$Z = \frac{X - np}{\sqrt{np(1 - p)}} = \frac{150 - 128}{\sqrt{(1{,}600)(.08)(.92)}} = \frac{22}{10.85} = +2.03$$

The probability of not more than $X = 150$ successes corresponds, on the standardized Z scale, to a value of not more than $+2.03$. This is depicted in Figure 7.20.

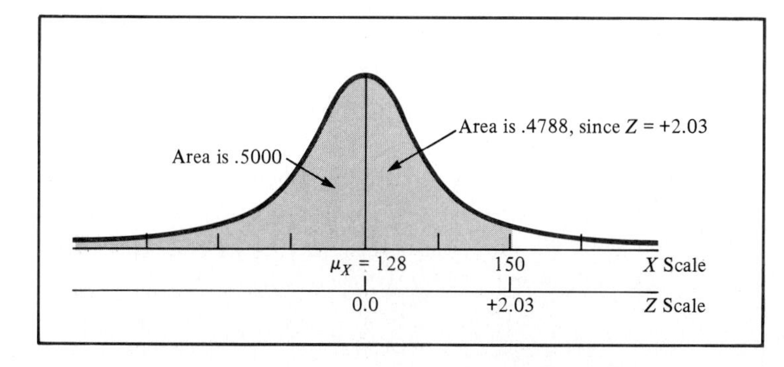

FIGURE 7.20 □ Approximating the binomial distribution.

Using Table C.2, we find that the area under the curve between the mean and $Z = +2.03$ is .4788, so that the approximate probability is given by .5000 + .4788 = .9788.

To appreciate the amount of labor saved by using the normal approximation to the binomial model in lieu of the exact probability computations, just imagine making the following 151 computations from Equation (7.4) and then summing up the results:

$$\frac{1{,}600!}{0!1{,}600!}(.08)^0(.92)^{1{,}600} + \frac{1{,}600!}{1!1{,}599!}(.08)^1(.92)^{1{,}599} + \cdots + \frac{1{,}600!}{150!1{,}450!}(.08)^{150}(.92)^{1{,}450}$$

7.6.2 Obtaining Individual Probability Approximations

Recall that with a continuous distribution (such as the normal), the probability of obtaining a *particular* value of a random variable is zero. On the other hand, when the normal distribution is used to approximate a discrete distribution, a **correction for continuity adjustment** can be employed so that we may approximate the probability of a specific value of the discrete distribution.

As a case in point, we refer to the product quality-control example. A random sample of $n = 1{,}600$ tires of the same type was selected from an ongoing production process in which 8% of all such tires produced are defective. Suppose that we now want to compute the probability of obtaining *exactly* 150 defectives. While a discrete random variable can have only a specified value (such as 150), a continuous random variable that is used to approximate it can take on any values whatsoever within an interval around that specified value, as demonstrated on the scale below:

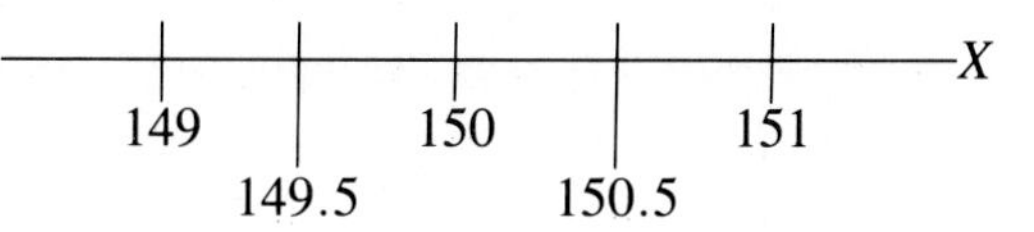

Therefore, the correction for continuity adjustment defines the integer value of interest (150) to range from one-half unit below it (149.5) to one-half unit above it (150.5). In our product quality-control example, the probability of obtaining 150 defective tires can now be approximated from the area under the normal curve between 149.5 and 150.5. Using Equation (7.10), we compute:

$$Z = \frac{X - np}{\sqrt{np(1-p)}}$$

$$= \frac{150.5 - 128}{\sqrt{1{,}600(.08)(.92)}} = +2.07$$

and

$$Z = \frac{149.5 - 128}{\sqrt{(1{,}600)(.08)(.92)}} = +1.98$$

From Table C.2 we note that the area under the normal curve from the mean to $X = 150.5$ is .4808, while the area under the curve from the mean to $X = 149.5$ is .4761. Thus, as depicted in Figure 7.21, the approximate probability of obtaining 150 defective tires is the difference in the two areas, .0047.

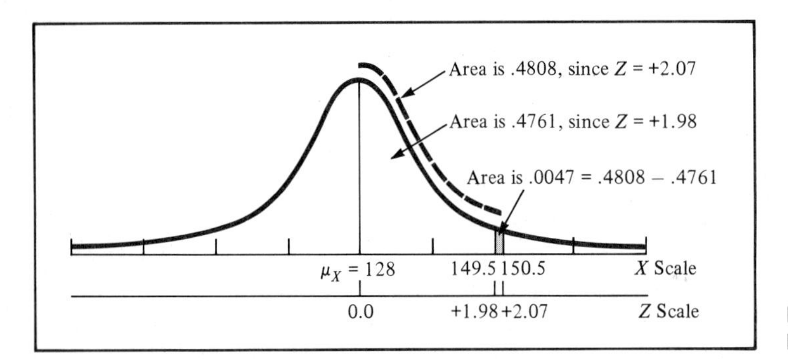

FIGURE 7.21 □ Approximating an exact binomial probability.

Problems

• 7.30 What is the (*approximate*) probability that a student could pass a 100-question true-false examination if the student were to guess on each question? (*Note*: To pass, the student must get at least 60 questions right.)

7.31 What is the (*approximate*) probability that a student would get exactly 60 questions right on a 100-question true-false examination if the student were to guess on each question?

• 7.32 Senator Phil E. Bluster believes that for a particular bill, each senator has a .75 probability of voting in favor of it. If 51 (out of 100) votes are needed for the bill to pass, and each senator decides independently how to vote, calculate:

(a) The probability that the bill will pass (i.e., that at least 51 senators will vote for it).
(b) The probability that at least 67 senators will vote for it.
(c) The probability that at least 76 senators will vote for it.

7.33 Little Johnny starts a lemonade stand near his home. Suppose that 100 people walk by every day, and that on the average one person out of five buys lemonade.

(a) Tell why each trial (whether each person buys or does not buy) may not really be independent; that is, why if one person buys lemonade, the next may be either more or less likely to buy it.
(b) If the trials really are independent, then use the normal approximation to the binomial to find
(1) The probability that Johnny will sell at least 20 glasses of lemonade.
(2) The probability that Johnny will sell less than 10 glasses of lemonade.
(3) The probability that Johnny will sell between 10 and 30 glasses of lemonade.
(c) Tell whether you think the normal approximation is warranted in this situation; justify your answer.

7.34 The text noted several conditions which, if they apply, indicate that the normal approximation to the binomial distribution will be a good approximation.

(a) For the case of $n = 9$ and $p = .5$, which of these conditions are met and which are not?
(b) Construct a relative frequency distribution for the binomial distribution for these values, and use the continuity correction formula to find the same relative frequency distribution using the normal approximation.
(c) Discuss the differences between the two distributions (i.e., how good does the normal approximation appear to be in this case?)

7.7 SUMMARY

In this chapter, we have developed the concept of mathematical expectation and we have examined two particularly important probability distributions—the (discrete) binomial and the (continuous) normal. In particular, the normal distribution was shown to be useful in its own right and also useful as an approximation of the binomial model. In the next chapter we shall investigate how the normal distribution provides the basis for classical statistical inference. In subsequent chapters pertaining to inference, we shall see that very often our data (that is, the outcomes of our phenomenon of interest) follow or can be approximated by the probability distribution functions considered here.

Supplementary Problems

Hapless Hal has decided to try to find a suitable mate through a video dating service. The service promises to arrange four dates for Hal with women who are matched with him on various personality characteristics and interests. Although the service claims a "95% success rate," meaning that 95% of their customers enjoy at least 2 of their 4 dates, Hal has been fooled by slick con artists in the past, so he decides to use his newly learned information on the binomial distribution to help him decide whether to sign up for the service. To do this, Hal knows he must assume some value for the probability of a date being "successful." He decides to try several different values, but to start with, he chooses $p = 1/3$.

(a) Find the probability distribution for Hal's chances of having 0, 1, 2, 3, and 4 successful dates.

(b) Using the information from (a), find the probability that at least 2 of the dates will be successful; that at least 3 of the dates will be successful.

(c) Recalculate the above probabilities using the more pessimistic estimate of $p = 1/10$, and the more optimistic estimate of $p = 1/2$.

(d) How *sensitive* is the expected probability of success to Hal's estimate of p? That is, how much does the probability of at least 2 successful dates vary when p changes from 1/10 to 1/3, and from 1/3 to 1/2?

7.36 The famous parapsychologist Professor Sy Klops decides to investigate whether people can read minds. He makes up five cards, each of which has a different symbol on it. When a subject comes to be tested, Klops picks a card at random and asks the subject to guess which symbol is on the card. Klops then records whether the subject is correct or not, shuffles the cards together again, and repeats the procedure. (Each such cycle on which one of two outcomes can occur is sometimes called a *trial*.)

(a) Suppose each subject is given 5 trials. Calculate what proportion of people tested will be expected to get none right, one right, and so on.

(b) In the situation described in (a), if the Professor tests 1,000 people and says that 5 "show promise for having telepathic powers" because they got all 5 trials correct, would you agree or disagree, and why?

(c) Now suppose that the Professor uses more trials for each subject, so that everyone is tested on 50 trials. Discuss whether or not you think he can use the normal approximation to the binomial distribution.

(d) Regardless of your answer to (c), use the normal approximation to the binomial to estimate the probability that someone will get at least 15 right by chance; at least 20 right by chance.

7.37 In the carnival game Under-or-Over-Seven a pair of fair dice are rolled once, and the resulting sum determines whether or not the player wins or loses his or her bet. For example, the player can bet \$1 that the sum is under 7—that is, 2, 3, 4, 5, or 6. For such a bet the player will lose \$1 if the outcome equals or exceeds 7 or will win \$1 if the result is under 7. Similarly, the player can bet \$1 that the sum is over 7—8, 9, 10, 11, or 12. Here the player wins \$1 if the result is over 7 but loses \$1 if the result is 7 or under. A third method of play is to bet \$1 on the outcome 7. For this bet the player will win \$4 if the result of the roll is 7 and lose \$1 otherwise.

(a) Form the probability distribution function representing the different outcomes that are possible for a \$1 bet on being under 7.

(b) Form the probability distribution function representing the different outcomes that are possible for a \$1 bet on being over 7.

(c) Form the probability distribution function representing the different outcomes that are possible for a \$1 bet on 7.

(d) Prove that the expected long-run profit (or loss) to the player is the same—no matter which method of play is used.

7.38 Using Table C.6 determine the following:

(a) If $n = 4$ and $p = .12$, then $P(X = 0|n = 4, p = .12) = ?$
(b) If $n = 10$ and $p = .40$, then $P(X = 9|n = 10, p = .40) = ?$
(c) If $n = 15$ and $p = .50$, then $P(X = 8|n = 15, p = .50) = ?$
(d) If $n = 7$ and $p = .83$, then $P(X = 6|n = 7, p = .83) = ?$
(e) If $n = 9$ and $p = .90$, then $P(X = 9|n = 9, p = .90) = ?$

7.39 In the carnival game of Chuck-a-Luck three fair dice are rolled after the player has placed a bet on the occurrence of a particular face of the dice, say ⚄. For every $1 bet that you place: you can lose the $1 if none of the three dice shows the face ⚄; you can win $1 if one die shows the face ⚄; you can win $2 if two of the dice show the face ⚄; or you can win $3 if all three dice show the face ⚄.

(a) Form the probability distribution function representing the different monetary values (winnings or losses) that are possible (from one roll of the three dice).
(b) Determine the mean of this probability distribution.
(c) What is the player's expected long-run profit (or loss) from a $1 bet? Interpret.
(d) What is the expected long-run profit (or loss) to the house? Interpret.
(e) Would you play Chuck-a-Luck and make a bet?

• **7.40** Refer to the probability distribution given in Problem 7.9 on page 170.

The game of Craps deals with the rolling of a pair of fair dice. A **field bet** in the game of Craps is a one-roll bet and is based on the outcome of the pair of dice. For every $1 bet you make: you can lose the $1 if the sum is 5, 6, 7, or 8; you can win $1 if the sum is 3, 4, 9, 10, or 11; or you can win $2 if the sum is either 2 or 12.

(a) Form the probability distribution function representing the different outcomes that are possible in a field bet.
(b) Determine the mean of this probability distribution.
(c) What is the player's expected long-run profit (or loss) from a $1 field bet? Interpret.
(d) What is the expected long-run profit (or loss) from a $1 field bet to the house? Interpret.
(e) Would you play this game and make a field bet?

7.41 Given a standardized normal distribution having mean of 0 and standard deviation of 1 (Table C.2):

(a) What is the probability that:
(1) Z is less than 1.57?
(2) Z exceeds 1.84?
(3) Z is between 1.57 and 1.84?
(4) Z is less than 1.57 or greater than 1.84?
(5) Z is between -1.57 and 1.84?
(6) Z is less than -1.57 or greater than 1.84?
(b) What is the value of Z if 50.0% of all possible Z values are larger?
(c) What is the value of Z if only 2.5% of all possible Z values are larger?
(d) Between what two values of Z (symmetrically distributed around the mean) will 68.26% of all possible Z values be contained?

7.42 The thickness of a batch of 10,000 brass washers of a certain type manufactured by a large company is normally distributed with a mean of .0191 inch and with a standard deviation of .000425 inch. Verify that 99.04% of these washers can be expected to have a thickness between .0180 and .0202 inch.

• **7.43** The length of time needed to service a car at a gas station is normally distributed with mean $\mu_X = 4.5$ minutes and standard deviation $\sigma_X = 1.1$ minutes.

(a) What is the probability that a randomly selected car will require:

(1) More than 6 minutes of service or under 5 minutes of service?

(2) Between 3.5 and 5.6 minutes of service?

(3) At most 3.5 minutes of service?

(b) What is the servicing time such that only 5% of all cars require more than this amount of time?

7.44 Based upon past experience, 15% of the bills of a large mail-order book company are incorrect. A random sample of three current bills is selected.

(a) What is the probability that:

(1) Exactly two bills are incorrect?

(2) No more than two bills are incorrect?

(3) At least two bills are incorrect?

(b) What assumptions about the probability distribution are necessary to solve this problem?

7.45 Verify the following:

(a) The area under the normal curve between 1.5 standard deviations above and below the mean is .8664.

(b) The area under the normal curve between 2.5 standard deviations above and below the mean is .9876.

7.46 The SL Trucking Company determined that on an annual basis the distance traveled per truck is normally distributed with a mean of 50.0 thousand miles and a standard deviation of 12.0 thousand miles.

(a) What proportion of trucks can be expected to travel between 34.0 and 50.0 thousand miles in the year?

(b) What is the probability that a randomly selected truck travels between 34.0 and 38.0 thousand miles in the year?

(c) What percentage of trucks can be expected to travel either below 30.0 or above 60.0 thousand miles in the year?

(d) How many of the 1,000 trucks in the fleet are expected to travel between 30.0 and 60.0 thousand miles in the year?

(e) How many miles will be traveled by at least 80% of the trucks?

7.47 A statistical analysis of 1,000 long-distance telephone calls made from a large business office indicates that the length of these calls is normally distributed with $\mu_X = 240$ seconds and $\sigma_X = 40$ seconds.

(a) What percentage of these calls lasted less than 180 seconds?

(b) What is the probability that a particular call lasted between 180 and 300 seconds?

(c) How many calls lasted less than 180 seconds or more than 300 seconds?

(d) What percentage of the calls lasted between 110 and 180 seconds?

(e) What is the length of a particular call such that only 1% of all calls are shorter?

(f) If the researcher could not assume that the data were normally distributed, what then would be the probability that a particular call lasted between 180 and 300 seconds? [*Hint*: Recall the Bienayme–Chebyshev rule in Section 3.7.]

(g) Discuss the differences in your answers to (b) and (f).

References

1. DERMAN, C., L. J. GLESER, AND I. OLKIN, *A Guide to Probability Theory and Application* (New York: Holt, Rinehart and Winston, 1973).
2. LARSEN, R. J., AND M. L. MARX, *An Introduction to Mathematical Statistics and Its Applications* (Englewood Cliffs, N.J.: Prentice-Hall, 1981).
3. SCARNE, J., *Scarne's New Complete Guide to Gambling* (New York: Simon and Schuster, 1974).
4. THORP, E. O., *Beat the Dealer* (New York: Random House, 1962).
5. WEAVER, W., ''Probability,'' in A. SHUCHMAN, ed., *Scientific Decision Making in Business* (New York: Holt, Rinehart and Winston, 1963).

Sampling Distributions

8.1 INTRODUCTION: WHAT'S AHEAD

A major goal of statistical analysis is to use statistics such as the sample mean and the sample proportion to estimate the corresponding parameters in the population. The process of generalizing these sample results to the population is referred to as **statistical inference.** In our previous two chapters we have examined basic rules of probability and investigated various probability distributions such as the binomial and normal distributions. In this chapter we shall use these rules of probability along with the normal distribution to begin focusing on how a sample statistic (such as the arithmetic mean) can be utilized in making inferences about a population parameter. In order to do this we turn to an illustrative example.

Case E—The Quality Control Director's Problem

The director of quality control at a local soft-drink bottling plant is interested in studying the distribution of the actual fill of two-liter bottles of soft drink. A production process is maintained in which the soft drink is bottled continuously during a fixed eight-hour period. The machines that pour the soft drink into the bottles are calibrated so that the population average amount of soft drink in each bottle is equal to two liters and the population standard deviation is equal to 0.05 liter.

Although the director recognizes that only a single sample of a certain size will be taken at any particular time, she realizes that in order to determine whether the filling process has been kept under control, samples must be taken at various times throughout the day. Since it is obviously more costly to take and evaluate large samples, she is interested in studying what distribution the sample average $\bar{X}$ would follow for different sample sizes. It so happens that recent records are available in which different sample sizes have been used in various areas of the bottling plant. The director has examined data that were compiled over a long period which seem to indicate that the bottle fill is approximately normally distributed. However, she wonders what the effect would be on the distribution of the sample mean if the bottle fill were evenly (i.e., uniformly) distributed between 1.90 and 2.10 liters.

The quality-control director realizes that an inference about the population average will be made on the basis of a *single* sample of a predetermined size n. However, in order to take the results of this sample and develop an inference about the population, we need to consider (in a conceptual if not actual manner) all the possible sample results that could conceivably occur. If the selections of all the possible samples were actually performed, and, for each sample, a summary measure such as the arithmetic mean were computed, its distribution would be referred to as a **sampling distribution.** This notion of a sampling distribution is crucial to providing statistical inferences about a population parameter based only on a sample statistic. It serves as the link between the ability to merely describe a statistical characteristic and the ability to draw a conclusion about an entire population.

8.2 THE SAMPLING DISTRIBUTION OF THE MEAN

8.2.1 Why the Arithmetic Mean

In Chapter 3 we discussed several measures of central tendency. Undoubtedly, the most widely used (if not always the best) measure of central tendency is the arithmetic mean. This is particularly the case if, as in the problem of interest to the quality-control director, the population can be assumed to be normally distributed.

Among several important mathematical properties (see Reference 2) of the arithmetic mean for a normal distribution are

1. Unbiasedness.
2. Consistency.
3. Efficiency.

The first property, **unbiasedness,** involves the fact that the average of all the possible sample means (of a given sample size n) will be equal to the population mean μ_X. We may demonstrate this property empirically by referring to the set of data displayed in Table 8.1. In the interest of evaluating integrated software packages for microcomputers (packages that combine several programs, such as graphics, word processing, electronic spreadsheet, and data base management), *Fortune* magazine commissioned a study of five such integrated software packages: Super Calc3, Lotus 1–2–3, Context MBA, Series One Plus, and Lisa. Among the tasks studied was the time it took to perform a new set of calculations on a financial model spreadsheet. The results are those shown in Table 8.1.

TABLE 8.1 □ Spreadsheet calculation time for a financial model

Package	Time (in seconds)
Super Calc3	229
Lotus 1–2–3	298
Context MBA	262
Series One Plus	162
Lisa	448

SOURCE: *Fortune*, December 26, 1983.

Let us assume that the five integrated software packages evaluated by *Fortune* constitute a population. As described in Section 3.7, when the data from a population are available, the mean can be computed from

$$\mu_X = \frac{\Sigma X}{N} \tag{3.9}$$

and the standard deviation can be obtained from

$$\sigma_X = \sqrt{\frac{\Sigma (X - \mu_X)^2}{N}} \tag{3.12}$$

Thus

$$\mu_X = \frac{229 + 298 + 262 + 162 + 448}{5}$$

$$= \frac{1,399}{5} = 279.8 \text{ seconds}$$

and

$$\sigma_X = \sqrt{\frac{(229 - 279.8)^2 + \cdots + (448 - 279.8)^2}{5}}$$

$$= \sqrt{\frac{45{,}396.8}{5}} = 95.286 \text{ seconds}$$

To illustrate the property of unbiasedness we can select all possible samples of two packages *without* replacement from this population. There would be ten possible samples of two packages, since

$$\frac{N!}{n!(N-n)!} = \frac{5!}{2!\,3!} = 10$$

These ten possible samples are listed in Table 8.2.

TABLE 8.2 □ All ten samples of $n = 2$ packages from a population of $N = 5$ packages when sampling without replacement

Sample	Packages	Sample outcomes	Sample mean $\bar{X}$
1	Super Calc3, Lotus 1–2–3	229, 298	263.5
2	Super Calc3, Context MBA	229, 262	245.5
3	Super Calc3, Series One Plus	229, 162	195.5
4	Super Calc3, Lisa	229, 448	338.5
5	Lotus 1–2–3, Context MBA	298, 262	280.0
6	Lotus 1–2–3, Series One Plus	298, 162	230.0
7	Lotus 1–2–3, Lisa	298, 448	373.0
8	Context MBA, Series One Plus	262, 162	212.0
9	Context MBA, Lisa	262, 448	355.0
10	Series One Plus, Lisa	162, 448	305.0
			$\mu_{\bar{X}} = 279.8 = \mu_X$

We observe that if all ten sample means are averaged, the mean of these values ($\mu_{\bar{X}}$) is equal to 279.8, the same as the mean of the population μ_X.

While unbiasedness is often desirable in an estimator of a parameter, it is not necessary. A more general concept is that of **consistency.** An estimator is consistent if, as the sample size increases, the difference between the estimate and the (true) population value approaches zero. As an example, consider formulas for the sample variance and standard deviation which have n instead of $n - 1$ in the denominator. The formulas for S^2 and S with $n - 1$ are unbiased (and also consistent), but those with n are consistent (but are biased). As the sample size gets larger, these estimates come closer and closer to the population value. As we pointed out, with a large enough sample it won't matter whether n or $n - 1$ is used in the denominator.

Commonly used unbiased estimators are consistent, but we see that we can have a consistent estimator which is biased.

The third property possessed by the mean, **efficiency,** refers to the precision of the sample statistic as an estimator of the population parameter. For distributions such as the normal, the arithmetic mean is considered to be more stable from sample to sample than are other measures of central tendency. For any size sample the mean will, on the average, come closer to the population mean than any other unbiased estimator, so that the sample mean is a "better" estimate of the population mean. In the terminology of the next section, no other unbiased estimator has a smaller standard error than does the arithmetic mean.

8.2.2 The Standard Error of the Mean

Now that we have discussed various properties of the arithmetic mean, we may return to our soft-drink bottling example. The quality-control director has collected data that include 100 different samples of sizes 5, 15, and 30 bottles. These data are summarized in Tables 8.3 to 8.5.

Figure 8.1 on page 222 illustrates these three sampling distributions (after the data have been condensed into a set of frequency distributions).

We observe from Tables 8.3–8.5 and Figure 8.1 that there is a good deal of fluctuation in the mean from sample to sample. However, there is less variation in this set of sample means than in the actual population. This fact follows directly from the so called **law of large numbers,** which can be explained as follows. The mean of a sample averages together all the values that occur in a particular sample. If an extremely high or low value falls into the sample, it will be averaged with all the other sample values. This averaging process reduces the influence of any

TABLE 8.3 □ Sampling distribution of the mean for 100 samples of n = 5 selected from a normal distribution where μ_X = 2.0 and σ_X = 0.05

Class interval	Frequency	Percent
1.945–under 1.955	2	2.0
1.955–under 1.965	3	3.0
1.965–under 1.975	11	11.0
1.975–under 1.985	8	8.0
1.985–under 1.995	17	17.0
1.995–under 2.005	17	17.0
2.005–under 2.015	18	18.0
2.015–under 2.025	9	9.0
2.025–under 2.035	5	5.0
2.035–under 2.045	5	5.0
2.045–under 2.055	3	3.0
2.055–under 2.065	2	2.0
	100	100.0

TABLE 8.4 □ Sampling distribution of the mean for 100 samples of $n = 15$ selected from a normal distribution where $\mu_X = 2.0$ and $\sigma_X = 0.05$

Class interval	Frequency	Percent
1.965–under 1.975	3	3.0
1.975–under 1.985	13	13.0
1.985–under 1.995	17	17.0
1.995–under 2.005	31	31.0
2.005–under 2.015	25	25.0
2.015–under 2.025	10	10.0
2.025–under 2.035	1	1.0
	100	100.0

TABLE 8.5 □ Sampling distribution of the mean for 100 samples of $n = 30$ selected from a normal distribution where $\mu_X = 2.0$ and $\sigma_X = 0.05$

Class interval	Frequency	Percent
1.965–under 1.975	1	1.0
1.975–under 1.985	3	3.0
1.985–under 1.995	18	18.0
1.995–under 2.005	47	47.0
2.005–under 2.015	23	23.0
2.015–under 2.025	8	8.0
	100	100.0

one particular value. As the sample size increases, the influence of any single value is further reduced, since it is being averaged with a larger number of observations.

The phenomenon we have just described is expressed statistically in the value of the standard deviation of the sample mean. This is the measure of variability in the mean from sample to sample and is referred to in statistics as the **standard error of the mean, $\sigma_{\overline{X}}$**. When sampling *with* replacement, the standard error of the mean is equal to the standard deviation of the population divided by the square root of the sample size. That is,

$$\sigma_{\overline{X}} = \frac{\sigma_X}{\sqrt{n}} \qquad \textbf{(8.1)}$$

Therefore, as the sample size increases, the standard error of the mean will decrease by a factor equal to the square root of the sample size. This relationship will be further examined in Chapter 9, when we address the issue of sample-size determination.

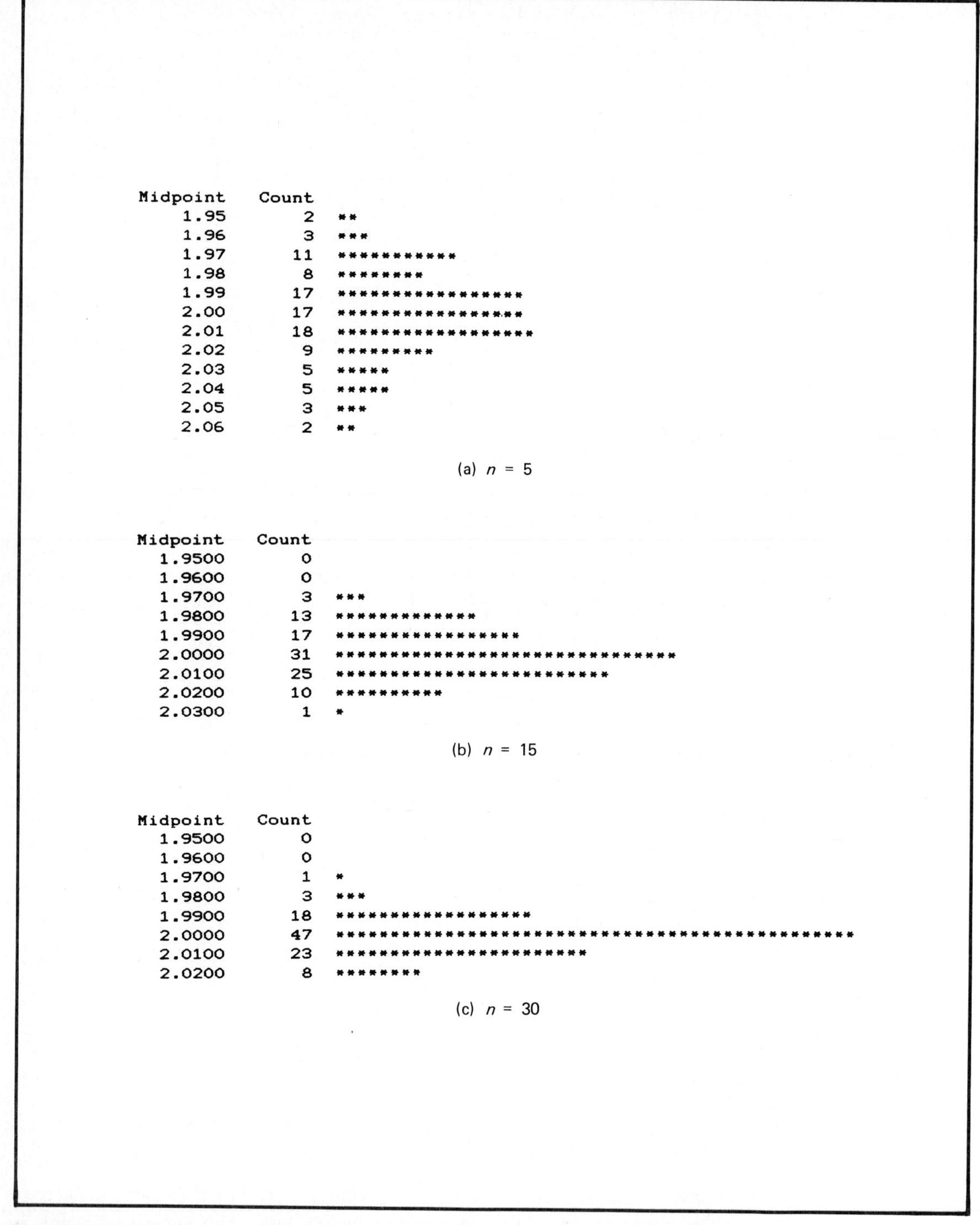

```
Midpoint   Count
    1.95       2  **
    1.96       3  ***
    1.97      11  ***********
    1.98       8  ********
    1.99      17  *****************
    2.00      17  *****************
    2.01      18  ******************
    2.02       9  *********
    2.03       5  *****
    2.04       5  *****
    2.05       3  ***
    2.06       2  **
```

(a) $n = 5$

```
Midpoint   Count
  1.9500       0
  1.9600       0
  1.9700       3  ***
  1.9800      13  *************
  1.9900      17  *****************
  2.0000      31  *******************************
  2.0100      25  *************************
  2.0200      10  **********
  2.0300       1  *
```

(b) $n = 15$

```
Midpoint   Count
  1.9500       0
  1.9600       0
  1.9700       1  *
  1.9800       3  ***
  1.9900      18  ******************
  2.0000      47  ***********************************************
  2.0100      23  ***********************
  2.0200       8  ********
```

(c) $n = 30$

FIGURE 8.1 □ **Sampling distribution of the mean for 100 samples of n = 5, 15, and 30 selected from a normal distribution where $\mu_X = 2.0$ and $\sigma_X = .05$.**

8.2.3 Sampling from Normal Populations

Now that we have introduced the idea of a sampling distribution and mentioned the standard error of the mean, we need to explore the question of what distribution the sample mean $\bar{X}$ will follow. It can be shown that if we sample *with* replacement from a population that is normally distributed with mean μ_X and standard deviation σ_X, the sampling distribution of the mean will also be normally distributed for *any size n* with mean $\mu_{\bar{X}} = \mu_X$ and have a standard error of the mean $\sigma_{\bar{X}} = \sigma_X/\sqrt{n}$.

In the most elementary case, if we draw samples of size $n = 1$, each possible sample mean is a single observation from the population, since

$$\bar{X} = \frac{\Sigma X}{n} = \frac{X}{1} = X$$

If we know that the population is normally distributed with mean μ_X and standard deviation σ_X, then of course the sampling distribution of $\bar{X}$ for samples of $n = 1$ must also follow the normal distribution with mean $\mu_{\bar{X}} = \mu_X$ and standard error of the mean $\sigma_{\bar{X}} = \sigma_X/\sqrt{n} = \sigma_X/\sqrt{1} = \sigma_X$. In addition, we note that as the sample size increases, the sampling distribution of the mean still follows a normal distribution with mean $\mu_{\bar{X}} = \mu_X$. However, as the sample size increases, the standard error of the mean decreases, so that a larger proportion of sample means are closer to the population mean. We may observe this by referring again to Figure 8.1.

Remember that the data displayed in Figure 8.1 represent the distributions of sample means for 100 different samples of size 5, 15, and 30. We can see clearly in Figure 8.1 that while the sampling distribution of the mean is approximately[1] normal for each sample size, the sample means are distributed more tightly around the population mean as the sample size increases.

We may obtain a deeper insight into the concept of the sampling distribution of the mean if we examine the following: Suppose the quality-control director intended to select a random sample of 25 bottles (from a population in which $\mu_X = 2.0$ and $\sigma_X = 0.05$). How could we determine the probability of obtaining a sample mean between 1.99 and 2.0 liters? We know from our study of the normal distribution (Section 7.5) that we can find the area between any value X and the population mean μ_X by converting to standardized Z units:

$$Z = \frac{X - \mu_X}{\sigma_X} \tag{7.8}$$

and finding the appropriate values in the table of the normal distribution (Table C.2). In the examples in Section 7.5 we were studying how any *single* value X

[1] We must remember that "only" 100 samples out of an infinite number of samples have been selected, so that the sampling distributions shown are only approximations of the true distributions.

differs from its mean. Now, in this soft-drink bottling example,[2] the value involved is a single sample mean $\overline{X}$, and we wish to determine the likelihood of obtaining a sample mean between 1.99 and the population mean of 2.0. Thus, by substituting $\overline{X}$ for X, $\mu_{\overline{X}}$ for μ_X, and $\sigma_{\overline{X}}$ for σ_X, we have:

$$Z = \frac{\overline{X} - \mu_{\overline{X}}}{\sigma_{\overline{X}}} = \frac{\overline{X} - \mu_X}{\frac{\sigma_X}{\sqrt{n}}} \tag{8.2}$$

Note that, based on the property of unbiasedness, it is always true that $\mu_{\overline{X}} = \mu_X$.

To find the area between 1.99 and 2.0 (Figure 8.2) we first calculate the Z value corresponding to an observed mean of 1.99:

$$Z = \frac{\overline{X} - \mu_X}{\frac{\sigma_X}{\sqrt{n}}} = \frac{1.99 - 2.0}{\frac{.05}{\sqrt{25}}}$$

$$= \frac{-.01}{.01} = -1.00$$

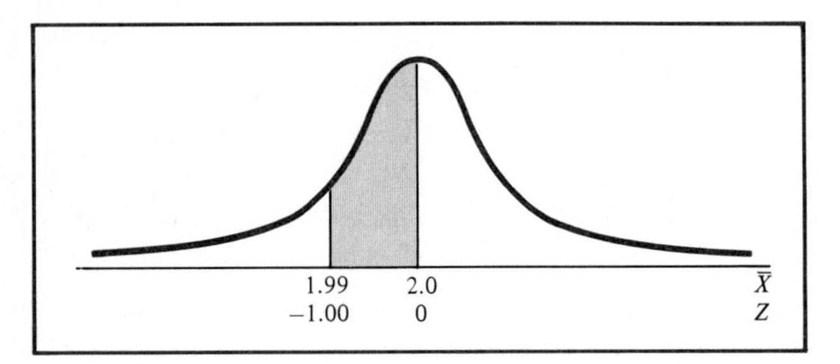

FIGURE 8.2 □ Diagram of normal curve needed to find area between 1.99 and 2.0 liters.

That is, a value of 1.99 is one standard error ($.05/\sqrt{25} = .01$) below the mean of 2.0. Looking up 1.00 in Table C.2, we find an area of .3413. Therefore, 34.13% of all possible samples of 25 bottles will have a *sample mean* between 1.99 and 2.0 liters.

We must realize that this is not the same as saying that a certain percentage of *individual* bottles will contain between 1.99 and 2.0 liters. In fact, that percentage can be computed from Equation (7.8) as follows:

$$Z = \frac{X - \mu_X}{\sigma_X} = \frac{1.99 - 2.0}{.05}$$

$$= \frac{-.01}{.05} = -.20$$

[2] In an ongoing production process the population size is usually considered to be infinite, so that there is no difference between sampling with replacement and sampling without replacement.

From Table C.2 the area under the normal curve from the mean to $Z = -.20$ is .0793. Therefore, 7.93% of the *individual* bottles are expected to contain between 1.99 and 2.0 liters. Comparing these two results, we may observe that many more *sample means* than *individual bottles* will be between 1.99 and 2.0 liters. This result can be explained by the fact that each sample consists of 25 different individual values, some small and some large. The averaging process dilutes the importance of any individual value, particularly when the sample size is large. Thus, the chance that the mean of a sample of 25 will be "close to" the population mean is greater than the chance that a *single individual* value will be.

How would these results be affected by using a different sample size, such as 100 bottles instead of 25? Here we would have the following:

$$Z = \frac{\overline{X} - \mu_{\overline{X}}}{\frac{\sigma_X}{\sqrt{n}}} = \frac{1.99 - 2.0}{\frac{.05}{\sqrt{100}}}$$

$$= \frac{-.01}{.005} = -2.00$$

From Table C.2, the area under the normal curve from the mean to $Z = -2.00$ is .4772. Therefore 47.72% of the samples of size 100 would be expected to have means between 1.99 and 2.0 liters, as compared to only 34.13% for samples of size 25.

Instead of determining the proportion of sample means that fall within a particular interval, we might be more interested in determining the interval within which a fixed proportion of sample means will fall. For example, suppose that the quality-control director wishes to find an interval around the population mean that will include 95% of the sample means based on samples of 25 bottles. The 95% could be divided into two equal parts, half below the mean and half above the mean (see Figure 8.3).

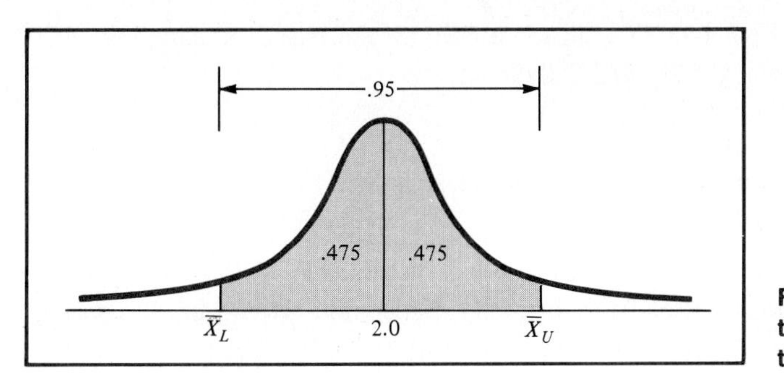

FIGURE 8.3 □ Diagram of normal curve needed to find upper and lower limits to include 95% of the sample means.

Analogous to Section 7.5, we are determining a distance below and above the population mean containing a specific area of the normal curve. From Equation (8.2) we have

$$Z_L = \frac{\overline{X}_L - \mu_{\overline{X}}}{\frac{\sigma_X}{\sqrt{n}}}$$

where $Z_L = -Z$ and

$$Z_U = \frac{\overline{X}_U - \mu_{\overline{X}}}{\frac{\sigma_X}{\sqrt{n}}}$$

where $Z_U = +Z$. Therefore the lower value of $\overline{X}$ is

$$\overline{X}_L = \mu_{\overline{X}} - Z\frac{\sigma_X}{\sqrt{n}} \quad \textbf{(8.3a)}$$

while the upper value for $\overline{X}$ is

$$\overline{X}_U = \mu_{\overline{X}} + Z\frac{\sigma_X}{\sqrt{n}} \quad \textbf{(8.3b)}$$

Since $\sigma_X = 0.05$ and $n = 25$ and the value of Z corresponding to an area of .475 under the normal curve is 1.96, the lower and upper values of $\overline{X}$ can be found as follows:

$$\overline{X}_L = 2.0 - (1.96)\left(\frac{.05}{\sqrt{25}}\right) = 2.0 - .0196 = 1.9804$$

$$\overline{X}_U = 2.0 + (1.96)\left(\frac{.05}{\sqrt{25}}\right) = 2.0 + .0196 = 2.0196$$

Our conclusion would be that 95% of all sample means based on samples of 25 bottles should fall between 1.9804 and 2.0196 liters.

8.2.4 Sampling from Nonnormal Populations

In the previous section we explored the sampling distribution of the mean for the case in which the population itself was normally distributed. However, we should realize that in many instances either we will know that the population is not normally distributed or we may believe that it is unrealistic to assume a normal distribution. Thus, we need to examine the sampling distribution of the mean for populations that are not normally distributed. This issue brings us to an important theorem in statistics, the **central limit theorem.**

> **Central Limit Theorem: As the sample size (number of observations in each sample) gets "large enough," the sampling distribution of the mean can be approximated by the normal distribution. This is true regardless of the shape of the distribution of the individual values in the population.**

What sample size is "large enough"? A great deal of statistical research has gone into this issue. As a general rule, statisticians have found that for most population distributions, once the sample size is at least 30, the sampling distribution of the mean will be approximately normal. However, we may be able to apply the central limit theorem for even smaller sample sizes if some knowledge of the population is available (for example, if the distribution is symmetric).

Recall that the quality-control director wondered what the effect would be if the distribution of bottle fill were uniformly rather than normally distributed. We can empirically evaluate this question by using a computer package such as Minitab (see Chapter 15) to select samples and generate sampling distributions from certain probability distributions.

Figure 8.4 on page 228 represents the sampling distribution that has been obtained by selecting 100 samples from a population that was uniformly distributed between 1.9 and 2.1 liters. These samples were selected for sizes of $n = 1, 2, 5, 15$, and 30 respectively. As depicted in part (a), for samples of size $n = 1$, each value in the population is approximately equally likely. However, when samples of only 2 are selected [part (b)], we already can notice that there are more sample means in the center of the distribution than in the tails. In parts (c), (d), and (e) (corresponding to sample sizes 5, 15, and 30) we observe that as the sample size increases, the sampling distribution of the mean rapidly approaches a normal distribution. In fact, once there are samples of at least 15 observations, the sampling distribution of the mean approximately follows a normal distribution.

We summarize our empirical findings about the sampling distribution of the mean with the following conclusions:

1. For most population distributions, regardless of their shape, the sampling distribution of the mean will be approximately normally distributed if samples of at least 30 observations are selected.
2. If the population distribution is symmetric, the sampling distribution of the mean will usually be approximately normal if samples of at least 15 observations are selected.
3. If the population is normally distributed, the sampling distribution of the mean will be normally distributed regardless of the sample size.

Midpoint	Count	
1.9000	1	*
1.9100	4	****
1.9200	6	******
1.9300	5	*****
1.9400	9	*********
1.9500	10	**********
1.9600	2	**
1.9700	3	***
1.9800	5	*****
1.9900	4	****
2.0000	3	***
2.0100	4	****
2.0200	5	*****
2.0300	6	******
2.0400	6	******
2.0500	7	*******
2.0600	1	*
2.0700	6	******
2.0800	4	****
2.0900	7	*******
2.1000	2	**

(a) $n = 1$

Midpoint	Count	
1.9100	2	**
1.9200	2	**
1.9300	4	****
1.9400	4	****
1.9500	3	***
1.9600	6	******
1.9700	1	*
1.9800	8	********
1.9900	8	********
2.0000	15	***************
2.0100	15	***************
2.0200	7	*******
2.0300	2	**
2.0400	6	******
2.0500	4	****
2.0600	4	****
2.0700	4	****
2.0800	4	****
2.0900	1	*

(b) $n = 2$

Midpoint	Count	
1.94	1	*
1.95	1	*
1.96	5	*****
1.97	9	*********
1.98	13	*************
1.99	21	*********************
2.00	14	**************
2.01	10	**********
2.02	14	**************
2.03	2	**
2.04	5	*****
2.05	5	*****

(c) $n = 5$

FIGURE 8.4 □ Sampling distribution of the mean for 100 samples of n = 1, 2, 5, 15, and 30 selected from a uniform population (between 1.9 and 2.1).

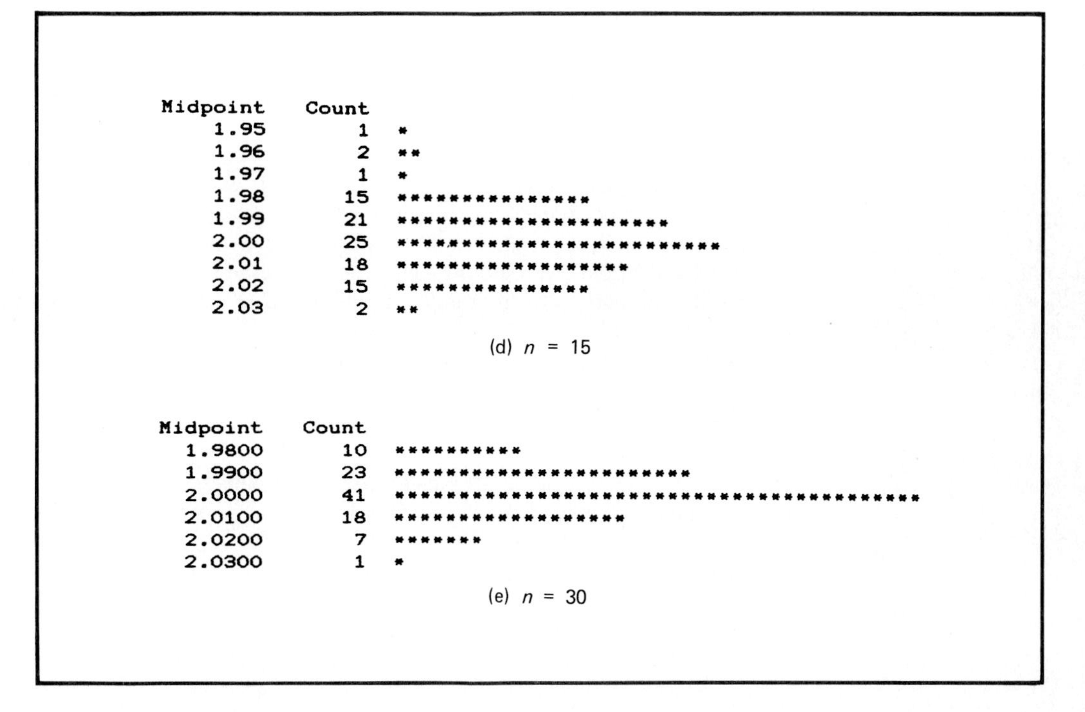

FIGURE 8.4 (continued)

Problems

8.1 Explain why a statistician would be interested in drawing conclusions about a population rather than merely describing the results of a sample.

8.2 Distinguish between a probability distribution and a sampling distribution.

8.3 The following data represent the number of days absent per year in a population of six employees of a small company:

1, 3, 6, 7, 7, 12

(a) Assuming that you sample *without* replacement:
  (1) select all possible samples of size 2 and set up the sampling distribution of the mean.
  (2) compute the mean of all the sample means and also compute the population mean. Are they equal? What property is this called?
  (3) do parts (1) and (2) for all possible samples of size 3.
  (4) compare the shape of the sampling distribution of the mean obtained in parts (1) and (3). Which sampling distribution seems to have the least variability? Why?

(b) Assuming that you sample *with* replacement, do parts (1)–(4) of part (a) and compare the results. Which sampling distributions seem to have the least variability, those in part (a) or part (b)? Why?

• 8.4 Paper bags used for packaging groceries are manufactured so that the breaking strength of the bag is normally distributed with a mean of 5 pounds per square inch and a standard deviation of 1 pound per square inch.

(a) What proportion of the bags produced have a breaking strength between 5 and 5.5 pounds per square inch?
(b) What proportion of the bags produced have a breaking strength between 4 and 4.1 pounds per square inch?
(c) If many random samples of 16 bags are selected:
(1) what would the mean and the standard error of the mean be expected to equal?
(2) what distribution would the sample means follow?
(3) what proportion of the sample means would be between 5 and 5.5 pounds per square inch?
(4) what proportion of the sample means would be between 4 and 4.1 pounds per square inch?
(5) compare the answers of (a) to (c)(3) and of (b) to (c)(4). Discuss.

8.5 Long-distance telephone calls are normally distributed with $\mu_X = 8$ minutes and $\sigma_X = 2$ minutes. If random samples of 25 calls were selected:

(a) Compute $\sigma_{\overline{X}}$.
(b) What proportion of the sample means would be between 7.8 and 8.2 minutes?
(c) What proportion of the sample means would be between 7.5 and 8 minutes?
(d) If random samples of 100 calls were selected:
(1) what proportion of the sample means would be between 7.8 and 8.2 minutes?
(2) explain the difference in the results of (b) and (d)(1).

8.6 The amount of time a bank teller spends with each customer has a population mean $\mu_X = 3.10$ minutes and standard deviation $\sigma_X = .40$ minutes. If a random sample of 16 customers is selected:

(a) What is the probability that the average time spent per customer will be at least 3 minutes?
(b) There is an 85% chance that the sample mean will be below how many minutes?
(c) What assumption must be made in order to solve (a) and (b)?
(d) If a random sample of 64 customers is selected:
(1) there is an 85% chance that the sample mean will be below how many minutes?
(2) what assumption must be made in order to solve (d)(1)?

8.7 In a certain country, the height of adult males is normally distributed with a mean of 69 inches and a variance of 36. If samples of 4 males are measured:

(a) What proportion of such samples will have average heights over 72 inches?
(b) What are the upper and lower limits between which 90% of the sample averages will lie?
(c) Repeat parts (a) and (b) for samples of size 9 and for samples of size 36. Comment on the effect of using larger samples.

• 8.8 IQ scores from the Risenshine Intelligence Test are normally distributed with a mean of 100 and a variance of 225. For samples of 16 people:

(a) What proportion of such samples will have mean IQ scores between 90 and 110?
(b) What proportion of such samples will have a mean IQ score greater than 110?
(c) What are the upper and lower limits between which 95% of the sample averages will lie?

(d) What are the upper and lower limits between which 95% of the IQs of *individuals* lie?

8.3 THE SAMPLING DISTRIBUTION OF THE PROPORTION

In our discussion of sampling distributions thus far we have focused on the distribution of the mean of a quantitative variable. However, as we observed in Chapter 2, many variables of interest are qualitative or categorical. In Chapter 6 we defined the probability of success as

$$\text{probability of success} = \frac{\text{number of favorable outcomes}}{\text{total number of outcomes}}$$

while in Chapter 7 we observed that the random variable "number of successes" followed a binomial distribution where the average number of successes μ_X was equal to np and the standard deviation of the number of successes σ_X was equal to $\sqrt{np(1-p)}$. Now, instead of expressing the variable in terms of the number of successes X, we can convert to the proportion of successes by dividing by n, the sample size. Thus we have

$$p_S = \frac{X}{n} = \frac{\text{number of successes}}{\text{sample size}} \qquad \textbf{(8.4)}$$

The average (or expected) proportion of successes μ_{p_S} is equal to p, while the standard deviation of the proportion of successes σ_{p_S} is equal to $\sqrt{p(1-p)/n}$. In Section 7.6.1 we said that when the sample size was large, the binomial distribution could be approximated by the normal distribution. The general rule of thumb was that if np and $n(1 - p)$ each were at least 5, the normal distribution provided a good approximation to the binomial distribution. In most cases in which inferences are being made about the proportion, the sample size is substantial enough to meet the conditions for using the normal approximation (see Reference 1). Thus, in many instances, we may use the normal distribution to evaluate the sampling distribution of the proportion.

We may develop this idea by referring to a hypothetical situation that involves a political pollster who is applying sample results in order to make predictions on election night. Assuming a two-candidate election, the pollster has decided that if, in a sample of 100 voters, a specific candidate receives at least 55% (i.e., $p_S \geq .55$) of the vote, then that candidate will be projected as the winner of the election. We would like to determine the probability that a candidate will be projected as the winner if, in fact, his or her true percentage in the population is 49% (i.e., he or she will really lose). Since the sampling distribution of the proportion can be assumed to be approximately normally distributed, we have the following:

$$Z = \frac{\bar{X} - \mu_{\bar{X}}}{\sigma_{\bar{X}}} \qquad \textbf{(8.2)}$$

and, because we are dealing with a sample proportion (not a sample mean),

$$p_S = \text{sample proportion}$$
$$p = \text{population proportion}$$
$$\sigma_{p_s} = \sqrt{\frac{p(1-p)}{n}}$$

and, substituting p_S for $\bar{X}$, $\mu_{p_s} = p$ for $\mu_{\bar{X}}$, and $\sigma_{p_s} = \sqrt{p(1-p)/n}$ for $\sigma_{\bar{X}}$, we have

$$Z = \frac{p_S - p}{\sqrt{\frac{p(1-p)}{n}}} \qquad \textbf{(8.5)}$$

Therefore,

$$Z = \frac{p_S - p}{\sqrt{\frac{p(1-p)}{n}}} = \frac{.55 - .49}{\sqrt{\frac{.49(.51)}{100}}} = \frac{.06}{\sqrt{\frac{.2499}{100}}}$$
$$= \frac{.06}{.05} = +1.20$$

Using Table C.2, the area under the normal curve from $Z = 0$ to $Z = +1.20$ is .3849. Since we need to find the probability of obtaining a sample proportion *above* .55 (i.e., 55%), we need to subtract .3849 from .50 to obtain the desired value of .1151 (see Figure 8.5). Thus, if the true population proportion is .49, there is an

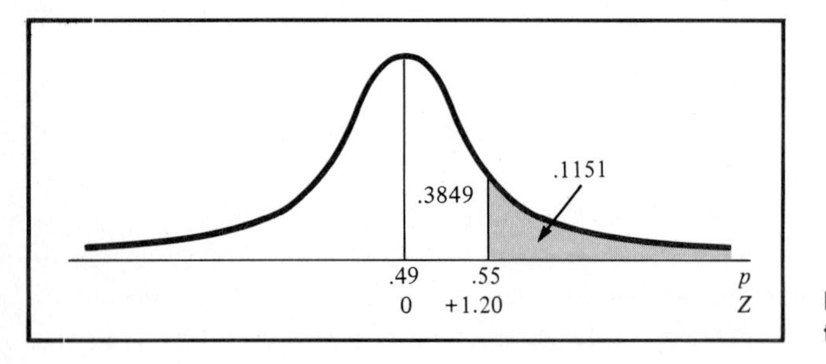

FIGURE 8.5 □ Diagram of normal curve needed to find the area above a proportion of .55.

11.51% chance of obtaining a sample proportion of .55 or above (and thereby incorrectly projecting the winner) if a sample of 100 voters is selected.

EXAMPLE 8.1

Suppose that the true population proportion were .60. What then would be the probability of correctly projecting the winner?

Solution

In this example $p_S = .55$ and $p = .60$. Using Equation (8.5),

$$Z = \frac{.55 - .60}{\sqrt{\frac{.60(.40)}{100}}} = \frac{-.05}{\sqrt{\frac{.24}{100}}}$$

$$= \frac{-.05}{.049} = -1.02$$

Using Table C.2, we find the area under the normal curve from $Z = 0$ to $Z = -1.02$ is .3461. Since we need to find the probability of obtaining a sample proportion *above* .55 (i.e. 55%), we need to add .3461 to .50 to obtain the desired value of .8461 (see figure). Thus if the true population proportion is .60, there is an 84.61% chance of obtaining a sample proportion of .55 or above and correctly projecting the winner if a sample of 100 voters is selected.

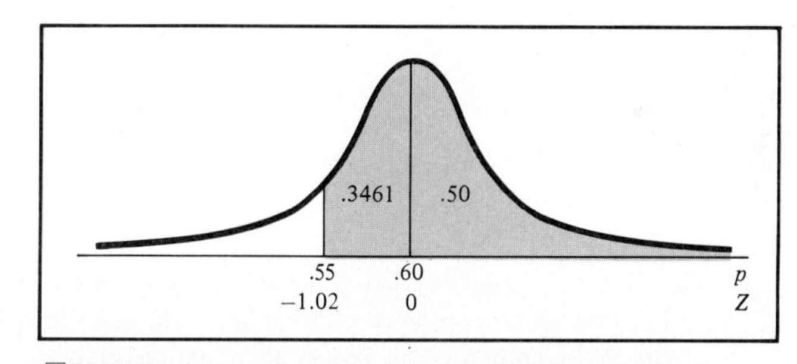

☐Diagram of normal curve needed to find the area above a proportion of .55.

Problems

8.9 Historically, 10% of a large shipment of machine parts are defective. If random samples of 400 parts are selected, what proportion of the samples will have:
(a) Between 9% and 10% defective parts?
(b) Less than 8% defective parts?

(c) If a sample size of only 100 had been selected, what would your answers have been in (a) and (b)?

8.10 Based on past data, 30% of the credit card purchases at a large department store are for amounts above $100. If random samples of 100 credit card purchases are selected:

(a) What proportion of samples are likely to have between 20% and 30% of the purchases over $100?

(b) Within what symmetrical limits of the population percentage will 95% of the sample percentages fall?

8.11 Suppose that a marketing experiment is to be conducted in which students are to taste two different brands of soft drink. Their task is to correctly identify the brand tasted. If random samples of 200 students are selected and it is assumed that the students have no ability to distinguish between the two brands:

(a) What proportion of the samples will have between 50% and 60% of the identifications correct?

(b) Within what symmetrical limits of the population percentage will 90% of the sample percentages fall?

(c) What is the probability of obtaining a sample percentage of correct identifications in excess of 65%?

[*Hint*: If an individual has no ability to distinguish between the two soft drinks, then each one is equally likely to be selected.]

• 8.12 Historically, 93% of the deliveries of an overnight mail service arrive before 10:30 on the following morning. If random samples of 500 deliveries are selected, what proportion of the samples will have:

(a) Between 93% and 95% of the deliveries arriving before 10:30 on the following morning?

(b) More than 95% of the deliveries arriving before 10:30 on the following morning?

(c) If a sample size of 1,000 were selected, what would your answers be in (a) and (b)?

8.13 The seeds of a certain variety of plant germinate 90% of the time. For samples of 64 seeds:

(a) In what proportion of such samples will more than 80% of the seeds germinate?

(b) What proportion of the samples will have between 85% and 95% of their seeds germinate?

(c) According to the rules in the book, should the normal distribution be nearly correct for samples such as this?

8.14 Suppose that, if nothing unusual happens to change the normal birth rate, 50% of births should be male. (In fact, the conception and birth rates of males are slightly higher than this, but more males die before and shortly after birth.)

(a) What proportion of samples of size 50 should have 35 or more (i.e., 70%) males?

(b) Compute the upper and lower bounds between which 90% of the proportions of males should lie for samples of 50 newborns.

(c) Recompute the results in parts (a) and (b) if 52% of the births are male.

8.4 SAMPLING FROM FINITE POPULATIONS

The central limit theorem and the standard error of the mean were based upon the premise that the samples selected were chosen *with* replacement. However, in virtually all survey research, sampling is conducted *without* replacement from populations that are of a finite size N. In these cases, particularly when the sample size n is not small as compared to the population size N ($n/N > .05$; i.e., more than 5% of the population is sampled), a **finite population correction factor (fpc)** should be used in defining the standard error of the mean. The finite population correction factor may be expressed as

$$\text{fpc} = \sqrt{\frac{N-n}{N-1}} \qquad \textbf{(8.6)}$$

The standard error of the mean becomes

$$\sigma_{\overline{X}} = \frac{\sigma_X}{\sqrt{n}}\sqrt{\frac{N-n}{N-1}} \qquad \textbf{(8.7)}$$

where n = sample size
N = population size

If we are referring to proportions, we have

$$\sigma_{p_s} = \sqrt{\frac{p(1-p)}{n}}\sqrt{\frac{N-n}{N-1}} \qquad \textbf{(8.8)}$$

Examining the formula for the finite population correction factor, we see that the numerator will always be smaller than the denominator, so the correction factor will be less than 1. Since this is multiplied by the standard error, the standard error becomes smaller when corrected. That is, we get more accurate estimates because we are sampling a large segment of the population.

We will illustrate the application of the finite population correction factor by referring back to the two problems discussed in this chapter. In Section 8.2.2, a sample of 25 bottles was selected from a production process. Suppose that the total number of two-liter bottles (i.e., our population) produced on this particular shift was 1,000. Using the finite population correction factor, we would have

$$\sigma_X = .05, \quad n = 25, \quad N = 1{,}000$$

$$\sigma_{\overline{X}} = \frac{\sigma_X}{\sqrt{n}} \sqrt{\frac{N-n}{N-1}}$$

$$= \frac{.05}{\sqrt{25}} \sqrt{\frac{1{,}000-25}{1{,}000-1}}$$

$$= (.01)\sqrt{.976} = .00988$$

The probability of obtaining a sample whose mean is between 1.99 and 2.00 liters is computed as follows:

$$Z = \frac{\overline{X} - \mu_{\overline{X}}}{\sigma_{\overline{X}}} = \frac{-.01}{.00988} = -1.01$$

From Table C.2 the approximate area under the normal curve is .3438.

It is evident in this example that the use of the finite population correction factor had a very small effect on the standard error of the mean and the subsequent area under the normal curve, since the sample was only 2.5% of the population size.

In the example concerning the political pollster, suppose that she was forecasting an election in a small town in which there were only 500 voters. Using the finite population correction factor, we have $p = .49$, $n = 100$, $N = 500$.

$$\sigma_{p_s} = \sqrt{\frac{p(1-p)}{n}} \sqrt{\frac{N-n}{N-1}}$$

$$= \sqrt{\frac{.49(.51)}{100}} \sqrt{\frac{500-100}{500-1}}$$

$$= \sqrt{\frac{.2499}{100}} \sqrt{\frac{400}{499}}$$

$$= .04476$$

The probability of obtaining a sample whose proportion is at least .55 is computed as follows:

$$Z = \frac{p_S - p}{\sigma_{p_s}} = \frac{.06}{.04476} = +1.34$$

From Table C.2, the area under the normal curve from $Z = 0$ to $Z = +1.34$ is .4099. Thus if the true proportion is .49, there is a 9.01% chance of obtaining a sample proportion which equals or exceeds .55 if a sample of 100 voters is selected.

EXAMPLE 8.2

Suppose that the population size for the election in the small town was only 200. (See pages 231–233.)

Using the finite population correction factor, we have

$$p = .49, \quad n = 100, \quad N = 200$$

$$\sigma_{p_S} = \sqrt{\frac{p(1-p)}{n}}\sqrt{\frac{N-n}{N-1}}$$

$$= \sqrt{\frac{.49(.51)}{100}}\sqrt{\frac{200-100}{200-1}}$$

$$= .0354$$

The probability of obtaining a sample whose proportion is at least .55 is computed as follows:

$$Z = \frac{p_S - p}{\sigma_{p_S}} = \frac{.06}{.0354} = +1.69$$

From Table C.2, the area under the normal curve from $Z = 0$ to $Z = +1.69$ is .4545. Thus, if the true population proportion is .49, there is a 4.55% chance of obtaining a sample proportion which equals or exceeds .55 if a sample of 100 voters is selected from a population of 200 voters.

Note that as we changed from an infinite population, to a finite population with 500 people, to one with only 200 people, the probability of making an error decreased. We became more certain of our conclusions, because we had sampled a larger proportion of the population; in Example 8.2 the sample of 100 was half of the population.

Problems

• 8.15 Referring to Problem 8.4 on page 230, if the population had consisted of 200 paper bags, what would be your answer to part (c) of that problem?

8.16 Referring to Problem 8.6 on page 230, if there was a population of 500 customers, what would be your answers to parts (a) and (b) of that problem?

8.17 Referring to Problem 8.9 on page 233, if the shipment included 5,000 machine parts, what would be your answers to parts (a) and (b) of that problem?

8.18 Referring to Problem 8.12 on page 234, if the population consisted of 10,000 deliveries, what would be your answers to parts (a) and (b) of that problem?

8.19 Suppose that the country described in Problem 8.7 on page 230 was a very small one, with only 400 adult males in the whole country. Redo the problem, and comment on how the estimates change from those in Problem 8.7.

8.5 SUMMARY AND OVERVIEW

In this chapter we have introduced the concept of the sampling distribution of the mean and the proportion. The importance of the normal distribution in statistics has been reinforced by the discussion of the central limit theorem. We have observed that knowledge of the population distribution is not always necessary in drawing conclusions based on the sample mean or proportion. These ideas are central to the development of statistical inference. In practice, a single sample is taken from a population. Although only a sample statistic is computed, knowledge of the sampling distribution will enable us to draw conclusions and make decisions about various population values. The statistical procedures relating to statistical inference (confidence intervals and tests of hypothesis) will be developed in the next three chapters.

Supplementary Problems

• **8.20** A cereal packaging machine is regulated so that the amount dispensed is normally distributed with $\mu_X = 13$ ounces and $\sigma_X = .5$ ounce. If samples of nine cereal boxes are taken, what value will be exceeded by 95% of the sample means?

8.21 The number of hours of life of a type of transistor battery is normally distributed with $\mu_X = 100$ hours and $\sigma_X = 20$ hours.

(a) What proportion of the batteries will last between 100 and 115 hours?

(b) If random samples of 16 batteries are selected, what proportion of the sample means will be:

(1) between 100 and 115 hours?

(2) more than 90 hours?

(3) Within what limits around the population mean will 90% of sample means fall?

(4) Is the central limit theorem necessary to answer (1), (2), and (3)? Explain.

8.22 The number of customers per week at each store of a supermarket chain has a population mean of $\mu_X = 5{,}000$ and standard deviation $\sigma_X = 500$. If a random sample of 25 stores is selected:

(a) What is the probability that the sample mean will be below 5,075 customers per week?

(b) Within what limits around the population mean can we be 95% certain that the sample mean will fall?

8.23 **(Class Project)** The table of random numbers is an example of a uniform distribution, since each digit is equally likely to occur. Starting in the row corresponding to the day of the month in which you were born, use the table of random numbers (Table C.1) to take *one digit* at a time. Select samples of size $n = 2$, $n = 5$, $n = 10$. Compute the sample mean $\bar{X}$ of each sample. For each sample size, each student should select five different samples so that a frequency distribution of the sample means can be developed for the results of the entire class. What can be said about the shape of the sampling distribution for each of these sample sizes?

8.24 **(Class Project)** A coin having one side "heads" and the other side "tails" is to be flipped 10 times and the number of heads obtained is to be recorded. If each student performs this experiment five times, a frequency distribution of the number of "heads" can be developed from the results of the entire class. Does this distribution seem to approximate the normal distribution?

8.25 **(Class Project)** The number of cars waiting in line at a car wash is distributed as follows:

Length of waiting line (number of cars)	Probability
0	.25
1	.40
2	.20
3	.10
4	.04
5	.01

The table of random numbers can be used to select samples from this distribution by assigning numbers as follows:

1. Start in the row corresponding to the day of the month in which you were born.
2. Select *two-digit* random numbers.
3. If a random number between 00 and 24 is selected, record a length of 0; if between 25 and 64, record a length of 1; if between 65 and 84, record a length of 2; if between 85 and 94, record a length of 3; if between 95 and 98, record a length of 4; if it is 99, record a length of 5.

Select samples of size $n = 2$, $n = 10$, $n = 25$. Compute the sample mean for each sample. For example, if a sample size 2 results in random numbers 18 and 46, these would correspond to lengths of 0 and 1, respectively, producing a sample mean of .5. If each student selects five different samples for each sample size, a frequency distribution of the sample means (for each sample size) can be developed from the results of the entire class. What conclusions can you draw about the sampling distribution of the mean as the sample size is increased?

References

1. Cochran, W. G., *Sampling Techniques*, 3d ed. (New York: Wiley, 1977).
2. Larsen, R. L., and M. L. Marx, *An Introduction to Mathematical Statistics and Its Applications* (Englewood Cliffs, N.J.: Prentice-Hall, 1981).

Chapter 9

Estimation

9.1 INTRODUCTION: WHAT'S AHEAD

Statistical inference is the process of using sample results to draw conclusions about the characteristics of a population. In this chapter we shall examine statistical procedures that will enable us to *estimate* the true population mean and the true population proportion.

There are two major types of estimates: **point estimates** and **interval estimates.** A point estimate consists of a single sample statistic that is used to estimate the true population parameter. For example, the sample mean $\overline{X}$ is a point estimate of the population mean μ_X, while the sample variance S^2 is a point estimate of the population variance σ_X^2. Recall from Section 8.2.1 that the sample mean $\overline{X}$ possessed the highly desirable properties of unbiasedness and efficiency. Although in practice only one sample is selected, we know that the average value of all possible sample means is the true population parameter (μ_X). Since we realize that the sample statistic ($\bar{X}$) varies from sample to sample (i.e., it depends on the elements selected in the sample), we need to provide for a more informative estimate of the true population characteristic. To accomplish this we shall develop an interval estimate of the true population mean by taking into account the sampling distribution of the mean. The interval that we construct will have a specified *confidence* or probability of correctly estimating the true value of the population parameter. Similar interval estimates will also be developed for the true population proportion p. In developing our discussion we will make use of an illustrative example.

Case F—The Personnel Director's Employee Benefits Study

The personnel director for a large corporation that employs 5,000 workers wanted to study the feasibility of providing an optional group dental insurance plan for employees. A survey was authorized in which the two most important questions were:

1. What were your actual family dental expenses in the last calendar year?
2. Would you purchase at the cost of $100/year an optional dental plan that covered 80% of your family's *actual* dental expenses in that year?

The personnel director realized that, owing to the costs involved, it would not be feasible to survey every employee of the company. However, a sample of 100 employees was selected and surveyed, and the results indicated a (sample) average family dental expense of $267.43 with a (sample) standard deviation of $191.74. Sixty-four of the 100 employees stated that they would purchase the optional group dental expense plan. The personnel director would like to use the sample results to draw conclusions about these characteristics in the entire population of employees. He wonders also whether a more "scientific" method of determining the proper sample size might be available to assist in the decision-making process. (He may also be concerned with possible "bias" in these estimates: Do employees tend to over- or underestimate their expenses or likelihood of joining? These questions are difficult to address statistically.)

9.2 CONFIDENCE-INTERVAL ESTIMATE OF THE MEAN (σ_X KNOWN)

In Section 8.2 we observed that from either the central limit theorem or knowledge of the population distribution we could determine the percentage of sample means that fell within certain distances of the population mean. For instance, in Section 8.2.3, in the example involving the quality control of bottle filling (in which $\mu_X = 2.0$, $\sigma_X = .05$, and $n = 25$) we observed that 95% of all sample means would fall between 1.9804 and 2.0196 liters.

The type of reasoning in this statement (*deductive reasoning*) is exactly opposite to the type of reasoning that is needed here (*inductive reasoning*). In statistical inference we must take the results of a single sample and draw conclusions about the population, not vice versa. In practice, the population mean is the unknown quantity that is to be estimated. Suppose, for example, in the quality-control director's situation, that the true population mean μ_X was unknown but the true population standard deviation σ_X was known to be .05 liter. Thus, rather than taking $\mu_X \pm (1.96)\,\sigma_X/\sqrt{n}$ to find the upper and lower limits around μ_X as in Section 8.2, let us determine the consequences of substituting the sample mean $\overline{X}$ for the unknown mean μ_X and using $\overline{X} \pm (1.96)\,\sigma_X/\sqrt{n}$ as an interval within which we estimate the unknown μ_X. Although in practice a single sample mean $\overline{X}$ is observed, we need to develop a *hypothetical* set of examples in order to understand the full meaning of the interval estimate that shall be obtained. In the first case, suppose that the sample mean was 2.01 liters. The interval developed to estimate μ_X would be $2.01 \pm (1.96)(.05)/\sqrt{25}$ or $2.01 \pm .0196$. That is, the estimate of μ_X would be

$$1.9904 \leq \mu_X \leq 2.0296$$

Since the true population mean μ_X (equal to 2.0) is included *within* the interval, we observe that this sample has led to a correct statement about μ_X (see Figure 9.1).

To continue our hypothetical example, suppose that for a different sample the

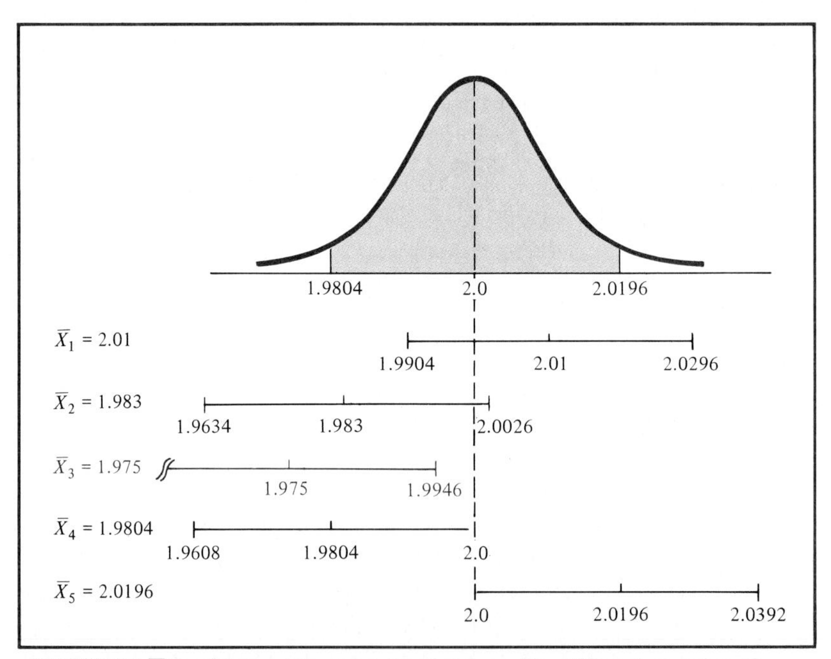

FIGURE 9.1 □ Confidence-interval estimates from five different samples of size n = 25 taken from a population where μ_X = 2.0 and σ_X = 0.05.

mean was 1.983. The interval developed from this sample would be 1.983 ± $(1.96)(.05)/\sqrt{25}$ or 1.983 ± .0196. That is, the estimate of μ_X would be

$$1.9634 \leq \mu_X \leq 2.0026$$

Since the true population mean μ_X (=2.0) is also included within this interval, we conclude that this statement about μ_X is correct.

Now, before we begin to think that we will *always* make correct statements about μ_X from the sample $\bar{X}$, let us examine a third hypothetical example, in which the sample mean is equal to 1.975 liters. The interval developed here would be 1.975 ± $(1.96)(.05)/\sqrt{25}$ or 1.975 ± .0196. In this case, the estimate of μ_X is

$$1.9554 \leq \mu_X \leq 1.9946$$

Observe that this estimate is *not* a correct statement, since the population mean μ_X is not included in the interval developed from this sample. Thus we are faced with a dilemma. For some samples the interval estimate of μ_X will be correct, while for others it will be incorrect. In addition, we must also realize that in practice we select only *one* sample, and since we do not know the true population mean, we cannot determine whether our particular statement is correct. What we can do in order to resolve this dilemma is to determine the proportion of samples producing intervals that result in correct statements about the population mean μ_X. In order to do this we need to examine two other hypothetical samples: the case in which $\bar{X}$ = 1.9804, and the case in which $\bar{X}$ = 2.0196. If $\bar{X}$ = 1.9804, the interval will be 1.9804 ± $(1.96)(.05)/\sqrt{25}$ or 1.9804 ± .0196. That is,

$$1.9608 \leq \mu_X \leq 2.00$$

Since the population mean of 2.0 is at the upper limit of the interval, the statement is a correct one.

Finally, if $\bar{X}$ = 2.0196, the interval will be 2.0196 ± $(1.96)(.05)/\sqrt{25}$ or 2.0196 ± .0196. That is,

$$2.00 \leq \mu_X \leq 2.0392$$

In this case, since the population mean of 2.0 is included at the lower limit of the interval, the statement is a correct one.

Thus, from these examples (see Figure 9.1) we can determine that if the sample mean falls anywhere between 1.9804 and 2.0196 liters, the population mean will be included *somewhere* within the interval. However, we know from our discussion of the sampling distribution in Section 8.2.3 that 95% of the sample means fall between 1.9804 and 2.0196 liters. Therefore, 95% of all sample means will include the population mean within the interval developed. The interval from 1.9804 to 2.0196 is referred to as a **95% confidence interval.**

In general a 95% confidence interval estimate can be interpreted to mean that if all possible samples were taken, 95% of them would include the true population mean somewhere within their interval, while only 5% of them would not.

Since only one sample is selected in practice and μ_X is unknown, we never know for sure whether the specific interval obtained includes the population mean. However, we can state that we have 95% confidence that we have selected a sample whose interval does include the population mean.

In our examples we had 95% confidence of including the population mean within the interval. In some situations we might desire a higher degree of assurance (such as 99%) of including the population mean within the interval. In other cases we might be willing to accept less assurance (such as 90%) of correctly estimating the true population mean.

In general, the level of confidence is symbolized by $(1 - \alpha) \times 100\%$, where α is the proportion in the tails of the distribution which are outside of the confidence interval [see Figure 9.2; the shaded area contains $(\alpha \times 100)\%$ of the distribution]. Therefore to obtain the $(1 - \alpha) \times 100\%$ confidence interval estimate of the mean with σ_X known, we have

$$\bar{X} \pm Z_{\alpha/2}\sigma_{\bar{X}} = \bar{X} \pm Z_{\alpha/2}\frac{\sigma_X}{\sqrt{n}} \qquad \textbf{(9.1a)}$$

or

$$\bar{X} - Z_{\alpha/2}\frac{\sigma_X}{\sqrt{n}} \leq \mu_X \leq \bar{X} + Z_{\alpha/2}\frac{\sigma_X}{\sqrt{n}} \qquad \textbf{(9.1b)}$$

where $Z_{\alpha/2}$ is the value of the standard normal variable for which only $(\alpha/2 \times 100)\%$ of the distribution's values are greater than this value.

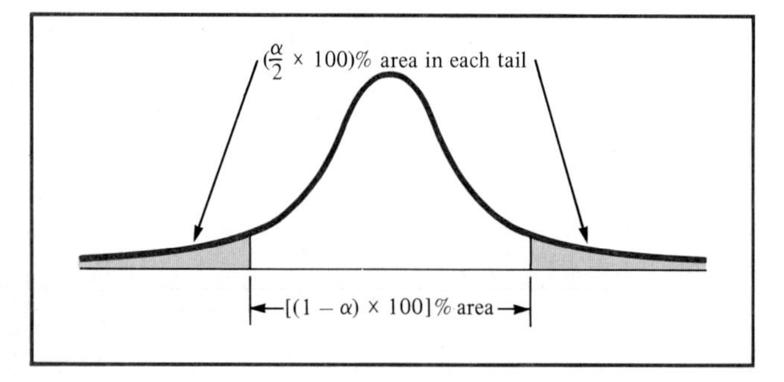

FIGURE 9.2 □ Relationship of α to the confidence interval.

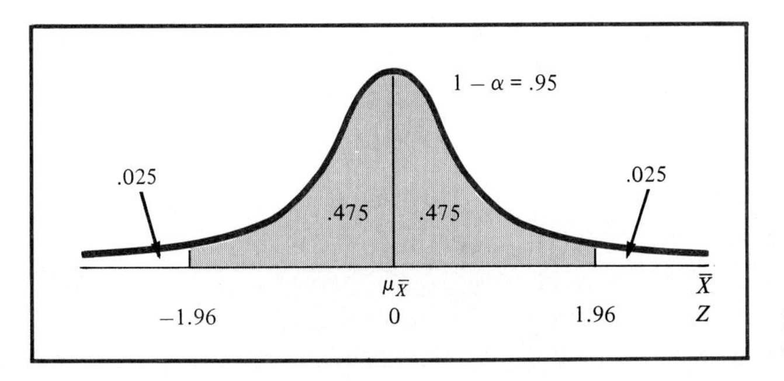

FIGURE 9.3 Normal curve for determining the Z values needed for 95% confidence.

For example, $.025 \times 100 = 2.5\%$ of the values of the standard normal distribution are greater than 1.96, which is why 1.96 is used for constructing a $(1 - .05) \times 100 = 95\%$ confidence interval. The value of $Z_{\alpha/2}$ selected for constructing such a confidence interval is called the **critical value** from the distribution. There is a different critical value for each level of α. However, since we usually use only one value of α in a given situation, we may omit the subscript $\alpha/2$ and simply write it as Z.

A level of confidence of 95% led to a Z value of ± 1.96 (see Figure 9.3). If 99% confidence were desired, the area of .99 would be divided in half, leaving .495 between each limit and μ_X (see Figure 9.4). The Z value corresponding to an area of .495 from the center of the normal curve is approximately 2.58.

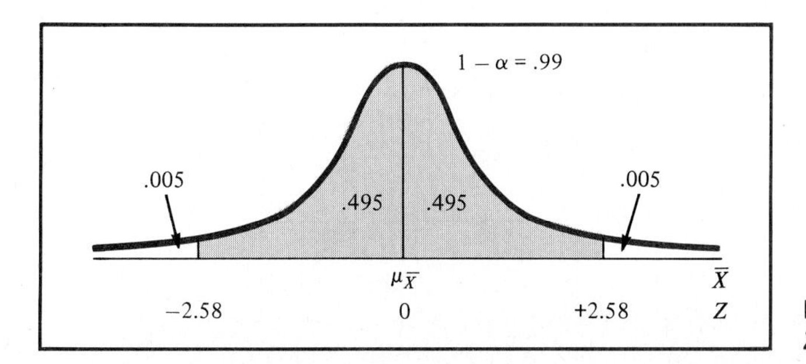

FIGURE 9.4 Normal curve for determining the Z values needed for 99% confidence.

EXAMPLE 9.1

What if 90% confidence were desired; what would be the corresponding Z value?

Solution

If 90% confidence were desired, the area of .90 would be divided in half, leaving .45 between each limit and μ_X (see accompanying figure on top of page 246). The Z value corresponding to an area of .45 from the center of the normal curve is 1.645 (halfway between 1.64 and 1.65).

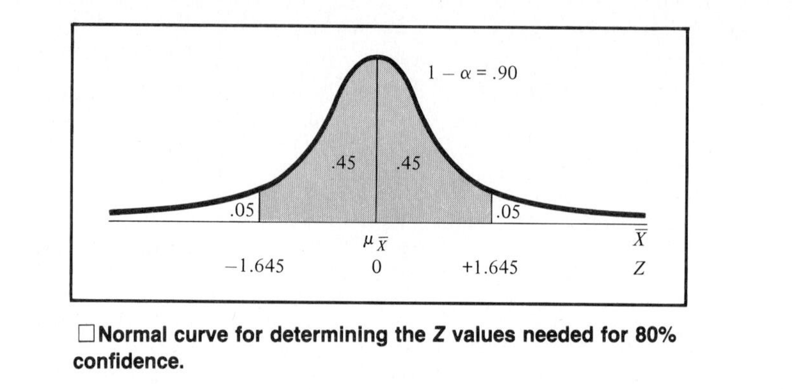

□ **Normal curve for determining the Z values needed for 80% confidence.**

EXAMPLE 9.2

If 80% confidence were desired, what would be the corresponding Z value?

Solution

If 80% confidence were desired, the area of .80 would be divided in half, leaving .40 between each limit and μ_X (see figure below). The Z value corresponding to an area of .40 from the center of the normal curve is 1.28.

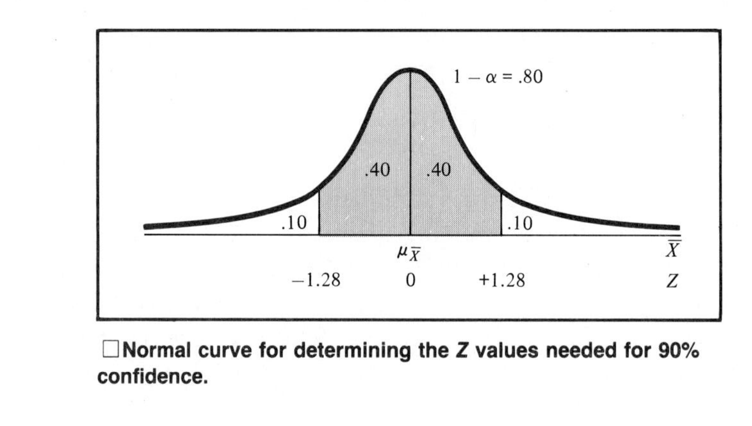

□ **Normal curve for determining the Z values needed for 90% confidence.**

Now that we have presented various levels of confidence, one might wonder why we wouldn't want to make the confidence level as close to 100% as possible. But any increase in confidence is achieved only by simultaneously widening (and making less precise and less useful) the confidence interval obtained. Thus we would have more confidence that the population mean is within a broader range of values. This tradeoff between the width of the confidence interval and the level of confidence will be discussed in greater depth when we investigate the determination of sample size (see Section 9.6).

Problems

9.1 Suppose that scores on an IQ test are known to be normally distributed and have a variance of 225. A random sample of nine people take the test; their mean score is 115. Compute 90%, 95%, and 99.9% confidence intervals for the population mean. What tradeoffs are involved in choosing which of the above to use?

• 9.2 Suppose that corn yield (in bushels per acre) is known to have a standard deviation of 18. If the average yield in a sample of one acre's production from each of 36 farms was 145 bushels per acre, construct 75%, 90%, and 99% confidence intervals for yield.

9.3 If the standard deviation of height (in inches) of adult females is known to be 3.6 inches, and a sample of 36 women have an average height of 63 inches, calculate 80%, 90%, and 95% confidence intervals for mean height. Does the population of heights have to be normally distributed here? Discuss.

9.4 Values on a test developed to detect marijuana in the blood are normally distributed and are known to have a standard deviation of 2.56 in people who have not smoked marijuana for at least one year. If a sample of 16 people who have not smoked marijuana for at least one year has a mean of 5.4, construct 70%, 90%, and 99.9% confidence intervals for the mean. Tell why an observed value for an individual would not necessarily be abnormal, even though it fell outside all of the computed confidence intervals.

9.5 Suppose that bacterial counts are approximately normally distributed, and that the variance of the bacterial count in the Smarmee River is 9,000,000. Counts for each of 25 days resulted in a mean count of 11,500. Calculate the 80%, 90%, and 99.9% confidence intervals for the mean bacterial count. Tell why an observed count of 15,000 on a particular day would not be unusual, even though it is outside the confidence intervals you calculated.

9.3 CONFIDENCE-INTERVAL ESTIMATE OF THE MEAN (σ_X UNKNOWN)

Now that we have developed the concept of a confidence-interval estimate, we are ready to return to the questions of interest to the personnel director. First, we may recall that the director wanted to estimate the true population mean annual family dental expenses for all employees. The survey results indicated that $\bar{X} = \$267.43$ and $S = \$191.74$ from a sample of 100 employees. We observe that for this set of data only the sample (not the population) standard deviation S is known. In most practical situations the population standard deviation σ_X will be unknown. Therefore we need to obtain a confidence-interval estimate of μ_X by using only the sample statistics $\bar{X}$ and S. To achieve this we turn to the work of William S. Gosset.

9.3.1 Student's *t* Distribution

At the turn of this century a statistician named William S. Gosset, an employee of Guinness Breweries in Ireland (see Reference 4), was interested in making inferences about the mean when σ_X was unknown. Since Guinness employees were not permitted to publish research work under their own names, Gosset adopted the pseudonym "Student." The distribution that he developed has come to be known as **Student's *t* distribution.** If the random variable X is normally distributed, then the statistic

$$\frac{\overline{X} - \mu_X}{\frac{S}{\sqrt{n}}}$$

has a t distribution with $n - 1$ *degrees of freedom*. Notice that this expression has the same form as Equation (8.2) on page 224, except that S is used to estimate σ_X, which is presumed unknown in this case.

9.3.2 Properties of the *t* Distribution

In appearance the t distribution is very similar to the normal distribution. Both distributions are bell-shaped and symmetric. However the Student t distribution has more area in the tails and less in the center than does the normal distribution (see Figure 9.5). This is because σ_X is unknown, and we are using S to estimate it. Since we are uncertain of the value of σ_X, the values of t which we observe will be more variable than for Z. Therefore, we must go a larger number of standard deviations from 0 to include a certain percentage of values from the t distribution than is the case with the standard normal.

However, as the number of degrees of freedom increases, the t distribution gradually approaches the normal distribution until the two are virtually identical. This happens because as the sample size gets larger, S becomes a better estimate of σ_X. With a sample size of about 120 or more, S estimates σ_X precisely enough that there is little difference between the t and Z distributions. For this reason, most statisticians will use Z instead of t when the sample size is over 120.

In practice, as long as the sample size is not too small and the population is not very skewed, the t distribution can be used in estimating the population mean when

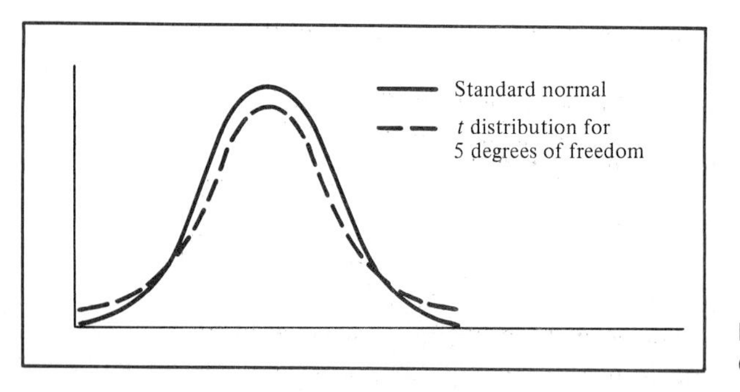

FIGURE 9.5 □ Standard normal distribution and *t* distribution for 5 degrees of freedom.

TABLE 9.1 □ Determining the critical value from the *t* table for an area of .025 in each tail with 99 degrees of freedom

Degrees of freedom	Upper tail areas					
	.25	.10	.05	.025	.01	.005
1	1.0000	3.0777	6.3138	12.7062	31.8207	63.6574
2	0.8165	1.8856	2.9200	4.3027	6.9646	9.9248
3	0.7649	1.6377	2.3534	3.1824	4.5407	5.8409
4	0.7407	1.5332	2.1318	2.7764	3.7469	4.6041
5	0.7267	1.4759	2.0150	2.5706	3.3649	4.0322
·	·	·	·	·	·	·
·	·	·	·	·	·	·
·	·	·	·	·	·	·
96	0.6771	1.2904	1.6609	1.9850	2.3658	2.6280
97	0.6770	1.2903	1.6607	1.9847	2.3654	2.6275
98	0.6770	1.2902	1.6606	1.9845	2.3650	2.6269
99	0.6770	1.2902	1.6604	1.9842	2.3646	2.6264
100	0.6770	1.2901	1.6602	1.9840	2.3642	2.6259

σ_X is unknown. The critical values of t for the appropriate degrees of freedom can be obtained from Table C.3. The top of each column of the t table indicates the area in the right tail of the t distribution (since positive entries are supplied, the values are for the upper tail), while each row represents the particular t value for each specific degree of freedom. For example, our personnel director's study contained a sample of 100 employees so that there are 99 degrees of freedom. If 95% confidence were desired, the appropriate value of t would be found as displayed in Table 9.1.

The 95% confidence level indicates that there would be an area of 2.5% (or in proportions .025) in each tail of the distribution. Referring to Table C.3 and looking in the column corresponding to an upper tail area of .025 and in the row corresponding to 99 degrees of freedom, we obtain a t value of 1.9842. A t value of 1.9842 means that the probability that t would exceed $+1.9842$ is .025 (see Figure 9.6). Since t is a symmetric distribution, if the upper t value is $+1.9842$, the value for the lower tail would be -1.9842.

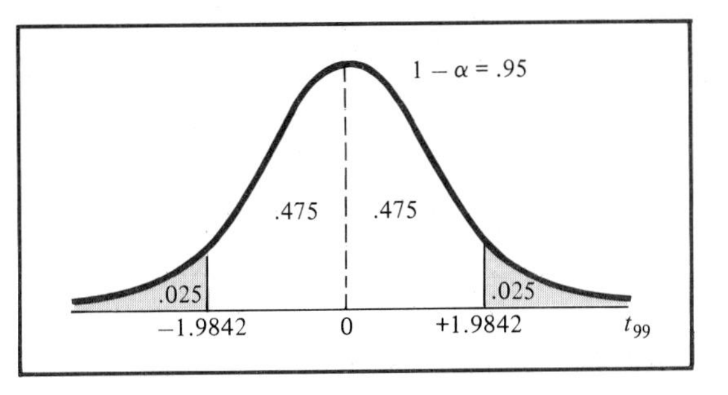

FIGURE 9.6 □ *t* distribution with 99 degrees of freedom.

9.3.3 The Concept of Degrees of Freedom

In Chapter 3 we learned that the sample variance S^2 requires the computation of $\Sigma (X - \overline{X})^2$. In order to compute this quantity we need to first know $\overline{X}$. Given the value of $\overline{X}$, only $n - 1$ of the sample values X are "free" to vary.

To see this, suppose that we had a sample of five values which had a mean of 10. How many distinct individual values would we need to know before we could obtain the remainder? The fact that $n = 5$ and $\overline{X} = 10$ tells us that $\Sigma X = 50$, since $\Sigma X/n = \overline{X}$. Therefore, once we know four of the values, the fifth one will *not* be "free" to vary, since the sum must add to 50. For example, if four of the values are 8, 14, 9, and 6, the fifth value can *only* be 13 if the sum is to equal 50. That is, there are $n - 1 = 5 - 1 = 4$ degrees of freedom for the observed values.

9.3.4 The Confidence-Interval Statement

The $(1 - \alpha) \times 100\%$ confidence-interval estimate for the mean with σ_X unknown is expressed as follows:

$$\overline{X} \pm t_{n-1} \frac{S}{\sqrt{n}}$$

or **(9.2)**

$$\overline{X} - t_{n-1} \frac{S}{\sqrt{n}} \leq \mu_X \leq \overline{X} + t_{n-1} \frac{S}{\sqrt{n}}$$

where t_{n-1} is the critical value at the $(\alpha/2 \times 100)\%$ level from the t distribution.

In this formula, the quantity $S/\sqrt{n}$ is the standard error of the mean $\overline{X}$. The confidence interval therefore extends from t_{n-1} standard errors below the mean to t_{n-1} standard errors above the mean. By choosing the value of t_{n-1} corresponding to the value of $\alpha/2$ we desire, this interval will include $(1 - \alpha) \times 100\%$ of the distribution. This is exactly what we did for the normal distribution [Equation (9.1)], where the population variance was known, except we used percentage points from the Z (standard normal) distribution instead of the t distribution.

Now we are ready to develop the desired confidence-interval estimate for the personnel director. Using Equation (9.2), we have

$$\bar{X} \pm t_{n-1} \frac{S}{\sqrt{n}}$$

$$= 267.43 \pm (1.9842) \frac{191.74}{\sqrt{100}}$$

$$= 267.43 \pm 38.045$$

so that

$$\$229.39 \leq \mu_X \leq \$305.48$$

We would conclude with 95% confidence that the average yearly family dental expense is between \$229.39 and \$305.48. The 95% confidence interval states that we are 95% sure that the sample we have selected is one in which the true population mean is located within the interval. This 95% confidence actually means that if all possible samples of size 100 were selected (something that would never be done in practice), 95% of the intervals developed would contain the true population mean while 5% would not.

EXAMPLE 9.3

A random sample of 40 college students was selected, and it was determined that the amount of money they spent on textbooks during the current year had a sample mean of \$328 and a sample standard deviation of \$55. Set up a 90% confidence-interval estimate of the true average amount spent for the population of students.

Solution

The 90% confidence level indicates that there would be an area of 5% (or in proportions .05) in each tail of the distribution. Referring to Table C.3, and looking in the column corresponding to an upper tail area of .05 and in the row corresponding to 39 degrees of freedom, we obtain a t value of 1.6849. A t value of 1.6849 means that the probability that t would exceed +1.6849 is .05 (see figure). Since t is a symmetric distribution, if the upper tail t value is +1.6849, the value for the lower tail is −1.6849.

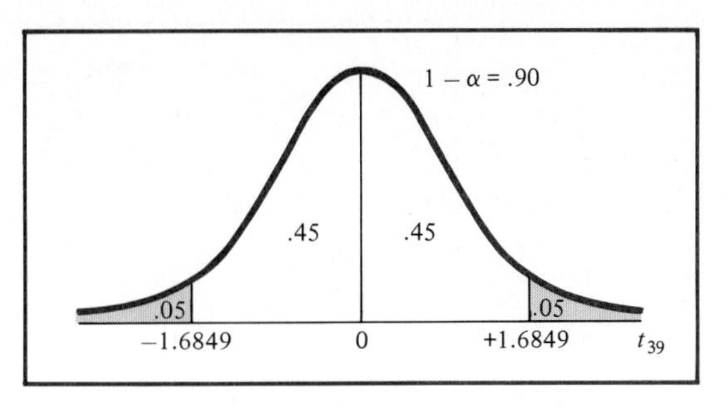

□ ***t* distribution with 39 degrees of freedom.**

Now that we have determined the critical values of t at the 90% confidence level for 39 degrees of freedom, we can develop the desired confidence-interval estimate. Using Equation (9.2), we have

$$\overline{X} \pm t_{n-1}\frac{S}{\sqrt{n}} = 328.00 \pm (1.6849)\frac{55.00}{\sqrt{40}}$$

$$= 328.00 \pm 14.65$$

so that

$$\$313.35 \leq \mu_X \leq \$342.65$$

We would conclude with 90% confidence that the average amount spent per student on textbooks this year is between \$313.35 and \$342.65.

EXAMPLE 9.4

A random sample of 300 undergraduate male students at a large university was selected and it was determined that the height of the students had a sample average of 71 inches and a sample standard deviation of 3 inches. Set up a 99% confidence-interval estimate of the true population average height of male undergraduates at this university.

Solution

Since the degrees of freedom which equal $300 - 1 = 299$ exceed 120, we may use the normal distribution instead of the t distribution, since the t distribution with more than 120 degrees of freedom is very close to the normal distribution. The 99% confidence level indicates that there would be an area of .5% (or in proportions .005) in each tail of the distribution (see figure). Referring to Table C.3, and looking in the column corresponding to an upper tail area of .005 and in the row corresponding to an infinite (∞) number of degrees of freedom (since we have more than 120), we obtain a t value of approximately 2.58. Of course,

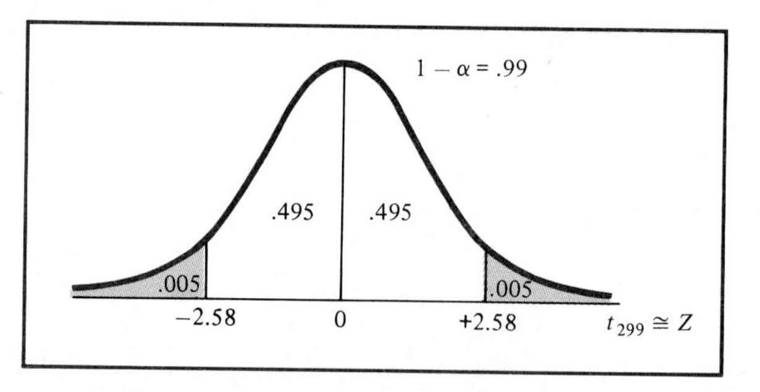

□ ***t* distribution with 299 degrees of freedom.**

had we used the normal distribution with an area of .005 in each tail of the distribution (.495 between each limit and the center), we also would have obtained a value of 2.58.

Now that we have determined the critical values of t at the 99% confidence level for 299 degrees of freedom, we can develop the desired confidence-interval estimate. Using Equation (9.2), we have

$$\bar{X} \pm t_{n-1}\frac{S}{\sqrt{n}} = 71 \pm (2.58)\frac{3}{\sqrt{300}}$$

$$= 71 \pm .45$$

so that

$$70.55 \leq \mu_X \leq 71.45$$

We would conclude with 99% confidence that the average height of male undergraduates is between 70.55 and 71.45 inches.

Problems

For Problems 9.6–9.16, assume that the population distribution is approximately normal.

9.6 Using the data on light bulbs from Problem 3.9 on page 46, calculate 80%, 90%, and 99% confidence intervals for the length of life of light bulbs. Compare these confidence intervals to similar confidence intervals calculated using (incorrectly) the normal distribution instead of the t distribution, and comment on the differences.

9.7 Using the data on mass of subatomic particles in Problem 3.11 on page 46, calculate 80%, 90%, and 95% confidence intervals for the mass of the particle. Calculate a 95% confidence interval using the (incorrect) normal distribution, and comment on the difference between the incorrect and correct confidence intervals.

9.8 Using the data from scores on a creativity test in Problem 3.12 on page 46, construct 80%, 90%, and 99% confidence intervals for the mean. Calculate a 90% confidence interval using the normal distribution (which is incorrect), and comment on the difference between this and the correct 90% confidence interval.

• 9.9 Using the data on batting averages in Problem 3.17 on page 47, calculate 80%, 90%, and 99% confidence intervals for the mean batting average. Give an example of a batting average that would be outside all of these confidence intervals but would not be unusual for an individual player, and tell why this is not a contradiction.

9.10 Using the data on vitamin C in oranges (Problem 3.23 on page 49) construct 80%, 90%, and 95% confidence intervals for the mean amount of vitamin C. If an individual orange had 118 milligrams of vitamin C, would this be considered unusual? (Explain your answer.)

9.11 From the data on tax revenues in Problem 3.62 on page 69, calculate the 80%, 95%, and 99% confidence intervals for the mean.

9.12 Using the data on GNP projections (Problem 3.74 on page 72) calculate 80% and 95% confidence intervals for the mean. Recalculate the 95% confidence interval (incorrectly) using the normal instead of the t distribution, and comment on the difference.

• 9.13 Calculate 95% confidence intervals for the data on acceleration of automobiles in Problem 3.75 on page 73. Do the German and Japanese cars separately.

9.14 Calculate 99% confidence intervals for the data on college tuition in Problem 3.76 on page 73. Do the colleges from the Northeast and Midwest separately.

9.15 Compute a 95% confidence interval for each of the following data:

(a) 1, 1, 1, 1, 1, 10, 10, 10, 10, 10

(b) 1, 2, 3, 4, 5, 6, 7, 8, 9, 10

Tell why they have different confidence intervals, even though they have the same mean and range.

9.16 Compute a 95% confidence interval for the numbers 1, 2, 3, 4, 5, 30. Change the number 30 to 6, and recalculate the confidence interval. Using these results, describe the effect of an outlier on the confidence interval.

9.4 CONFIDENCE-INTERVAL ESTIMATE OF THE MEAN DIFFERENCE IN MATCHED OR PAIRED SAMPLES

9.4.1 Development

Suppose that instead of a sample of individual observations we have a sample of pairs of observations. For example, suppose a shoe company wanted to test a new material for the soles of shoes. It might make up a number of pairs of shoes with the new material on one shoe of each pair and the usual material on the other shoe. People wear the shoes for three months, and the wear is measured on each shoe. There will be variation from person to person in how much wear each pair of shoes gets, because some people will walk (or run) quite a bit, while others will use the shoes less often. We are interested in calculating the difference between the mean wear of the new material and of the old material and in constructing a confidence interval for that difference. That is, we want to construct a confidence interval for

$$\overline{X}_n - \overline{X}_o$$

where $\overline{X}_n$ is the mean wear in the shoes with the new material and $\overline{X}_o$ is the mean wear in the shoes with the old material.

To see how we can do this, using exactly the same methods we used in the previous sections, we will express the desired outcome in a different but equivalent way. For each pair of shoes, we can calculate the difference in the measure of wear between the shoe with the new material and the shoe with the old material. Now we have a single measure for each pair of shoes: the difference between new and old material. Using the methods of the previous sections, we can construct a confidence interval for the average of the difference scores. This can be shown (see Section 9.4.2) to be equivalent to the confidence interval for the difference in means; therefore, two situations which appear different are, in fact, the same in structure.

Suppose that six pairs of shoes were used in the test and that wear was measured on a 1-to-10 scale, with 1 being the least wear and 10 the most wear. If the data were as follows:

	Pair number					
Material	**I**	**II**	**III**	**IV**	**V**	**VI**
New	2	4	5	7	7	5
Old	4	5	3	8	9	4
Differences	−2	−1	+2	−1	−2	+1

then the mean wear of the new materials is 5.0, and, for the old material, it is 5.5. Note that the difference between the means is $5.0 - 5.5 = -.5$. The differences between the pairs of wear measurements are −2, −1, 2, −1, −2, and 1. The mean of these differences is −.5, which is the same as the difference in means. The standard deviation of these six differences is 1.6432 (see Problem 9.17). Assuming that these differences are approximately normally distributed, the 95% confidence interval for the true mean difference is

$$-.5 \pm (2.571)(1.6432)/\sqrt{6}$$

or from −2.22 to 1.22.

★ 9.4.2 Demonstrating the Equivalence Between Difference in Means and Mean Difference (Optional Subsection)

Let us write out the statement of equivalence in means using the individual observations for a small example. Suppose that there are three pairs of scores, so that we want to find a confidence interval for

$$\frac{X_{a1} + X_{a2} + X_{a3}}{3} - \frac{X_{b1} + X_{b2} + X_{b3}}{3}$$

where the subscript a denotes the first member of a pair and the subscript b the second member. For example, X_{b3} is the second member of the third pair. If we rearrange the terms, putting the pairs together, we get

★ Note that this is an optional subsection that may be skipped without any loss of continuity.

$$\frac{(X_{a1} - X_{b1}) + (X_{a2} - X_{b2}) + (X_{a3} - X_{b3})}{3}$$

which is just the average of the difference scores, as we wished to demonstrate. It is obvious that the same method can be used to prove the equivalence of the average difference and the difference of the averages for any sample size.

Problems

9.17 Verify that the standard deviation of the numbers -2, -1, 2, -1, -2, and 1 is 1.6432.

9.18 The Gigantic Conglomerate Drug Company Inc. wants to test a new drug called E-Z-Sleep that one of its scientists invented to help people sleep. The company pays people to come into its sleep laboratory, where on one night they take the new drug and on another night they take a placebo (that is, a pill which, unknown to the participants in the study, contains no active ingredients). The scientists measure how many hours each person sleeps under each condition. The results for 12 people are:

E-Z-Sleep:	7	4	6	2	6	7	4	9	6	4	10	8
Placebo:	5	4	7	1	4	6	5	6	4	3	9	6

Set up a 95% confidence interval estimate of the mean difference in sleep hours between the drug and the placebo.

• 9.19 A public interest group wants to see the effects of alcohol on driving. They arrange a winding path marked with pylons (plastic cones about 2 feet high) in a parking lot and have 12 people drive through the path at 30 miles per hour. The number of pylons which they knock over is taken to be an indicator of the quality of their driving; a perfect score means no pylons are knocked over. Each person drives the course twice—first while sober, and then after drinking a six-pack of Iron Gullet Beer. The number of pylons knocked over by each person under each condition is:

Sober:	2	4	3	1	3	5	3	2	0	2	1	4
Drunk:	4	7	2	2	7	5	2	5	2	3	2	3

Set up a 99% confidence interval estimate of the mean difference in the number of pylons knocked over when sober and when drunk.

9.20 Researchers at White Kernel College want to test a new variety of corn seed which they have developed. They get nine farmers to plant the new seed on half of their land, and their usual seed on the other half. The number of bushels per acre which each farmer obtained is:

New seed:	132	154	121	163	159	138	143	149	136
Old seed:	139	145	134	153	167	139	148	142	133

Set up a 95% confidence interval estimate of the mean difference in yield between the new seed and the old seed.

9.21 A group of nine children wanted to see whether the amount of air in their bicycle tires made a difference in how easy it was to pedal their bikes. They decided to ride a particular route around town under two different conditions: in one condition, they would have 40 pounds per square inch (psi) of pressure in the tires, and in the other condition they would have 65 psi. Each child would ride independently of the others. Some would ride first with 65 psi, others with 40 psi, so that the fatigue or warm up effects (whichever occurred) from the first ride would have equal effects on both conditions (this is called *counterbalancing*). The length of time (in minutes) it took each child to ride the circuit under each condition is:

40 psi:	34	54	23	67	46	35	49	51	27
65 psi:	32	45	21	63	37	40	51	39	23

Set up a 95% confidence interval estimate of the mean difference in time between 40 psi and 65 psi.

9.5 CONFIDENCE-INTERVAL ESTIMATE OF THE PROPORTION

In addition to estimating the average annual family dental expenses, the personnel director also wanted to estimate the true proportion of employees who say they would purchase the optional dental insurance plan. In this section we extend the concept of the confidence interval to qualitative data to estimate the population proportion p from the sample proportion $p_S = X/n$. Recall from Chapters 7 and 8 that when both np and $n(1 - p)$ were at least 5, the binomial distribution could be approximated by the normal distribution. Hence, we could set up the following $(1 - \alpha) \times 100\%$ confidence interval estimate for the population proportion p:

$$p_S \pm Z\sqrt{\frac{p_S(1-p_S)}{n}}$$

or **(9.3)**

$$p_S - Z\sqrt{\frac{p_S(1-p_S)}{n}} \leq p \leq p_S + Z\sqrt{\frac{p_S(1-p_S)}{n}}$$

In the personnel director's survey, 64 out of 100 employees sampled said that they would purchase the optional group dental insurance plan. If 90% confidence were desired, the Z value obtained from Table C.2 would be 1.645 (since .05 would be in each tail of the normal distribution). Since $n = 100$ and $p_S = 64/100 = .64$, using Equation (9.3) we would have

$$p_S \pm Z\sqrt{\frac{p_S(1-p_S)}{n}} = .64 \pm (1.645)\sqrt{\frac{.64(.36)}{100}}$$

$$= .64 \pm (1.645)(.048)$$

$$= .64 \pm .079$$

or

$$.561 \leq p \leq .719$$

Thus the personnel director would estimate with 90% confidence that between 56.1% and 71.9% of the employees would choose to participate in the optional dental insurance plan.[1]

For a given sample size, confidence intervals for proportions often seem to be wider relative to those for continuous variables. With continuous variables, the measurement on each subject contributes more information than for a dichotomous variable. In other words, a variable with only two possible values is a very crude measure compared to a continuous variable, so each subject contributes only a little information about the parameter we are estimating.

EXAMPLE 9.5

A random sample of 200 registered voters was selected in a small city to determine the proportion who intended to vote in an upcoming school board election. The results of the sample indicated that 28 stated they intended to vote. Set up a 95% confidence-interval estimate of the true population proportion of registered voters that intend to vote in the school board election.

Solution

If 95% confidence were desired, the Z value obtained from Table C.2 would be 1.96 (since .025 would be in each tail of the normal distribution). Since $n = 200$ and $p_S = 28/200 = .14$, using Equation (9.3) we would have

$$p_S \pm Z\sqrt{\frac{p_S(1-p_S)}{n}} = .14 \pm (1.96)\sqrt{\frac{.14(.86)}{200}}$$
$$= .14 \pm (1.96)(.0245)$$
$$= .14 \pm .048$$

or

$$.092 \leq p \leq .188$$

Thus we would estimate with 95% confidence that between 9.2% and 18.8% of the voters intend to vote in the upcoming school board election.

---

[1] Since the number of successes is 64 and the number of failures is 36, the normal distribution provides an excellent approximation for the binomial distribution. However, if the sample size is not large or the proportion of successes is either very low or very high, the binomial distribution should be used (see Reference 1). Exact confidence intervals for various sample sizes and proportions of successes have been tabled by Fisher and Yates (see Reference 2).

Problems

9.22 Out of a sample of 50 flower bulbs, 35 grew into flowering plants and 15 did not. Construct 75% and 90% confidence intervals for the proportion of plants that will flower.

• 9.23 A computer program is written to generate random sequences of zeros and ones, with the ones occurring a fixed proportion of the time (determined by the user of the computer program). If the program generated a sequence with 16 zeros and 24 ones, construct a 95% confidence interval for the proportion of digits which will be ones.

9.24 Wanda is working her way through college selling magazines door to door. Suppose that out of 50 houses visited, 15 people buy magazines from her. Construct 80%, 90%, and 95% confidence intervals for the proportion of households making a purchase.

• 9.25 Suppose you wanted to determine the proportion of books being published which were nonfiction. If you randomly sampled 100 recently published books, and 34 were nonfiction books, construct 80%, 90%, and 95% confidence intervals for the proportion which are nonfiction.

9.26 A suburban bus company is considering the institution of a commuter bus route from a particular suburb into the central business district of the city. A random sample of 50 commuters is selected, and 18 indicate that they would use this bus route. Set up a 95% confidence-interval estimate of the true proportion of commuters who would utilize this new bus route.

9.27 An automobile dealer would like to estimate the proportion of customers who still own the same cars they purchased five years earlier. A random sample of 200 customers selected from the automobile dealer's records indicated that 82 still owned the cars that had been purchased five years earlier. Set up a 95% confidence-interval estimate of the true proportion of all customers who still own the same cars five years after they were purchased.

9.28 A stationery supply store receives a shipment of a certain brand of inexpensive ballpoint pens from the manufacturer. The owner of the store wishes to estimate the proportion of pens that are defective. A random sample of 300 pens is tested, and 30 are found to be defective. Set up a 90% confidence-interval estimate of the proportion of defective pens in the shipment. If the shipment can be returned if it is more than 5% defective, based on the sample results, can the owner return this shipment?

9.6 SAMPLE-SIZE DETERMINATION FOR THE MEAN

Now that the personnel director has obtained confidence interval estimates for the mean and the proportion, he realizes that he is not completely satisfied with the precision that he has obtained. Specifically he feels that the intervals obtained are too wide (and therefore not precise enough) for his needs. Thus, for future surveys he would like to be able to determine the sample size given his requirements for desired confidence level and width of the confidence interval. In determining the sample size for estimating the population mean, these requirements must be considered along with information about the standard deviation. Had σ_X been known, the confidence-interval estimate for the population mean would be obtained from Equation (9.1b):

$$\overline{X} \pm Z\frac{\sigma_X}{\sqrt{n}}$$

To develop a formula for determining the sample size, we begin by considering the quantity[2] $e = \overline{X} - \mu_X$. This difference between the sample mean $\overline{X}$ and the population mean μ_X is called the **sampling error.** We recall from Equation (8.2) that:

$$Z = \frac{\overline{X} - \mu_X}{\dfrac{\sigma_X}{\sqrt{n}}}$$

Multiplying each side by $\sigma_X/\sqrt{n}$, we have

$$\frac{Z\sigma_X}{\sqrt{n}} = \overline{X} - \mu_X$$

Therefore, the sampling error e can be expressed as

$$e = \frac{Z\sigma_X}{\sqrt{n}} \qquad \textbf{(9.4)}$$

where $e = (\overline{X} - \mu_X)$.

Solving this equation for n, we multiply each side by $\sqrt{n}/e$ and then we square each side to obtain

$$n = \frac{Z^2\sigma_X^2}{e^2} \qquad \textbf{(9.5)}$$

Therefore, in order to determine the sample size, we must define three unknowns:

1. **The confidence level desired,** which determines the value of Z, the critical value from the normal distribution.[3]
2. **The sampling error permitted, e.**
3. **The standard deviation, σ_X.**

In practice the determination of these three quantities may not be easy. How is one to know what level of confidence to use and what sampling error is desired?

---

[2] In this context, the letter e is *not* the same as the constant 2.718 . . . used in the formula for the normal distribution.

[3] We use Z instead of t because (1) to determine the critical value of t we would need to know the sample size, which is what we don't know yet, and (2) for most studies the sample size needed will be large enough that the normal is a good approximation to the t distribution.

Typically these questions can be answered only by the individual who is familiar with the variables to be analyzed. Although 95% is the most common confidence level used (in which case $Z = 1.96$), if one desires greater confidence, 99% might be more appropriate, while if less confidence is deemed acceptable, then 90% (or even 80%) might be utilized.

For the sampling error, the personnel director should be thinking not of how much sampling error he would like to have (he really doesn't want any error) but of how much he can "live with" and still be able to provide adequate conclusions for the data. Even when the confidence level and the sampling error are specified, an estimate of the standard deviation must be available. Unfortunately, in very few cases is the population standard deviation σ_X known. In some instances the standard deviation can be estimated from past data. In other situations, one can develop an "educated guess" by taking into account the range and distribution of the variable. For example, if one assumes a normal distribution, the range is approximately equal to $6\sigma_X$ (that is, $\pm 3\sigma_X$ around the mean), so that σ_X can be estimated as range/6 for a normally distributed variable. If σ_X cannot be estimated in this manner, a **pilot study** can be conducted to estimate the standard deviation.

Returning to the problem of interest to the personnel director, suppose that he desired 95% confidence for estimating the population mean annual family dental expense to within $\pm\$25$ of the true value. Based upon information about the distribution of dental expenses and the previous study, he has arrived at an estimate of \$200 for the standard deviation. Thus, using Equation (9.5) with 95% confidence ($Z = 1.96$), $e = \pm\$25$, and $\sigma_X = \$200$,

$$n = \frac{Z^2 \sigma_X^2}{e^2} = \frac{(1.96)^2(200)^2}{(25)^2}$$

$$= \frac{3.8416(40{,}000)}{625} = 245.86$$

Rounding up to the nearest integer, $n = 246$. (The rule of thumb used in determining sample size is to always round up to the next highest integer value in order to slightly oversatisfy the criteria desired.) Therefore, our conclusion is that if the personnel director desired to estimate the population average dental expense to within $\pm\$25$ with 95% confidence and he assumed a standard deviation of \$200, then a sample size of 246 employees would be required.

Problems

9.29 In Problem 9.1 on page 247, you estimated a confidence interval for the mean of a set of IQ scores. If you wanted a 90% confidence interval for the mean of such scores, and wanted the estimate to be within ± 2 IQ points, what sample size would you need? If you wanted 99% confidence, what sample size would you need?

• 9.30 In Problem 9.2 on page 247, you estimated confidence intervals for the yield of corn. If you wanted a 90% confidence interval for the mean, and you wanted the estimate to be accurate to plus or minus 5 bushels per acre, how many farms would you have to sample? How many farms would be necessary for 99% confidence?

• 9.31 In Problem 9.5 on page 247, you estimated confidence intervals for bacterial counts in the Smarmee River. If you wanted 75% confidence that you estimate the mean within ±1,000, what sample size would you need? What sample size would you need if you wanted 90% confidence?

9.32 In Problem 9.7 on page 253, you estimated confidence intervals for the mass of a subatomic particle. If you want 90% confidence in estimating the mass to within ±.01, what sample size would you need? If you wanted 99% confidence, what sample size would you need? [*Hint*: Treat the original data as a pilot study and use the standard deviation to estimate σ_X.]

9.33 In Problem 9.12 on page 254, you estimated confidence intervals for GNP projections. If you wanted to know the average projection within ±.1% with 90% confidence, how many projections would you need? [*Hint*: Treat the original data as a pilot study and use the standard deviation to estimate σ_X.]

9.34 A survey is planned to determine the average annual family medical expenses of employees of a large company. The management of the company wishes to be 95% confident that the sample average is correct to within ±$50 of the true average family expenses. A pilot study indicates that the standard deviation can be estimated as $400. How large a sample size is necessary?

9.35 A consumer group would like to estimate the average monthly electric bills for the month of July for one-family homes in a large city. Based upon studies conducted in other cities, the standard deviation is assumed to be $25. The group would like to estimate the average bill for July to within ±$5 of the true average with 99% confidence. What sample size is needed?

9.7 SAMPLE-SIZE DETERMINATION FOR THE PROPORTION

In the previous section we discussed the determination of the sample size necessary for estimating the population mean. Since the personnel director also wishes to estimate the proportion of employees who would purchase the optional dental plan, we need to determine the sample size necessary for estimating the population proportion. The methods of sample-size determination for the proportion are similar to those utilized for the mean.

The confidence-interval estimate of the true proportion p obtained from Equation (9.3) is

$$p_S \pm Z\sqrt{\frac{p_S(1-p_S)}{n}}$$

To determine the sample size, recall from Equation (8.5) that since

$$Z = \frac{p_S - p}{\sqrt{\frac{p(1-p)}{n}}}$$

we have

$$Z\sqrt{\frac{p(1-p)}{n}} = p_S - p$$

In this case we define the sampling error e as the difference between the sample estimate p_S and the population parameter p. That is,

$$e = Z\sqrt{\frac{p(1-p)}{n}} \quad \textbf{(9.6)}$$

Solving this equation for n, we obtain

$$n = \frac{Z^2 p(1-p)}{e^2} \quad \textbf{(9.7)}$$

where $e = p_S - p$.

In determining the sample size for estimating a proportion, three unknowns must be defined:

1. **The level of confidence desired.**
2. **The sampling error permitted, *e*.**
3. **The true proportion of success, *p*.**

In practice the selection of these three quantities is often difficult. Once we determine the desired level of confidence, we will be able to obtain the appropriate Z value from the normal distribution. The sampling error e indicates the amount of error that we are willing to accept in estimating the population proportion. The third quantity—the true proportion of success, p—is actually the population parameter that we are trying to find! Thus, how can we state a value for the very thing that we are taking a sample in order to determine?

Here there are two alternatives. First, in many situations past information or relevant experience may be available that enables us to provide an ''educated'' estimate of p. Second, if past information or relevant experience is not available, we try to provide a value for p that would never *underestimate* the sample size needed. Referring to Equation (9.7), we observe that the quantity $p(1 - p)$ appears in the numerator. Thus, we need to determine the value of p that will make $p(1 - p)$ as large as possible. It can be shown that when $p = .5$, then the product $p(1 - p)$ achieves its maximum result. Several values of p along with the accompanying products of $p(1 - p)$ are:

$$\begin{aligned}
p = .1: &\quad p(1-p) = .1(.9) = .09 \\
p = .4: &\quad p(1-p) = .4(.6) = .24 \\
p = .5: &\quad p(1-p) = .5(.5) = .25 \\
p = .7: &\quad p(1-p) = .7(.3) = .21 \\
p = .99: &\quad p(1-p) = .99(.01) = .0099
\end{aligned}$$

Therefore, when we have no prior knowledge or estimate of the true proportion, p, we should use $p = .5$ as the most conservative way of determining the sample size. However, the use of $p = .5$ may result in an *overestimate* of the sample size. If the actual sample proportion is very different from .50, the width of the real confidence interval may be substantially narrower than the estimate we get using this method.

Returning to the problem of interest to the personnel director, suppose that he desired 90% confidence of estimating the true population proportion of employees who would purchase the optional dental plan, to within $\pm.05$. Since no information is available concerning the population proportion, p is set equal to .50. Thus, using Equation (9.7) with 90% confidence ($Z = 1.645$), $e = \pm.05$ and $p = .50$:

$$n = \frac{Z^2 p(1-p)}{e^2} = \frac{(1.645)^2(.5)(.5)}{(.05)^2} = 270.6$$

Therefore, $n = 271$.

In order to be 90% confident of estimating the population proportion to within $\pm.05$, we would need a sample size of 271 employees.

Problems

9.36 In Problem 9.22 on page 259 you calculated confidence intervals for the proportion of flower bulbs which would grow into flowering plants. If you wanted to estimate this proportion to within $\pm.03$, with 90% confidence, what sample size would be necessary? For 99% confidence, what sample size would be necessary?

• 9.37 In Problem 9.23 on page 259 you calculated confidence intervals for the proportion of digits generated by a computer which were ones (rather than zeros). If you wanted to estimate the proportion of digits which are ones to within $\pm.02$, with 95% confidence, how long would the string of digits you examine have to be? What if you wanted to be within $\pm.005$ with 99% confidence?

• 9.38 If you wanted to estimate the proportion of births which are girls to within $\pm.01$ with 90% confidence, what sample size would be necessary? What sample size would be necessary for 99% confidence?

9.39 Suppose that less than 10% of all widgets manufactured by the Wooly Widget Company are defective. If you wanted to know the exact proportion to within $\pm.005$ with 95% confidence, how many widgets would you have to test?

9.40 A political pollster would like to estimate the proportion of voters who will vote for the Democratic candidate in a presidential campaign. The pollster would like 90% confidence that her prediction is correct to within ±.04 of the true proportion. What sample size is needed?

9.41 A cable television company would like to estimate the proportion of its customers that would purchase a cable television program guide. The company would like to have 95% confidence that its estimate is correct to within ±.05 of the true proportion. Past experience in other areas indicates that 30% of the customers will purchase the program guide. What sample size is needed?

9.8 ESTIMATION AND SAMPLE-SIZE DETERMINATION FOR FINITE POPULATIONS

9.8.1 Estimating the Mean

In discussing confidence-interval estimation and sample-size determination in the case concerning the personnel director, we have not considered the fact that there is a finite population of 5,000 employees in the company. In Section 8.4 we introduced the *finite population correction factor*, which served to reduce the standard error by a factor equal to $\sqrt{(N-n)/(N-1)}$. As a rule of thumb, whenever the sample size, n, is at least 5% of the population size, N (that is, $n/N \geq .05$), this correction factor should be employed to reduce the width of any confidence interval developed (provided that we are sampling without replacement—as is the typical case). Therefore the confidence interval for the mean would become:

$$\overline{X} \pm t_{n-1} \frac{S}{\sqrt{n}} \sqrt{\frac{N-n}{N-1}} \qquad \textbf{(9.8)}$$

In the personnel director's study, the sample of 100 employees was selected from a population of 5,000 employees. Since the sample was only 2% of the population (that is, $n/N = 100/5{,}000$), the correction factor will have only a slight effect on the confidence-interval estimate (as it was computed in Section 9.3). Using Equation (9.8) with $\overline{X} = \$267.43$, $S = \$191.74$, $n = 100$, $N = 5{,}000$, and $t_{99} = 1.9842$, we have

$$\overline{X} \pm t_{n-1} \frac{S}{\sqrt{n}} \sqrt{\frac{N-n}{N-1}} = 267.43 \pm (1.9842) \frac{191.74}{\sqrt{100}} \sqrt{\frac{5{,}000-100}{5{,}000-1}}$$

$$= 267.43 \pm 38.045(.99)$$

$$= 267.43 \pm 37.665$$

or

$$\$229.77 \leq \mu_X \leq \$305.10$$

These limits differ by less than \$0.50 from the uncorrected limits we originally computed.

9.8.2 Estimating the Proportion

For the population proportion, the $(1 - \alpha) \times 100\%$ confidence interval estimate using the finite population correction factor would be

$$p_S \pm Z \sqrt{\frac{p_S(1 - p_S)}{n}} \sqrt{\frac{N - n}{N - 1}} \tag{9.9}$$

Using Equation (9.9) for the personnel director's study in which he wishes to estimate the proportion of employees who would purchase the optional group dental expense plan, we have $p_S = 64/100 = .64$, $n = 100$, $N = 5{,}000$, and $Z = 1.645$. Thus

$$p_S \pm Z \sqrt{\frac{p_S(1 - p_S)}{n}} \sqrt{\frac{N - n}{N - 1}} = .64 \pm (1.645) \sqrt{\frac{.64(.36)}{100}} \sqrt{\frac{5{,}000 - 100}{5{,}000 - 1}}$$

$$= .64 \pm .079(.99)$$

$$= .64 \pm .0782$$

or

$$.562 \leq p \leq .718$$

Once again, because the sample is a small fraction of the population, the correction factor has a negligible effect on the confidence-interval estimate (as was previously computed in Section 9.5).

9.8.3 Determining the Sample Size

Just as the correction factor was used in developing confidence-interval estimates, it also should be used in determining sample size when sampling without replacement. For example, in estimating means the sampling error would be

$$e = Z\frac{\sigma_X}{\sqrt{n}}\sqrt{\frac{N-n}{N-1}} \tag{9.4a}$$

while in estimating proportions the sampling error would be

$$e = Z\sqrt{\frac{p(1-p)}{n}}\sqrt{\frac{N-n}{N-1}} \tag{9.6a}$$

The sample size needed could then be determined in a two-stage procedure. First, the sample size would be determined as in Sections 9.6 and 9.7 without regard to the correction factor. Then the correction factor would be applied to obtain the final sample size.

In determining the sample size in estimating the mean, we would have from Equation (9.5),

$$n_0 = \frac{Z^2\sigma_X^2}{e^2}$$

where n_0 is the sample size without considering the finite population correction factor.

Applying the correction factor gives

$$n = \frac{n_0 N}{n_0 + (N-1)} \tag{9.10}$$

In the personnel director's study, the sample size needed in order to be 95% confident of being correct to within ±$25 (assuming a standard deviation of $200) was 246, since n_0 was computed as 245.86. Using Equation (9.10), we have

$$n = \frac{(245.86)(5{,}000)}{245.86 + (5{,}000 - 1)} = 234.38$$

Thus $n = 235$.

Here, the use of the correction factor had a slight effect on the required sample size, since n/N was approximately 5%. A similar effect is noted when estimating proportions. Previously it was determined that a sample size of 271 was needed ($n_0 = 270.6$), since the personnel director desired 90% confidence and a sampling error of ±.05. Using Equation (9.10), we have

$$n = \frac{(270.6)(5{,}000)}{270.6 + (5{,}000 - 1)} = 256.76$$

Thus $n = 257$.

Here, once again the use of the finite population correction factor had only a small effect on the sample size required. Since the personnel director would certainly include several questions in a *single* questionnaire, if he wanted to *simultaneously* satisfy his requirements concerning the average dental expense and the proportion of employees who would choose the optional plan, the larger sample size of 257 would have to be utilized for the survey.

EXAMPLE 9.6

Suppose that the population size was only 750. What sample size would be needed?

Solution

Using Equation (9.10), we have

$$n = \frac{(270.6)(750)}{270.6 + (750 - 1)} = 199.05$$

Thus $n = 200$. In this instance the use of the finite population correction factor had a substantial effect on the sample size.

Problems

• **9.42** Matthew bought 400 firecrackers to use for the July 4th celebration. He decided to test them to be sure that they were good. He wanted to know with 95% confidence, within ±.05, what proportion of the firecrackers were bad.

(a) How many firecrackers does he need to test? (This is called *destructive testing*; that is, the product being tested is destroyed by the test and is then unavailable to be used. Obviously, manufacturers prefer to have methods of nondestructive testing.)

(b) If he actually tested 40 firecrackers, out of which 8 were bad, construct a 95% confidence interval for the proportion that were bad.

• **9.43** Suppose that you want to estimate the average IQ of a group of 300 people.

(a) If the standard deviation of the test is 15 points, how many people must you test if you want your estimate to be within ±5 points, with 95% confidence?

(b) What sample size would be necessary if the group had 150 people instead of 300?

(c) If 30 out of 300 were tested, and the average was 95, construct a 99% confidence interval for the mean IQ. Assume σ_X is 15 points.

9.44 There are 200 farms in Deutsch County on which corn is grown.

(a) How many must be sampled to estimate within ±.03, with 95% confidence, the proportion which use a particular method of cultivation?

(b) How many must be sampled to determine within ± 5 bushels with 95% confidence, the average yield of the corn crop (assuming a variance of 400)?

(c) If a sample of 40 farms was actually taken, with 16 using the new method of cultivation, and with an average yield of 144 bushels per acre, calculate 95% confidence intervals for (i) the proportion using the new method of cultivation and (ii) the average yield per acre (assuming a variance of 400).

9.45 There are 250 students at Lurie High in Fort Dodge.

(a) How many must be sampled to estimate within $\pm .02$, with 95% confidence, the proportion who plan to go to college?

(b) How many must be sampled to determine the mean grade-point index (on a 4-point scale) within $\pm .1$ with 95% confidence? [*Hint*: Assume grade-point index has a normal distribution and estimate σ_X based on the range.]

(c) If a sample of 50 students had a mean grade-point index of 2.75 and a sample standard deviation of .48, and if 42 of them plan to go to college, compute 90% confidence intervals for the true mean grade-point index and the proportion going to college.

9.46 Sixty law firms each hired attorneys who just graduated from law school. Suppose that, within a firm, everyone newly hired is paid the same amount.

(a) How many firms must be sampled to determine within $\pm$\$2,000 with 95% confidence the average starting salary at these firms? (Assume a population standard deviation of \$5,000.)

(b) How many firms must be sampled to determine within $\pm 2\%$, with 95% confidence, the percentage of firms paying over \$60,000 to newly hired attorneys?

(c) Suppose that 15 firms are sampled and that the average salary for those firms is \$57,000 with a sample standard deviation of \$4,000. Furthermore, suppose that 5 of the firms pay over \$60,000. Compute 95% confidence intervals for the salary and the proportion paying over \$60,000.

Supplementary Problems

9.47 A market researcher states that she has 95% confidence that the true monthly sales of a product will be between \$170,000 and \$200,000. Explain the meaning of this statement.

9.48 Suppose that a paint supply store wanted to estimate the actual amount of paint contained in one-gallon cans purchased from a nationally known manufacturer. It is known from the manufacturer's specifications that the standard deviation of the amount of paint is equal to .08 gallon. A random sample of 50 gallons is selected, and the average amount of paint per one-gallon can is 0.97 gallon. Set up a 99% confidence-interval estimate of the true amount of paint contained in a one-gallon can. Based on your results, do you think the store owner has a right to complain to the manufacturer? Why?

9.49 Determine the critical value of t in each of the following circumstances:

(a) $1 - \alpha = .95$, $n = 10$.
(b) $1 - \alpha = .99$, $n = 10$.
(c) $1 - \alpha = .95$, $n = 32$.
(d) $1 - \alpha = .95$, $n = 65$.
(e) $1 - \alpha = .90$, $n = 16$.

9.50 If σ_X were known and the random variable were normally distributed and a sample size of 10 were selected, which distribution would be more appropriate in estimating the true population mean, the normal or the t? Why?

9.51 A new breakfast cereal is to be test-marketed for one month at stores of a large supermarket chain. The results for a sample of 16 stores indicated average sales of $1,200 with a sample standard deviation of $180. Set up a 99% confidence-interval estimate of the true average sales of this new breakfast cereal.

9.52 The personnel department of a large corporation would like to estimate the family medical expenses of its employees in order to determine the feasibility of providing a medical insurance plan. A random sample of 10 employees revealed the following family medical expenses in the previous year:

$110, 362, 246, 85, 510, 208, 173, 425, 316, 179

(a) Set up a 90% confidence interval estimate of the average family medical expenses for all employees of this corporation?
(b) What assumption about the population distribution must be made in part (a)?

9.53 The manager of a branch of a local savings bank wanted to estimate the average amount held in passbook savings accounts by depositors at the bank. A random sample of 30 depositors was selected, and the results indicated a sample average of $4,750 and a sample standard deviation of $1,200. Set up a 95% confidence-interval estimate of the population average amount held in passbook savings accounts.

9.54 An auditor for a consumer protection agency would like to determine the proportion of claims that are paid by a health insurance company within two months of receipt of the claim. A random sample of 200 claims is selected, and it is determined that 80 were paid out within two months of the receipt of the claim. Set up a 99% confidence-interval estimate of the true proportion of the claims paid within two months.

• **9.55** A market researcher for a large consumer electronics company wanted to study television viewing habits of residents of a particular small city. A random sample of 40 respondents was selected, and each respondent was instructed to keep a detailed record of all television viewing in a particular week. The results were as follows:

Amount of viewing per week: $\overline{X} = 15.3$ hours, $S = 3.8$ hours;
27 respondents watched the Evening News on at least three weeknights.

Set up 95% confidence-interval estimates for each of the following:
(a) The average amount of television watched per week in this city.
(b) The proportion of respondents who watch the Evening News at least three nights per week.

If the market researcher wanted to take another survey in a different city:
(c) What sample size is required if he wishes to be 95% confident of being correct to within ± 2 hours and assumes the population standard deviation is equal to 5 hours?
(d) What sample size is needed if he wishes to be 95% confident of being within $\pm .035$ of the true proportion who watch the Evening News on at least three weeknights and no estimate is available based on past experience?

9.56 The real estate assessor for a county government wished to study various characteristics concerning single-family houses in the county. A random sample of 70 houses revealed the following:

Heating area of the house: $\overline{X} = 1{,}759$ sq ft, $S = 380$ sq ft;
42 houses had central air conditioning

(a) Set up a 99% confidence-interval estimate of the population average heating area of the house.
(b) Set up a 95% confidence-interval estimate of the population proportion of houses that had central air conditioning.

9.57 The personnel director of a large corporation wished to study absenteeism among clerical workers at the corporation's central office during last year. A random sample of 25 clerical workers revealed the following:

$\overline{X} = 9.7$ days, $S = 4.0$ days;
12 employees were absent more than 10 days

Set up 95% confidence-interval estimates of each of the following:
(a) The average number of days absent for clerical workers last year.
(b) The proportion of clerical workers absent more than 10 days last year.

If the personnel director also wishes to take a survey in a branch office:
(c) What sample size is needed if the director wishes to be 95% confident of being correct to ± 1.5 days and the population standard deviation is assumed to be 4.5 days?
(d) What sample size is needed if the director wishes to be 90% confident of being correct to within $\pm .075$ of the true proportion of workers who are absent more than 10 days and no estimate is available based upon previous experience?

Database Problems

The following problems refer to the sample data obtained from the questionnaire of Figure 2.5 and presented in Figure 2.10. They should be solved with the aid of an available computer package.

□ 9.58 Set up a 99% confidence interval for the proportion of alumni who went on to graduate school (question 5).

□ 9.59 State a 95% confidence interval for the average salary of all alumni (question 7).

□ 9.60 Construct a 90% confidence interval for the proportion of alumni who are female (question 18).

□ 9.61 Calculate a 95% confidence interval for the average reported grade-point index of alumni (question 15).

□ 9.62 Estimate a 95% confidence interval for the proportion of alumni who would attend the college again (question 16).

□ 9.63 Show the 95% confidence interval for the average age of alumni (question 20).

□ 9.64 Set up a 95% confidence-interval estimate of the average number of years since graduation (question 1).

□ 9.65 Set up a 99% confidence-interval estimate of the average number of years worked in the current job activity (question 10).

□ 9.66 Set up a 95% confidence-interval estimate of the average number of areas worked in during a professional career (question 12).

□ 9.67 Set up a 90% confidence-interval estimate of the average number of full-time jobs held in the last ten years (question 13).

□ 9.68 Set up a 95% confidence-interval estimate of the proportion of alumni who majored in business (question 2).

□ 9.69 Set up a 99% confidence-interval estimate of the proportion of alumni who chose the college primarily because of cost (question 3).

□ 9.70 Set up a 90% confidence-interval estimate of the proportion of alumni whose current political ideology is more conservative (question 4).

□ 9.71 Set up a 95% confidence-interval estimate of the proportion of alumni whose present job is at least somewhat related to their undergraduate major (question 6, codes 4 and 5).

□ 9.72 Set up a 99% confidence-interval estimate of the proportion of alumni who are employed by a corporation/company with under 100 employees (question 8).

□ 9.73 Set up a 95% confidence-interval estimate of the proportion of alumni whose current job activity is Marketing and Retailing (question 9).

□ 9.74 Set up a 90% confidence-interval estimate of the proportion of alumni whose first full-time job after graduation was in Accounting (question 11).

□ 9.75 Set up a 99% confidence-interval estimate of the proportion of alumni who found Accountancy to be the most valuable subject area for their career (question 14A).

□ 9.76 Set up a 95% confidence-interval estimate of the proportion of alumni who found Psychology to be the most important subject area in their daily nonworking activities (question 14B).

□ 9.77 Set up a 90% confidence-interval estimate of the proportion of alumni who would recommend or highly recommend the college to a close friend or relative (question 17, codes 4 and 5).

References

1. COCHRAN, W. G., *Sampling Techniques*, 3d ed. (New York: Wiley, 1977).
2. FISHER, R. A., AND F. YATES, *Statistical Tables for Biological, Agricultural and Medical Research*, 5th ed. (Edinburgh: Oliver & Boyd, 1957).
3. LARSEN, R. L., AND M. L. MARX, *An Introduction to Mathematical Statistics and Its Applications* (Englewood Cliffs, N.J.: Prentice-Hall, 1981).
4. PEARSON, E. S., AND KENDALL, M. *Studies in the History of Statistics and Probability* (Darien, Conn.: Hafner Press, 1970).
5. SNEDECOR, G. W., AND W. G. COCHRAN, *Statistical Methods*, 7th ed. (Ames, Iowa: Iowa State University Press, 1980).

Chapter 10

Hypothesis Testing for Quantitative Variables

10.1 INTRODUCTION: WHAT'S AHEAD

In Chapter 9 we focused on the *confidence interval* as a way of estimating the true value of a mean or proportion in a population. In this chapter we will begin to focus upon another aspect of statistical inference, *hypothesis testing*, in which decisions are made about population characteristics based only on sample information.

Case G—The University President's Evaluation of Student Grades and Employee Information

Suppose that as part of an annual report to be made to the trustees of a university, the president must provide detailed status reports concerning both academic and administrative aspects of the education process. In this case we shall consider four different issues:

1. The president would like to know whether the average grade-point index of undergraduate students has changed in the last twenty years. Twenty years ago the population mean was 2.50 with a population standard deviation of .40. A random sample of 100 students will be selected and their cumulative grade-point index will be computed at the end of a particular semester. Based on the sample results, the president must conclude that the average GPI has or has not changed from its previous value of twenty years past. In doing this sampling, she must weigh the consequences of making incorrect decisions based on sample evidence.
2. The president is particularly concerned about the amount of job experience possessed by administrative employees. In the past, the university has attempted to maintain an average employee job experience of six years. University records are to be sampled on the first day of next month to determine the amount of job experience for 30 employees. Based on the results of the sample, the president must report to the board of trustees on whether the average job experience of employees has changed from the previous norm.
3. In maintaining a smoothly functioning administrative process, it is important that the absenteeism level of employees be kept to a minimum. In order to probe the variability that exists in different employees' absenteeism rates (measured as number of days absent) the president wants to see whether there is any difference in yearly absenteeism levels between males and females.
4. Finally, the university has set up an experimental plan to (hopefully) *reduce* the amount of absenteeism among workers. That is, workers are to be paid an attendance bonus that has a graduated scale of incentives. At the end of the experimental year, a decision is to be made based on a sample of employees as to whether the incentive plan should be continued. The president will need to determine whether there is evidence that the set of incentives led to a decrease in absenteeism.

10.2 THE HYPOTHESIS-TESTING PROCEDURE

In each of the aforementioned situations, the president must make a decision based only on sample evidence. In Chapter 9 we used the concept of a sampling distribution to develop the confidence-interval estimate. Now we will extend this concept to another phase of statistical inference, the hypothesis-testing procedure.

Referring to the first question of interest to the president, one of the following two conclusions will be reached

1. The average grade-point index (GPI) is equal to 2.5.
2. The average grade-point index (GPI) is not equal to 2.5.

From the perspective of statistical hypothesis testing, these two conclusions would be represented as follows:

$$\text{(null hypothesis)} \quad H_0: \ \mu_X = 2.5$$
$$\text{(alternative hypothesis)} \quad H_1: \ \mu_X \neq 2.5$$

In this example, the **null hypothesis, H_0**, represents the conclusion that would be drawn if the average GPI had not changed. The null hypothesis can also be viewed from the perspective of the American legal process, in which innocence is presumed until guilt is proven. Thus we assume that the average is ''not different'' unless evidence of ''guilt'' is found showing that the average has changed. The **alternative hypothesis, H_1**, which is usually the negation of the null hypothesis, represents the conclusion that would be drawn if evidence of ''guilt'' were found. In our example, ''guilt'' refers to the average grade-point index being different from 2.5.

We may develop the logic underlying the hypothesis-testing procedure by contemplating how we can determine, based only on a sample, whether or not the population average is 2.5. If our sample average is 2.493, for example, we will be inclined to conclude that the average has not changed (that $\mu_X = 2.5$), since the sample mean is ''close to'' the hypothesized value of 2.5. Intuitively, we will be thinking that it is not unlikely that we could obtain a sample mean of 2.493 from a population whose mean was 2.5. On the other hand, if our sample average is 3.4, our instinct will be to conclude that the average is not 2.5 (that $\mu_X \neq 2.5$), since the sample mean is far away from the hypothesized value of 2.5. In this case we will be reasoning that it would be very unlikely that a sample mean of 3.4 could be obtained if the population mean was 2.5; therefore it is more reasonable to conclude that the mean is not equal to 2.5.

Unfortunately, the decision-making process is not always so clear-cut. It would be arbitrary for us to determine what is ''close'' and what is ''far away'' from the population mean. For this reason statistical hypothesis testing is employed to quantify the decision-making process so that the probability of obtaining a given sample result can be found. This is achieved by first determining the sampling distribution that the sample statistic (i.e., the mean) follows and then computing the appropriate **test statistic.** This test statistic measures how close the sample value has come to the null hypothesis. The test statistic often follows a well-known statistical distribution (such as the normal, t, etc.).

The appropriate distribution of the test statistic is divided into two regions, a **region of rejection** (sometimes called the **critical region**) and a **region of nonrejection** (see Figure 10.1). If the test statistic falls into the region of nonrejection, the null hypothesis cannot be rejected, and the president will conclude that the average GPI has not changed. If the test statistic falls into the rejection region, the null

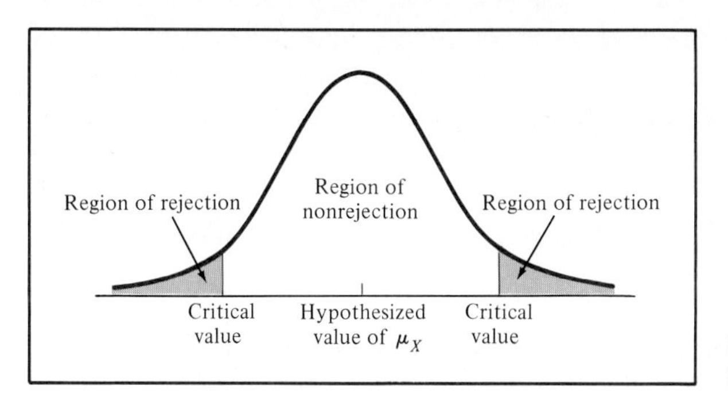

FIGURE 10.1 □ Regions of rejection and nonrejection in hypothesis testing.

hypothesis will be rejected and the president will conclude that the true mean GPI is not 2.5 (and therefore has changed).

The region of rejection may be thought of as consisting of the values of the test statistic which are unlikely to occur if the null hypothesis is true. On the other hand, they are not so unlikely to occur if the null hypothesis is false. Therefore, if we observe a value of the statistic which falls into the critical region, we reject the null hypothesis because the observation of that value would be unlikely to occur if the null hypothesis were true.

In order to make this decision concerning the null hypothesis, we must first determine the **critical value** for the statistical distribution of interest. The critical value divides the nonrejection region from the rejection region. However, the determination of this critical value depends on the size of the rejection region. As we will see in the next section, the size of the rejection region is directly related to the risks involved in using only sample evidence to make decisions about population parameters.

10.3 TYPE I AND TYPE II ERRORS

In using a sample to draw conclusions about a population there is a risk that an incorrect conclusion will be reached. Indeed, two different errors can occur in applying the hypothesis-testing approach to decision making concerning population parameters.

The first error is called the **Type I error**; the probability of a Type I error is called $\boldsymbol{\alpha}$. (α is the lower-case Greek letter alpha.)

A Type I error occurs if the null hypothesis H_0 is rejected when, in fact, it is true.

In our student-grades example, the Type I error would occur if we concluded (based on sample data) that the population GPI was *not* 2.5 when in fact it was 2.5. The probability of a Type I error (α) is also called the **level of significance**. Traditionally, the statistician controls the Type I error rate by establishing the risk level he or she is willing to tolerate in terms of rejecting a true null hypothesis. The selection of a particular risk level α is, of course, dependent on the cost of making a Type I error. Once the value for α is specified, the size of the rejection region is known, since α is the probability of rejection under the null hypothesis. From this fact the critical value or values that divide the rejection and nonrejection regions can be determined.

The second error, called the **Type II error**, occurs with a probability called $\boldsymbol{\beta}$. (β is the lower-case Greek letter beta.)

A Type II error occurs if the null hypothesis H_0 is not rejected when it is false and should be rejected.

In our student-grades example, a Type II error would occur if we concluded that the average GPI had not changed (that it remained at 2.5) when in fact the true population average GPI was different from 2.5. Unlike the Type I error rate α, which is preset at a specified value, the magnitude of the Type II error rate β is dependent on the actual population value of the mean. For example, if the true population average (which is unknown to us) were 3.4, there would be a small chance (β) of concluding that the average had not changed from 2.5. However, if the true population average were 2.49, we would have a high probability of concluding that the population GPI had not changed (and we would be making a Type II error).

The complement $(1 - \beta)$ of the probability of a Type II error is called the **power** of a statistical test.

The power of a statistical test $(1 - \beta)$ is the probability of rejecting the null hypothesis when it is false (and should be rejected).

In our student-grades example, the power of the test is the probability of concluding that the average GPI has changed if in fact it has actually changed (and the population mean is no longer 2.5).

An easy way to remember which probability goes with which type of error is to note that α is the first letter of the Greek alphabet, and it is used to represent the probability of a Type I error. The letter β is the second letter of the Greek alphabet and is used to represent the probability of a Type II error. (If you have trouble remembering the Greek alphabet, note that the word "alphabet" tells you its first two letters.)

Table 10.1 illustrates the two types of errors, the probability of each, and the power of the test.

TABLE 10.1 □ Hypothesis-testing outcomes and their probabilities

Statistical decision	Actual situation: H_0 True	Actual situation: H_0 False
Do not reject H_0	$1 - \alpha$	β (Type II error)
Reject H_0	α (Type I error)	$1 - \beta$ (power)

The usual procedure in research is to choose a level of α; this (along with sample size) will determine the level of β for a particular study design. Very often, α is chosen to be .05; sometimes it is .01; other values of α are less common. One way in which a researcher can control the probability of these errors is to increase the sample size. For a given level of α, increasing the sample size will decrease β and therefore increase the power to detect that H_0 is false.

For a given sample size the decision maker must balance the two types of errors. If α is to be decreased, then β will be increased. If β is to be decreased, then α will be increased. The values for α and β depend on the importance of each risk in a particular problem. The risk of a Type I error in the student-grades study involves concluding that the average GPI has changed when it has not changed. The risk of a Type II error involves concluding that the average GPI has not changed when in truth it has changed.

The choice of reasonable levels of α and β depends on the costs inherent in each type of error. For example, if it were very costly to make changes from the status quo, then we would want to be very sure that a change would be beneficial, so the risk of a Type I error might be most important and α would be kept very low. On the other hand, if we wanted to be very certain of detecting changes from a hypothesized mean, the risk of a Type II error would be most important and we might choose a higher level of α. Of course, by increasing the sample size, we could control both α and β, but there is always a limit on our resources, so we still must consider the tradeoffs between making Type I and Type II errors.

Problems

10.1 Explain in your own words what the terms null hypothesis and alternative hypothesis mean.

10.2 Draw a picture illustrating the concepts of region of rejection, region of nonrejection, and critical value, relative to a null hypothesis such as H_0: $\mu_X = 7.3$.

10.3 Explain what Type I and Type II errors are.

10.4 What is the probability of a Type I error called? Of a Type II error?

10.5 Which type of error (I or II) usually has its probability directly controlled by the choice of the person analyzing the data? What determines the probability of the other type of error?

10.6 How is power related to the probability of making a Type II error?

10.7 Tell why the following statement is false: The probability of making a Type II error (β) is equal to $1 - \alpha$.

10.8 Why is it possible for the null hypothesis to be rejected when in fact it is true? Why is it possible that the null hypothesis will not always be rejected when it is false?

• 10.9 For a given sample size, if the α risk is reduced from .05 to .01, what will happen to the β risk?

10.10 For H_0: $\mu_X = 100$, H_1: $\mu_X \neq 100$, and for a sample of size n, β will be larger if the actual value of μ_X is 90 than if the actual value of μ_X is 75. Why?

10.11 In the American legal system, a defendant is presumed innocent until proven guilty. Consider then a null hypothesis H_0 that the defendant is innocent and an alternative hypothesis H_1 that the defendant is guilty. A jury has two possible decisions: convict the defendant (i.e., reject the null hypothesis) or do not convict the defendant (i.e., do not reject the null hypothesis). Explain the meaning of the α and β risks in this example.

10.12 Suppose the defendant in Problem 10.11 were presumed guilty until proven innocent. How would the null and alternative hypotheses differ from those in Problem 10.11? What would be the meaning of the α and β risks?

10.4 TEST OF HYPOTHESIS FOR THE MEAN (σ_X KNOWN)

Now that we have defined the Type I and Type II errors, we may return to the first question of interest to the university president. For the student-grades problem, the null and alternative hypotheses were:

$$H_0: \quad \mu_X = 2.5$$
$$H_1: \quad \mu_X \neq 2.5$$

If we assume that the standard deviation σ_X is known, then based on the central limit theorem the sampling distribution of the mean would follow the normal distribution and the test statistic would be

$$Z = \frac{\overline{X} - \mu_X}{\frac{\sigma_X}{\sqrt{n}}} \tag{10.1}$$

In this formula, the numerator measures how far (in an absolute sense) the observed mean $\overline{X}$ is from the hypothesized mean μ_X. The denominator is the standard error, so Z represents how many standard errors $\overline{X}$ is from μ_X. With σ_X known, Z has a standard normal distribution.

If the president decided to choose a level of significance of 5%, the size of the rejection region would be .05, and the critical values of the normal distribution could be determined. Since the rejection region is divided into the two tails of the distribution, the 5% is divided into two equal parts of 2.5% each.

Since we have a normal distribution, the critical values can be expressed in standard-deviation units. A rejection region of .025 in each tail of the normal distribution results in an area of .475 between the hypothesized mean and each critical value. Looking up this area in the normal distribution (Table C.2), we find that the critical values that divide the rejection and nonrejection regions are $+1.96$ and -1.96. Figure 10.2 on page 280 illustrates this case: it shows that if the mean is actually 2.5, as H_0 claims, then the values of the test statistic Z will have a standard normal distribution centered at $\mu_X = 2.5$. Observed values of Z greater than 1.96 or less than -1.96 indicate that $\overline{X}$ is so far from 2.5 that it is unlikely that such a value would occur if H_0 were true.

Therefore the rejection rule would be

$$\text{Reject } H_0 \text{ if } Z > +1.96$$
$$\text{or if } Z < -1.96;$$
$$\text{otherwise do not reject } H_0.$$

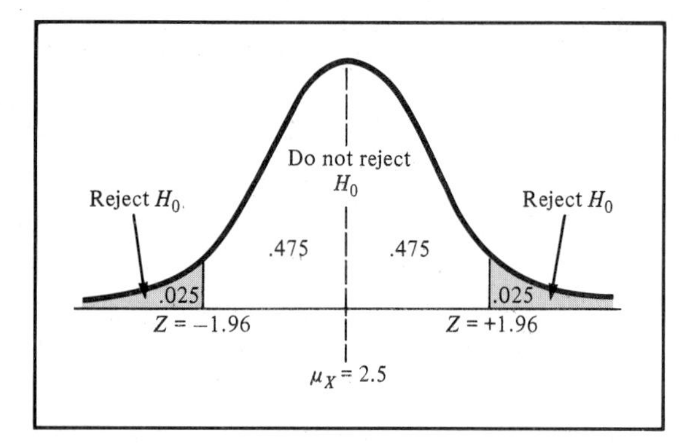

FIGURE 10.2 □ Testing a hypothesis about the mean (σ_X known) at the 5% level of significance.

Suppose that a sample of 100 student grade-point indexes are to be selected from all undergraduate students currently enrolled full time at the university. The standard deviation in the population was previously equal to .40; for illustrative purposes we assume that it has not changed, so that $\sigma_X = .40$. If the sample of 100 students had a mean GPI of 2.59, what conclusion would we reach? Using Equation (10.1), we have

$$Z = \frac{\bar{X} - \mu_X}{\dfrac{\sigma_X}{\sqrt{n}}}$$

$$= \frac{2.59 - 2.50}{\dfrac{.40}{\sqrt{100}}}$$

$$= \frac{+.09}{.04}$$

$$= +2.25$$

Since $Z = +2.25$, we can see that $+2.25 > +1.96$, so we reject H_0. The decision to reject H_0 means we are concluding that there is evidence that the average grade-point index is different from 2.5. Thus we may conclude that there is evidence that the population average GPI has changed in the last twenty years. However, we realize that there is a risk α (of .05) that we have rejected the null hypothesis when in fact the population average GPI has not changed in the last twenty years.

Problems

10.13 For the situation described in Problem 9.1 on page 247, test the null hypothesis that the average IQ in the population is 120. Use $\alpha = .05$.

• 10.14 For the situation described in Problem 9.2 on page 247, test the null hypothesis that the average yield is 140 bushels per acre. Use $\alpha = .01$.

10.15 For the situation described in Problem 9.3 on page 247, test the null hypothesis that the average height of women is 64 inches. Use $\alpha = .05$.

10.16 For the situation described in Problem 9.4 on page 247, test the null hypothesis that the average value on the blood test is 5.0. Use $\alpha = .10$.

10.17 For the situation described in Problem 9.5 on page 247, test the null hypothesis that the average bacteria count is 12,000. Use $\alpha = .01$.

10.5 SUMMARIZING THE STEPS OF HYPOTHESIS TESTING

Now that we have used the hypothesis-testing procedure to draw a conclusion about the population mean, it will be useful to summarize the steps involved in this procedure.

1. State the null hypothesis, H_0.
2. State the alternative hypothesis, H_1.
3. Choose the level of significance, α.
4. Choose the sample size, n.
5. Determine the appropriate statistical technique and corresponding test statistic to use.
6. Set up the critical values that divide the rejection and nonrejection regions.
7. Collect the data and compute the sample value of the appropriate test statistic.
8. Determine whether the test statistic has fallen into the rejection or the nonrejection region.
9. Make the statistical decision.
10. Express the statistical decision in terms of the problem.

Steps 1 and 2. The null and alternative hypotheses must be stated in statistical terms. In testing whether the average GPI was 2.5, the null hypothesis was that μ_X equals 2.5, while the alternative hypothesis was that μ_X was not equal to 2.5.

Step 3. The level of significance is specified according to the relative importance of Type I and Type II errors in the problem. We chose $\alpha = .05$; along with the sample size, this determines β, which we did not explicitly determine here.

Step 4. The sample size is determined after taking into account the desired α and β risks and considering budget constraints in carrying out the study. Here 100 students were selected.

Step 5. The statistical technique that will be used to test the null hypothesis must be chosen. Since σ_X was assumed to be known, a Z test was selected.

Step 6. Once the null and alternative hypotheses are specified and the level of significance and sample size are determined, the critical values for the appropriate statistical distribution can be found so that the rejection and nonrejection regions can be indicated. Here the values 1.96 and -1.96 were used to define these regions.

Step 7. The data are collected and the value of the test statistic is computed. Here, $\bar{X} = 2.59$, so $Z = +2.25$.

Step 8. The value of the test statistic is compared to the critical values for the appropriate distribution to determine whether it falls into the rejection or nonrejection region. Here, $Z = 2.25$ is in the rejection region, since $2.25 > 1.96$.

Step 9. The hypothesis-testing decision is made. If the test statistic falls into the nonrejection region, the null hypothesis H_0 cannot be rejected. If the test statistic falls into the rejection region, the null hypothesis is rejected. Here, H_0 is rejected.

Step 10. The consequences of the hypothesis-testing decision must be expressed in terms of the actual problem involved. Thus in our student-grades problem we concluded that, since there was evidence that the population mean was different from 2.5, the president should conclude that the average GPI had changed in the last twenty years.

10.6 TEST OF HYPOTHESIS FOR THE MEAN (σ_X UNKNOWN)

Now that we have studied a situation in which the population standard deviation was known, let us turn to the second question of interest to the university president. Recall that the board of trustees wanted to know whether the average amount of employee experience had shifted from a previous norm of 6 years. For this example the null and alternative hypotheses may be stated as

$$H_0: \mu_X = 6$$
$$H_1: \mu_X \neq 6$$

However, in this example only the standard deviation of the sample (S) is known. Section 9.3 stated that if the population is assumed to be normally distributed, the sampling distribution of the mean will follow a t distribution with $n - 1$ degrees of freedom. The test statistic for determining the difference between the sample mean $\bar{X}$ and the population mean μ_X when the sample standard deviation S is used is given by

$$t = \frac{\bar{X} - \mu_X}{\frac{S}{\sqrt{n}}} \quad \textbf{(10.2)}$$

TABLE 10.2 ☐ Determining the critical value from the *t* table for an area of .05 in each tail with 29 degrees of freedom

Degrees of freedom	Upper tail areas					
	.25	.10	.05	.025	.01	.005
1	1.0000	3.0777	6.3138	12.7062	31.8207	63.6574
2	0.8165	1.8856	2.9200	4.3027	6.9646	9.9248
3	0.7649	1.6377	2.3534	3.1824	4.5407	5.8409
4	0.7407	1.5332	2.1318	2.7764	3.7469	4.6041
5	0.7267	1.4759	2.0150	2.5706	3.3649	4.0322
⋮	⋮	⋮		⋮	⋮	⋮
26	0.6840	1.3150	1.7056	2.0555	2.4786	2.7787
27	0.6837	1.3137	1.7038	2.0518	2.4727	2.7707
28	0.6834	1.3125	1.7011	2.0484	2.4671	2.7633
29	0.6830	1.3114	1.6991	2.0452	2.4620	2.7564
30	0.6828	1.3104	1.6973	2.0423	2.4573	2.7500

In practice it has been found that as long as the sample size is not small and the population is not very skewed, the *t* distribution gives a good approximation to the sampling distribution of the mean.

Suppose that the sample of 30 employees reveals that $\bar{X} = 5.7$ and $S = 2.2$. If a level of significance of .10 is selected, the critical values of the *t* distribution with $30 - 1 = 29$ degrees of freedom can be obtained from Table C.3, as illustrated in Table 10.2 and Figure 10.3.

The rejection region of .10 is divided into two equal parts of .05 each. Using the *t* table given in Table C.3, the critical values are -1.6991 and $+1.6991$. The decision rule is:

$$\text{Reject } H_0 \text{ if } t > t_{29} = +1.6991$$
$$\text{or if } t < -t_{29} = -1.6991;$$
$$\text{otherwise do not reject } H_0.$$

Using Equation (10.2), we have

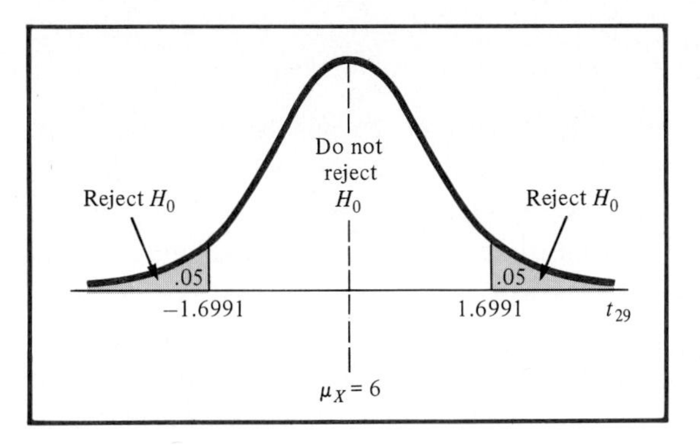

FIGURE 10.3 ☐ Testing a hypothesis about the mean (σ_X unknown) at the 10% level of significance with 29 degrees of freedom.

$$t = \frac{5.7 - 6.0}{\frac{2.2}{\sqrt{30}}}$$

$$= \frac{-.30}{.40} = -.75$$

Since $t = -.75$, we see that $-1.6991 \le -.75 \le +1.6991$. Thus our decision is to not reject H_0. We would conclude that the average experience is still 6 years. Alternatively, to take into account the possibility of a Type II error, we may phrase this conclusion as, "There is no evidence that the average is different from 6 years."

Problems

10.18 The manager of the credit department for an oil company would like to determine whether the average monthly balance of credit card holders is equal to $75. An auditor selects a random sample of 100 accounts and finds that the average owed is $83.40 with a sample standard deviation of $23.65. Using the .05 level of significance, should the auditor conclude that there is evidence that the average balance is different from $75?

• 10.19 Referring to the data of Problem 9.6 on page 253, is there evidence that the average life of light bulbs produced is different from 500 hours? Use $\alpha = .01$.

10.20 Referring to the data of Problem 9.7 on page 253, is there evidence that the average mass of the riton is different from 2.10×10^{-24} gram? Use $\alpha = .01$.

• 10.21 Referring to the data of Problem 9.10 on page 253, is there evidence that the average amount of Vitamin C in oranges is different from 120 milligrams? Use $\alpha = .05$.

10.22 Referring to the data of Problem 9.11 on page 253, is there evidence that the average tax revenue per return is different from $12,000? Use $\alpha = .05$.

10.23 Referring to the data of Problem 9.14 on page 254, is there evidence that the average tuition at colleges in the Northeast is different from $10,000? Use $\alpha = .05$.

10.7 ONE-TAILED TESTS

In Section 10.4 we examined the question of whether the population average GPI was equal to 2.5. The alternative hypothesis contained two possibilities: either the average could be less than 2.5, or the average could be more than 2.5. For this reason the rejection region was divided into two tails of the distribution. In many instances the alternative hypothesis focuses in a particular direction. For instance, the board of trustees might be interested only in whether the average GPI has *increased* above 2.5 in the last twenty years. This would mean that unless the sample average was "significantly above" 2.5, the average GPI would be considered to

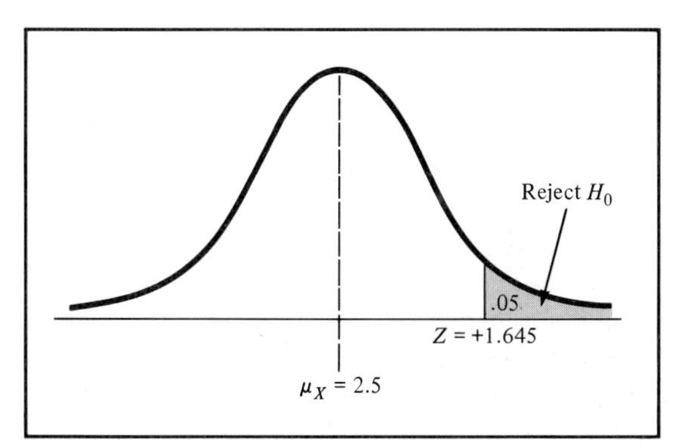

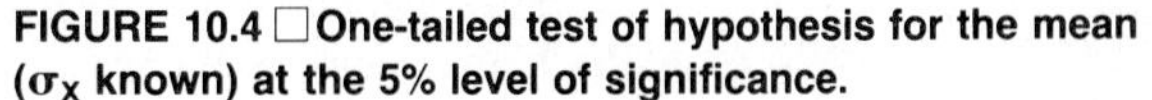
FIGURE 10.4 □ One-tailed test of hypothesis for the mean (σ_X known) at the 5% level of significance.

have either fallen or stayed the same. The null and alternative hypotheses would be stated as follows:

$$H_0: \quad \mu_X \leq 2.5 \text{ (GPI has not increased)}$$
$$H_1: \quad \mu_X > 2.5 \text{ (GPI has increased)}$$

The rejection region would be entirely contained in the upper tail of the distribution, since we want to reject H_0 only when the sample mean is significantly above 2.5. If we again use an α level of .05, the critical value on the normal distribution can be determined. Since the entire rejection region of .05 is in the upper tail of the distribution, the critical value (see Table C.2) will be 1.645 (see Figure 10.4).

Thus the decision would be to reject the null hypothesis if the sample statistic is above +1.645, otherwise do not reject the null hypothesis. For the data of Section 10.4, since $Z = +2.25 > +1.645$, our conclusion would be to reject H_0 and state that there is evidence that the average GPI is above 2.5.

This conclusion happens to be the same as we reached when we used the two-tailed test in Section 10.4. However, this will not always be the case. Assuming that we have correctly specified the direction of the alternative hypothesis, the one-tailed test will provide a higher chance (i.e., power) of rejecting the null hypothesis. Thus, when the test for the mean involves a *particular direction*, a one-tailed test should be utilized.

Problems

10.24 Referring to the data of Problem 9.1 on page 247, is there evidence that the mean IQ is greater than 110? Use $\alpha = .05$.

10.25 Referring to the data of Problem 9.5 on page 247, is there evidence that the mean bacteria count is greater than 10,000? Use $\alpha = .01$.

• 10.26 Referring to the data of Problem 9.6 on page 253, is there evidence that the average length of the life of the bulbs is less than 600 hours? Use $\alpha = .05$.

• **10.27** Referring to the data of Problem 9.10 on page 253, is there evidence that the average orange has less than 120 milligrams of Vitamin C? Use $\alpha = .01$.

10.28 Referring to the data of Problem 9.11 on page 253, is there evidence that the average amount paid in taxes is greater than \$10,000? Use $\alpha = .05$.

10.29 The Glen Valley Steel Company manufactures steel bars. The production process turns out steel bars with an average length of at least 2.8 feet when the process is working properly. A sample of 25 bars is selected from the production line. The sample indicates an average length of 2.43 feet and a standard deviation of .20 foot. The company wishes to determine whether the machine needs any adjustment.

(a) State the null and alternative hypotheses.

(b) If the company wishes to test the hypothesis at the .05 level of significance, what decision would it make?

• **10.30** A large nationwide chain of home improvement centers is having an end-of-season clearance of lawnmowers. The number of lawnmowers sold during this sale at a sample of 10 stores was as follows:

8 11 0 4 7 8 10 5 8 3

(a) At the .05 level of significance, is there evidence that an average of more than 5 lawnmowers per store have been sold during this sale?

(b) What assumption is necessary to perform this test?

10.8 TESTING THE DIFFERENCE BETWEEN THE MEANS OF TWO INDEPENDENT POPULATIONS

Thus far in this chapter we have studied situations in which we were testing a hypothesis about the mean of a single population. We may now extend our hypothesis-testing concepts to situations in which we would like to determine whether there is any difference between the means of two independent populations. This will enable us to investigate the third question of interest to the university president: whether or not there is a difference in absenteeism between males and females. The test to be performed can be either two-tailed or one-tailed, depending on whether we are testing whether the two population means are merely *different* or whether one population mean is *greater than* the other mean. A two-tailed test is used if we are testing to see whether the two population means are significantly different. A one-tailed test is used if we are testing to see whether one population mean is significantly greater than the other.

Two-tailed test	One-tailed test	One-tailed test
H_0: $\mu_1 = \mu_2$ H_1: $\mu_1 \neq \mu_2$	H_0: $\mu_1 \geq \mu_2$ H_1: $\mu_1 < \mu_2$	H_0: $\mu_1 \leq \mu_2$ H_1: $\mu_1 > \mu_2$

where μ_1 = mean of population 1
μ_2 = mean of population 2

The statistic used to determine the difference between the population means is based upon the difference between the sample means $(\bar{X}_1 - \bar{X}_2)$. Because of the central limit theorem, this statistic $(\overline{X}_1 - \overline{X}_2)$ will approximately follow the normal distribution for large enough sample sizes. If σ_1 and σ_2 are known, the test statistic is

$$Z = \frac{\overline{X}_1 - \overline{X}_2}{\sqrt{\dfrac{\sigma_1^2}{n_1} + \dfrac{\sigma_2^2}{n_2}}} \tag{10.3}$$

where $\overline{X}_1$ is the sample mean in population 1
$\overline{X}_2$ is the sample mean in population 2
σ_1^2 is the population variance for group 1
σ_2^2 is the population variance for group 2
n_1 is the sample size for population 1
n_2 is the sample size for population 2

EXAMPLE 10.1

The quality-control manager at an electric light bulb factory would like to determine whether there is any difference in the average life of bulbs manufactured on two different machines. Machine I is known to have a population standard deviation of 110 hours, Machine II a population standard deviation of 125 hours. A random sample of 25 light bulbs obtained from Machine I indicated a sample mean of 375 hours, a sample of 25 light bulbs from Machine II indicated a sample mean of 362 hours. Using the .01 level of significance, is there any evidence of a difference in the average life of bulbs produced by the two machines?

Solution

For this problem the null and alternative hypotheses would be

$$H_0: \mu_1 = \mu_2$$
$$H_1: \mu_1 \neq \mu_2$$

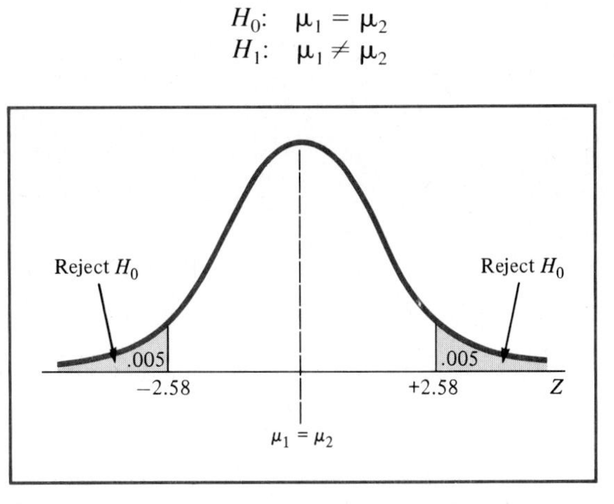

Two-tailed test of hypothesis for the difference between two means at the 1% level of significance.

Since the test is being conducted at the $\alpha = .01$ level of significance, from Table C.2 the critical values would be -2.58 and $+2.58$. Thus the decision rule would be (see figure on page 287):

$$\text{Reject } H_0 \text{ if } Z > +2.58 \text{ or } Z < -2.58; \text{ otherwise do not reject } H_0.$$

For our data we have

$$Z = \frac{\bar{X}_1 - \bar{X}_2}{\sqrt{\frac{\sigma_1^2}{n_1} + \frac{\sigma_2^2}{n_2}}} = \frac{375 - 362}{\sqrt{\frac{110^2}{25} + \frac{125^2}{25}}} = +.39 < +2.58$$

Therefore we do not reject H_0. We conclude that there is no evidence of a difference in the average life of light bulbs produced by the two types of machines.

As mentioned previously, in most cases we do not know the standard deviation of either of the two populations (σ_1, σ_2). Usually we know only the sample means and the sample standard deviations ($\bar{X}_1$, $\bar{X}_2$; S_1, S_2). If the assumptions are made that each population is normally distributed and that the population variances are equal ($\sigma_1^2 = \sigma_2^2$), the t distribution with $n_1 + n_2 - 2$ degrees of freedom can be used to test for the difference between the means of the two populations. Since we are assuming equal variances in the two populations, the variances of the two samples can be pooled together (i.e., combined) to form one estimate (S_p^2) of the common population variance. The test statistic will be

$$t = \frac{\bar{X}_1 - \bar{X}_2}{\sqrt{S_p^2\left(\frac{1}{n_1} + \frac{1}{n_2}\right)}} \qquad \textbf{(10.4)}$$

where[1]

$$S_p^2 = \frac{(n_1 - 1)S_1^2 + (n_2 - 1)S_2^2}{n_1 + n_2 - 2}$$

[1] For the special case where the two sample sizes n_1 and n_2 are equal, the formula for the pooled variance can be simplified. Thus, if $n_1 = n_2$, then

$$S_p^2 = \frac{S_1^2 + S_2^2}{2}$$

S_P^2 is the pooled variances of the two groups
$\overline{X}_1$ is the sample mean in population 1
S_1^2 is the sample variance in population 1
n_1 is the sample size for population 1
$\overline{X}_2$ is the sample mean in population 2
S_2^2 is the sample variance in population 2
n_2 is the sample size for population 2

Since the president wanted to determine whether there was a *difference* in average absenteeism between males and females, we have a two-tailed test in which the null and alternative hypotheses are

$$H_0: \mu_1 = \mu_2$$
$$H_1: \mu_1 \neq \mu_2$$

Suppose the results of the sample of 30 employees revealed the following summary information for the 18 males and 12 females on the number of days they were absent:

Males	**Females**
$n_M = 18$	$n_F = 12$
$\overline{X}_M = 8.4$	$\overline{X}_F = 7.6$
$S_M = 3.0$	$S_F = 2.5$

Since there are 18 males and 12 females, the t distribution with $18 + 12 - 2 = 28$ degrees of freedom would be utilized. If the test was conducted at the .05 level of significance, the critical values would be -2.0484 and $+2.0484$ (see Figure 10.5 on page 290) and the decision rule is:

Reject H_0 if $t > +2.0484$
or if $t < -2.0484$;
otherwise do not reject H_0.

For our data we have

$$t = \frac{\overline{X}_1 - \overline{X}_2}{\sqrt{S_P^2\left(\frac{1}{n_1} + \frac{1}{n_2}\right)}}$$

where

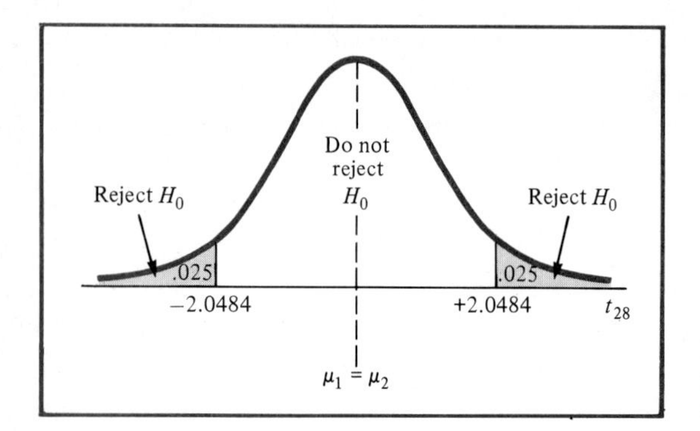

FIGURE 10.5 □ Two-tailed test of hypothesis for the difference between two means at the 5% level of significance.

$$S_P^2 = \frac{(n_1 - 1)S_1^2 + (n_2 - 1)S_2^2}{n_1 + n_2 - 2}$$

$$= \frac{17(3.0)^2 + 11(2.5)^2}{18 + 12 - 2}$$

$$= \frac{221.75}{28} = 7.92$$

so that

$$t = \frac{8.4 - 7.6}{\sqrt{7.92\left(\frac{1}{18} + \frac{1}{12}\right)}} = \frac{.8}{\sqrt{1.10}} = \frac{.8}{1.049} = .763$$

Since $-2.0484 \leq +.763 \leq +2.0484$, we do not reject H_0. Thus the president would conclude that there is no evidence of a significant difference in the average absenteeism between males and females.

In testing for the difference between the means, we have assumed that we were sampling from normal distributions and we have assumed equality of variances in the two populations. We must examine the consequences on the t test of departures from each of these assumptions. With respect to the assumption of normality, the t test is a "robust" test; it is not sensitive to modest departures from normality. As long as the sample sizes are not very small, the assumption of normality can be violated without serious effect on the power of the test. When the second assumption, equality of variances, is violated, it creates what is called the **Behrens-Fisher problem.** When σ_1^2 is significantly different from σ_2^2, an approximate t test can be utilized (see Reference 1).

Problems

10.31 Referring to the data of Problem 3.68 on page 70, is there evidence that the average rent is higher in Manhattan than Brooklyn? Use $\alpha = .05$.

• **10.32** Referring to the data of Problem 3.75 on pager 73, is there evidence of a difference between German and Japanese cars in their rate of acceleration? Use $\alpha = .10$.

10.33 Referring to the data of Problem 3.76 on pager 73, is there evidence of a difference in the average tuition rate between colleges in the Northeast and the Midwest? Use $\alpha = .05$.

10.34 Referring to the data of Problem 4.11 on pager 83, is there evidence that the average noise is higher for gas-powered chainsaws than for electric-powered chainsaws? Use $\alpha = .01$. (*Note*: The first 13 observations are from the gas-powered chain saws.)

• **10.35** Referring to the data of Problem 4.12 on pager 83, is there evidence that the average number of calories is higher for the 4% milkfat cottage cheeses than for the 2% milkfat cottage cheeses? Use $\alpha = .05$. (*Note*: The first 24 observations are from 4% milkfat cottage cheeses.)

• **10.36** A consumer testing agency found that a sample of 44 White Plume pens had enough ink to write an average of 3,421 feet, with a sample standard deviation of 521 feet, while 37 Purple Peacock pens wrote an average of 4,284 feet, with a sample standard deviation of 633 feet.

(a) At the .10 significance level, is there evidence of a difference in how far the pens of the two manufacturers will write?

(b) If you (incorrectly) used the formula which assumed that the standard deviations were known population values instead of estimates, would your conclusion be any different?

10.37 Management of the Sycamore Steel Co. wishes to determine if there is any difference in performance between the day shift of workers and the evening shift of workers. A sample of 100 day-shift workers reveals an average output of 74.3 parts per hour with a sample standard deviation of 16 parts per hour. A sample of 100 evening-shift workers reveals an average output of 69.7 parts per hour with a sample standard deviation of 18 parts per hour. At the .10 level of significance, is there evidence of a difference in output between the day shift and the evening shift?

10.38 An independent testing agency has been contracted to determine whether there is any difference in gasoline mileage output of two different gasolines on the same model automobile. Gasoline A was tested on 200 cars and produced a sample average of 18.5 miles per gallon with a sample standard deviation of 4.6 miles per gallon. Gasoline B was tested on a sample of 100 cars and produced a sample average of 19.34 miles per gallon with a sample standard deviation of 5.2 miles per gallon. At the .05 level of significance, is there evidence of a difference in performance of the two gasolines?

10.9 TESTING THE MEAN DIFFERENCE IN MATCHED OR PAIRED SAMPLES

10.9.1 Rationale

Now that we have discussed the test for the difference between the means of two independent samples, we may turn to the fourth and last question of interest to the university president. We may recall that the board of trustees wishes to determine whether the experimental incentive policy has reduced absenteeism from the previous year. The president realizes that the samples from year 1 and year 2 could be selected in two different ways. The first method would simply be to select a sample from the personnel records in year 1 and a different sample in year 2. The second method would be to take a sample of the employees and study their absenteeism records in both year 1 and year 2.

The first method of selection is an example of taking two *independent* samples. Unfortunately, this approach suffers from the fact that there is variability between the two different samples as well as between the two years. The second method of selection is an example of taking *related* samples; the results from year 1 (the first sample) are not independent of the results from year 2 (the second sample). We examined such a method in the last chapter when we constructed a confidence interval for the average difference in wear of pairs of shoes. In this instance the related feature occurs because *repeated measurements* have been obtained from the same individuals. This is advantageous here because it serves to reduce the variability between the employees and allows us to focus on the *difference* in absenteeism between year 1 and year 2 for each employee.[2]

10.9.2 Development

In order to determine whether any difference exists between the two related samples, we must obtain the individual values for each group (as shown in Table 10.3 on page 293), as was done for the confidence interval for related samples in Section 9.4.

From the central limit theorem, the average difference $\overline{D}$ follows the normal distribution when the population standard deviation of the difference σ_D is known and the sample size is large enough. In practice, however, only the sample standard deviation of the difference S_D is usually available. If it can be assumed that the population of D values is normally distributed, then the test statistic t with $(n - 1)$ degrees of freedom takes the form

$$t = \frac{\overline{D}}{\frac{S_D}{\sqrt{n}}} \qquad \textbf{(10.5)}$$

[2] A second approach to the related-samples problem involves the matching of items or individuals according to some characteristic of interest. For example, if a market researcher wished to evaluate the effect of two product packaging designs on the sales of a laundry product, the supermarkets studied could be matched according to their sales volume and other characteristics, and then randomly assigned to one package design or the other (see Example 10.2). Thus, regardless of whether matched (paired) samples or repeated measurements are utilized, the objective is to study the difference between two groups by reducing the effect of the variability due to the items or individuals themselves.

TABLE 10.3 □ Determining the difference between two related groups

	Group A	Group B	Difference, D	Squared difference, D^2
1	X_{A1}	X_{B1}	$D_1 = X_{A1} - X_{B1}$	D_1^2
2	X_{A2}	X_{B2}	$D_2 = X_{A2} - X_{B2}$	D_2^2
.	.	.	.	.
.	.	.	.	.
.	.	.	.	.
i	X_{Ai}	X_{Bi}	$D_i = X_{Ai} - X_{Bi}$	D_i^2
.	.	.	.	.
.	.	.	.	.
.	.	.	.	.
n	X_{An}	X_{Bn}	$D_n = X_{An} - X_{Bn}$	D_n^2

where X_{Ai} = ith value in group A
X_{Bi} = ith value in group B
$D_i = X_{Ai} - X_{Bi}$ = difference between the ith value in group A and the ith value in group B
D_i^2 = squared difference between the ith value in group A and the ith value in group B

where

$$\bar{D} = \frac{\sum D}{n}$$

$$S_D = \sqrt{\frac{\sum D^2 - \frac{(\sum D)^2}{n}}{n - 1}}$$

and n = sample size.

The three panels of Figure 10.6 indicate the null and alternative hypothesis and rejection regions for the possible one-tailed and two-tailed tests. If, as shown in

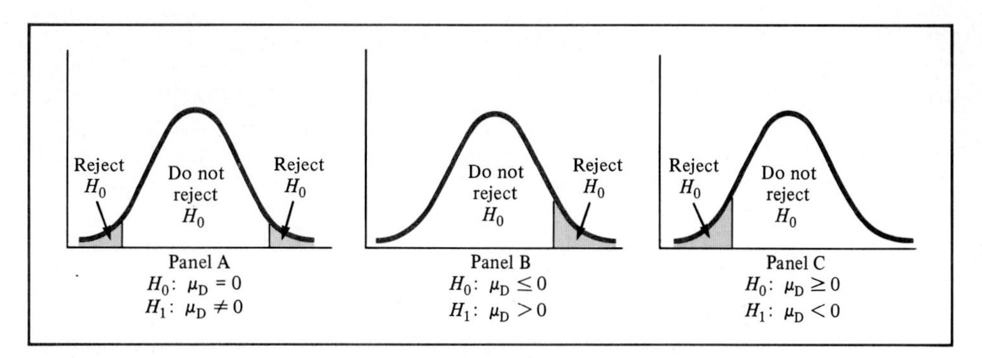

FIGURE 10.6 □ Testing for the difference between two means in related samples. Panel A, two-tailed test. Panel B, one-tailed test. Panel C, one-tailed test.

panel A, the test of hypothesis is two-tailed, the rejection region is split into the lower and upper tails of the t distribution. If, however, the test is one-tailed, the rejection region is in either the upper tail (panel B) or the lower tail (panel C) of the t distribution, depending on the direction of the alternative hypothesis.

10.9.3 Application

To apply the test for the difference between the means of two related samples, we may refer back to the absenteeism problem of interest to the university president. Suppose a sample of 15 employees was selected and their absenteeism (in days and parts of days) for year 1 and year 2 was recorded as illustrated in Table 10.4.

TABLE 10.4 □ Absenteeism for a random sample of 15 employees in year 1 and year 2

	Year		Difference	
Employee	1	2	D	D^2
1	3.0	2.8	+0.2	0.04
2	6.7	5.1	+1.6	2.56
3	11.3	8.4	+2.9	8.41
4	5.0	5.0	0.0	0.00
5	9.4	6.2	+3.2	10.24
6	15.7	12.2	+3.5	12.25
7	8.0	10.0	−2.0	4.00
8	10.0	6.8	+3.2	10.24
9	9.7	6.0	+3.7	13.69
10	4.5	4.2	+0.3	0.09
11	7.3	5.0	+2.3	5.29
12	3.1	3.0	+0.1	0.01
13	1.0	0.5	+0.5	0.25
14	8.6	5.9	+2.7	7.29
15	7.0	4.8	+2.2	4.84
Totals			$\Sigma D = 24.4$	$\Sigma D^2 = 79.2$

For these data,

$$\Sigma D = 24.4, \qquad \Sigma D^2 = 79.2, \qquad n = 15$$

Thus

$$\bar{D} = \frac{\Sigma D}{n} = \frac{24.4}{15} = 1.627$$

and

$$S_D^2 = \frac{\sum D^2 - \frac{(\sum D)^2}{n}}{n-1}$$

$$= \frac{79.2 - \frac{(24.4)^2}{15}}{14}$$

$$= \frac{39.509}{14}$$

$$= 2.822$$

and therefore $S_D = 1.68$.

Since the president is interested in determining whether there is evidence that average absenteeism is *less* in year 2 (i.e., *greater* in year 1), we have a one-tailed test in which the null and alternative hypotheses can be stated as follows:

H_0: $\mu_D \leq 0$ (that is, $\mu_1 - \mu_2 \leq 0$; absenteeism is not higher in year 1)
H_1: $\mu_D > 0$ (that is, $\mu_1 - \mu_2 > 0$; absenteeism is higher in year 1)

Since a sample of 15 employees was chosen, if a level of significance of 1% is selected, the decision rule can be stated as follows (see Figure 10.7):

Reject H_0 if $t > t_{14} = 2.6245$;
otherwise do not reject H_0.

From Equation (10.5) the test statistic is

$$t = \frac{\bar{D}}{\frac{S_D}{\sqrt{n}}}$$

and therefore we have

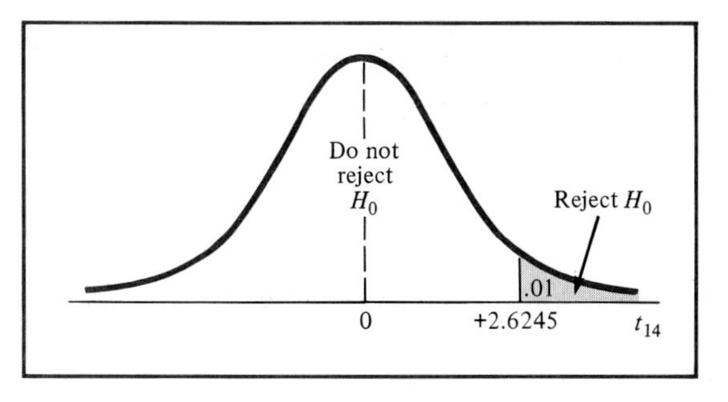

FIGURE 10.7 □ One-tailed test for the paired difference at the 1% level of significance with 14 degrees of freedom.

$$t = \frac{1.627}{\frac{1.68}{\sqrt{15}}} = 3.75$$

Since 3.75 > 2.6245, we reject H_0. We would conclude that there is evidence that absenteeism is higher in year 1 than in year 2. Thus we may state that the experimental attendance incentive plan instituted in year 2 resulted in a significant decrease in absenteeism as compared to year 1.

EXAMPLE 10.2

The project director for a large consumer products company was interested in studying the effect of a new product packaging design on sales of a laundry stain remover. The new package design was to be test-marketed over a period of a month in a sample of supermarkets in a particular city. A random sample of 10 pairs of supermarkets were matched according to weekly sales volume and other demographic characteristics. One store of each pair (the experimental group) was to sell the laundry stain remover in its new package, while the other member of the pair (the control group) was to sell the laundry stain remover in its old package. At the .05 level of significance, the project director would like to know whether sales (in thousands of units sold) are higher for the new package design. The following data indicate the results over the test period of one month:

Pair	New package	Old package	Difference (D)
1	4.58	3.97	+0.61
2	5.19	4.88	+0.31
3	3.94	4.09	−0.15
4	6.32	5.87	+0.45
5	7.68	6.93	+0.75
6	3.48	4.00	−0.52
7	5.72	5.08	+0.64
8	7.04	6.95	+0.09
9	5.27	4.96	+0.31
10	5.84	5.13	+0.71

***Solution*:**

For these data

$$\Sigma D = 3.20, \qquad \Sigma D^2 = 2.544, \qquad n = 10$$

Thus

$$\overline{D} = \frac{\Sigma D}{n} = \frac{3.20}{10} = .32$$

and

$$S_D^2 = \frac{2.544 - \frac{(3.2)^2}{10}}{9}$$

$$S_D^2 = \frac{1.52}{9} = .1689$$

$$S_D = \sqrt{.1689} = .411$$

Since the project director wants to determine whether there is evidence that the new package design has increased sales, we have a one-tailed test in which the null and alternative hypotheses can be stated as follows:

$$H_0\text{: } \mu_D \leq 0 \qquad (\mu_{\text{new}} \leq \mu_{\text{old}})$$
$$H_1\text{: } \mu_D > 0 \qquad (\mu_{\text{new}} > \mu_{\text{old}})$$

Since a sample of 10 stores was chosen, if a level of significance of 5% is selected, the decision rule can be stated as follows (see figure):

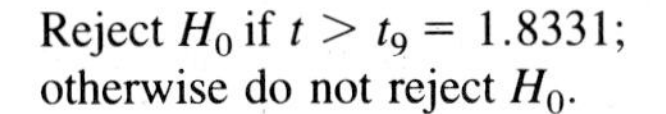

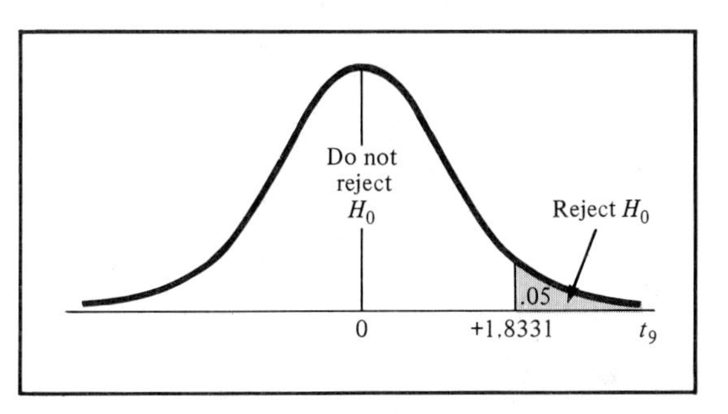

☐**One-tailed test for the paired difference at the 1% level of significance with 9 degrees of freedom.**

Using Equation (10.5), we have

$$t = \frac{.32}{\frac{.411}{\sqrt{10}}} = +2.462$$

Since $+2.462 > +1.8331$, we reject H_0. We conclude that there is evidence that sales are higher with the new package design. In the light of these findings, the project director might want to change the package design for the laundry stain remover and adopt the new package.

Problems

• 10.39 Referring to the data of Problem 9.18 on page 256:

(a) Is there evidence that E-Z-Sleep is more effective (i.e., that the number of hours of sleep is higher) than a placebo? Use $\alpha = .05$.

(b) Suppose that in part (a) you merely wanted to determine whether there was any evidence of a difference between E-Z-Sleep and a placebo.
(1) What then would be the null and alternative hypotheses?
(2) Compare the results of testing (b)(1) with those obtained in part (a). Explain.
(3) Which approach [(a) or (b)(1)] seems to make the most sense here? Why?

• 10.40 Referring to the data of Problem 9.19 on page 256, is there any evidence that the average driving skills are higher for people driving when they are sober? Use $\alpha = .01$.

10.41 Referring to the data of Problem 9.20 on page 256, is there any evidence of a difference in the average yield between the two varieties of corn seed? Use $\alpha = .05$.

10.42 Referring to the data of Problem 9.21 on page 256.
(a) Is there evidence that the average length of time to complete the circuit is higher at 40 psi than at 65 psi? Use $\alpha = .05$.
(b) What conclusions can you reach about the effect of tire pressure on the ability to pedal a bicycle?

10.43 A group of engineering students decide to see whether cars which supposedly do not need high-octane gasoline get more miles per gallon using regular or high-octane gas. They test several cars, using both types of gas in each car at different times. The mileage in each condition for each car is:

	Car									
Gas Type	**A**	**B**	**C**	**D**	**E**	**F**	**G**	**H**	**I**	**J**
Regular	15	23	21	35	42	28	19	32	31	24
High-octane	18	21	25	34	47	30	19	27	34	20

Is there any evidence of a difference in the average gasoline mileage between regular and high-octane gas? Use $\alpha = .05$.

10.44 A systems analyst is testing the feasibility of using a new computer system. The analyst will switch the processing to the new system only if there is evidence that the new system uses less processing time than the old system. In order to make a decision, a sample of seven jobs was selected and the processing time in seconds was recorded on the two systems with the following results:

Job	Old	New
1	8	6
2	4	3
3	10	7
4	9	8
5	8	5
6	7	8
7	12	9

(a) At the .01 level of significance, is there evidence that the old system uses more processing time?
(b) What assumption is necessary to perform this test?

10.45 The personnel director of a large company would like to know whether it will take less time to type a standard monthly summary report with a word processing typewriter than with a standard electric typewriter. A random sample of 7 secretaries was selected and the amount of typing time (in hours) was recorded with the following results:

	Secretary						
Equipment	**A**	**B**	**C**	**D**	**E**	**F**	**G**
Word processing typewriter	6.3	7.5	6.8	6.0	5.3	7.4	7.2
Standard electric typewriter	7.0	7.4	7.8	6.7	6.1	8.1	7.5

(a) At the .05 level of significance, is there evidence to enable the personnel director to conclude that the typing time is less for the word processing typewriter?
(b) What assumption is necessary to perform this test?

10.46 A financial analyst wishes to know whether there had been a significant change in earnings per share from one period of time to another among the largest industrial corporations in the United States. A random sample of 15 companies selected from the *Fortune* 500 yielded the following results:

	Earnings per share	
	1985	**1984**
Ford Motor	$13.63	$15.79
Borden	3.75	3.56
American Cyanimid	2.68	4.41
Whirlpool	4.98	5.18
Libbey-Owens-Ford	5.71	5.82
International Minerals and Chem.	4.39	3.04
Gould	−3.94	0.40
Pennwalt	−2.80	3.31
Dayco	2.04	2.35
Nalco Chemical	1.87	1.94
Macmillan	2.07	1.72
Nashua	3.46	4.22
American Bakeries	2.56	2.36
Revere Copper and Brass	2.00	−0.79
Morton Thiokol	4.01	2.17

SOURCE: *Fortune*, April 28, 1986.

(a) At the .01 level of significance, is there evidence of a difference in earnings per share between the two years?
(b) What assumption is necessary to perform this test?

10.10 A CONNECTION BETWEEN CONFIDENCE INTERVALS AND HYPOTHESIS TESTING

In the last two chapters we have examined the two major areas of statistical inference: confidence intervals and hypothesis testing. They are based on the same set of concepts, but we have used them for different purposes. We used confidence intervals to estimate parameters, while we used hypothesis testing in making decisions about specified values of population parameters.

In many situations we can use confidence intervals to do a test of a null hypothesis. This can be illustrated for the test of a hypothesis for a mean. Referring back to the student grades problem, we first attempted to determine whether the true average GPI had changed from 2.5. We tested this in Section 10.4 using the formula

$$Z = \frac{\overline{X} - \mu_X}{\frac{\sigma_X}{\sqrt{n}}}$$

We could also solve the problem by obtaining a confidence-interval estimate of μ_X. If the hypothesized value of $\mu_X = 2.5$ fell into the interval, the null hypothesis would not be rejected. That is, the value 2.5 would not be considered unusual for the data observed. On the other hand, if it did not fall into the interval, the null hypothesis would be rejected, because 2.5 would then be an unusual value. Using Equation (9.1), the confidence-interval estimate could be set up from the following data:

$$n = 100, \qquad \sigma_X = .40, \qquad \overline{X} = 2.59$$

For a confidence level of 95% (corresponding to a 5% level of significance—that is, $\alpha = .05$) we have

$$\overline{X} \pm Z\frac{\sigma_X}{\sqrt{n}} = 2.59 \pm 1.96\frac{(.40)}{\sqrt{100}}$$

$$= 2.59 \pm .0784$$

so that

$$2.5116 \leq \mu_X \leq 2.6684$$

Since the interval does not include the hypothesized value of 2.5, we would reject the null hypothesis. This, of course, was the same decision we reached by using the hypothesis-testing technique.

We may also illustrate the connection between confidence intervals and hypothesis testing in the case of the means of two independent samples. Referring back to the absenteeism of workers, we attempted to determine whether there was a difference in the average absenteeism between males and females. This was tested using Equation (10.4):

$$t = \frac{\overline{X}_1 - \overline{X}_2}{\sqrt{S_P^2 \left(\frac{1}{n_1} + \frac{1}{n_2}\right)}}$$

This problem could also be solved by obtaining a confidence-interval estimate of $\mu_1 - \mu_2$. If the hypothesized difference of zero were to fall into the interval, the null hypothesis would not be rejected. From Equation (10.4), the confidence-interval estimate for the difference between the means can be set up as follows:

$$(\overline{X}_1 - \overline{X}_2) \pm t_{n_1+n_2-2} \sqrt{S_P^2 \left(\frac{1}{n_1} + \frac{1}{n_2}\right)} \quad \textbf{(10.6)}$$

For a confidence level of 95% (corresponding to a 5% level of significance—that is, $\alpha = .05$) we have

$$(8.4 - 7.6) \pm (2.0484) \sqrt{7.92 \left(\frac{1}{18} + \frac{1}{12}\right)}$$

$$(+.8) \pm (2.0484)(1.049)$$

$$(+.8) \pm 2.149$$

so that

$$-1.349 \leq (\mu_1 - \mu_2) \leq 2.949$$

Since the interval includes zero, the null hypothesis cannot be rejected. This, of course, is the same decision that we reached using the hypothesis-testing approach of Section 10.8.

Problems

10.47 Show how the confidence intervals you computed in Problem 9.1 on page 247 could be used to test the null hypothesis in Problem 10.13 on page 280. (Choose any one of the confidence levels you used in Problem 9.1.)

• **10.48** Show how the confidence intervals you computed in Problem 9.2 on page 247 could be used to test the null hypothesis in Problem 10.14 on page 281. (Choose any one of the confidence levels you used in Problem 9.2.)

10.49 Show how the confidence intervals you computed in Problem 9.4 on page 247 could be used to test the null hypothesis in Problem 10.16 on page 281. (Choose any one of the confidence levels you used in Problem 9.4.)

10.50 Show how the confidence intervals you computed in Problem 9.5 on page 247 could be used to test the null hypothesis in Problem 10.17 on page 281. (Choose any one of the confidence levels you used in Problem 9.5.)

10.51 Show how the confidence intervals you computed in Problem 9.7 on page 253 could be used to test the null hypothesis in Problem 10.20 on page 284. (Choose any one of the confidence levels you used in Problem 9.7.)

10.52 Show how the confidence intervals you computed in Problem 9.10 on page 253 could be used to test the null hypothesis in Problem 10.21 on page 284. (Choose any one of the confidence levels you used in Problem 9.10.)

• **10.53** Compare the results obtained from Problems 9.18 and 10.39(b) on pages 256 and 297. Are the conclusions the same? Why?

10.54 Compare the results obtained from Problems 9.20 and 10.41 on pages 256 and 298. Are the conclusions the same? Why?

10.55 Referring to Problem 10.43 on page 298:

- **(a)** Set up a 95% confidence-interval estimate of the average difference in the mileage of the two gasolines.
- **(b)** Is the value of zero included within the interval?
- **(c)** Use the results of (b) to determine whether there is evidence that the average difference is different from zero.

10.56 Why shouldn't we compare the results obtained from the confidence interval in Problem 9.19 to the results of the test of hypothesis in Problem 10.40 (see pages 256 and 298)?

• **10.57** Referring to Problem 3.75 on page 73:

- **(a)** Set up a 90% confidence-interval estimate of the average difference in acceleration between German and Japanese cars.
- **(b)** Is the value of zero included within the interval?
- **(c)** Use the results of (b) to determine whether the average difference is equal to zero.
- **(d)** Compare the conclusions obtained in (c) to those of Problem 10.32 on page 291. Are the conclusions the same? Why?

10.58 Referring to Problem 3.76 on page 73:

- **(a)** Set up a 95% confidence-interval estimate of the average difference in tuition between colleges in the Northeast and the Midwest.
- **(b)** Is the value of zero included within the interval?
- **(c)** Use the results of (b) to determine whether the average difference is equal to zero.
- **(d)** Compare the conclusions obtained in (c) to those of Problem 10.33 on page 291. Are the conclusions the same? Why?

10.59 Referring to Problem 10.37 on page 291:

- **(a)** Set up a 90% confidence-interval estimate of the average difference in output between the day and evening shifts.
- **(b)** Is the value of zero included within the interval?
- **(c)** Use the results of (b) to determine whether the average difference is equal to zero.
- **(d)** Compare the conclusions obtained in (c) to those of Problem 10.37 on page 291. Are the conclusions the same? Why?

10.60 Referring to Problem 10.38 on page 291:

- **(a)** Set up a 95% confidence-interval estimate of the average difference in mileage between the two gasolines.

(b) Is the value of zero included within the interval?
(c) Use the results of (b) to determine whether the average difference is equal to zero.
(d) Compare the conclusions obtained in (c) to those of Problem 10.38 on page 291. Are the conclusions the same? Why?

10.61 Referring to Problem 4.11 on page 83:
(a) Set up a 99% confidence-interval estimate of the average difference in noise between gas-powered and electric-powered chainsaws.
(b) Is the value of zero included within the interval?
(c) Use the results of (b) to determine whether the average difference is equal to zero.
(d) Why shouldn't we compare the results obtained from the confidence interval in (a) to the results of the test of hypothesis in Problem 10.34 on page 291?

10.11 THE p-VALUE APPROACH TO HYPOTHESIS TESTING

In recent years an approach to hypothesis testing that has increasingly gained acceptance deals with the concept of the ***p* value**.

> **The *p* value is the probability of obtaining a test statistic equal to or more extreme than the result obtained, given that H_0 is true.**

The p value is often referred to as the ''observed level of significance.'' If the p value is greater than the selected values of α, the null hypothesis is not rejected; if the p value is smaller than α, the null hypothesis can be rejected.

We obtain a better understanding of the p value by referring back to the student-grades problem. In Section 10.4 we tested whether or not the average GPI was 2.5. We obtained a Z value of $+2.25$ and rejected the null hypothesis, since this was greater than the upper critical value of $+1.96$. Using the p-value approach, we could determine the probability of obtaining a Z value larger than $+2.25$. From Table C.2 this probability is $.5000 - .4878 = .0122$. Since we are performing a two-tailed test, we need also to find the probability of obtaining a value less than -2.25. Since the normal distribution is symmetric, this value is also .0122. Thus the p value for the two-tailed test is .0244. This may be interpreted to mean that if H_0 is true, the probability of obtaining a *more extreme* result than the one observed is only .0244. Since this is less than $\alpha = .05$, the null hypothesis is rejected. Had a one-tailed test been specified instead of a two-tailed test, the p value would have been the probability in one tail (in this case .0122). This value would then have been compared to the critical α value.

Unless we are dealing with a test statistic that follows the normal distribution, the computation of the exact p value may be difficult. Thus, it is fortunate that many statistical computer software packages now present the p value as part of the output for most hypothesis-testing procedures.

Problems

• **10.62** Find the exact p value for the test statistic you calculated in Problem 10.13 on page 280.

10.63 Find the exact p value for the test statistic you calculated in Problem 10.17 on page 281.

• **10.64** In Problem 10.19 on page 284 approximate the upper and lower limits of the p value and interpret.

10.65 In Problem 10.20 on page 284 approximate the upper and lower limits of the p value and interpret.

10.66 In Problem 10.27 on page 286 approximate the upper and lower limits of the p value and interpret.

10.67 In Problem 10.29 on page 286 approximate the upper and lower limits of the p value and interpret.

10.68 In Problem 10.34 on page 291 approximate the upper and lower limits of the p value and interpret.

10.69 In Problem 10.38 on page 291 approximate the upper and lower limits of the p value and interpret.

Supplementary Problems

• **10.70** The personnel department of a large company would like to determine the amount of time it takes for employees to arrive at work. A random sample of 12 employees is selected and the time in minutes to arrive at work is recorded, with the following results:

15 30 50 60 25 65 45 90 75 50 50 20

(a) At the .01 level of significance, is there evidence that the average travel time of employees is less than 60 minutes?
(b) What assumption is necessary to perform this test?

10.71 An auditor for the Department of Energy wishes to study the price of unleaded gasoline per gallon in New York City. A random sample of 50 gas stations was selected, with the following results: $\bar{X}$ = \$1.364, S = \$.073. Is there evidence that the average price of unleaded gasoline is different from \$1.30 per gallon? (α = .05.)

10.72 What is the relationship of α to the Type I error? What is the relationship of β to the Type II error?

10.73 The purchasing director for an industrial parts factory is investigating the possibility of purchasing a new type of milling machine. She has determined that the new machine will be bought if there is evidence that the parts produced have a higher breaking strength than those from the old machine. The process standard deviation of the breaking strength for the old machine is 10 kilograms and for the new machine is 9 kilograms. A sample of 100 parts taken from the old machine indicated a sample mean of 65 kilograms, while a similar sample of 100 from the new machine indicated a sample mean of 72 kilograms. Using the

.01 level of significance, is there evidence that the purchasing director should buy the new machine?

10.74 The budget director of a large retailing company wanted to compare luncheon expense vouchers for executives in the sales department and in the production department. A random sample of 15 vouchers in each department was selected with the following results:

Sales department	Production department
$\bar{X}_1 = \$42.50$	$\bar{X}_2 = \$31.75$
$S_1 = \$\ 9.50$	$S_2 = \$\ 8.20$

(a) Using the .01 level of significance, is there any evidence of a difference in the average luncheon expense voucher between the sales and production departments?
(b) Set up a 99% confidence-interval estimate of the average difference in luncheon expense vouchers between the sales and production departments.
(c) Compare the conclusions obtained in parts (a) and (b). Are they the same? Why?

10.75 The advertising manager of a breakfast cereal company would like to determine whether a new package shape would improve sales of the product. In order to test the feasibility of the new package shape, a sample of 40 equivalent stores was selected and 20 were randomly assigned as the test market of the new package shape, while the other 20 were to continue receiving the old package shape. The weekly sales during the time period studied were as follows:

New	Old
$\bar{X}_1 =$ 130 boxes	$\bar{X}_2 =$ 117 boxes
$S_1 =$ 10 boxes	$S_2 =$ 12 boxes

(a) Using the .05 level of significance, is there evidence that the new package shape resulted in increased sales?
(b) Set up a 95% confidence-interval estimate of the average difference in sales between the new package shape and the old package shape.
(c) Why shouldn't we compare the results obtained in parts (a) and (b)?

• **10.76** A consumer reporting agency wished to determine whether an ''unknown brand'' calculator sells at a lower price than the ''famous brand'' calculator of the same type. A random sample of eight stores was selected, and the prices (at the stores) of each of the two calculators were recorded with the following results:

Store	Unknown brand	Famous brand
1	10	11
2	8	11
3	7	10
4	9	12
5	11	11
6	10	13
7	9	12
8	8	10

(a) At the .01 level of significance, is there evidence that the unknown brand sells for a lower price?
(b) What assumption is necessary to perform this test?
(c) Set up a 99% confidence-interval estimate of the average difference in price between the unknown brand and the famous brand.
(d) Why shouldn't we compare the results obtained in parts (a) and (c)?

Database Problems

The following problems refer to the sample data obtained from the questionnaire of Figure 2.5 and presented in Figure 2.10. They should be solved (to the greatest extent possible) with the aid of an available computer package.

□ **10.77** At the .05 level of significance, is there evidence that the average salary (question 7) is greater than $25,000?

□ **10.78** At the .01 level of significance, is there evidence that the average number of years worked in the current job activity (question 10) is different from 5?

□ **10.79** At the .05 level of significance, is there evidence that the average grade-point index (question 15) of graduates is greater than 2.8?

□ **10.80** At the .10 level of significance, is there evidence that the average height of males is less than 70 inches (questions 18 and 19)?

□ **10.81** At the .05 level of significance, is there evidence that males have worked in fewer job areas than females (questions 12 and 18)?

□ **10.82** At the .01 level of significance, is there evidence of a difference in the average grade-point index (question 15) of Business versus Social Science/Humanities majors (question 2, codes 1 and 2)?

□ **10.83** At the .05 level of significance, is there evidence of a difference in the average grade-point index between males and females (questions 15 and 18)?

□ **10.84** At the .05 level of significance, is there evidence of a difference in the average number of years worked in the current job activity (question 10) between males and females (question 18)?

□ **10.85** At the .10 level of significance, is there evidence that Business majors have worked in fewer areas than Social Science/Humanities majors (questions 2 and 12)?

□ **10.86** At the .01 level of significance, is there evidence that the average salary (question 7) is higher for those graduates who consider themselves more conservative (question 4, code 1) than for all other graduates (question 4, codes 2 and 3)?

□ **10.87** At the .05 level of significance, is there evidence that the average age (question 20) is higher for those graduates who consider themselves more conservative (question 4, code 1) than for all other graduates (question 4, codes 2 and 3)?

□ **10.88** At the .05 level of significance, is there evidence of a difference between the average grade-point indices (question 15) of those who attended graduate school and those who did not (question 5)?

☐ 10.89 At the .10 level of significance, is there evidence of a difference in the average annual salary (question 7) between those who are employed at a company with over 100 employees (question 8, code 1) and those who are employed at a company with fewer than 100 employees (question 8, code 2)?

Reference

1. SNEDECOR, G. W., AND W. G. COCHRAN, *Statistical Methods*, 7th ed. (Ames, Iowa: Iowa State University Press, 1980).

Chapter 11

Hypothesis Testing for Qualitative Variables

11.1 INTRODUCTION: WHAT'S AHEAD

In Chapter 10 we used hypothesis-testing techniques to test hypotheses about the mean of one group and about differences between the means of two groups. In this chapter we shall extend these procedures to test hypotheses about categorical (qualitative) variables.

Case H—The Marketing Director's Microcomputer Survey

The increasing availability of the personal computer has drawn into the information-processing revolution thousands of people who are using such equipment for business, educational, and entertainment purposes. As a result of this phenomenon, periodic "computer shows" have been organized at various locations around the country. As marketing director of the company that runs a series of these computer shows, it is your job to obtain data that might enable you to develop a profile of those individuals who attend. With this in mind you have constructed a questionnaire that will be administered to a sample of 400 attendees as they exit from a forthcoming computer show. The attendees will be asked whether or not they currently own a personal computer. If they answer *No*, they will be asked:

Do you plan to purchase a personal computer in the next six months?
(1)Yes (2) No or Not Sure

If they answer the original question *Yes*, they will be asked the following set of questions:

1. For what purpose is your personal computer primarily used?
   (1) Business (2) Education (3) Entertainment
2. Among the brands of microcomputers that you currently own which one do you use most often?
   (1) IBM or compatible (2) Apple or compatible (3) Commodore or Atari (4) Other
3. Did you purchase any computer software, hardware, or supplies at this show?
   (1) Yes (2) No
4. Would you return for a similar show next year?
   (1) Yes (2) No or Not Sure

Using the results of this questionnaire, the marketing director is primarily interested in drawing conclusions about the following five issues:

1. Whether there is evidence that more than 40% of those who attend the show already own a personal computer.
2. Whether there is evidence of a difference in the willingness of personal computer owners to return to a similar show next year based upon purchasing decisions made at this show.
3. Whether there is evidence of a difference in the intention of personal computer owners to return to a similar show next year based on how they primarily use their computers (i.e., for business, education, or entertainment).
4. Whether there is evidence of a relationship between the type of personal computer owned and how it is primarily used.
5. Based on a follow-up study that was to be conducted among the computer nonowners six months after the show, whether there was evidence of a difference in the proportion that planned to purchase and the proportion that actually purchased a home computer.

11.2 TEST OF HYPOTHESIS FOR A PROPORTION (ONE SAMPLE)

In order to draw conclusions concerning each of these issues, the marketing director must test hypotheses concerning one or more qualitative variables. Referring to the first question of interest, the marketing director wanted to determine whether there was evidence that more than 40% of the attendees of the show owned a personal computer. In terms of proportions rather than percentages, the null and alternative hypothesis can be stated as follows:

$$H_0: \; p \leq .40$$
$$H_1: \; p > .40$$

We recall that although the *number of successes* follows a binomial distribution, if the sample size is large enough (both $np \geq 5$ and $n(1 - p) \geq 5$), the normal distribution provides a good approximation to the binomial distribution (see Section 8.3). The test statistic can be stated in two forms, in terms of either the *proportion* of successes [see Equation (11.1)] or the *number* of successes [see Equation (11.2)]:

$$Z = \frac{p_S - p}{\sqrt{\frac{p(1-p)}{n}}} \qquad \textbf{(11.1)}$$

where $p_S = \frac{X}{n} = \frac{\text{number of successes in sample}}{\text{sample size}} = \text{observed proportion of successes}$

p = proportion of successes from the null hypothesis

or

$$Z = \frac{X - np}{\sqrt{np(1-p)}} \qquad \textbf{(11.2)}$$

Aside from possible rounding errors, both formulas will provide the same solution for Z.

Suppose the results of the questionnaire indicate that 188 out of the 400 respondents owned personal computers. Since the marketing director wants to be able to make a national advertising claim, he decides to use the .01 level of significance to reduce the possibility of a false claim. Thus we can set up the rejection and nonrejection regions as displayed in Figure 11.1, since the decision rule would be:

Reject H_0 if $Z > +2.33$;
otherwise do not reject H_0.

For the survey data, $n = 400$, $X = 188$, and $p_S = .47$. Using Equation (11.1), we have

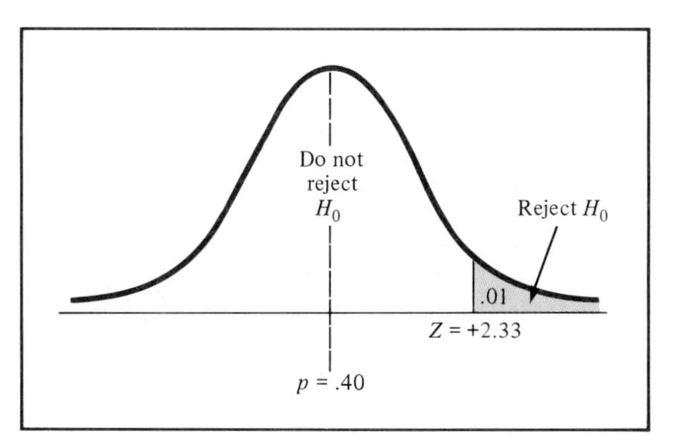

FIGURE 11.1 □ One-tailed test of hypothesis for a proportion at the 1% level of significance.

$$Z = \frac{p_S - p}{\sqrt{\frac{p(1-p)}{n}}} = \frac{.47 - .40}{\sqrt{\frac{.4(.6)}{400}}} = \frac{.07}{\sqrt{.0006}} = \frac{.07}{.0245} = +2.858$$

or, using Equation (11.2), we have

$$Z = \frac{X - np}{\sqrt{np(1-p)}} = \frac{188 - 400(.4)}{\sqrt{400(.4)(.6)}} = \frac{188 - 160}{\sqrt{96}} = \frac{28}{9.798} = +2.858$$

Since $Z = +2.858 > +2.33$, the statistical decision is to reject H_0.

Thus, the marketing director may conclude there is evidence that more than 40% of the attendees of this show own a personal computer.

Problems

11.1 Prove the statement on page 310 of the text that the formula on the right-hand side of Equation (11.1) is equivalent to the formula on the right-hand side of Equation (11.2).

11.2 Problem 9.22 on page 259 described a situation in which 35 out of 50 flower bulbs grew into flowering plants. Is there evidence that less than 90% of the bulbs grew into flowering plants? Use $\alpha = .05$.

• 11.3 In Problem 9.24 on page 259 we found Wanda working her way through college selling magazines; 15 out of 50 people she visited bought magazines. Is there evidence that more than 20% of all people will buy magazines? Use $\alpha = .01$.

• 11.4 In a poll of registered voters, 354 out of 809 said they were voting for the Republocrat candidate. Is there evidence that the proportion of voters who will vote Republocrat is different from .50? Use $\alpha = .05$.

11.5 The advertising director for a fast-food chain claims that at least 75% of all respondents are familiar with a particular commercial that has been broadcast on radio and television during the past month. A random sample of 400 respondents indicated that 260 were familiar with the commercial. Is the claim valid, or is there evidence that it is not valid? Use $\alpha = .10$.

11.6 A television manufacturer has claimed in its warranty that in the past only 10% of its television sets needed any repair during their first two years of operation. In order to test the validity of this claim, a government testing agency selects a sample of 100 sets and finds that 14 sets required some repair within their first two years of operation. Is the manufacturer's claim valid, or is there evidence that it is not valid? Use $\alpha = .01$.

11.3 TESTING FOR A DIFFERENCE BETWEEN PROPORTIONS FROM TWO INDEPENDENT POPULATIONS: USING THE NORMAL APPROXIMATION

Now that we have tested a hypothesis concerning the proportion of successes in one group, we can turn to the marketing director's second concern. Here two groups of personal computer owners are to be compared with respect to their response to a particular question (whether or not they would return to a similar show next year). This is the first of several issues of interest to the marketing director that involve a comparison of two or more groups. In this section we will examine how we can test for the difference between two proportions obtained from independent groups, while in subsequent sections we will examine situations either involving more than two groups or involving related samples.

The method we shall apply in this section once again involves the normal distribution for large sample sizes. When comparing two population proportions, we may be interested in determining either whether there is any *difference* in the proportion of successes in the two groups (two-tailed test) or whether one group had a *higher* proportion of successes than the other group (one-tailed test).

Two-tailed test	One-tailed test	One-tailed test
H_0: $p_1 = p_2$ H_1: $p_1 \neq p_2$	H_0: $p_1 \geq p_2$ H_1: $p_1 < p_2$	H_0: $p_1 \leq p_2$ H_1: $p_1 > p_2$

where p_1 is the proportion of successes in population 1
p_2 is the proportion of successes in population 2

The test statistic would be

$$Z = \frac{p_{S_1} - p_{S_2}}{\sqrt{\bar{p}(1 - \bar{p})\left(\frac{1}{n_1} + \frac{1}{n_2}\right)}} \quad \textbf{(11.3)}$$

where

$$p_{S_1} = \frac{X_1}{n_1}, \qquad p_{S_2} = \frac{X_2}{n_2}$$

$$\bar{p} = \frac{X_1 + X_2}{n_1 + n_2} = \frac{n_1 p_{S_1} + n_2 p_{S_2}}{n_1 + n_2}$$

and p_{S_1} is the sample proportion from population 1
p_{S_2} is the sample proportion from population 2
X_1 is the number of successes in sample 1
X_2 is the number of successes in sample 2
n_1 is the sample size from population 1
n_2 is the sample size from population 2
$\bar{p}$ is the pooled estimate of the population proportion

Note that $\bar{p}$, the pooled estimate for the population proportion, is based on the null hypothesis. Under the null hypothesis it is assumed that the two population proportions are equal. Therefore we may obtain an overall estimate of the common population proportion by pooling together the two sample proportions. The estimate $\bar{p}$ is simply the number of successes in the two samples combined $(X_1 + X_2)$ divided by the total sample size from the two groups $(n_1 + n_2)$.

We may now apply this test for the difference between two proportions to the second problem of interest to the marketing director. He would like to determine whether there is any evidence of a difference in the proportion of computer owners who intend to return to next year's show based upon whether or not they purchased anything at this show. We may summarize the results for the 188 microcomputer owners as follows:

YES, Purchased Items at the Show	*NO*, Did Not Purchase Items
$n_1 = 84$	$n_2 = 104$
$X_1 = 58$ would return	$X_2 = 61$ would return

Of course, these results pertain only to those respondents who own microcomputers, since the nonowners answered a shorter questionnaire.

For this situation, the null and alternative hypotheses would be:

$$H_0: \; p_1 = p_2$$
$$H_1: \; p_1 \neq p_2$$

If the test were to be carried out at the .05 level of significance, the critical values would be -1.96 and $+1.96$ (see Figure 11.2 on page 314) and our decision rule would be:

Reject H_0 if $Z > +1.96$
or if $Z < -1.96$;
otherwise do not reject H_0.

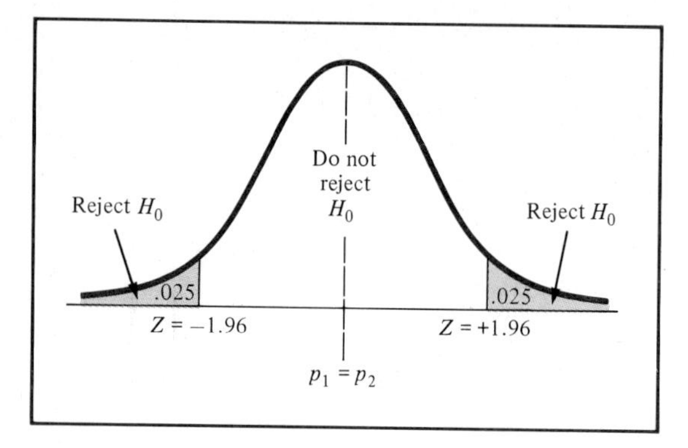

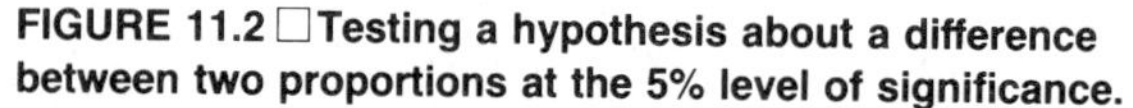
FIGURE 11.2 □ Testing a hypothesis about a difference between two proportions at the 5% level of significance.

For these data we compute

$$p_{S_1} = \frac{X_1}{n_1} = \frac{58}{84} = .691, \qquad p_{S_2} = \frac{X_2}{n_2} = \frac{61}{104} = .587$$

$$\bar{p} = \frac{X_1 + X_2}{n_1 + n_2} = \frac{58 + 61}{84 + 104} = .633$$

and, using Equation (11.3),

$$Z = \frac{p_{S_1} - p_{S_2}}{\sqrt{\bar{p}(1-\bar{p})\left(\frac{1}{n_1} + \frac{1}{n_2}\right)}}$$

$$= \frac{.691 - .587}{\sqrt{.633(.367)\left(\frac{1}{84} + \frac{1}{104}\right)}} = \frac{.104}{\sqrt{.232(.022)}}$$

$$= \frac{.104}{\sqrt{.005}} = \frac{.104}{.071} = +1.47$$

Since $Z = 1.47$, we see that $-1.96 < 1.47 < +1.96$, and the decision is not to reject H_0. The marketing director would conclude that there is no evidence of a difference in the proportion of microcomputer owners who would return to next year's show based upon whether they had made any purchases at the present show.[1]

Problems

• **11.7** A psychologist studying handedness found that 12 out of 76 males in his study were left-handed, while 6 out of 89 females were left-handed. Is there evidence of a difference in the proportion of males and females who are left-handed? Use $\alpha = .05$.

[1] In this problem the rejection region has been divided up into both tails of the distribution. Of course, if the alternative hypothesis had been in a specific direction (such as $p_1 > p_2$), then the rejection region would be entirely contained in one tail of the normal distribution.

• 11.8 A sociologist found that 84 out of 124 Republicans he interviewed favored capital punishment, while 45 out of 98 Democrats did. Is there evidence of a difference in the proportion of Republicans and Democrats who favor capital punishment? Use $\alpha = .05$.

• 11.9 Students entering the University of Northwest South Dakota who scored below 50 on an entrance examination were randomly divided into two groups; one group got remedial tutoring the summer before starting their college career. Of the 23 students in the special program, 15 eventually got their degrees; of the 47 students who did not get the special program, 19 got degrees. Is there evidence that the program was effective? Use $\alpha = .01$.

11.10 In the original experiment to evaluate the clinical utility of penicillin, 8 rats were infected with a virulent bacterium. Of the 4 rats that were then given penicillin, none died; of the 4 that were not given penicillin, all died. Is there evidence that the penicillin was effective? Use $\alpha = .05$. (Notice that the conditions for a good approximation to the normal distribution are not present here, but do the test anyway just to see what happens.)

11.11 We wish to determine if there is any difference in the popularity of football between college-educated males and non-college-educated males. A sample of 100 college-educated males revealed 55 who considered themselves football fans. A sample of 200 non-college-educated males revealed 125 who considered themselves football fans. Is there any evidence of a difference in football popularity between college-educated and non-college-educated males? Use $\alpha = .01$.

11.12 We wish to determine whether there is a sex difference in preference for margarine versus butter. A sample of 80 males indicated that 28 preferred margarine to butter; a sample of 120 females indicated that 52 preferred margarine. Is there any evidence of a difference in preference for margarine between males and females? Use $\alpha = .05$.

11.4 TESTING FOR A DIFFERENCE BETWEEN PROPORTIONS FROM TWO INDEPENDENT POPULATIONS: USING THE CHI-SQUARE TEST

For the example of the previous section, we may also view the data in terms of a two-way cross-tabulation called a **contingency table** (see Sections 4.7 and 6.3). The "cells" in the table indicate the frequency count for each possible combination of responses to the two questions of interest. This contingency table is displayed as Table 11.1.

TABLE 11.1 ☐ Contingency table for intention to return to next year's show and current purchasing decisions

	Did you purchase anything?		
Do you intend to return to next year's show?	**Yes**	**No**	**Totals**
Yes	58	61	119
No or Not Sure	26	43	69
Totals	84	104	188

11.4.1 Test for a Difference Between Two Proportions

This contingency table is the basis for testing the difference between the two proportions using a second method of analysis, the **chi-square test.** The method begins from the null hypothesis that the proportion of successes is the same in the two populations ($p_1 = p_2$). If this null hypothesis were true, the frequency that theoretically should be expected in each cell could be computed. For example, we can determine how many personal computer owners would *theoretically* be expected to return to next year's show for each of the two subgroups (those who purchased and those who did not purchase). A total of 119 out of 188 personal computer owners (63.3%) stated that they intend to return to next year's show. If there was no difference in this proportion between those who did and did not purchase something at the present show, then theoretically 63.3% in each group should intend to return to next year's show. This would mean a theoretical frequency (of returning) of $.633 \times 84 = 53.17$ for the group that purchased something and $.633 \times 104 = 65.83$ for the group that made no purchases. The number of microcomputer owners who theoretically would not return to next year's show can be found by subtracting the theoretical number who would return from the total number in each group. The actual (observed) frequencies and theoretical (expected) frequencies (circled numbers in the table) are displayed in Table 11.2.

TABLE 11.2 □ Observed and theoretical frequencies for the intention to return–current purchase problem

Do you intend to return to next year's show?	Did you purchase anything? Yes		No		
	Observed	Theoretical	Observed	Theoretical	Total
Yes	58	53.17	61	65.83	119
No or Not Sure	26	30.83	43	38.17	69
Total	84		104		188

The theoretical (or expected) frequencies f_t in each cell can also be computed by using a simple formula:

$$f_t = \frac{n_R n_C}{n} \qquad (11.4)$$

where n_R is the total number in the row
n_C is the total number in the column
n is the total sample size

Thus the theoretical frequency of 53.17 for the cell "intends to return and made current purchase" can be arrived at by taking the total number in the appropriate row (119), multiplying by the total number in the appropriate column (84), and dividing by the overall number of personal computer owners (188). The computation of the theoretical frequencies for each cell is illustrated as follows:

$$f_t = \frac{119(84)}{188} = 53.17, \qquad f_t = \frac{119(104)}{188} = 65.83$$

$$f_t = \frac{69(84)}{188} = 30.83, \qquad f_t = \frac{69(104)}{188} = 38.17$$

The chi-square method of analysis uses the squared difference between the observed frequency f_o and the theoretical frequency f_t in each cell. If there is no difference in the population proportions of success in the two groups, then the squared difference between the observed and theoretical frequencies should be small. If the proportions for the two groups are significantly different, then the squared difference between the observed and theoretical frequencies should be large. Of course, what is a "large" difference is relative; the same difference would mean more in a cell with only a few observations than in a cell where there are many observations. Therefore an adjustment is made for the size of the cell: The squared difference is divided by the expected frequency for the cell.

The test statistic is $(f_o - f_t)^2/f_t$ summed over all cells of the table. That is,

$$\chi^2 = \Sigma \frac{(f_o - f_t)^2}{f_t} \qquad \textbf{(11.5)}$$

where f_o is the observed frequency in a cell
f_t is the theoretical frequency in a cell

This statistic approximately follows a chi-square (χ^2) distribution, with the degrees of freedom[2] equal to the number of rows in the contingency table minus 1 (that is, $R - 1$) times the number of columns in the contingency table minus 1 (that is, $C - 1$):

$$\text{degrees of freedom} = (R - 1)(C - 1)$$

[2] To see why the degrees of freedom is $(R - 1)(C - 1)$, consider how we calculated the expected frequencies. There are RC expected frequencies to calculate in the table. To calculate these we needed to know (1) the total sample size, (2) the marginal frequencies for the rows (row totals), and (3) the marginal frequencies for the columns (column totals). The expected frequency for each cell is the row total times the column total divided by the overall total. How many of these values are independent? Starting with the overall total, we need $R - 1$ additional row totals (since the final one can be calculated by subtraction from the overall total) and $C - 1$ additional column totals (again the final one can be found by subtraction). We have therefore used up $1 + (R - 1) + (C - 1)$ independent pieces of information in calculating the expected frequencies. This leaves
$RC - [1 + (R - 1) + (C - 1)] = RC - R - C + 1 = R(C - 1) - (C - 1) = (R - 1)(C - 1)$
degrees of freedom for the chi-square test.

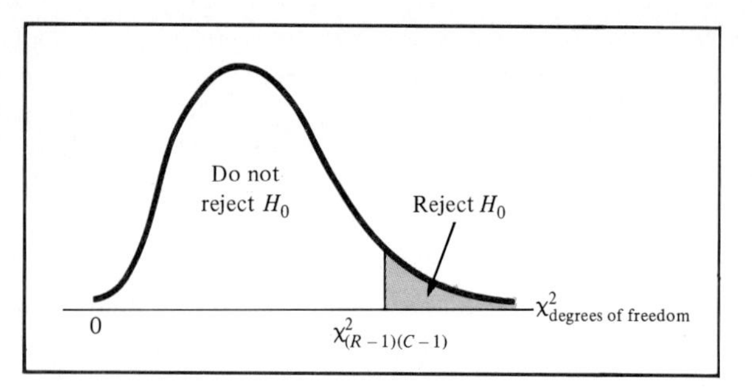

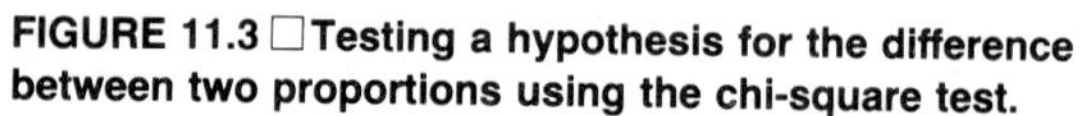
FIGURE 11.3 □ Testing a hypothesis for the difference between two proportions using the chi-square test.

The null hypothesis of equality between the two proportions ($p_1 = p_2$) will be rejected when the computed value of the test statistic is greater than the critical value of chi-square with the appropriate number of degrees of freedom (see Figure 11.3).

The chi-square distribution is a skewed distribution whose shape depends solely on the number of degrees of freedom. As the number of degrees of freedom increases, the chi-square distribution becomes more symmetrical. Table C.4 contains various upper tail areas of the chi-square distribution for different degrees of freedom (see Table 11.3).

The value at the top of each column indicates the area in the upper portion (or right side) of the chi-square distribution. For example, with one degree of freedom the value for an upper tail area of .05 is 3.841 (see Figure 11.4). This means that for 1 degree of freedom the probability of exceeding a chi-square value of 3.841 is .05.

Returning to the intention to return–current purchase problem (see Table 11.1),

TABLE 11.3 □ Obtaining the critical value from the chi-square distribution for $\alpha = .05$ and 1 degree of freedom

Degrees of freedom	Upper tail areas (α)											
	.995	.99	.975	.95	.90	.75	.25	.10	.05	.025	.01	.005
1			0.001	0.004	0.016	0.102	1.323	2.706	3.841	5.024	6.635	7.879
2	0.010	0.020	0.051	0.103	0.211	0.575	2.773	4.605	5.991	7.378	9.210	10.697
3	0.072	0.115	0.216	0.352	0.584	1.213	4.108	6.251	7.815	9.348	11.345	12.838
4	0.207	0.297	0.484	0.711	1.064	1.923	5.385	7.779	9.488	11.143	13.277	14.860
5	0.412	0.554	0.831	1.145	1.610	2.675	6.626	9.236	11.071	12.833	15.086	16.750
6	0.676	0.872	1.237	1.635	2.204	3.455	7.841	10.645	12.592	14.449	16.812	18.548
7	0.989	1.239	1.690	2.167	2.833	4.255	9.037	12.017	14.067	16.013	18.475	20.278
8	1.344	1.646	2.180	2.733	3.490	5.071	10.219	13.362	15.507	17.535	20.090	21.955
9	1.735	2.088	2.700	3.325	4.168	5.899	11.389	14.684	16.919	19.023	21.666	23.589
10	2.156	2.558	3.247	3.940	4.865	6.737	12.549	15.987	18.307	20.483	23.209	25.188

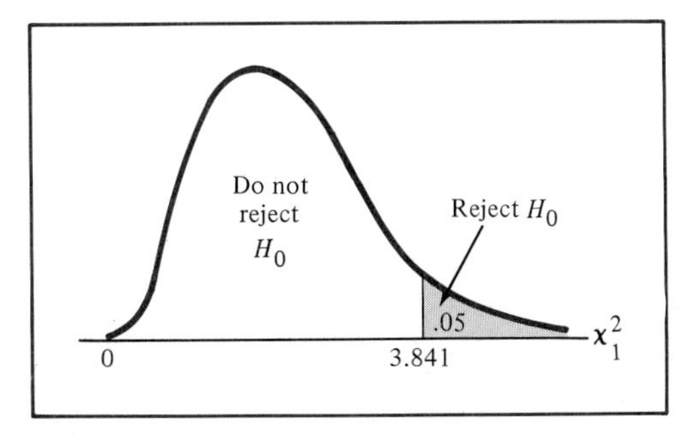

FIGURE 11.4 □ Finding the critical value of chi-square at the .05 level of significance with 1 degree of freedom.

since there are two rows and two columns there will be $(2 - 1) \times (2 - 1) = 1$ degree of freedom. Thus the decision rule will be

Reject H_0 if $\chi^2 > \chi_1^2 = 3.841$;
otherwise do not reject H_0.

Using Equation (11.5), we may compute the chi-square test statistic as summarized in Table 11.4.

TABLE 11.4 □ Computation of the chi-square test statistic for the intention to return–current purchase problem

f_o	f_t	$(f_o - f_t)$	$(f_o - f_t)^2$	$(f_o - f_t)^2/f_t$
58	53.17	4.83	23.3289	.43876
26	30.83	−4.83	23.3289	.75669
61	65.83	−4.83	23.3289	.35438
43	38.17	4.83	23.3289	.61118
				$\chi^2 = 2.16101$

From Table 11.4 we observe that the computed chi-square value is 2.16101. Since 2.16101 < 3.841, the null hypothesis is not rejected. Thus, the marketing director may again conclude that there is no evidence of a difference in the intention to return to next year's show between those personal computer owners who purchase and those who do not purchase something at the current show.

The chi-square test is most accurate when the expected frequency in each cell of the contingency table is at least five. This is important particularly for the 2 × 2 contingency table, which has only 1 degree of freedom. Other procedures, such as Fisher's exact test (see Reference 1) can be utilized if this is not true.

11.4.2 Test for Independence

Since the contingency table is cross-classifying two qualitative variables (intention to return next year and current purchase behavior), we may state the null and alternative hypotheses in an alternative form. The statement of no difference in intention between the two groups (those who did and did not make a purchase) also means that the two variables are *independent* (that is, there is *no relationship* between intention to return and purchase behavior). When the conclusions are stated in this manner, the chi-square test is called a **test of independence.** The null and alternative hypotheses would be:

H_0: There is independence (no relationship) between intention to return and current purchase behavior

H_1: There is a relationship between intention to return and current purchase behavior

If the null hypothesis cannot be rejected, the conclusion is drawn that there is no evidence of a relationship between the two variables. If the null hypothesis is rejected, the conclusion is drawn that there is evidence of a relationship between the two variables. Regardless of whether the chi-square test is considered a test for the difference between two proportions or a test of independence, the computations and results are exactly the same. Hence for our example the marketing director can conclude that there is no evidence of a relationship between intention to return and current purchase behavior.

11.4.3 Testing the Equality of Proportions by Z and by χ^2: A Comparison of Results

We have seen from this particular problem of interest to the marketing director that both the Z test and chi-square test have led to the same conclusion. This can be explained by the interrelationship between the value for the normal distribution and that for the chi-square distribution with 1 degree of freedom. The chi-square statistic will be the square of the two-sided test statistic based on the normal distribution (that is, $\chi^2 = Z^2$). Thus, for our example, we observe that $Z = 1.47$ while the chi-square value was $(1.47)^2 = 2.161$. Also, if we compare the critical values of the two distributions, we see that at the .05 level of significance the chi-square value of 3.841 is the square of the normal value of 1.96 (that is, $\chi^2 = Z^2$). Therefore it is clear that in testing the *difference* between two proportions, the Z test and chi-square test are equivalent methods. However, if we are interested in determining a *directional difference*, such as $p_1 > p_2$, then the Z test must be used, with the entire rejection region located in one tail of the normal distribution.

Problems

11.13 **(a)** Do Problem 11.7 on page 314 using the chi-square test.
(b) Show that the square of the Z value you calculated in Problem 11.7 is equal to the chi-square statistic.

• 11.14 **(a)** Do Problem 11.8 on page 315 using the chi-square test.
(b) Show that the square of the Z value you calculated in Problem 11.8 is equal to the chi-square statistic.

• 11.15 **(a)** For the data in Problem 11.9 on page 315, use the chi-square test to determine whether there is evidence of a difference between the two groups. ($\alpha = .01$.)
(b) Show that the square of the Z value you calculated in Problem 11.9 is equal to the chi-square statistic.
(c) How do the null and alternative hypothesis of Problem 11.9 differ from those used in part (a)?

11.16 Refer to Problem 11.10 on page 315. Provide two reasons why the chi-square test is inappropriate for solving this problem.

11.17 **(a)** Do Problem 11.11 on page 315 using the chi-square test.
(b) Show that the square of the Z value you calculated in Problem 11.11 is equal to the chi-square statistic.

11.18 **(a)** Do Problem 11.12 on page 315 using the chi-square test.
(b) Show that the square of the Z value you calculated in Problem 11.12 is equal to the chi-square statistic.

11.5 TESTING THE DIFFERENCE AMONG PROPORTIONS FROM C INDEPENDENT POPULATIONS

We may extend the chi-square method to the general case in which there are C independent populations. The null and alternative hypotheses would be:

$$H_0: \; p_1 = p_2 = p_3 = \cdots = p_c$$
H_1: At least one proportion is different from the others

The contingency table would have two rows and C columns, so that there would be $(2 - 1) \times (C - 1) = (C - 1)$ degrees of freedom in the chi-square test. Referring back to the survey, we may recall that as the third issue of concern, the marketing director wanted to determine whether there was a difference in the intention of personal computer owners to return based on their primary use of the computer (business, education, entertainment). This information can be analyzed by first cross-classifying these two variables as shown in Table 11.5.

TABLE 11.5 □ Cross-classification of intention to return and primary use

Do you intend to return to next year's show	Primary use			
	Business	Education	Entertainment	Total
Yes	87	25	7	119
No or Not Sure	33	19	17	69
Total	120	44	24	188

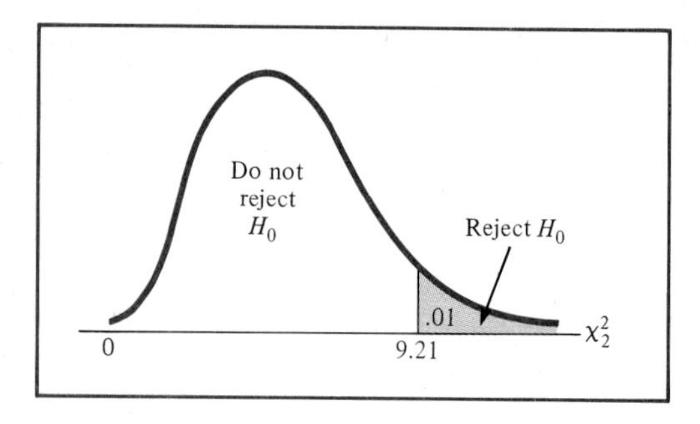

FIGURE 11.5 □ Testing for a relationship between intention to return and primary use at the 1% level of significance with 2 degrees of freedom.

Since there are three groups, the null and alternative hypotheses can be stated as follows:

H_0: $p_{\text{Bus}} = p_{\text{Ed}} = p_{\text{Ent}}$ (There is no relationship between intention to return and primary use.)

H_1: At least one primary use has a different proportion than the others. (There is a relationship between intention to return and primary use.)

If H_0 were to be tested at the .01 level of significance, the critical value (from Table C.4) would be 9.21, since there are $(2 - 1)(3 - 1) = 2$ degrees of freedom (see Figure 11.5). The decision rule would be

Reject H_0 if $\chi^2 > \chi_2^2 = 9.21$;
otherwise do not reject H_0.

The theoretical frequencies are computed using Equation (11.4). For the six cells we have:

$\frac{119(120)}{188} = 75.96$	$\frac{119(44)}{188} = 27.85$	$\frac{119(24)}{188} = 15.19$
$\frac{69(120)}{188} = 44.04$	$\frac{69(44)}{188} = 16.15$	$\frac{69(24)}{188} = 8.81$

The chi-square test statistic would be computed as shown in Table 11.6.

TABLE 11.6 □ Computation of the chi-square test statistic for the intention to return–primary use problem

f_o	f_t	$f_o - f_t$	$(f_o - f_t)^2$	$(f_o - f_t)^2/f_t$
87	75.96	+11.04	121.8816	1.60455
33	44.04	−11.04	121.8816	2.76752
25	27.85	− 2.85	8.1225	.29165
19	16.15	+ 2.85	8.1225	.50294
7	15.19	− 8.19	67.0761	4.41581
17	8.81	+ 8.19	67.0761	7.61363
				$\chi^2 = 17.1961$

Since $\chi^2 = 17.1961 > \chi_2^2 = 9.21$, the null hypothesis can be rejected and the marketing director can conclude that there is evidence of a difference in the intention to return among business, education, and entertainment personal computer users. Alternatively, he may conclude that there is evidence of a relationship between intention to return and primary computer use.

Problems

11.19 Teachers in an elementary school were asked to say whether each child in their class "never tells a lie," "lies sometimes," or "lies quite frequently." The results were:

	Gender	
Lies	**Boys**	**Girls**
Never	12	21
Sometimes	32	28
Frequently	10	3

Is there evidence at the .05 level of significance that a tendency to be classified as a liar is related to gender? If it is, list as many reasons as you can think of why this might be so.

• 11.20 The Sudso Soap Company tested its new brand of detergent by having families try one box of the new brand and one box of the old brand and expressing a preference for one or the other. Families were classified by the softness of their water, resulting in the following table:

	Preference	
Water softness	**New**	**Old**
Soft	19	29
Medium	23	47
Hard	24	43

Is there evidence that water softness is related to preference for the new detergent? Use $\alpha = .01$.

• 11.21 Cancer researchers wanted to see whether the patient's psychological attitude toward his or her cancer was related to survival. Attitude was classified as being either a desire to fight the cancer or an attitude of resignation. The results after a 10-year follow-up were:

	Patient's attitude	
Results after 10 years	**Fighting**	**Resignation**
Alive, no cancer	23	12
Alive, with cancer	6	5
Dead	19	29

(a) Is there evidence at the .05 level of significance that attitude is related to results after 10 years?
(b) If it is, discuss possible reasons why these variables might be related.

11.22 Elementary school students were randomly assigned to one of four types of training in how to study (one, labeled Control, consisted of no specific training in study techniques but was included as a comparison condition). At the end of the semester, each student's teacher rated the student as being either improved or not improved in classroom performance. The results were:

	Classroom performance	
Training group	**Improved**	**Not improved**
A	17	7
B	13	11
C	11	13
Control	6	18

Is there evidence at the .10 level of significance that the proportion of students showing improvement differs over training groups?

/11.23 Why can't the normal approximation be used for determining differences in the proportion of successes in more than two groups?

11.6 CHI-SQUARE TEST OF INDEPENDENCE IN THE $R \times C$ TABLE

We have just seen that the chi-square test can be used to determine differences between the proportion of successes in any number of populations. For a contingency table that has R rows and C columns the chi-square test can be generalized as a test of independence. As an example, we may refer to the fourth issue of interest to the marketing director. He wanted to determine whether there was a relationship between the type of computer owned and its primary use. This information can be analyzed by first cross-classifying these two variables as illustrated in Table 11.7.

TABLE 11.7 □ Cross-classification of type of computer and primary use

	Primary use			
Computer type	**Business**	**Education**	**Entertainment**	**Total**
IBM or compatible	57	0	1	58
Apple or compatible	34	15	0	49
Commodore or Atari	3	26	23	52
Other	26	3	0	29
Total	120	44	24	188

The null and alternative hypothesis would be:

H_0: There is no relationship between type of computer and primary use
H_1: There is a relationship between type of computer and primary use

If this were to be tested at the .05 level of significance, the critical value (from Table C.4) would be 12.592, since there are $(4 - 1)(3 - 1) = 6$ degrees of freedom (see Figure 11.6). The decision rule would be

Reject H_0 if $\chi^2 > \chi^2_6 = 12.592$;
otherwise don't reject H_0.

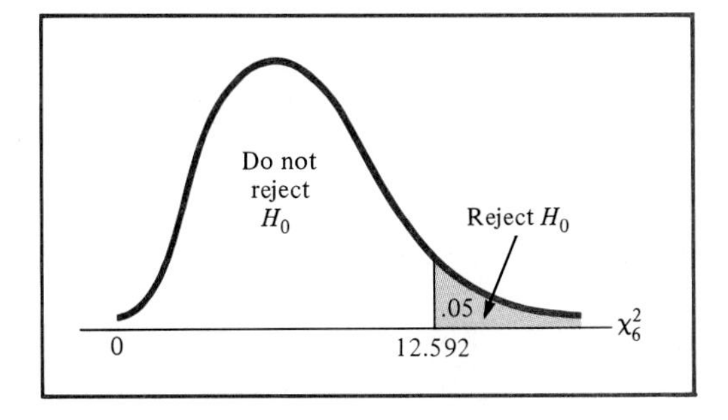

FIGURE 11.6 □ Testing for the relationship between type of computer owned and primary use at the 5% level of significance with 6 degrees of freedom.

The theoretical frequencies for the twelve cells, computed from Equation (11.4), are

$\frac{58(120)}{188} = 37.02$	$\frac{58(44)}{188} = 13.57$	$\frac{58(24)}{188} = 7.41$
$\frac{49(120)}{188} = 31.28$	$\frac{49(44)}{188} = 11.47$	$\frac{49(24)}{188} = 6.25$
$\frac{52(120)}{188} = 33.19$	$\frac{52(44)}{188} = 12.17$	$\frac{52(24)}{188} = 6.64$
$\frac{29(120)}{188} = 18.51$	$\frac{29(44)}{188} = 6.79$	$\frac{29(24)}{188} = 3.70$

One simple check on the computations of the expected frequencies is that the marginal totals of the expected frequencies should equal the marginal totals of the observed frequencies (except for possible rounding error). For example, for the first column we have

$$37.02 + 31.28 + 33.19 + 18.51 = 120$$

which is the same as the first-column marginal total for the observed frequencies.

We note that one cell (computer type = *other*; primary use = *entertainment*)

TABLE 11.8 □ Computation of the chi-square test statistic for the type of computer–primary use problem

f_o	f_t	$f_o - f_t$	$(f_o - f_t)^2$	$(f_o - f_t)^2/f_t$
57	37.02	19.98	399.2004	10.78337
34	31.28	2.72	7.3984	.23652
3	33.19	−30.19	911.4361	27.46117
26	18.51	7.49	56.1001	3.03080
0	13.57	−13.57	184.1449	13.57000
15	11.47	3.53	12.4609	1.08639
26	12.17	13.83	191.2689	15.71643
3	6.79	− 3.79	14.3641	2.11548
1	7.41	− 6.41	41.0881	5.54495
0	6.25	− 6.25	39.0625	6.25000
23	6.64	16.36	267.6496	40.30867
0	3.70	3.70	13.6900	3.70000
				$\chi^2 = 129.80378$

has a theoretical frequency below 5. However, in a large contingency table (such as the resulting 4 × 3 table) the existence of a single cell with a theoretical frequency below 5 does not seriously affect the chi-square test, unless that theoretical frequency is below 1 (see Reference 1). Thus, the chi-square test statistic may be computed as shown in Table 11.8.

Since $\chi^2 = 129.80378 > \chi^2_6 = 12.592$, the null hypothesis can be rejected, and the marketing director can conclude that there is evidence of a relationship between the type of computer owned and the primary use of the computer.

Problems

• 11.24 During the Viet Nam War a lottery system was instituted to choose males to be drafted into the military. Numbers representing days of the year were ''randomly'' selected; men born on days of the year with low numbers were drafted first, while those with high numbers were not drafted. The following shows how many low (1–122), medium (123–244), and high (245–366) numbers were drawn for birthdates in each quarter of the year:

Number set	Jan.–Mar.	Apr.–Jun.	Jul.–Sep.	Oct.–Dec.
Low	21	28	35	38
Medium	34	22	29	37
High	36	41	28	17

(a) Is there evidence that the numbers drawn were related to the time of year ($\alpha = .05$).
(b) Would you conclude that the lottery drawing appears to have been random?

• **11.25** To see whether people appear to find mates on the basis of height, a sociologist classified each partner in married couples as being short, average, or tall. The results of her survey are in the table below:

	Husband's height		
Wife's height	**Short**	**Medium**	**Tall**
Tall	6	23	42
Medium	13	43	59
Short	21	23	9

Is there evidence ($\alpha = .01$) that the height of wives is related to the height of husbands?

11.26 A scientist studying various inherited characteristics examined the relationship between hair color and eye color. The following results were obtained:

	Eye color		
Hair color	**Blue**	**Brown**	**Other**
Black	12	21	9
Brown	21	54	13
Red	5	3	3
Blonde	19	8	4

(a) Is there evidence that hair color and eye color are related? ($\alpha = .10$.)

(b) Since several expected frequencies are below 5, what do you think should be done: eliminate redheads, or combine them with some other hair color? Discuss.

11.27 Referring to Problem 6.52 on page 153, the statistics association at a large state university would like to determine whether there is a relationship between student interest in statistics and ability in mathematics. At the .05 level of significance, is there evidence of a relationship between interest in statistics and ability in mathematics?

11.28 Suppose that a survey has been undertaken to determine if there is a relationship between place of residence and automobile preference. A random sample of 200 car owners from large cities, 150 from suburbs, and 150 from rural areas was selected with the following results:

	Automobile preference					
Residence	**GM**	**Ford**	**Chrysler**	**American**	**Foreign**	**Totals**
Large city	64	40	26	8	62	200
Suburb	53	35	24	6	32	150
Rural	53	45	30	6	16	150
Totals	170	120	80	20	110	500

At the .05 level of significance, is there evidence of a relationship between place of residence and automobile preference?

11.29 For any of the tables in Problems 11.24–11.28, show how the degrees of freedom are counted by putting as many observed values as possible into the table without determining any of the row or column marginal totals, and noticing that the number of cells filled in is equal to the number of degrees of freedom. [*Hint*: See footnote 2 on page 317.]

11.7 TESTING FOR A DIFFERENCE BETWEEN TWO PROPORTIONS IN MATCHED OR PAIRED POPULATIONS: THE McNEMAR TEST

11.7.1 Rationale

In discussing the chi-square test in the last three sections, we have been concerned with situations in which the data have been obtained from groups that were separate or independent from each other. Often, however, as in the case of quantitative variables (see Section 10.9), qualitative data are obtained from related samples. We might wish to determine whether there is a difference between two groups that have been matched according to some control characteristic. We could also examine a single group to see whether its responses on two variables (or on one variable measured at different times or under different conditions) are different. For example, in the fifth issue of interest to the marketing director, he wanted to determine whether there was any evidence of a difference in (1) the proportion of computer nonowners who said they *intended to purchase* a personal computer in the next six months and (2) the proportion who *actually had purchased* a personal computer when contacted at the end of the six-month period. Since the data have been obtained from repeated measurements on the same individuals, the chi-square test of independence as presented in the previous sections in this chapter should not be used. (Later we show a correct way to do the chi-square test.) In such situations, when two proportions (each with two categories) are involved, a test developed by McNemar (see Reference 2) may be utilized.

11.7.2 Development

We may be interested in determining either (a) whether there is a *difference* between the two proportions (two-tailed test) or (b) whether one group has a *higher* proportion than the other group (one-tailed test). Thus, we may set up a 2 × 2 contingency table (see Table 11.9).

The sample proportions are

$p_{S_1} = \dfrac{A + B}{n}$ = the proportion of respondents in the sample that answer yes on variable 1

TABLE 11.9 ☐ 2 × 2 contingency table for the McNemar test

	Variable 2		
Variable 1	**Yes**	**No**	**Total**
Yes	A	B	$A + B$
No	C	D	$C + D$
Total	$A + C$	$B + D$	n

where A is the number of respondents that answer yes on variable 1 and variable 2
B is the number of respondents that answer yes on variable 1 and no on variable 2
C is the number of respondents that answer no on variable 1 and yes on variable 2
D is the number of respondents that answer no on variable 1 and variable 2
n is the number of respondents in the sample (that is, the sample size)

$p_{s_2} = \frac{A + C}{n}$ = the proportion of respondents in the sample that answer yes on variable 2

Notice that $p_{s_1} = p_{s_2}$ if and only if $(A + B)/n = (A + C)/n$, which in turn is true if and only if $B = C$. But $B = C$ means that $B - C = 0$, which is the basis of the McNemar test.

The test statistic for the McNemar test is

$$Z = \frac{B - C}{\sqrt{B + C}} \tag{11.6}$$

If the proportions are approximately equal, then $B - C$ will be small, making Z small; if the proportions differ greatly, then $B - C$ will be large, making Z large.

The data obtained from the computer nonowners at the show and through a followup six months later are summarized in Table 11.10.

Since the marketing director wanted to determine whether there was *any difference*

TABLE 11.10 ☐ Purchase intent and actual purchases of microcomputers

	Actual purchase		
Intended to purchase	**Yes**	**No**	**Total**
Yes	91	25	116
No	32	64	98
Total	123	89	212

in the proportion that intended to purchase and the proportion that actually purchased, the null and alternative hypotheses would be:

$$H_0: \; p_1 = p_2$$
$$H_1: \; p_1 \neq p_2$$

The test statistic is referred to critical values from the standard normal distribution (Table C.2). If the test was carried out at the .05 level of significance, the critical values would be -1.96 and $+1.96$ (see Figure 11.7) and our decision rule would be

Reject H_0 if $Z > +1.96$
or $Z < -1.96$;
otherwise do not reject H_0.

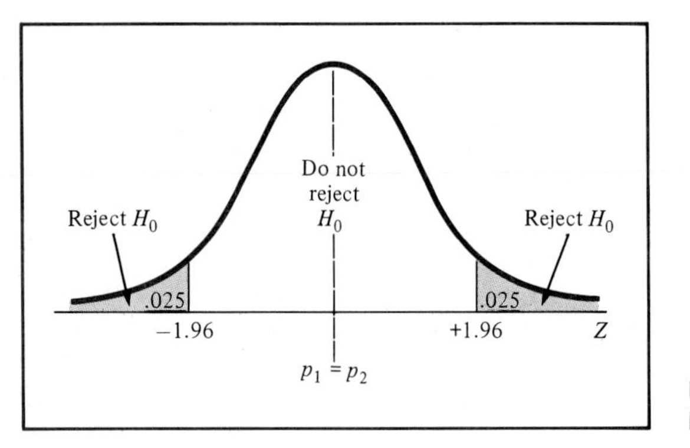

FIGURE 11.7 □ Testing a hypothesis about the difference between two proportions at the .05 level of significance.

For our data

$$A = 91, \quad B = 25, \quad C = 32, \quad D = 64$$

$$p_{S_1} = \frac{91 + 25}{212} = .5472, \qquad p_{S_2} = \frac{91 + 32}{212} = .5801$$

and, using Equation (11.6),

$$Z = \frac{25 - 32}{\sqrt{25 + 32}} = \frac{-7}{\sqrt{57}} = \frac{-7}{7.5498} = -.927$$

Since $Z = -.927$, we see that $-1.96 < -.927 < +1.96$, and the statistical decision is to not reject H_0. The marketing director can conclude that there is no evidence of a difference in the proportion of computer nonowners who intended to purchase and the proportion that actually purchased within the six-month period.

The McNemar test is equivalent to doing a chi-square test in which the theoretical frequencies are computed differently than for the tests of independence we have done so far. If we consider the two cells with the values B and C above, we noted that the test of equality of proportions is equivalent to the test that $B = C$. Using this as the null hypothesis, we may compute expected values of these two cells as $(B + C)/2$ for each cell. The chi-square statistic computed using these expected values is the square of the value computed for the McNemar test.

For the data in Table 11.10 on page 329 the expected values are $(32 + 25)/2 = 28.5$, so the chi-square value is calculated as

$$\begin{aligned}\chi^2 &= \frac{(32 - 28.5)^2}{28.5} + \frac{(25 - 28.5)^2}{28.5} \\ &= \frac{(3.5)^2}{28.5} + \frac{(-3.5)^2}{28.5} \\ &= .8596\end{aligned}$$

The statistic for the McNemar test was $-.927$, and $(-.927)^2 = .859$, the value of the chi-square statistic.

Since the square of a standard normal variable is a chi-square variable with one degree of freedom, the tests will always give the same result for a two-tailed test.

Problems

11.30 Two antibiotics, code-named W and Z, are tested by seeing how many common bacteria strains they will kill. The results are:

	Killed by W?	
Killed by Z?	**Yes**	**No**
Yes	42	21
No	8	16

(a) Is there evidence ($\alpha = .05$) that W and Z are equally effective in killing bacteria?

(b) Tell why the following table gives less information than the table above, how it can be derived from the table above, and why the information in it is insufficient to test the hypothesis of interest:

	Effect on bacteria strain	
Antibiotic type	**Kill**	**No kill**
W	50	37
Z	63	24

• 11.31 A researcher wanted to know what type of remedy people try most often for colds. He divided answers into two broad categories: natural remedies and drugs. The results of his survey are:

	Natural remedies?	
Drugs?	**Yes**	**No**
Yes	9	46
No	23	13

(a) Is there evidence ($\alpha = .05$) that drugs and natural remedies are not equally likely to be used?

(b) Tell why the following table gives less information than the table above, how it can be derived from the table above, and why the information in it is insufficient to test the hypothesis of interest:

	Use this remedy?	
Type of remedy	**Yes**	**No**
Natural	32	59
Drugs	55	36

11.32 A market researcher wanted to study the effect of an advertising campaign for Brand A cola to determine whether the proportion of cola drinkers who preferred Brand A would increase as a result of the campaign. A random sample of 200 cola drinkers was selected and they indicated their preference for either Brand A or Brand B prior to the beginning of the advertising campaign and at its completion. The results were as follows:

	Preference after completion of advertising campaign		
Preference prior to advertising campaign	**Brand A**	**Brand B**	**Totals**
Brand A	101	9	110
Brand B	22	68	90
Totals	123	77	200

At the .05 level of significance, is there evidence that the proportion of cola drinkers who prefer Brand A is higher at the end of the advertising campaign than at the beginning?

• 11.33 The personnel director of a large manufacturing company would like to reduce excessive absenteeism among assembly-line workers. She has decided to institute an experimental incentive plan that will provide financial rewards to employees who are absent less than 5 days in a given calendar year. A sample of 100 workers is selected at the end of the one-year trial period. For each of the two years, information is obtained for each employee selected that indicates whether the worker was absent less than 5 days in that year. The results were as follows:

	Year 2		
Year 1	**<5 days absent**	**≥5 days absent**	**Totals**
<5 days absent	32	4	36
≥5 days absent	25	39	64
Totals	57	43	100

(a) At the .01 level of significance, is there evidence that the proportion of employees absent less than 5 days is lower in year 1 than in year 2?

(b) What conclusion should the personnel director reach regarding the effect of the incentive plan?

11.34 Children in the third grade were tested on whether they knew the meaning of the words "democracy" and "ambiguous." The results were:

	Know "democracy?"	
Know "ambiguous?"	**Yes**	**No**
Yes	21	5
No	14	41

(a) Is there evidence ($\alpha = .05$) that a different proportion of students know each word?

(b) Tell why the following representation of the data is incomplete, and why it is insufficient to test the hypothesis:

	Yes	No
Know "Democracy?"	35	46
Know "Ambiguous?"	26	55

(c) Tell how a study might be designed so that the type of display in (b) would be the correct way to display the data.

Supplementary Problems

11.35 An auditor from the Department of Energy wishes to study the availability of diesel fuel in New York City. A random sample of 50 gas stations was selected; 11 sold diesel fuel. Is there evidence that less than 25% of the gas stations sell diesel fuel? (Use $\alpha = .05$.)

• 11.36 A marketing study conducted in a large city showed that of 100 married women who worked full time, 43 ate dinner at a restaurant at least one night during a typical workweek, while a sample of 100 married women who did not work full time indicated that 27 ate dinner at a restaurant at least one night during the typical workweek. Using a Z test with a .05 level of significance, is there evidence of a difference between the two groups of married women in the proportion that eat dinner in a restaurant at least once during the typical workweek?

11.37 Redo Problem 11.36 using the chi-square test.

11.38 The faculty council of a large university would like to determine the opinion of various groups toward a proposed trimester academic calendar. A random sample of 100 undergraduate students, 50 graduate students, and 50 faculty members is selected with the following results:

	Group		
Opinion	**Undergraduate**	**Graduate**	**Faculty**
Favor trimester	63	27	30
Oppose trimester	37	23	20
Totals	100	50	50

At the .01 level of significance, is there evidence of a difference in attitude toward the trimester among the various groups?

11.39 An agronomist is studying three different varieties of tomato to determine whether there is a difference in the proportion of seeds that germinate. Random samples of 100 seeds of each variety (beefsteak, plum, and cherry) are subjected to the same starting conditions with the following results:

	Tomato variety		
Number of seeds	**Beefsteak**	**Plum**	**Cherry**
Germinated	82	70	58
Did not germinate	18	30	42
Totals	100	100	100

At the .10 level of significance, is there evidence of a difference among the varieties of tomatoes in the proportion of seeds that germinate?

• 11.40 A large corporation was interested in determining whether an association exists between commuting time of their employees and the level of stress-related problems observed on the job. A study of 116 assembly-line workers revealed the following:

	Stress			
Commuting time	**High**	**Moderate**	**Low**	**Totals**
Under 15 min.	9	5	18	32
15 min. to 45 min.	17	8	28	53
Over 45 min.	18	6	7	31
Totals	44	19	53	116

At the .01 level of significance, is there evidence of a relationship among commuting time and stress?

11.41 **(a)** What is the difference between the McNemar test and the chi-square test for the difference between proportions?
(b) Under what circumstances should the McNemar test be used?

11.42 **(a)** Why shouldn't the chi-square test for independence be used when the theoretical frequencies in some of the cells are too small?

(b) What action can be taken in such circumstances to enable one to analyze that data from the contingency table?

Database Exercises

The following problems refer to the sample data obtained from the questionnaire of Figure 2.5 and presented in Figure 2.10. They should be solved with the aid of a computer package.

□ 11.43 **(a)** At the .05 level of significance, is there evidence that more than one-half of all alumni were business majors (question 2)?

(b) Determine the p value and interpret its meaning.

□ 11.44 **(a)** At the .05 level of significance, is there evidence that more than 40% of the alumni were attracted to the college primarily due to cost (question 3)?

(b) Determine the p value and interpret its meaning.

□ 11.45 **(a)** At the .01 level of significance, is there evidence that the proportion of alumni whose current political philosophy is about the same as in college days (question 4) is different from .50?

(b) Determine the p value and interpret its meaning.

□ 11.46 **(a)** At the .10 level of significance, is there evidence that the proportion of alumni who have ever attended graduate school (question 5) is greater than .35?

(b) Determine the p value and interpret its meaning.

□ 11.47 **(a)** At the .05 level of significance, is there evidence that more than 25% of the alumni are employed by a corporation with more than 100 employees (question 8)?

(b) Determine the p value and interpret its meaning.

□ 11.48 **(a)** At the .01 level of significance, is there evidence that the proportion of alumni who are currently employed in accounting (question 9) is different from .35?

(b) Determine the p value and interpret its meaning.

□ 11.49 **(a)** At the .10 level of significance, is there evidence that the proportion of alumni whose first full-time job after graduation from college was in accounting (question 11) is greater than .40?

(b) Determine the p value and interpret its meaning.

□ 11.50 **(a)** At the .05 level of significance, is there evidence that more than 50% of the alumni would attend the college if they had to make the choice again (question 16)?

(b) Determine the p value and interpret its meaning.

□ 11.51 **(a)** At the .01 level of significance, is there evidence that more than 50% of the alumni would either recommend or highly recommend the college to a close relative or friend (question 17)?

(b) Determine the p value and interpret its meaning.

□ 11.52 At the .05 level of significance, is there evidence of a difference in the proportion of males and females (question 18) who attended graduate school (question 5)? Use $\alpha = .05$.

□ 11.53 Is there evidence of a relationship between undergraduate major[3] (question 2) and graduate school attendance (question 5)? Use $\alpha = .01$.

□ 11.54 Is there evidence of a relationship between undergraduate major[3] (question 2) and gender (question 18)? Use $\alpha = .10$.

□ 11.55 Is there evidence of a relationship between undergraduate major (question 2) and the type of organization in which the alumni are employed[3] (question 8)? Use $\alpha = .01$.

□ 11.56 Is there evidence of a relationship between undergraduate major (question 2) and the advice the alumni would give concerning the college[3] (question 17)? Use $\alpha = .05$.

□ 11.57 Is there evidence of a relationship between the reason for attending the college (question 3) and whether the alumni would attend again[3] (question 16)? Use $\alpha = .10$.

□ 11.58 Is there evidence of a relationship between the reason for attending the college (question 3) and the advice the alumni would give[3] (question 17)? Use $\alpha = .05$.

□ 11.59 Is there evidence of a relationship between the reason for attending the college[3] (question 3) and gender (question 18)? Use $\alpha = .01$.

□ 11.60 Is there evidence of a relationship between current political ideology[3] (question 4) and gender (question 18)? Use $\alpha = .05$.

□ 11.61 Is there evidence of a relationship between current political ideology (question 4) and type of organization in which the respondent is currently employed[3] (question 8)? Use $\alpha = .01$.

□ 11.62 Is there evidence of a relationship between current political ideology (question 4) and whether the respondent would attend the college again[3] (question 16)? Use $\alpha = .05$.

□ 11.63 Is there evidence of a relationship between graduate school attendance (question 5) and current job activity[3] (question 9)? Use $\alpha = .05$.

□ 11.64 Is there evidence of a relationship between graduate school attendance (question 5) and advice concerning the college[3] (question 17)? Use $\alpha = .10$.

□ 11.65 Is there evidence of a relationship between type of organization at which currently employed (question 8) and current job activity[3] (question 9)? Use $\alpha = .01$.

□ 11.66 Is there evidence of a relationship between type of organization at which currently employed (question 8) and advice concerning the college[3] (question 17)? Use $\alpha = .05$.

References

1. CONOVER, W. J., *Practical Nonparametric Statistics,* 2d ed. (New York: Wiley, 1980).
2. DANIEL, W., *Applied Nonparametric Statistics* (Boston: Houghton Mifflin, 1978).

[3] The categories of certain variables may have to be combined and recoded in order to meet the assumptions of the chi-square test.

Chapter 12

The Analysis of Variance

12.1 INTRODUCTION: WHAT'S AHEAD

In many fields of application we need to compare alternative methods, treatments, or materials, according to some predetermined criteria. A consumer organization, for example, may wish to determine which brand of tires lasts longer under highway conditions; an agricultural researcher would like to know which type of green beans provides the highest yield; a medical researcher would like to evaluate the effect of different brands of prescription medicine on the reduction of diastolic blood pressure.

In Chapter 10 we used statistical inference to draw conclusions about differences between the means of two groups. In this chapter we shall examine techniques that have been developed to test for differences in the means of several groups. This methodology is classified under the general title **the analysis of variance.** The procedure covered in this text pertains only to the **one-way analysis of variance,** since only one **factor** (such as type of tire, variety of green bean, or brand of drug) is to be considered. However, the analysis of variance can easily be extended to handle cases where more than one factor at a time is studied in one experiment (see References 1–5).

Case I—The Educational Researcher's Problem

Suppose that the assistant superintendent for curriculum affairs wishes to evaluate three alternative sets of mathematics materials so that one set can be selected to be purchased by the entire school district. A third-grade teacher in the district has volunteered to make the comparison. The 24 students in her class are homogeneous with respect to their academic abilities. They are to be divided randomly into three groups, containing seven, nine, and eight students, respectively. The first group is assigned to set I, the second to set II, and the third to set III.

At the end of the year all 24 students are given the same standardized mathematics test. The scores on a scale of 0–100 are displayed in Table 12.1.

TABLE 12.1 □ Score on standardized mathematics exam according to sets of material used

Material set		
I	II	III
87	58	81
80	63	62
74	64	70
82	75	64
74	70	70
81	73	72
97	80	92
	62	63
	71	
$\bar{X}_I = 82.14$	$\bar{X}_{II} = 68.44$	$\bar{X}_{III} = 71.75$

Based on these results, the assistant superintendent would like to know whether there is any difference among the sets in the scores achieved and, if so, which set(s) are superior to the others.

12.2 THE LOGIC BEHIND THE ANALYSIS OF VARIANCE

For the data of Table 12.1, we must test whether the various groups all have the same population mean. Suppose we have c groups. In general, the null and alternative hypotheses would be stated as:

$$H_0: \mu_1 = \mu_2 = \mu_3 = \cdots = \mu_c$$
$$H_1: \text{Not all means are equal}$$

For our educational researcher's example, since there are three sets, the null and alternative hypotheses would be:

$$H_0: \mu_{\text{I}} = \mu_{\text{II}} = \mu_{\text{III}}$$
$$H_1: \text{Not all the sets have equal means}$$

We have seen in Table 12.1 that there are differences in the *sample* means of the three sets ($\bar{X}_{\text{I}} = 82.14$, $\bar{X}_{\text{II}} = 68.44$, $\bar{X}_{\text{III}} = 71.75$). The question that must be answered is whether the differences among the *sample* means of the groups are large enough to lead to the conclusion that the *population* means differ; or, alternatively, that the sample differences can reasonably be explained merely by random variation. Since under the null hypothesis the population means of the three sets are presumed equal, a measure of the **total variation** among the scores can be obtained by summing up the squared differences between each observation and an overall combined mean, $\bar{\bar{X}}$, based on all the observations. The total variation would be computed as

$$\text{total variation} = \sum_i \sum_j (X_{ij} - \bar{\bar{X}})^2 \qquad \textbf{(12.1)}$$

where X_{ij} = ith observation in group j

$$\bar{\bar{X}} = \frac{\sum_i \sum_j X_{ij}}{n} \text{ is called the } \textbf{grand mean}$$

n = total number of observations in the sample

Since this total variation is actually the sum of squared deviations, it is also referred to as the **total sum of squares, or SST.** In our study the total sum of squares would be computed as follows:[1]

$$\bar{\bar{X}} = \frac{\sum_i \sum_j X_{ij}}{n}$$

$$= \frac{87 + 80 + \cdots + 58 + \cdots + 63}{24}$$

$$= \frac{1{,}765}{24} = 73.54$$

[1] All calculations are carried out to more decimal places than are recorded here; using the numbers here will produce some slight discrepancies due to rounding error.

$$\text{SST} = \sum_i \sum_j (X_{ij} - \overline{\overline{X}})^2$$

$$= \begin{pmatrix} (87 - 73.54)^2 \\ +(80 - 73.54)^2 \\ +(74 - 73.54)^2 \\ +(82 - 73.54)^2 \\ +(74 - 73.54)^2 \\ +(81 - 73.54)^2 \\ +(97 - 73.54)^2 \end{pmatrix} + \begin{pmatrix} (58 - 73.54)^2 \\ +(63 - 73.54)^2 \\ +(64 - 73.54)^2 \\ +(75 - 73.54)^2 \\ +(70 - 73.54)^2 \\ +(73 - 73.54)^2 \\ +(80 - 73.54)^2 \\ +(62 - 73.54)^2 \\ +(71 - 73.54)^2 \end{pmatrix} + \begin{pmatrix} (81 - 73.54)^2 \\ +(62 - 73.54)^2 \\ +(70 - 73.54)^2 \\ +(64 - 73.54)^2 \\ +(70 - 73.54)^2 \\ +(72 - 73.54)^2 \\ +(92 - 73.54)^2 \\ +(63 - 73.54)^2 \end{pmatrix}$$

$$= \begin{pmatrix} 181.13 \\ +\ 41.71 \\ +\ 0.21 \\ +\ 71.54 \\ +\ 0.21 \\ +\ 55.63 \\ +550.29 \end{pmatrix} + \begin{pmatrix} 241.54 \\ +111.13 \\ +\ 91.04 \\ +\ 2.13 \\ +\ 12.54 \\ +\ 0.29 \\ +\ 41.71 \\ +133.21 \\ +\ 6.46 \end{pmatrix} + \begin{pmatrix} 55.63 \\ +133.21 \\ +\ 12.54 \\ +\ 91.04 \\ +\ 12.54 \\ +\ 2.38 \\ +340.71 \\ +111.13 \end{pmatrix}$$

$$= 2{,}299.96$$

This total variation measures the sum of the squared differences between each value X_{ij} and the overall (grand) mean $\overline{\overline{X}}$.

Why is there variation among the values; that is, why are the observations not all the same? One reason is that by treating people differently (in this case, giving them different instructional material), we affect their scores. This would explain part of the reason why the groups have different means: the bigger the effect of the treatment, the more variation in group means we will find. But there is another reason for variability in the scores, which is that people are naturally variable whether we treat them alike or not. So even within a particular group, where everyone got the same treatment, there is variability. Because it occurs within each group, it is called **within-group variation.**

The differences among the group means are called the **between-group variation** (even when there are more than two groups, statisticians usually use "between" and not "among"). Part of the between-group variation, as we noted above, is due to the effect of being in different groups. But even if there is no effect of being in different groups (that is, the null hypothesis is true), there will likely be differences among the group means. This is because variability among subjects will make the sample means different just because we have different samples. Therefore, if the null hypothesis is true, then the between-group variation will estimate the population variability just as well as the within-group variation. But if the null

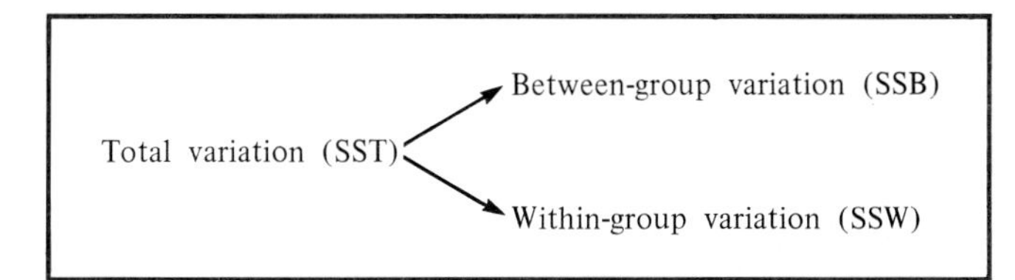

FIGURE 12.1 □ Partitioning the total variation

hypothesis is false, then the between-group variation will be larger. This fact forms the basis for the statistical test of group differences.

The total variation can therefore be subdivided (i.e., "analyzed") into two separate components (see Figure 12.1); one part consists of variation between the groups, and the other part consists of variation within the groups. It is always true that

$$\text{total variation (SST)} = \text{between-group variation (SSB)} + \text{within-group variation (SSW)} \qquad \textbf{(12.2)}$$

The between-group variation is measured by the sum of the squared differences between the sample mean of each group $\bar{X}_j$ and the grand mean $\bar{\bar{X}}$, weighted by the sample size n_j in each group. The between-group variation, which is usually called the **sum of squares between** (or **SSB**), may be computed from:

$$\text{SSB} = \sum_j n_j(\bar{X}_j - \bar{\bar{X}})^2 \qquad \textbf{(12.3)}$$

where n_j is the number of observations in group j
$\bar{X}_j$ is the sample mean of group j
$\bar{\bar{X}}$ is the grand mean

For our data the between-group variation would be computed as follows:

$$\begin{aligned}
\text{SSB} &= 7(82.14 - 73.54)^2 + 9(68.44 - 73.54)^2 + 8(71.75 - 73.54)^2 \\
&= 7(8.60)^2 + 9(-5.10)^2 + 8(-1.79)^2 \\
&= 7(73.98) + 9(25.98) + 8(3.21) \\
&= 517.86 + 233.84 + 25.68 \\
&= 777.39
\end{aligned}$$

The within-group variation, usually called the **sum of squares within** (or **SSW**), measures the difference between each value and the mean of its own group, and cumulates the squares of these differences over all groups. The within-group variation may be computed as:

$$\text{SSW} = \sum_{i}\sum_{j}(X_{ij} - \bar{X}_j)^2 \qquad \textbf{(12.4)}$$

where $\bar{X}_j$ is the mean of group j
X_{ij} is the ith observation in group j

For the data of Table 12.1 the within-group variation would be computed as follows:

$$\begin{pmatrix} (87 - 82.14)^2 \\ +(80 - 82.14)^2 \\ +(74 - 82.14)^2 \\ +(82 - 82.14)^2 \\ +(74 - 82.14)^2 \\ +(81 - 82.14)^2 \\ +(97 - 82.14)^2 \end{pmatrix} + \begin{pmatrix} (58 - 68.44)^2 \\ +(63 - 68.44)^2 \\ +(64 - 68.44)^2 \\ +(75 - 68.44)^2 \\ +(70 - 68.44)^2 \\ +(73 - 68.44)^2 \\ +(80 - 68.44)^2 \\ +(62 - 68.44)^2 \\ +(71 - 68.44)^2 \end{pmatrix} + \begin{pmatrix} (81 - 71.75)^2 \\ +(62 - 71.75)^2 \\ +(70 - 71.75)^2 \\ +(64 - 71.75)^2 \\ +(70 - 71.75)^2 \\ +(72 - 71.75)^2 \\ +(92 - 71.75)^2 \\ +(63 - 71.75)^2 \end{pmatrix}$$

$$= \begin{pmatrix} 23.59 \\ + \ 4.59 \\ + \ 66.31 \\ + \ 0.02 \\ + \ 66.31 \\ + \ 1.31 \\ +220.73 \end{pmatrix} + \begin{pmatrix} 109.09 \\ + \ 29.64 \\ + \ 19.75 \\ + \ 42.98 \\ + \ 2.42 \\ + \ 20.75 \\ +133.53 \\ + \ 41.53 \\ + \ 6.53 \end{pmatrix} + \begin{pmatrix} 85.56 \\ + \ 95.06 \\ + \ 3.06 \\ + \ 60.06 \\ + \ 3.06 \\ + \ 0.06 \\ + \ 410.06 \\ + \ 76.56 \end{pmatrix}$$

$$= 1{,}522.58$$

Referring back to the previous computations for the total variation and the between-group variation, we check to determine that

$$\text{SST} = \text{SSB} + \text{SSW}$$
$$2{,}299.96 = 777.39 + 1{,}522.58 \qquad \text{(except for rounding error)}$$

12.3 THE *F* DISTRIBUTION

In order to determine whether the means of the various groups are all equal, we can examine two different estimates of the population variance. One estimate is based on the sum of squares within groups (SSW); the other is based on the sum of squares between groups (SSB). If the null hypothesis is true, these estimates should be nearly the same; if it is false, the estimate based on the sum of squares between groups should be larger.

In Section 3.5.2 we said that a variance is estimated by dividing the sum of squared deviations by its appropriate degrees of freedom. In dividing by the degrees of freedom, we calculated the variance as a sort of average squared deviation. Since another word for "average" is "mean," we could call this the **mean squared deviation;** in statistics, this is shortened to the term **mean square.**

In the analysis of variance, the sum of squared deviations is represented by the component sums of squares we have computed. Within each group, the degrees of freedom is equal to the number of subjects in that group minus one. Since there are c groups, if we add the degrees of freedom within the groups together, we get $df_W = n - c$; this is the degrees of freedom for computing the variance estimate using the within sum of squares.[2] That is, the within-group mean square (also called the **mean square within,** or **MSW**), measures variability around the particular sample mean of each group. Since this variability is not affected by group differences, it can be considered a measure of the random variation of values within a group.

The between-group variance estimate (**mean square between** or **MSB**) is calculated by dividing the between-group sum of squares by the degrees of freedom between groups. The degrees of freedom is one less than the number of groups; i.e., $df_B = c - 1$. This is because SSB was calculated from the difference between the c group means and the overall mean. Using the same logic as we did in calculating degrees of freedom for the variance in Chapter 3, we see that only $c - 1$ of the group means can vary; the last is determined because the overall mean is known.

The variance between groups, MSB, is estimated by

$$\text{MSB} = \frac{\begin{array}{c}\text{sum of squares}\\ \text{between groups}\end{array}}{\begin{array}{c}\text{degrees of freedom}\\ \text{between groups}\end{array}} = \frac{\text{SSB}}{df_B} = \frac{\sum_j n_j(\bar{X}_j - \bar{\bar{X}})^2}{c - 1} \qquad \textbf{(12.5)}$$

The variance within groups, MSW, is estimated by

$$\text{MSW} = \frac{\begin{array}{c}\text{sum of squares}\\ \text{within groups}\end{array}}{\begin{array}{c}\text{degrees of freedom}\\ \text{within groups}\end{array}} = \frac{\text{SSW}}{df_W} = \frac{\sum_i \sum_j (X_{ij} - \bar{X}_j)^2}{n - c} \qquad \textbf{(12.6)}$$

[2] Within each group there are $(n_j - 1)$ degrees of freedom. When these are summed over all c groups, there are $n - c$ degrees of freedom within groups:

$$(n_1 - 1) + (n_2 - 1) + \cdots + (n_c - 1) = (n_1 + n_2 + \cdots + n_c) + (-1 - 1 - \cdots - 1) = n - c$$

The between-group variance estimate (MSB) not only takes into account random fluctuations from observation to observation, but also measures differences from one group to another. If there is no difference from group to group, any sample mean differences will be explainable by random variation, and the variance between groups, MSB, should be close to the variance within groups, MSW. If there is really a difference between the groups, however, the variance between groups, MSB, will be significantly *larger* than the variance within groups, MSW. The test statistic is based upon the ratio of the two variances, MSB/MSW. If the null hypothesis is true, this ratio should be near 1; if the null hypothesis is false, then the numerator should be bigger than the denominator, and the ratio should be bigger than 1. To know how big the ratio must be before we reject the null hypothesis, we need to know what the distribution of this statistic would be if the null hypothesis were true; this distribution is called the ***F* distribution** (see Table C.5), named after the famous statistician R. A. Fisher. That is,

$$F = \frac{\text{MSB}}{\text{MSW}} \qquad \textbf{(12.7)}$$

From Table C.5, we can see that the shape of the F distribution depends on two types of degrees of freedom: the degrees of freedom between the group means in the numerator, (called df_1 in most tables of the F distribution) and the degrees of freedom within groups (df_2) in the denominator.

The decision rule is to reject the null hypothesis of no difference between the groups if at the α level of significance

$$F = \frac{\text{MSB}}{\text{MSW}} > F_{c-1,n-c}$$

where $F_{c-1,n-c}$ is the critical value of the F distribution with $c - 1$ and $n - c$ degrees of freedom. This is depicted in Figure 12.2.

Returning to the data of Table 12.1, since SSB = 777.39, SSW = 1522.58, $n = 24$, and $c = 3$, we have

$$\text{MSB} = \frac{777.39}{3-1} = \frac{777.39}{2} = 388.69$$

$$\text{MSW} = \frac{1{,}522.58}{24-3} = \frac{1{,}522.58}{21} = 72.50$$

Therefore we have

$$F = \frac{\text{MSB}}{\text{MSW}} = \frac{388.69}{72.50} = 5.36$$

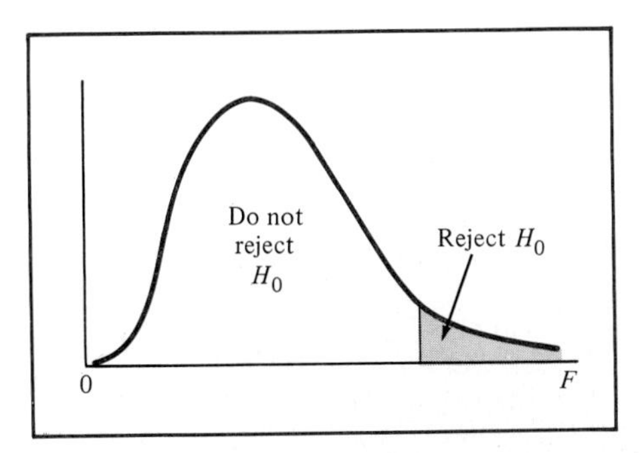

FIGURE 12.2 ☐ Regions of rejection and nonrejection in the analysis of variance.

If a significance level of .05 was selected, the critical value of the F distribution could be determined from Table C.5. The values in the body of this table refer to selected *upper* percentage points of the F distribution. In this example, since there are 2 degrees of freedom in the numerator and 21 degrees of freedom in the denominator, the critical value of F at the .05 level of significance is 3.47 (see Table 12.2).

Thus, the decision rule would be to reject the null hypothesis (H_0: $\mu_I = \mu_{II} = \mu_{III}$) if the calculated F value exceeds 3.47 (see Figure 12.3 on page 346).

In this example, since $F = 5.36 > F_{2,21} = 3.47$, the statistical decision is to

TABLE 12.2 ☐ Obtaining the critical value of F with 2 and 21 degrees of freedom and α = .05

Denominator, df_2	Numerator, df_1													
	1	2	3	4	5	6	7	8	9	10	12	15	20	24
1	161.4	199.5	215.7	224.6	230.2	234.0	236.8	238.9	240.5	241.9	243.9	245.9	248.0	249.1
2	18.51	19.00	19.16	19.25	19.30	19.33	19.35	19.37	19.38	19.40	19.41	19.43	19.45	19.45
3	10.13	9.55	9.28	9.12	9.01	8.94	8.89	8.85	8.81	8.79	8.74	8.70	8.66	8.64
4	7.71	6.94	6.59	6.39	6.26	6.16	6.09	6.04	6.00	5.96	5.91	5.86	5.80	5.77
5	6.61	5.79	5.41	5.19	5.05	4.95	4.88	4.82	4.77	4.74	4.68	4.62	4.56	4.53
6	5.99	5.14	4.76	4.53	4.39	4.28	4.21	4.15	4.10	4.06	4.00	3.94	3.87	3.84
7	5.59	4.74	4.35	4.12	3.97	3.87	3.79	3.73	3.68	3.64	3.57	3.51	3.44	3.41
8	5.32	4.46	4.07	3.84	3.69	3.58	3.50	3.44	3.39	3.35	3.28	3.22	3.15	3.12
9	5.12	4.26	3.86	3.63	3.48	3.37	3.29	3.23	3.18	3.14	3.07	3.01	2.94	2.90
10	4.96	4.10	3.71	3.48	3.33	3.22	3.14	3.07	3.02	2.98	2.91	2.85	2.77	2.74
11	4.84	3.98	3.59	3.36	3.20	3.09	3.01	2.95	2.90	2.85	2.79	2.72	2.65	2.61
12	4.75	3.89	3.49	3.26	3.11	3.00	2.91	2.85	2.80	2.75	2.69	2.62	2.54	2.51
13	4.67	3.81	3.41	3.18	3.03	2.92	2.83	2.77	2.71	2.67	2.60	2.53	2.46	2.42
14	4.60	3.74	3.34	3.11	2.96	2.85	2.76	2.70	2.65	2.60	2.53	2.46	2.39	2.35
15	4.54	3.68	3.29	3.06	2.90	2.79	2.71	2.64	2.59	2.54	2.48	2.40	2.33	2.29
16	4.49	3.63	3.24	3.01	2.85	2.74	2.66	2.59	2.54	2.49	2.42	2.35	2.28	2.24
17	4.45	3.59	3.20	2.96	2.81	2.70	2.61	2.55	2.49	2.45	2.38	2.31	2.23	2.19
18	4.41	3.55	3.16	2.93	2.77	2.66	2.58	2.51	2.46	2.41	2.34	2.27	2.19	2.15
19	4.38	3.52	3.13	2.90	2.74	2.63	2.54	2.48	2.42	2.38	2.31	2.23	2.16	2.11
20	4.35	3.49	3.10	2.87	2.71	2.60	2.51	2.45	2.39	2.35	2.28	2.20	2.12	2.08
21	4.32	3.47	3.07	2.84	2.68	2.57	2.49	2.42	2.37	2.32	2.25	2.18	2.10	2.05

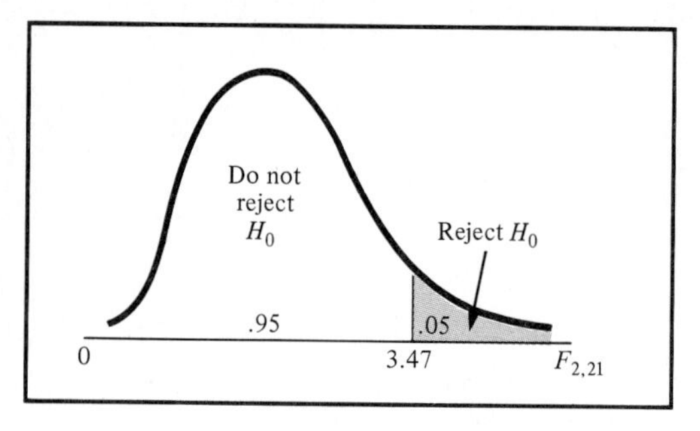

FIGURE 12.3 □ Regions of rejection and nonrejection for the analysis of variance at the 5% level of significance with 2 and 21 degrees of freedom.

reject the null hypothesis. The educational researcher may conclude that there is a significant difference in the average scores for the three sets of mathematics material.

12.4 THE ANALYSIS-OF-VARIANCE TABLE

Since several steps are involved in the computation of both the between- and within-group variances, the entire set of results may be organized into an *analysis-of-variance (ANOVA) table*. This table, presented as Table 12.3, includes the sources of variation, the sums of squares, the degrees of freedom, the variances, and the calculated F value.

In our educational researcher's example, the computations can be summarized as in Table 12.4.

TABLE 12.3 □ Analysis-of-variance table

Source	Sum of squares	Degrees of freedom	Mean square (variance)	F
Between	$SSB = \sum_j n_j(\bar{X}_j - \bar{\bar{X}})^2$	$c - 1$	$MSB = SSB/(c - 1)$	MSB/MSW
Within	$SSW = \sum_i \sum_j (X_{ij} - \bar{X}_j)^2$	$n - c$	$MSW = SSW/(n - c)$	
Total	$SST = \sum_i \sum_j (X_{ij} - \bar{\bar{X}})^2$	$n - 1$		

TABLE 12.4 □ Analysis-of-variance table for the educational research case study

Source	Sum of squares	Degrees of freedom	Mean square	F
Between	777.39	2	388.69	5.36
Within	1,522.58	21	72.50	
Total	2,299.97	23		

12.5 COMPUTATIONAL METHODS

So far in this chapter the formulas utilized in computing the between-group sum of squares, the within-group sum of squares, and the total sum of squares may be considered as conceptual (or definitional). Unfortunately, such formulas involve a large amount of tedious computations. However, as we saw in Section 3.5 when the variance was computed, a "short-cut" computational formula was derived from the definitional formula. There are similar computational formulas for the analysis of variance; they require the calculation of the following quantities:

1. $\sum_i \sum_j X_{ij}^2$, the sum of the squared values of each observation in the data.
2. T_j, the total (or sum) of the values in group j.
3. Grand total (GT) $= \sum_i \sum_j X_{ij}$, the sum of all values.
4. $\sum_j T_j^2$, the sum of the squared totals of the values in each group.

Using computational formulas, we can then compute the various sums of squares in the following way:

$$\text{SSB} = \sum_j \frac{T_j^2}{n_j} - \frac{(\text{GT})^2}{n} \quad \textbf{(12.3a)}$$

$$\text{SSW} = \sum_i \sum_j X_{ij}^2 - \sum_j \frac{T_j^2}{n_j} \quad \textbf{(12.4a)}$$

$$\text{SST} = \sum_i \sum_j X_{ij}^2 - \frac{(\text{GT})^2}{n} \quad \textbf{(12.1a)}$$

The format for the analysis-of-variance table using the computational formulas is given in Table 12.5 on page 348.

Returning to the educational researcher's study, the following quantities must be computed:

$$\sum_i \sum_j X_{ij}^2, \qquad T_j, \qquad \sum_j \frac{T_j^2}{n_j}, \qquad \frac{(\text{GT})^2}{n}$$

TABLE 12.5 □ Analysis-of-variance table using computational formulas

Source	Sum of squares	Degrees of freedom	Mean square	F
Between	$SSB = \sum_j \frac{T_j^2}{n_j} - \frac{(GT)^2}{n}$	$c - 1$	$\frac{SSB}{c-1}$	$\frac{MSB}{MSW}$
Within	$SSW = \sum_i \sum_j X_{ij}^2 - \sum_j \frac{T_j^2}{n_j}$	$n - c$	$\frac{MSW}{n-c}$	
Total	$SST = \sum_i \sum_j X_{ij}^2 - \frac{(GT)^2}{n}$	$n - 1$		

Referring back to Table 12.1, there were three sets of mathematics material (I, II, and III). The results, giving the total and the mean for each set, are summarized below:

	I	II	III
Sample size, n_j	$n_1 = 7$	$n_2 = 9$	$n_3 = 8$
Total, T_j	$T_1 = 575$	$T_2 = 616$	$T_3 = 574$
Sample mean, $\bar{X}_j$	$\bar{X}_1 = 82.14$	$\bar{X}_2 = 68.44$	$\bar{X}_3 = 71.75$

Therefore, we compute the following:

$$GT = \sum_i \sum_j X_{ij} = 575 + 616 + 574 = 1{,}765$$

$$\frac{(GT)^2}{n} = \frac{(1{,}765)^2}{24} = 129{,}801.04$$

$$\sum_i \sum_j X_{ij}^2 = 87^2 + 80^2 + \cdots + 63^2 = 132{,}101$$

$$\sum_j \frac{T_j^2}{n_j} = \frac{(575)^2}{7} + \frac{(616)^2}{9} + \frac{(574)^2}{8} = 130{,}578.42$$

With this information, the analysis-of-variance table for the educational research data can be set up using the computational formulas (Table 12.6).

If we compare the entries given in Table 12.4 with those of Table 12.6, we can see, of course, that the two methods (definitional and computational) have produced exactly the same results. Therefore, it is recommended that the computational method be utilized for all calculations needed for the analysis-of-variance table, since these computations are less tedious and simpler to perform.

TABLE 12.6 ☐ Analysis-of-variance table for the educational researcher's data using computational formulas

Source	Sum of squares	Degrees of freedom	Mean square	F
Between	130,578.42 − 129,801.04 = 777.39	2	388.69	5.36
Within	132,101 − 130,578.42 = 1,522.58	21	72.50	
Total	132,101 − 129,801.04 = 2,299.96	24 − 1 = 23		

Problems

• 12.1 A gerontologist wanted to see whether staying "lean and mean" (i.e., being under normal weight) would lengthen life. He randomly assigned newborn rats to get either (i) unlimited access to food, (ii) 90% of the amount a rat of that size would normally eat, or (iii) 80% of the amount a rat of that size would normally eat. He then measured how long they lived (in years), with the following results:

Unlimited Food:	2.5, 3.1, 2.3, 1.9, 2.4
90%:	2.7, 3.1, 2.9, 3.7, 3.5
80%:	3.1, 2.9, 3.8, 3.9, 4.0, 3.6

(a) Construct an appropriate graph, plot, or chart of the data.
(b) Calculate the mean length of life for each group.
(c) Using your answers to (a) and (b), guess whether there is any difference among the groups.
(d) Do an analysis of variance of the data, and describe the results. (α = .05.)

12.2 In order to see whether marijuana has any effect on cognitive functioning, drug users were recruited by a researcher and were asked to smoke two cigarettes. The cigarettes had either no THC (the active ingredient in marijuana), 4.5 milligrams of THC, or 18 milligrams of THC. Their scores on a test of cognitive ability (high scores are better) were:

No THC:	74, 86, 93, 69, 85, 78
4.5 mg THC:	85, 73, 82, 88, 90
18 mg THC:	83, 65, 72, 78, 70, 76

(a) Construct an appropriate graph, plot, or chart of the data.
(b) Calculate the mean test score for each group.
(c) Using your answers to (a) and (b), guess whether there is any difference among the groups.
(d) Do an analysis of variance of the data, and describe the results. (α = .05.)

12.3 A metallurgist tested five different alloys for tensile strength. He tested several samples of each alloy; the tensile strength of each sample was:

Alloy 1:	12.4, 19.8, 15.2, 14.8, 18.5
Alloy 2:	8.9, 11.6, 10.0, 10.3
Alloy 3:	10.5, 13.8, 12.1, 11.9, 12.6
Alloy 4:	12.8, 14.2, 15.9, 14.1
Alloy 5:	16.4, 15.9, 17.8, 20.3

(a) Construct an appropriate graph, plot, or chart of the data.
(b) Calculate the mean tensile strength for each group.
(c) Using your answers to (a) and (b), guess whether there is any difference among the groups.
(d) Do an analysis of variance of the data, and describe the results. ($\alpha = .05$.)

12.4 In Problem 3.11 on page 46, a study was described in which physicists at the University of Gorgon measured the mass of the riton, a new subatomic particle. These masses ($\times 10^{-24}$ grams) are listed below:

2.11, 2.13, 2.21, 2.19, 2.09, 2.14,
2.17, 2.12, 2.20, 2.16, 2.14, 2.15

Physicists at Holy Cow College in Rizzutoville, New Jersey decided to reproduce the original experiment; their measurements for the masses were:

2.05, 2.09, 2.10, 2.07, 2.09, 2.04,
2.08, 2.12, 2.07, 2.06, 2.11

(a) Construct an appropriate graph, plot, or chart of the data.
(b) Calculate the mean particle mass for each group.
(c) Using your answers to (a) and (b), guess whether there is any difference between the groups.
(d) Do an analysis of variance of the data, and tell whether the two groups of physicists are getting results which are consistent with each other. ($\alpha = .05$.)
(e) What procedure that you learned in Chapter 10 could have been used to solve part (d)?

12.5 In Problem 10.32 on page 291 you used a t test to compare the acceleration of German and Japanese cars.
(a) Do an analysis of variance on this data set. ($\alpha = .01$.)
(b) Square the value of t you computed in Problem 10.32; notice that it is the same (except for rounding error) as the F value.

12.6 In Problem 10.34 on page 291 you used a t test to compare the noise made by gasoline-powered and electric-powered chainsaws.
(a) Do an analysis of variance on this data set. ($\alpha = .01$.)
(b) Square the value of t you computed in Problem 10.34; notice that it is the same (except for rounding error) as the F value.

• **12.7** In Problem 10.35 on page 291 you used a t test to compare the caloric content of low- and high-fat cottage cheeses.
(a) Do an analysis of variance on this data set. ($\alpha = .05$.)
(b) Square the value of t you computed in Problem 10.35; notice that it is the same (except for rounding error) as the F value.

• **12.8** A consumer magazine was interested in determining whether any difference existed in the average life of four different brands of transistor radio batteries. A random sample of four batteries of each brand was tested with the following results (lifetimes are in hours):

Brand 1	Brand 2	Brand 3	Brand 4
12	14	21	14
15	17	19	21
18	12	20	25
10	19	23	20

Is there evidence of a difference in the average life of these four brands of transistor radio batteries? ($\alpha = .05$.)

12.9 A consumer organization wanted to compare the price of a particular toy in three types of stores in a suburban county: discount toy stores, department stores, and variety stores. A random sample of 4 discount toy stores, 6 department stores, and 5 variety stores was selected. The results were as follows:

Discount Toy	Department	Variety
12	15	19
14	18	16
15	14	16
16	18	18
	18	15
	15	

Is there evidence of a difference in average price among the types of stores? ($\alpha = .05$.)

12.6 ASSUMPTIONS OF THE ANALYSIS OF VARIANCE

In Chapters 10 and 11 we mentioned the assumptions made in applying each particular statistical procedure to a set of data and the consequences of departures from these assumptions. The analysis-of-variance technique also makes certain assumptions about the data being investigated. There are three major assumptions in the analysis of variance:

1. Normality.
2. Homogeneity of variance.
3. Independence of errors.

The first assumption, normality, states that the values in each group are normally distributed. Just as in the case of the t test, the analysis-of-variance test is "robust" against departures from the normal distribution; that is, as long as the distributions are not extremely different from the normal distribution, the level of significance of the analysis-of-variance test is not greatly affected by lack of normality, particularly for large samples.

The second assumption, homogeneity of variance, states that the variance within each population should be equal for all populations ($\sigma_1^2 = \sigma_2^2 = \cdots = \sigma_c^2$). This assumption is needed in order to combine or pool the variances within the groups into a single "within-group" source of variation SSW. If there are equal sample sizes in each group, inferences based upon the F distribution may not be seriously affected by unequal variances. If, however, there are unequal sample sizes in different groups, unequal variances from group to group can have serious effects on drawing inferences made from the analysis of variance. Thus, from the point of view of computational simplicity, robustness, and power, there should be equal sample sizes in all groups whenever possible.

The third assumption, independence of errors, refers not to mistakes, but to the difference of each value from its own group mean. The assumption is that these should be independent for each value. That is, the error for one observation should not be related to the error for any other observation. This assumption might be violated, for example, in the evaluation of the teaching material described in this chapter if one student copied answers from another. The assumption would also be violated if two of the subjects in one group were twins: their behavior would likely be more similar than would the behavior of any other two subjects in the study.

This assumption is often violated when data are collected over a period of time, because observations made at adjacent time points are often more alike than those made at very different times. Consider, for example, temperature recorded every day for a month. The temperature on a given day is likely to be near what it was the day before, but less likely to be close to the temperature several weeks later. Departures from this assumption can seriously affect inferences from the analysis of variance. These problems are discussed more thoroughly in References 1 and 4.

★ 12.7 CONTRASTS AMONG GROUP MEANS (OPTIONAL SECTION)

On page 346, we have answered the first question raised by the educational researcher. By rejecting the null hypothesis, we showed that the group means are not all equal. But which means are different? To give us more information, we want to be able to compare individual groups with each other. Furthermore, we may want to do other kinds of comparisons, perhaps comparing one group with the mean of two other groups. These kinds of comparisons are called **contrasts** among the group means.

We may decide to do contrasts before we even collect our data, because certain comparisons are of theoretical importance. If we decide ahead of time, the contrasts are called **a priori,** or preplanned, contrasts, because we decide which contrasts to use *prior* to doing the study. On the other hand, we may wait until after we collect our data, and then notice that certain groups differ from others. At that time we may decide to test whether these apparent differences among groups are significant. Contrasts which are devised after looking at the results are called **post hoc** (after the fact) contrasts.

★ Note that this is an optional section that may be skipped without any loss of continuity.

The simplest contrast compares the means of two groups. In the educational researcher's problem, for example, we might want to test the hypothesis that the mean score for the group using material set II is the same as the mean score for the group using material set III. We could express this as

$$\mu_{II} = \mu_{III}$$

which is equivalent to testing whether

$$\mu_{II} - \mu_{III} = 0$$

Therefore, if we want to test this hypothesis, we could look at the difference in sample means between groups II and III. If it is near zero, it is consistent with the null hypothesis above; if it is not near zero, it is not consistent with the null hypothesis. A hypothesis such as this, where we are testing whether an expression involving the group means is equal to zero, is called a contrast among the means.

As a shorthand way of expressing the contrasts, the numbers which are multiplied by the means, called the **coefficients of the contrast,** are listed. For example, the coefficients of the contrast above are 1, −1. But we have three groups in the experiment, so how can we tell which groups the coefficients refer to? In the example above, the contrast is between groups II and III; group I is ignored. We can think of group I as having a coefficient of 0 (so that no matter what the group mean for I is, it doesn't contribute to the contrast). Then the coefficients for the contrast above are 0, 1, −1. To contrast group I with group III, the coefficients would be 1, 0, −1.

Let us consider another hypothesis we might be interested in testing. Suppose that material set II and material set III both involved using many visual aids, while set I consisted mostly of verbal material. We might want to know if using visual aids produced different results than using mostly verbal material. This would mean that we want to compare group I with the average performance by groups II and III. In other words, is the following hypothesis true?

$$\mu_{I} = \frac{\mu_{II} + \mu_{III}}{2}$$

Rewriting this expression,

$$\mu_{I} - \frac{(\mu_{II} + \mu_{III})}{2} = 0$$

That is, if the sample mean for group I is close to the average of the sample means of groups II and III, then the difference above should be close to zero; if it is not, then there is evidence that the hypothesis is false.

The coefficients for this particular contrast are 1, $-\frac{1}{2}$, $-\frac{1}{2}$; that is, the contrast is formed by taking the mean of the first group, minus half the mean of the second group, minus half the mean of the third group. Notice that the coefficients for the contrast add up to zero; the same was true for the first contrast we devised. This is a good way to check that we have set up a meaningful contrast among the group means.

An F test can be used to test whether each contrast is zero. The between sum of squares for testing a contrast is

$$SS_{CON} = \frac{\left(\sum_j C_j \bar{X}_j\right)^2}{\sum_j \frac{C_j^2}{n_j}} \qquad \textbf{(12.8)}$$

where C_j is the contrast coefficient for group number j
n_j is the number of observations in group j

The numerator is the square of the value calculated for the contrast. The denominator is the sum of one term for each coefficient C_j: each term is the square of the coefficient for group j divided by the number of observations in group j.

For the first contrast we wanted to test, the sum of squares is

$$\frac{[(0)(82.14) + (1)(68.44) + (-1)(71.75)]^2}{[(0)^2(\frac{1}{7}) + (1)^2(\frac{1}{9}) + (-1)^2(\frac{1}{8})]} = \frac{10.956}{.2361} = 46.40$$

For the second contrast, the sum of squares is

$$\frac{[(1)(82.14) + (-\frac{1}{2})(68.44) + (-\frac{1}{2})(71.75)]^2}{[(1)^2(\frac{1}{7}) + (-\frac{1}{2})^2(\frac{1}{9}) + (-\frac{1}{2})^2(\frac{1}{8})]} = \frac{145.082}{.2019} = 718.58$$

Each contrast (no matter how many means are involved in the comparison) has 1 degree of freedom. So the mean square for each contrast is equal to its sum of squares. That is, $MS_{CON} = SS_{CON}$.

The within mean square for testing each contrast is the same within mean square (MSW) we calculated in the analysis-of-variance table, 72.50. Therefore, the F ratio for testing a contrast is

$$F_{CON} = \frac{MS_{CON}}{MSW} \qquad \textbf{(12.9)}$$

Hence, for our first contrast

$$F_{CON} = \frac{46.40}{72.50} = .64$$

while for our second contrast

$$F_{CON} = \frac{718.58}{72.50} = 9.91$$

For testing a priori contrasts, we use the critical value of F, at the .05 level, with 1 degree of freedom in the numerator and $n - c = 21$ degrees of freedom in the denominator. At the .05 level of significance, this value is 4.32. The first contrast is not significant, so that groups II and III do not differ from each other. The second contrast is significant, so group I differs from the average of groups II and III. We can see from the sign of the contrast, or from direct inspection of the group means, that group I scores higher than the average of groups II and III.

The above tests were calculated as a priori contrasts, because we had theoretical reasons for making the comparisons we made. That is, by the nature of the design of the study, we expected that groups II and III might perform similarly because the materials were both constructed using visual aids heavily. On the other hand, we expected group I to differ from II and III, because its materials were designed much differently.

Suppose we had not made any such preplanned hypotheses before collecting the data, but rather we looked at the results (after performing an ANOVA) and noticed that groups II and III performed similarly, and that group I seemed to perform differently. Now we are doing a **post hoc test.**

Even with only the three groups we have in our study, there are a great many post hoc contrasts which might be devised. With more groups, there is an even larger number of contrasts. With so many possible contrasts, it would not be unusual in any study to be able to find at least one contrast which would be significant if done as a preplanned comparison. If *each* of a large number of contrasts is tested at the .05 level, the probability that *at least one* of the contrasts is significant can become very large. Because of this, statisticians have developed ways to control the Type I error level for the number of possible comparisons which might be done (see Reference 3).

Many procedures have been devised for testing specific kinds of post hoc contrasts; most, for example, test all possible pairs of means (see Reference 3). We will demonstrate the **Scheffé procedure,** because (1) it can be used to test any contrast, not just contrasts of pairs of means, and (2) it is very similar to doing a priori contrasts. In fact, it differs from a priori contrasts only in the critical value used to test the calculated F values. The F values themselves are calculated in the same way as for the a priori contrasts.

To use the Scheffé procedure to do post hoc tests, we calculate a new critical value using (1) the number of groups in the study, c, and (2) the critical value for the usual ANOVA, $F_{(c-1,n-c)}$. The critical value (F_S) for the Scheffé procedure is the product of the between-group degrees of freedom from the ANOVA, $(c - 1)$, and the critical value from the one-way ANOVA:

$$F_S = (c - 1)F_{(c-1),(n-c)} \tag{12.10}$$

In the study we have been analyzing in this chapter, the critical value for the ANOVA was 3.47, and there were 2 degrees of freedom between groups. The critical value for the Scheffé test is therefore (2)(3.47) = 6.94. Compare this with the critical value of 4.32 which we could use for preplanned contrasts. The F value for a contrast must be bigger to be significant in a post hoc test than in a preplanned test, because we have essentially used a smaller value of α to control the overall chance of a Type I error in the study. In our example, there would be no difference in the conclusions reached if we had done the tests post hoc; this is because the second contrast was so large that it is significant even using a conservative test, and the first so small that it is clearly not significant.

Problems

• 12.10 Referring to Problem 12.1 on page 349, if appropriate, use Scheffé's method ($\alpha = .05$) to determine which diets are different.

12.11 Referring to Problem 12.2 on page 349, if appropriate, use Scheffé's method ($\alpha = .05$) to determine which levels of THC differ in their effect on cognitive ability.

12.12 Referring to Problem 12.3 on page 349, if appropriate, use Scheffé's method ($\alpha = .05$) to determine which alloys differ in tensile strength.

• 12.13 Referring to Problem 12.8 on page 350, if appropriate, use Scheffé's method ($\alpha = .05$) to determine which brands are different.

12.8 OVERVIEW OF EXPERIMENTAL DESIGN

In the example that we discussed in this chapter, we used the analysis of variance to test for the difference in the means of three groups. This was an example of the most elementary experimental design model, the *one-factor* or *completely randomized design* in which the researcher is evaluating only one factor of interest over several levels or groups (in our case the different sets of mathematics materials).

We can extend our development of experimental design along several different lines (each of which is beyond the scope of this textbook). First, researchers often either wish to evaluate several factors simultaneously or are restricted in some fashion in how the subjects are assigned to groups. For example, a study may involve the factors Gender (male and female) and Treatment Group (such as material sets I, II, and III in this chapter). There are then six groups of subjects: males using set I, males using set II, males using set III; females using set I, females using set II, and females using set III. The general issues in such a study are: (1) Whether males and females scored the same; (2) whether people using sets I, II, and III scored the same; and (3) whether the difference between males and females is the same for each set of materials. The first two questions involve testing what is known as **main effects,** while the third tests for an **interaction** between the Gender and Material factors. The techniques for testing contrasts may be used to analyze data from this type of study.

Second, researchers may want to measure variables which they cannot control experimentally, and somehow control for them statistically. For example, the educational researcher may measure prior knowledge of mathematics in the study described in this chapter, because that may explain some of the variation among subjects who got the same treatment. Such variables are sometimes called **covariates,** or **control variables.** Such situations lead to experimental designs that are more complicated than the completely randomized model. For a more detailed discussion of these and other experimental design topics, see References 1, 2, and 5.

Supplementary Problems

Note: A star indicates a problem or part of a problem that requires a study of an optional section or subsection of this chapter.

12.14 Explain the purpose of an analysis of variance.

12.15 Explain what the mean square between and mean square within are estimating. What is expected to happen to these quantities if there is no difference among the groups in the population?

12.16 Explain the assumptions of the analysis of variance.

12.17 Do the assumptions of homogeneity of variance and independence of observations appear to be true for the situation described in Problem 12.1 on page 349? Discuss.

✓12.18 Do the assumptions of homogeneity of variance and independence of observations appear to be true for the situation described in Problem 12.7 on page 350? (*Note*: Part of this—the independence of observations—is tricky. Just do the best job of reasoning about it that you can. The point here is to think, not necessarily to get the "right" answer.)

✓12.19 Refer to Problem 12.9 on page 351. If the stores were not a sample, but the only such stores in the county, would it make any sense to do a statistical analysis? (You should be able to give arguments for and against the sensibility of doing an analysis). [*Hint*: Consider to what populations you might want to generalize, and whether the sample could be considered a random sample for each possible population of interest.]

12.20 Explain how the graphical methods of Chapters 4 and 5 might be used to evaluate the validity of the assumptions of the analysis of variance.

12.21 A medical researcher decided to compare several over-the-counter sleep medications. People who had trouble sleeping came to the researcher's sleep laboratory and took either a placebo (a pill with no active ingredients), Nighty-night, Snooze-Away, or Mr. Sandman. The number of hours each one slept is given below:

Placebo:	2, 4, 3, 5, 2, 4, 3
Nighty-night:	3, 4, 5, 3, 5, 4, 6, 5
Snooze-Away:	6, 5, 7, 4, 8, 6
Mr. Sandman:	5, 4, 7, 5, 8, 6, 7

(a) Construct an appropriate graph, plot, or chart of the data.
(b) Calculate the mean time slept for each group.
(c) Using your answers to (a) and (b), guess whether there is any difference among the groups.
(d) Do an analysis of variance of the data, and describe the results. ($\alpha = .05$.)
★**(e)** Does it appear that some drugs are more useful than others? Test. ($\alpha = .05$.)

• 12.22 Researchers in the area of human factors wanted to test several designs for the dials used in airplanes to see which could be read most accurately. They tested 21 pilots, each of whom ''flew'' one of three simulators. Each simulator had a different type of dial. The researchers recorded the number of errors made in reading the dials:

Digital dial:	4, 3, 2, 5, 3, 2, 4, 3
Circular analog dial:	8, 6, 7, 5, 4, 6
Vertical analog dial:	5, 3, 4, 5, 4, 2, 3

(a) Construct an appropriate graph, plot, or chart of the data.
(b) Calculate the mean number of errors for each group.
(c) Using your answers to (a) and (b), guess whether there is any difference among the groups.
(d) Do an analysis of variance of the data, and describe the results. ($\alpha = .05$.)
★**(e)** Does it appear that one type of dial is easier to read then the others? Test. ($\alpha = .05$.)

• 12.23 The retailing manager of a food chain wishes to determine whether product location has any effect on the sale of pet toys. Three different aisle locations are to be considered: front, middle, and rear. A random sample of 18 stores was selected with 6 stores randomly assigned to each aisle location. At the end of a one-week trial period, the sales of the product in each store were as follows:

Aisle location		
Front	**Middle**	**Rear**
86	20	46
72	32	28
54	24	60
40	18	22
50	14	28
62	16	40

(a) At the .01 level of significance, is there evidence of a difference in average sales among the various aisle locations?

★**(b)** If appropriate, use the Scheffé method to determine which aisle locations are different in average sales.

12.24 An auditor for the Internal Revenue Service would like to compare the efficiency of four regional tax processing centers. A random sample of five returns was selected at each center, and the number of days between receipt of the tax return and final processing was determined. The results (in days) were as follows:

Regional center			
East	**Midwest**	**South**	**West**
49	47	39	52
54	56	55	42
40	40	48	57
60	51	43	46
43	55	50	50

At the .05 level of significance:

(a) Is there evidence of a difference in processing time among the four regional centers?

★**(b)** If appropriate, use the Scheffé method to determine the regional centers that differ in average processing time.

12.25 As a consulting industrial engineer, you are hired to perform a "human factors experiment" at a large law firm. A pool of 12 typists of similar ability and experience is selected to participate. The 12 subjects are randomly assigned to one of three working conditions—Very noisy atmosphere (90 db constant), Somewhat noisy atmosphere (65 db constant), and Pleasant atmosphere (40 db constant). The subjects are then asked to type a technical manuscript. The data below represent the number of mistakes on the manuscript made by the typists under the various working conditions.

Group		
Very noisy (90 db)	**Somewhat noisy (65 db)**	**Pleasant (40 db)**
14	2	2
12	5	6
13	8	6
13	5	2

At the .01 level of significance:

(a) Is there evidence of a difference in the average number of errors between the three groups?

★**(b)** If appropriate, use the Scheffé method to determine which groups differ in average number of errors.

★12.26 An agronomist primarily wishes to compare the yield (in pounds) from winter versus summer squash. Four types of squash were to be studied:

1. Butternut (winter) squash.
2. Acorn (winter) squash.
3. Zucchini (summer) squash.
4. Scallop (summer) squash.

A field was divided into 16 plots, with 4 plots randomly assigned to each type of squash. The results of the experiment (yield in pounds) were as follows:

Type			
1	2	3	4
86	40	30	48
74	48	36	54
88	54	42	42
76	46	34	56

(a) Using a .05 level of significance, is there evidence of a difference in the yield from winter squash versus summer squash?
(b) Explain why it is appropriate to consider this as a problem in preplanned or a priori contrasts.

Database Exercises

The following problems refer to the sample data obtained from the questionnaire of Figure 2.5 and presented in Figure 2.10. They should be solved with the aid of a computer package. Use $\alpha = .05$ throughout.

□ 12.27 Is there evidence of a difference in salary among alumni who had different undergraduate majors (questions 7 and 2)?

□ 12.28 Is there evidence of a difference in height among male alumni who work in different types of organizations (questions 18, 19, and 8)?

□ 12.29 Is there evidence of a difference in salary based on current philosophical ideology (questions 7 and 4)?

□ 12.30 Is there evidence of a difference in age based on current philosophical ideology (questions 20 and 4)?

□ 12.31 Is there evidence of a difference in grade-point average among alumni who had different undergraduate majors (questions 15 and 2)?

□ 12.32 Is there evidence of a difference in salary among alumni who work in different types of organizations (questions 7 and 8)?

□ 12.33 Is there evidence of a difference in salary based on type of advice given to prospective college students (questions 7 and 17)?

□ 12.34 Is there evidence of a difference in age based on primary reason for attending the college (questions 20 and 3)?

References

1. BERENSON, M. L., D. M. LEVINE, AND M. GOLDSTEIN, *Intermediate Statistical Methods and Applications: A Computer Package Approach* (Englewood Cliffs, N.J.: Prentice-Hall, 1983).
2. HICKS, C. R., *Fundamental Concepts in the Design of Experiments*, 3d ed. (New York: Holt, Rinehart and Winston, 1982).
3. KIRK, R. *Experimental Design*, 2d ed. (Belmont, Calif.: Brooks-Cole, 1982).
4. NETER, J., W. WASSERMAN, AND M. H. KUTNER, *Applied Linear Statistical Models*, 2d ed. (Homewood, Ill.: Richard D. Irwin, 1985).
5. WINER, B., *Statistical Principles in Experimental Design*, 2d ed. (New York: McGraw-Hill, 1971).

Chapter 13

Simple Linear Regression and Correlation

13.1 INTRODUCTION: WHAT'S AHEAD

In previous chapters we have been concerned primarily with situations involving a single quantitative response variable, such as annual salary, quarterly sales tax receipts, verbal SAT score, soft-drink bottle fill, and worker absenteeism. In analyzing the resulting data we studied various measures of description (see Chapters 3 through 5) and applied different techniques of statistical inference (that is, confidence intervals and tests of hypothesis) to make estimates and draw conclusions about the underlying parameters (see Chapters 9 through 12). In this chapter we will concern ourselves with both descriptive and inferential situations involving two quantitative variables as a means of viewing the relationships that exist between them. Two techniques will be discussed: regression and correlation.

Regression analysis is used for the purpose of prediction. Our objective in regression analysis is the development of a statistical model that can *predict the values* of a **dependent** or **response variable** Y based upon the values of a **predictor** or **independent variable** X. As an example, an economist might want to develop a statistical model that would predict how much money we will spend (Y) based on how much money we earn (X). In this chapter we shall focus on such a "simple" regression model—one which would utilize a single quantitative independent variable X to predict the quantitative dependent variable Y.

Correlation analysis, on the other hand, is used to measure the *strength of association* between variables X and Y. As an example, an educational psychologist might wish to determine the correlation between verbal ability and IQ in sixth-grade children. In this instance, the objective is not to use one variable to predict another, but rather to measure the strength of the association or "covariation" that exists between the two variables.

The goals of this chapter are:

1. To highlight the essential concepts of regression and correlation.
2. To demonstrate how regression and correlation methods can be used in data analysis by focusing on the fundamental aspects of model development and evaluation through statistical inference.

13.2 SIMPLE LINEAR REGRESSION

In a *simple* linear regression analysis we attempt to develop a linear model from which the values of a dependent (i.e., response) variable can be predicted based on particular values of a *single* independent variable. To develop the model, we assume that a sample of n independent pairs of observations (X_1, Y_1), (X_2, Y_2), . . . , (X_n, Y_n) is obtained, where X_i represents the ith value of the independent or predictor variable X and where Y_i represents the corresponding response—that is, the ith value of the dependent variable Y.

In order to motivate the discussion concerning the main ideas of this chapter let us consider an illustrative example.

Case J—The Placement Officer's Problem

Suppose the placement officer at a university wishes to investigate the relationship between the achieved grade-point index and the starting salary of recent graduates majoring in business so that when advising students she may build a model to predict the starting salary of business majors based on the grade-point index. A random sample of 30 recent graduates from the College of Business is drawn, and the data pertaining to the grade-point index (GPI) and starting salary (in thousands of dollars) are recorded for each individual as in Table 13.1.

TABLE 13.1 □ Starting salary and grade-point index for a random sample of 30 recent graduates majoring in business

Individual no.	(X) GPI	(Y) Starting salary ($000)
1	2.7	17.0
2	3.1	17.7
3	3.0	18.6
4	3.3	20.5
5	3.1	19.1
6	2.4	16.4
7	2.9	19.3
8	2.1	14.5
9	2.6	15.7
10	3.2	18.6
11	3.0	19.5
12	2.2	15.0
13	2.8	18.0
14	3.2	20.0
15	2.9	19.0
16	3.0	17.4
17	2.6	17.3
18	3.3	18.1
19	2.9	18.0
20	2.4	16.2
21	2.8	17.5
22	3.7	21.3
23	3.1	17.2
24	2.8	17.0
25	3.5	19.6
26	2.7	16.6
27	2.6	15.0
28	3.2	18.4
29	2.9	17.3
30	3.0	18.5

In particular, for the placement officer to be provided with useful information to advise the business majors, the following questions must be addressed:

1. What model best represents the underlying form of the relationship between starting salary and grade-point index?
2. For a business student who has achieved a particular grade-point index, what starting salary should he/she expect to be offered?
3. Is the model useful?
4. What inferences can be drawn from the sample data?

The exact mathematical form of the model expressing the relationship, as well as methods for predicting starting salary for a given grade-point index, will be discussed in subsequent sections of this chapter.

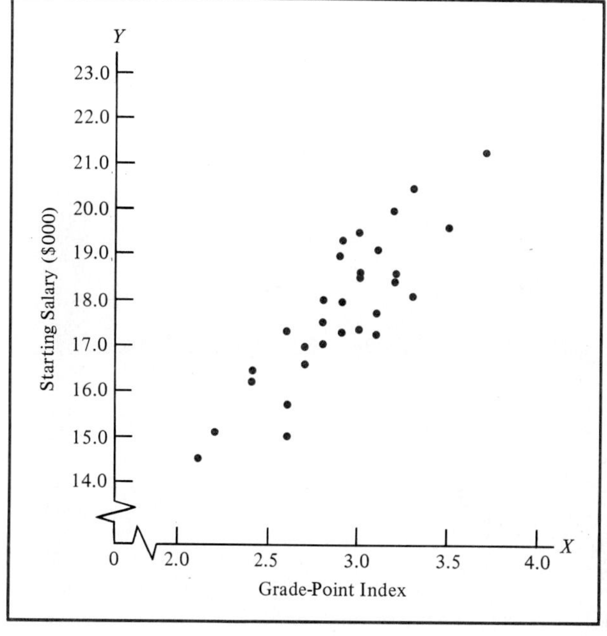

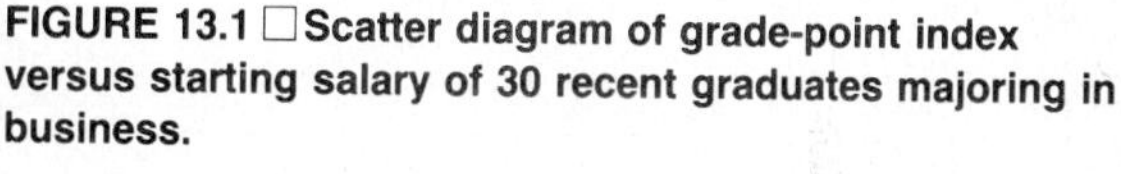

FIGURE 13.1 □ Scatter diagram of grade-point index versus starting salary of 30 recent graduates majoring in business.

SOURCE: Data are taken from Table 13.1.

13.3 THE SCATTER DIAGRAM

To visually study the possible underlying relationship between X (grade-point index) and Y (starting salary), the $n = 30$ individual pairs of observations can be plotted on a two-dimensional graph called a **scatter diagram.** Each individual pair of observations is plotted at its particular X and Y coordinates. The dependent variable Y is plotted on the vertical axis, while the independent variable X is plotted on the horizontal axis.

The scatter diagram of the data for the placement officer's problem is displayed in Figure 13.1. A brief examination of the figure indicates an increasing (i.e., a positive) linear relationship between the grade-point index X and the starting salary Y. As the grade-point index increases, the starting salary also tends to increase, and the points appear to fall approximately in a straight line.

13.4 DEVELOPING THE SIMPLE LINEAR REGRESSION EQUATION

The scatter diagram aids the researcher in selecting an appropriate regression model. By examining the plotted sample points, the researcher attempts to project the underlying mathematical relationship that may exist between X and Y. The nature of the relationship can take many forms, ranging from simple mathematical functions to extremely complicated ones (see References 2 and 6). The simplest relationship between X and Y consists of a straight-line or *linear* relationship, as shown in Figure 13.2.

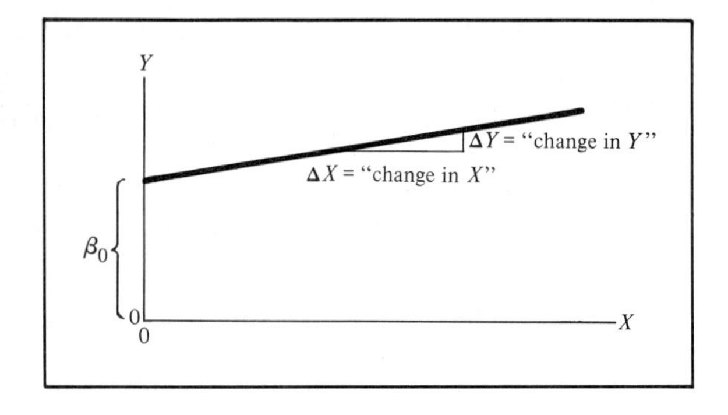

FIGURE 13.2 □ A positive straight-line relationship.

The straight-line (linear) model can be represented as

$$Y_i = \beta_0 + \beta_1 X_i + \epsilon_i \qquad \textbf{(13.1)}$$

where β_0 is the true Y intercept for the population
β_1 is the true slope for the population
ϵ_i is the random error in Y for observation i

In this model, the slope of the line (β_1) represents the unit change in Y per unit change in X; that is, it represents the amount that Y changes (either positively or negatively) for a particular unit change in X. The Y intercept (β_0) represents a constant factor that is included in the equation. It represents the value of Y when X equals zero. The last component of the model, ϵ_i, represents the random error in Y for each observation i that occurs. This term is included because the statistical model is only an approximation of the exact relationship between the two variables.

Since we do not have access to the entire population, we cannot compute the parameters β_0 and β_1 and obtain the population regression model. The objective then becomes one of obtaining estimates b_0 (for β_0) and b_1 (for β_1) from the sample. The sample regression equation for representing a straight-line model is

$$\hat{Y}_i = b_0 + b_1 X_i \qquad \textbf{(13.2)}$$

where $\hat{Y}_i$ is the predicted value of Y for observation i
X_i is the value of X for observation i

To see what properties we would like the estimates b_0 and b_1 to have, start by trying to draw a straight line through the data points in Figure 13.1 so that the line comes as close as possible to all the points. Of course, you cannot draw a line which will actually go through all the points, and you might even pick a line which does not go through any of them. Some lines are obviously better than others, but how should we define more exactly what we mean by a line being "close" to all of the points?

One way is the following: for any particular line (i.e., for specific values of b_0 and b_1), calculate the Y value on the line for each X value in the data set. These predicted values of Y, which we have called $\hat{Y}$ (pronounced "Y hat"), will differ depending on the values of b_0 and b_1. If we subtract $\hat{Y}$ from Y for each point, we get the error of prediction (ϵ) for each point.

How can we summarize the sizes of the various errors? Of course, a negative value (overprediction) for an error is as bad as a positive value (underprediction). To remove the sign, we could square the values. If we add these squared values, we get the sum of squared errors as a measure of how close the line comes to the data points. Using this measure, we would like to find estimates b_1 and b_0 which make the sum of squared errors as small as possible.

To obtain the statistics b_1 and b_0 which do minimize the sum of squared errors in a sample, we use a technique known as the **method of least squares.** Derived using the mathematical techniques from calculus, this method tells us to compute the sample slope b_1 and sample intercept b_0 as follows:

$$b_1 = \frac{(\Sigma XY) - \dfrac{(\Sigma X)(\Sigma Y)}{n}}{(\Sigma X^2) - \dfrac{(\Sigma X)^2}{n}} \qquad \textbf{(13.3)}$$

$$b_0 = \overline{Y} - b_1\overline{X} \qquad \textbf{(13.4)}$$

where $\overline{Y} = \dfrac{\Sigma Y}{n}$ and $\overline{X} = \dfrac{\Sigma X}{n}$

The denominator in Equation (13.3) should look familiar: it was used in computing the sum of squares in previous chapters. The numerator is similar, except that both X and Y values are used at the same time; it is called the **sum of cross products.** If we abbreviate these quantities as SS_X and SS_{XY}, respectively, we have the simplified formula

$$b_1 = \frac{SS_{XY}}{SS_X} \qquad \textbf{(13.3a)}$$

Examining Equations (13.3) and (13.4), we see that we must determine five quantities in order to obtain b_1 and b_0 (and the sample regression equation). These are:

- ΣX, the sum of the X values,
- ΣY, the sum of the Y values,
- ΣXY, the sum of the cross products of X and Y,
- ΣX^2, the sum of the squared X values, and
- n, the sample size.

In addition, for subsequent calculations we must also compute the quantity ΣY^2, the sum of the squared Y values.

Using the data in Table 13.1, the summary computations recorded in Table 13.2 are needed not only for computing the intercept b_0 and slope b_1 but also for studying other aspects of regression and correlation to be discussed throughout this chapter.

TABLE 13.2 □ Summary computations needed for simple regression and correlation analysis

$\Sigma X = 87.0$	$\bar{X} = 2.90$
$\Sigma Y = 534.3$	$\bar{Y} = 17.81$
$\Sigma XY = 1{,}564.24$	$\Sigma X^2 = 256.06$
$n = 30$	$\Sigma Y^2 = 9{,}593.41$

SOURCE: Table 13.1.

By using Equations (13.3) and (13.4), the slope is

$$b_1 = \frac{(\Sigma XY) - \dfrac{(\Sigma X)(\Sigma Y)}{n}}{(\Sigma X^2) - \dfrac{(\Sigma X)^2}{n}} = \frac{1{,}564.24 - \left[\dfrac{87.0(534.3)}{30}\right]}{256.06 - \left[\dfrac{(87.0)^2}{30}\right]} = +3.92819$$

while the intercept is

$$b_0 = \bar{Y} - b_1\bar{X} = 17.81 - (3.92819)(2.90) = +6.41825$$

Thus the equation we want for these data is (rounded to two places)[1]

$$\hat{Y}_i = 6.42 + 3.93\,X_i$$

INTERPRETING THE SLOPE AND INTERCEPT

The slope $b_1 = 3.93$ indicates that for each full unit (point) increase in the grade-point index, the starting salary is predicted to increase by approximately 3.93 thousands of dollars (i.e., \$3,930). The Y intercept, $b_0 = 6.42$, represents the predicted value if $X = 0$. As we can see from this case, this predicted value is not necessarily meaningful, since no one could really graduate with a 0.0 grade-point index.

[1] To make the estimates of the slope and intercept accurate, we have used five decimal places in our calculations.

PREDICTION

The sample regression equation can now be used for prediction purposes. For example, if the placement officer wished to advise a senior from the College of Business about expected starting salary, she would substitute the student's cumulated grade-point index into her model [Equation (13.2)] to make the prediction. As a case in point, suppose a senior had achieved a 3.0 grade-point index. Then,

$$\hat{Y} = 6.42 + 3.93(3.0) = 18.19$$

and the expected starting salary for such a senior would be 18.19 thousands of dollars.

When using a regression model for prediction purposes, it is important that we consider only the relevant range of the independent variable in making our predictions. This relevant range encompasses all values from the smallest to the largest X used in developing the regression model. Hence, when predicting Y for a given value of X, we may *interpolate* within this relevant range of the X values, but we may not *extrapolate* beyond the range of X values. For example, in predicting starting salaries of recent graduates of the College of Business, we note from Table 13.1 that the grade-point indexes varied from 2.1 through 3.7. Predictions of starting salaries should be made only for grade-point indexes in this interval. Any predictions outside this range assume that the fitted relationship holds outside the interval 2.1 through 3.7 (and there is no justification for such an assumption).

The sample regression line for the placement officer's data is plotted on the scatter diagram in Figure 13.3. As it should, it seems to come close to most of the

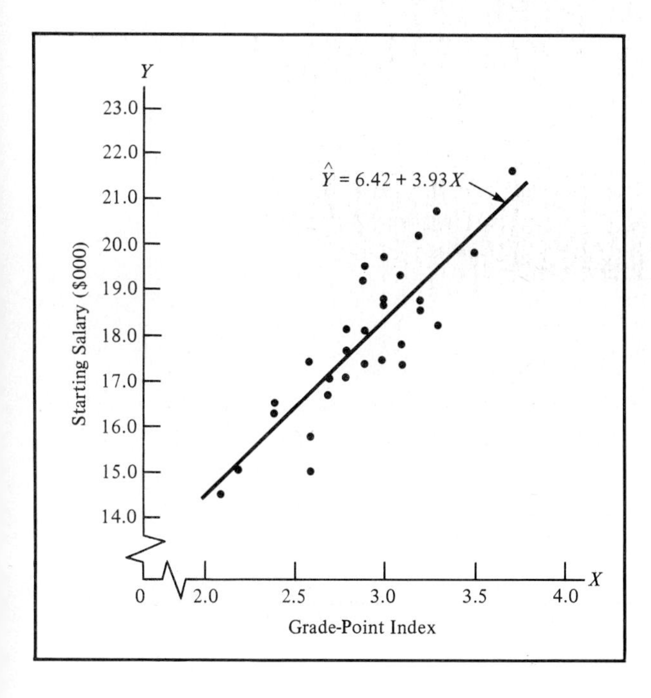

FIGURE 13.3. □ **Plotting the sample regression line.**

data points. Such a plot is one way of checking the calculations; any gross errors will be easily detected if the line does not go through the points.

EXAMPLE 13.1

An industrial psychologist wishes to predict the *time* it takes to complete a task (in minutes) based on the *level of alcohol consumed* (in ounces) during the previous hour. The following data are obtained for a sample of $n = 5$ college seniors.

Subject	Level of alcohol consumed (ounces), X	Time to complete task (minutes), Y
Elana	2	3
Irene	1	2
Kathy	3	5
Micki	5	7
Roni	4	8

1. Plot the scatter diagram.
2. Fit a straight line to the data and plot the line on the scatter diagram.
3. Predict the time needed to complete the task if alcoholic consumption is:
   (a) 2.0 ounces.
   (b) 4.5 ounces.

Solution

1. The scatter diagram is plotted as follows:

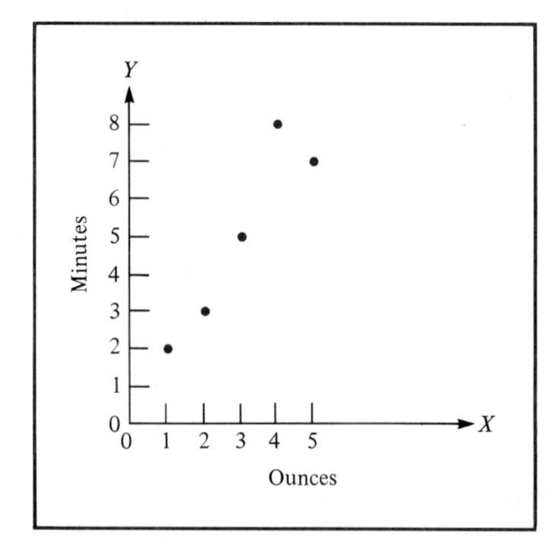

2. Fitting the straight line:

X	Y	XY	X^2	Y^2	
2	3	6	4	9	
1	2	2	1	4	
3	5	15	9	25	
5	7	35	25	49	
4	8	32	16	64	
15	25	90	55	151	$n = 5$
↕	↕	↕	↕	↕	
ΣX	ΣY	ΣXY	ΣX^2	ΣY^2	

$$b_1 = \frac{(\Sigma XY) - \dfrac{(\Sigma X)(\Sigma Y)}{n}}{(\Sigma X^2) - \dfrac{(\Sigma X)^2}{n}}$$

$$= \frac{90 - \left[\dfrac{(15)(25)}{5}\right]}{55 - \left[\dfrac{(15)^2}{5}\right]}$$

$$= \frac{90 - 75}{55 - 45}$$

$$= \frac{15}{10}$$

$$= 1.5$$

For each additional ounce of alcohol consumed, the time to complete the task increases by 1.5 minutes.

$$b_0 = \overline{Y} - b_1\overline{X}$$

$$\overline{Y} = \frac{\Sigma Y}{n} = \frac{25}{5} = 5.0$$

$$\overline{X} = \frac{\Sigma X}{n} = \frac{15}{5} = 3.0$$

$$b_0 = 5.0 - (1.5)(3.0) = 0.5$$

The constant $b_0 = .5$ minutes is the predicted time needed to complete the task when no alcohol is consumed (i.e., when $X = 0$).

Thus,

$$\hat{Y}_i = b_0 + b_1X_i$$

$$= 0.5 + 1.5X_i$$

2. Plotting the line: To plot the line, we first plot two points. We then use a ruler to draw the line (through the points) from the Y axis to the largest value of X.
   (a) When $X = 0$,

$$\hat{Y} = 0.5 + 1.5(0) = 0.5 = b_0$$

   Plot the coordinates ($X = 0$, $Y = 0.5$).
   (b) When $X = 5$,

$$\hat{Y} = 0.5 + 1.5(5) = 8.0$$

   Plot the coordinates ($X = 5$, $Y = 8$)
   (c) Draw the line through these plotted points.

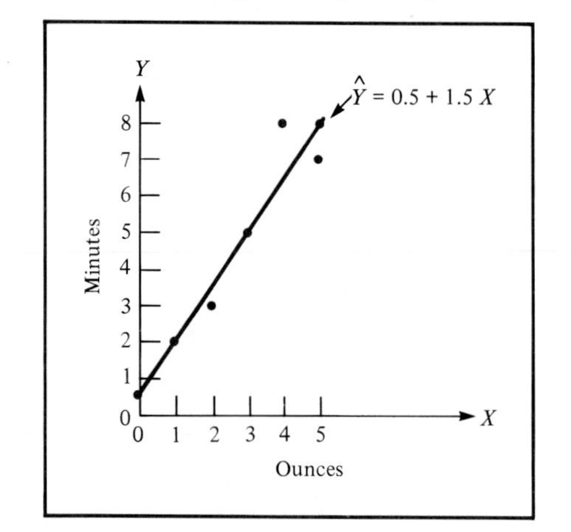

3. **(a)** Predicting Y when X is 2 ounces:

$$\hat{Y} = 0.5 + 1.5(2) = 3.5 \text{ minutes}$$

   (b) Predicting Y when X is 4.5 ounces:

$$\hat{Y} = 0.5 + 1.5(4.5) = 7.25 \text{ minutes}$$

Problems

13.1 Given the following data:

X	Y
2	8
1	10
3	6
5	2
4	4

(a) Plot the scatter diagram.
(b) Fit a straight line to the data and plot the line on the scatter diagram.
(c) What can you conclude about the model?
(d) Predict Y if $X = 1.5$.
(e) Predict Y if $X = 4.0$.

13.2 A computer software testing company evaluated every major piece of software which performed a certain task. The ratings (on a scale from 1 to 10, with 10 being highest) and the retail price (in dollars) of each piece of software are presented below:

Software	Rating	Price
A	7.4	200
B	6.8	445
C	6.7	145
D	6.2	645
E	6.0	49
F	5.8	345
G	5.8	545
H	5.0	149
I	4.9	345
J	4.6	1,100
K	4.4	945
L	4.1	645
M	4.0	450
N	2.0	545

(a) Plot a scatter diagram of rating (Y) versus price (X). Does there appear to be a relationship between price and the rating of the software package?
(b) Calculate the regression line predicting rating from price. Draw this line on the scatter diagram.
(c) What is the predicted rating of a product costing $200?
(d) Why is it improper to predict a rating of a product costing $2,000?

13.3 A measure of air pollution and the mortality rate (deaths per 100,000 population) for several metropolitan areas were:

Area	Pollution level	Mortality
1	190	1020
2	100	880
3	240	1000
4	150	840
5	100	915
6	80	790

(a) Plot the scatter diagram of mortality by pollution level.
(b) Calculate the regression equation for the straight line which fits the data best, and plot this line on the scatter diagram.
(c) What is the mortality level predicted for a city with a pollution level of 200?

13.4 A cricket was caught by a little boy, who measured how many times the cricket chirped in 15 seconds. The boy measured the cricket at several different temperatures, with the following results:

Temperature	Number of chirps
45	7
55	14
65	25
75	31
85	48
95	53

(a) Plot the number of chirps (as the Y variable) by the temperature (as the X variable). Does it appear that a straight line describes the relationship fairly well?
(b) Compute the coefficients in the linear equation for predicting the number of chirps in 15 seconds from the temperature. Draw this line on the scatter diagram.
(c) Using the equation, estimate the number of chirps per minute which would be expected at a temperature of 65 degrees.
(d) Now reverse the roles of chirping rate and temperature, and develop an equation for predicting temperature from number of chirps in 15 seconds. What temperature is it if a cricket chirps 20 times in 15 seconds?
(e) Which makes more sense to you: predicting temperature from chirps, or chirps from temperature? Why?

13.5 The government of Lower Slivovich wanted to study the relationship between capital punishment and the number of homicides committed. Each of its provinces had developed its own laws about whether to use capital punishment, and judges in provinces in which capital punishment was allowed had a wide latitude in deciding what punishment to assign when defendants were convicted. The number of executions in a recent year, and the number of homicides per 100,000 population the following year, were as follows for the provinces:

Province	Executions	Homicide rate
A	5	25.3
B	0	19.0
C	12	23.9
D	8	15.6
E	15	28.1
F	3	17.5

(a) Plot the homicide rate (Y) by number of executions (X). Tell whether the plot looks like a straight line.
(b) Calculate the regression equation for predicting the homicide rate from the number of executions.
(c) Interpret the coefficients of the equation in terms of the context of the problem. That is, what do the slope and intercept mean in this context?
(d) What would the person who did this study probably want you to conclude? What other reasons for the relationship (if one is found) might there be?

• 13.6 The weight and gas mileage of several cars were measured, and are presented below:

Weight (lb)	Mileage (miles/gallon)
2,100	19.5
1,450	31.7
1,890	25.3
2,340	20.9
1,760	28.1
1,910	26.3
1,790	27.1
1,980	23.6

(a) Plot the mileage (Y) by weight (X). Does the scatter diagram seem linear?
(b) Calculate the linear regression equation for predicting mileage from weight.
(c) Interpret the slope coefficient in the equation; that is, what does it signify here?
(d) What is the predicted mileage of a car which weighs 1,800 pounds?
(e) If you were asked to predict the mileage of a car which weighs 500 pounds using the regression equation, why should you refuse to do so?

• 13.7 A state agency wants to calibrate a machine which its police can use to measure the alcohol level in drivers suspected of driving while intoxicated. They have a machine which has a digital readout, and they test people who have consumed a known amount of beer. The results of the tests on eight people are:

Driver	Machine reading	Beers consumed
S.F.	4.8	6
B.Z.	8.1	9
M.F.	2.1	3
A.G.	7.1	8
P.S.	5.9	7
P.R.	2.0	2
C.T.	9.2	10
B.K.	3.1	4

(a) Plot a scatter diagram of the number of beers consumed (Y) by machine reading (X). Does the plot look as if a straight line should fit it well?
(b) Calculate the regression line for predicting the number of beers consumed from the machine reading. Draw the line on the scatter diagram.
(c) What is the predicted number of beers consumed if the machine reading is 5.0? 7.0? 9.0?

13.8 A random sample of people in a recent census was used to establish whether there is a relationship between the amount of education (in number of years of schooling) a person has and his or her income (in thousands of dollars). The data for a portion of this sample are shown on page 376.

Person	Education	Income
Daisy	12	21
Debby	15	34
Heidi	10	19
Jonathan	17	40
Matthew	16	42
Oliver	9	16
Penny	11	22

(a) Draw a scatter diagram of income (Y) by education (X). Do the data appear to be in approximately a straight line?
(b) Find the regression equation for predicting income from education, and plot the regression line on the scatter diagram.
(c) Predict the income of a person with 12 years of education; 16 years of education.
(d) What are the meaning of the slope and intercept in this problem?
(e) When this type of data is presented, it is usually done to show that "it pays to get more education." What other explanation can you think of for the relationship you see in these data?

13.9 An automobile club wants to find a simple rule which will help its members estimate the length of time it will take to travel between two cities if they know the distance between the cities. Of course, the distance on the map will be less than the actual distance travelled on roads, and the length of time to travel a particular distance will depend on characteristics of the road itself, traffic patterns, and individual drivers. The auto club timed some "average" drivers travelling several different routes between cities. The distance between the cities (point to point, not road distance) and the length of time to drive these distances (in hours and minutes) are:

Distance	Time
350	7:15
125	2:45
250	5:25
310	6:20
230	4:40
280	5:15

(a) Plot a scatter diagram of time (Y) versus distance (X). Does the plot look approximately linear?
(b) Calculate the regression equation for predicting time from distance. Plot the line on the scatter diagram.
(c) If you have to make a trip of 250 miles, how long will it take, according to the formula?
(d) How would you interpret the slope and intercept in the equation?

13.10 In order to establish standards for height and weight of healthy people, an insurance company weighed and measured the height of people applying for life insurance. The data for a small portion of the males measured in their sample are:

Person	Weight (pounds)	Height (inches)
Alan	235	76
Bennett	164	66
Benson	185	70
Errol	172	69
Ken	194	73
Mike	155	67
Richard	178	72
Robert	182	71
Ronald	216	74

(a) Plot the scatter diagram of the data; use weight as the Y variable and height as the X variable.
(b) Find the linear regression equation for predicting weight from height. Draw the line of the equation on the scatter diagram.
(c) Predict the weight of a person 70 inches tall.
(d) What is the interpretation of the slope in the equation?

13.5 THE STANDARD ERROR OF ESTIMATE

In the previous section the least-squares method was used to develop a linear regression equation to predict a starting salary based on a grade-point index. The least-squares method results in a line that fits the data with a *minimum* amount of error in prediction. Moreover, as illustrated in Figure 13.3, the least-squares method results in a line such that the *vertical distances* between the observed Y values and the predicted Y values on the (fitted) regression line ''balance out.'' That is, the Y values *above* the line cancel the Y values *below* the line, so that, for all individuals in the sample,

$$\Sigma (Y - \hat{Y}) = 0$$

To measure the amount of variability or scatter around the regression line we compute $S_{Y|X}$, the **standard error of estimate.**[2] It is defined as

$$S_{Y|X} = \sqrt{\frac{\Sigma (Y - \hat{Y})^2}{n - 2}} \qquad \textbf{(13.5)}$$

where Y is the observed value of Y for a given X
$\hat{Y}$ is the predicted value of Y for a given X

[2] $S_{Y|X}$ is also known as the standard deviation of the regression equation.

In words, the standard error of the estimate is the standard deviation of the observed Y values around the predicted Y values ($\hat{Y}$), rather than around the mean $\overline{Y}$ in the usual standard deviation.

The computation of the standard error of estimate using Equation (13.5) is cumbersome. It would first require the calculation of the predicted value of Y for each X value in the sample. The computation can be simplified, however, because of the following identity:

$$\sum (Y - \hat{Y})^2 = \sum Y^2 - b_0 (\sum Y) - b_1 (\sum XY)$$

The standard error of estimate $S_{Y|X}$ can then be obtained using the following computational formula:

$$S_{Y|X} = \sqrt{\frac{\sum Y^2 - b_0 (\sum Y) - b_1 (\sum XY)}{n - 2}} \qquad \textbf{(13.6)}$$

By using the summary calculations in Table 13.2 on page 368 as well as the regression coefficients obtained in the problem of interest to the placement officer, we can compute the standard error of estimate $S_{Y|X}$ from Equation (13.6) as

$$S_{Y|X} = \sqrt{\frac{9{,}593.41 - (6.41825)(534.3) - (3.92819)(1{,}564.24)}{30 - 2}}$$

$$= \sqrt{\frac{19.5071}{28}}$$

$$= \sqrt{.69668}$$

$$= .835$$

The standard error of estimate is measured in units of the dependent variable Y. Hence this standard error of estimate, equal to .835 (in thousands of dollars), represents the average variation around the fitted regression line.[3] The interpretation of the standard error of estimate, as we noted earlier, is analogous to that of the standard deviation. Just as the standard deviation measured variability around the arithmetic mean, the standard error of estimate measures variability around the fitted line of regression.

As we shall observe in the later sections of this chapter, the standard error of estimate will be used extensively in statistical inference. Intuitively, the closer $S_{Y|X}$ is to zero, the better the model fits the observed data.

[3] Note that in computing the standard error of estimate we divide the sum of squared residuals by $n - 2$. Two degrees of freedom are lost because two statistics—b_0 and b_1—are used in Equation (13.6) in order to compute the statistic $S_{Y|X}$.

Problems

13.11 What does the standard error of the estimate measure?

13.12 How does the standard error of the estimate differ from the standard deviation of the Y values?

13.13 Based on the results from Example 13.1 on page 370, compute $S_{Y|X}$ two ways. [*Hint*: Use Equation (13.5) and use Equation (13.6). Show that the results are identical.]

13.14 Using Equation (13.6), compute $S_{Y|X}$ for the data in Problem 13.2 on page 373 which dealt with software price and quality.

13.15 Using Equation (13.6), compute $S_{Y|X}$ for the data in Problem 13.3 on page 373 which dealt with pollution and mortality.

13.16 Using Equation (13.6), compute $S_{Y|X}$ for the data in Problem 13.4 on page 374 which dealt with temperature and cricket chirps.

13.17 Using Equation (13.6), compute $S_{Y|X}$ for the data in Problem 13.5 on page 374 which dealt with executions and the homicide rate.

• 13.18 Using Equation (13.6), compute $S_{Y|X}$ for the data in Problem 13.6 on page 375 which dealt with car weight and gas mileage.

• 13.19 Using Equation (13.6), compute $S_{Y|X}$ for the data in Problem 13.7 on page 375 which dealt with the reading on a machine for testing alcohol consumption and the number of beers consumed.

13.20 Using Equation (13.6), compute $S_{Y|X}$ for the data in Problem 13.8 on page 375 which dealt with education and income.

13.21 Using Equation (13.6), compute $S_{Y|X}$ for the data in Problem 13.9 on page 376 which dealt with distance between cities and the time needed to drive between the cities.

13.22 Using Equation (13.6), compute $S_{Y|X}$ for the data in Problem 13.10 on page 376 which dealt with weight and height of adult males.

13.6 PARTITIONING THE TOTAL VARIATION

To determine how well the independent variable predicts the dependent variable in our regression equation, we need to develop several measures of variation. The first measure, the **total variation**, is a measure of variation of the observed Y values around their mean $\overline{Y}$. It measures the variability in the Y values without taking into consideration the X values at all.

The total variation or **total sum of squares, SST,** can be computed from

$$\text{SST} = \Sigma\,(Y - \overline{Y})^2 = \Sigma\,Y^2 - \frac{(\Sigma\,Y)^2}{n} \qquad \textbf{(13.7)}$$

This calculation is the same as the short-cut formula for ANOVA in Chapter 12.

The total variation can now be partitioned into two components: the **explained variation** (that which is attributable to the model relating X and Y) and the **unexplained variation** (that which is not accounted for by the model relating X and Y). That is,

$$\text{total variation} = \text{explained variation} + \text{unexplained variation} \qquad \textbf{(13.8)}$$

These different measures of variation are depicted in Figure 13.4.

The explained variation or **sum of squares due to regression, SSR,** is obtained from

$$\text{SSR} = \Sigma\,(\hat{Y} - \overline{Y})^2 = b_0\,(\Sigma\,Y) + b_1\,(\Sigma\,XY) - \frac{(\Sigma\,Y)^2}{n} \qquad \textbf{(13.9)}$$

From Figure 13.4 and the middle term of Equation (13.9) we observe that SSR represents the differences between the $\hat{Y}$ (the values of Y that would be predicted from the regression relationship) and $\overline{Y}$ (the average value of Y that exists even without considering the regression relationship). Intuitively, if a good-fitting regression equation is obtained, then SSR, the variability explained by regression, will represent a large portion of the total variation SST. Of course, a *perfect fit* occurs if $Y = \hat{Y}$ at each value X. All the observed Y values would then lie on the computed regression line, and SSR = SST.

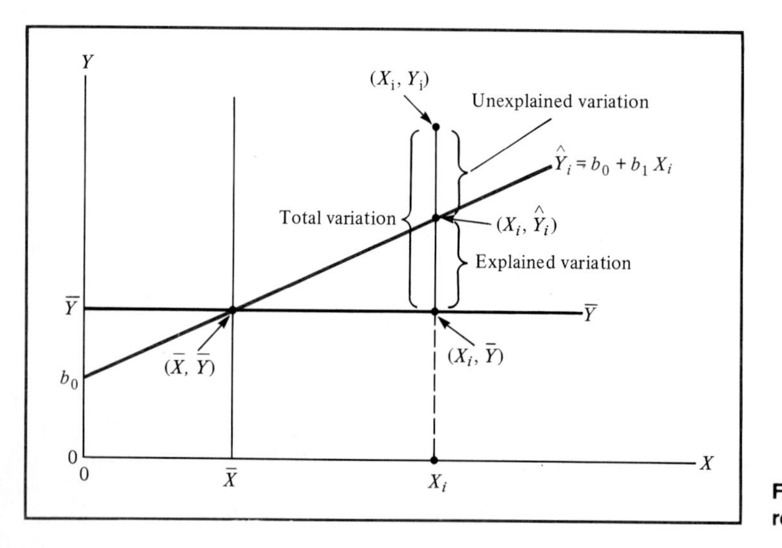

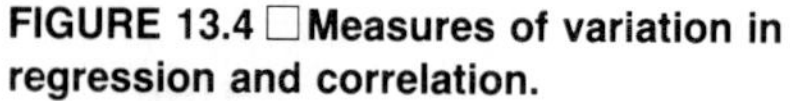
FIGURE 13.4 □ Measures of variation in regression and correlation.

The unexplained variation or **error sum of squares, SSE,** is obtained from

$$\text{SSE} = \sum (Y - \hat{Y})^2 = \sum Y^2 - b_0\left(\sum Y\right) - b_1\left(\sum XY\right) \qquad \textbf{(13.10)}$$

From Figure 13.4 and the middle term of Equation (13.10) we see that SSE represents the sum of squared **residuals** (that is, differences between the observed and predicted Y values). Hence the unexplained variation is a measure of scatter about the regression line. Moreover, from Equations (13.5) and (13.6) we see that SSE is the numerator under the square root in the computation of the standard error of estimate $S_{Y|X}$. Intuitively, if a good-fitting regression equation is obtained, then SSE, the variability unexplained by the regression, would represent a small portion of SST. If a perfect fit were obtained, there would be no scatter about the regression line and SSE would be zero (as would, of course, $S_{Y|X}$).

For the example of interest to the placement officer, the unexplained variation was already obtained when we computed $S_{Y|X}$ on page 378.

Unexplained Variation (SSE):

$$\begin{aligned}\text{SSE} &= \sum Y^2 - b_0\left(\sum Y\right) - b_1\left(\sum XY\right)\\ &= 9{,}593.41 - 6.41825(534.3) - 3.92819(1{,}564.24) = 19.5071\end{aligned}$$

By using Equations (13.7) and (13.9), the other two measures of variation are computed from the summary data in Table 13.2 on page 368:

Total Variation (SST):

$$\begin{aligned}\text{SST} &= \sum Y^2 - \frac{\left(\sum Y\right)^2}{n}\\ &= 9{,}593.41 - \frac{(534.3)^2}{30} = 77.5270\end{aligned}$$

Explained Variation (SSR):

$$\begin{aligned}\text{SSR} &= b_0\left(\sum Y\right) + b_1\left(\sum XY\right) - \frac{\left(\sum Y\right)^2}{n}\\ &= 6.41825(534.3) + 3.92819(1{,}564.24) - \frac{(534.3)^2}{30}\\ &= 58.0199\end{aligned}$$

Note that from Equation (13.8) the explained variation could have been easily obtained through subtraction. That is,

$$\begin{aligned}\text{explained variation} &= \text{total variation} - \text{unexplained variation}\\ &= 77.5270 - 19.5071\\ &= 58.0199\end{aligned}$$

DESCRIPTIVE MEASURES

Now that the total variation SST has been partitioned into two additive components—explained variation SSR and unexplained variation SSE—we may examine the strength of the linear relationship between X and Y as follows:

The **coefficient of determination,** denoted by r^2, is defined as

$$r^2 = \frac{\text{explained variation}}{\text{total variation}} = \frac{\text{SSR}}{\text{SST}}$$

$$= \frac{\sum(\hat{Y} - \bar{Y})^2}{\sum(Y - \bar{Y})^2} = \frac{b_0(\sum Y) + b_1(\sum XY) - \frac{(\sum Y)^2}{n}}{\sum Y^2 - \frac{(\sum Y)^2}{n}} \qquad \textbf{(13.11)}$$

That is:

The coefficient of determination, r^2, represents the portion of the total variation in Y that is "explained" or "accounted for" by the fitted simple linear regression model.

Using Equation (13.11) for the example of interest to the placement officer, we compute

$$r^2 = \frac{\text{SSR}}{\text{SST}} = \frac{58.0199}{77.5270} = .7484$$

Thus, 74.8% of the variation in starting salaries is explained by the simple linear regression model based on the grade-point index. On the other hand, 25.2% of the variability in starting salaries cannot be explained by the linear regression model. This, then, is an example demonstrating a strong linear relationship between the two variables, since the use of the simple linear regression model has reduced the variability in predicting starting salaries by 74.8%.

EXAMPLE 13.1 (CONTINUED FROM PAGE 372)

Because the data set is smaller, we will illustrate some of these concepts using the data on alcohol (X) and time to complete a task (Y) in Example 13.1. The slope was 1.5, and the intercept was .5 for the regression line. The accompanying table shows the X and Y values, the predicted Y values ($\hat{Y}$), the errors (ϵ), and the squared errors. (The data have been reordered from small to large values of X.)

X	Y	$\hat{Y}$	$\epsilon = Y - \hat{Y}$	ϵ^2
1	2	2.0	.0	.00
2	3	3.5	−.5	.25
3	5	5.0	.0	.00
4	8	6.5	1.5	2.25
5	7	8.0	−1.0	1.00
				$\Sigma \epsilon^2 = 3.50$

Solution:

The sum of squared errors, calculated directly from the errors themselves, is 3.50. Using Equation (13.10), we calculate the sum of squared errors as

$$\begin{aligned}\Sigma Y^2 - b_0 \Sigma Y - b_1 \Sigma XY &= 151 - (.5)(25) - (1.5)(90)\\ &= 151 - 12.5 - 135\\ &= 3.5\end{aligned}$$

which agrees with the direct estimate.

Now suppose that we did not have the X variable to help us predict Y. In that case, the best estimate $\hat{Y}$ of Y would be the mean $\bar{Y}$. Since $\bar{Y} = 5.0$, using this estimate would give us the following table:

Y	$\hat{Y}$	$\epsilon = Y - \hat{Y}$	ϵ^2
2	5.0	−3.0	9.0
3	5.0	−2.0	4.0
5	5.0	.0	.0
8	5.0	3.0	9.0
7	5.0	2.0	4.0
			$\Sigma \epsilon^2 = 26.0$

Now the sum of squared errors is 26.0; when X was used as a predictor, the sum of squared errors was reduced by $26.0 - 3.5 = 22.5$. So the proportion of variation accounted for by X is $22.5/26.0 = .8654$. This is the value of r^2, as calculated directly from the conceptual formula instead of the computational formula.

Problems

13.23 Using the computational formulas [(13.7), (13.9), and (13.10)] to calculate SST, SSR, and SSE, demonstrate that they are the same as shown in the continuation of Example 13.1.

13.24 For the data in Problem 13.2 on software price and quality on page 373:
(a) Calculate SST, SSR, and SSE using the computational formulas.

(b) Calculate the coefficient of determination for these data.
(c) Redo parts (a) and (b) using quality to predict price, and notice that even though the answers to (a) are different, the answer to (b) remains the same.

13.25 For the data in Problem 13.3 on pollution and mortality on page 373:
(a) Calculate SST, SSR, and SSE using the computational formulas.
(b) Calculate the coefficient of determination and interpret its meaning.

13.26 For the data in Problem 13.4 on temperature and the rate of cricket chirps on page 374:
(a) Calculate SST, SSR, and SSE using the computational formulas.
(b) Calculate the coefficient of determination and interpret its meaning.
(c) Redo parts (a) and (b) using chirps to predict temperature, and notice that even though the answers to (a) are different, the answer to (b) remains the same.

13.27 For the data in Problem 13.5 on executions and homicide rate on page 374:
(a) Calculate SST, SSR, and SSE using the computational formulas.
(b) Calculate the coefficient of determination and interpret its meaning.

• 13.28 For the data in Problem 13.6 on car weight and mileage on page 375:
(a) Calculate SST, SSR, and SSE using the computational formulas.
(b) Calculate the coefficient of determination and interpret its meaning.
(c) Redo parts (a) and (b) using mileage to predict weight, and notice that even though the answers to (a) are different, the answer to (b) remains the same.

• 13.29 For the data in Problem 13.7 on the reading of a machine to measure alcohol consumption and the number of beers consumed on page 375:
(a) Calculate SST, SSR, and SSE using the computational formulas.
(b) Calculate the coefficient of determination and interpret its meaning.

13.30 For the data in Problem 13.8 on education and income on page 375:
(a) Calculate SST, SSR, and SSE using the computational formulas.
(b) Calculate the coefficient of determination and interpret its meaning.
(c) Redo parts (a) and (b) using income to predict education, and notice that even though the answers to (a) are different, the answer to (b) remains the same.

13.31 For the data in Problem 13.9 on distance between cities and the length of time to drive between the cities on page 376:
(a) Calculate SST, SSR, and SSE using the computational formulas.
(b) Calculate the coefficient of determination and interpret its meaning.

13.32 For the data in Problem 13.10 on weight and height on page 376:
(a) Calculate SST, SSR, and SSE using the computational formulas.
(b) Calculate the coefficient of determination and interpret its meaning.
(c) Redo parts (a) and (b) using weight to predict height, and notice that even though the answers to (a) are different, the answer to (b) remains the same.

13.7 CORRELATION ANALYSIS: MEASURING THE STRENGTH OF THE ASSOCIATION

13.7.1 Introduction

To this point of the chapter we have extensively analyzed a simple linear relationship between X and Y using the least-squares method of regression. As such, we have formulated and developed the appropriate prediction model. For completeness, however, we shall now examine this relationship from a slightly different perspective. That is, using methods of **correlation,** we shall focus our attention on measuring, evaluating, and understanding the *strength of this association* between X and Y.

13.7.2 Interpreting r

From our previous section we note that the coefficient of determination, r^2, ranges from 0 to 1 inclusive. If a perfect linear relationship between X and Y exists, all the variability in Y will be explained by the variation in X, and r^2 will be 1. At the other extreme, if no relationship exists, none of the variability in Y will be explained by the variation in X, and r^2 will be 0.

Figure 13.5 illustrates four different types of relationships between X and Y: perfect negative linear relationship; no relationship; perfect positive linear relationship; and a perfect relationship which is nonlinear. For the latter, linear measures such as the correlation coefficient can give misleading results.

The **coefficient of correlation,** denoted by r, measures the strength of the linear relationship between X and Y. It is obtained as the square root of the coefficient of

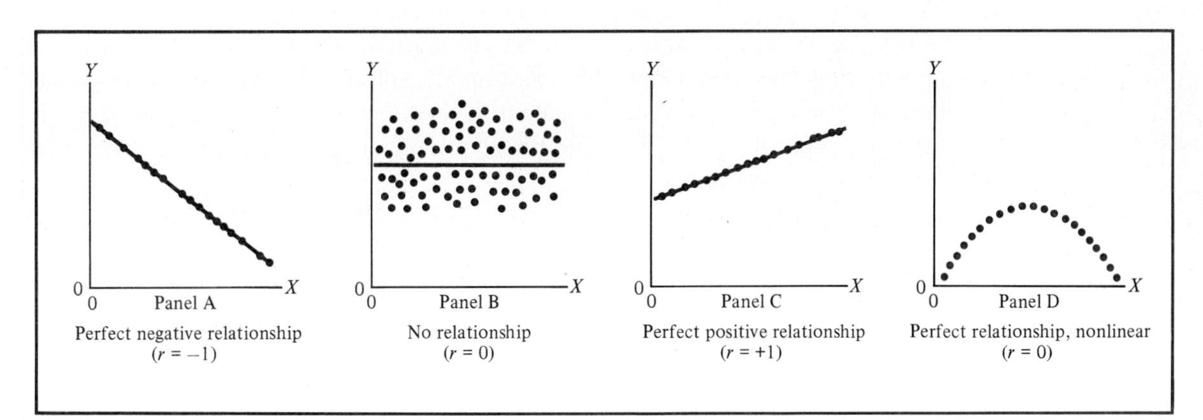

FIGURE 13.5 □ Types of relationships between variables *X* and *Y*.

determination with the appropriate sign (+ or −) affixed, according to the sign of b_1. That is,

$$r = \pm\sqrt{r^2} \tag{13.12}$$

The coefficient of correlation ranges from −1 (for perfect negative correlation as in Panel A of Figure 13.5) to +1 (for perfect positive correlation as in Panel C of Figure 13.5). Thus, for any given sample size n, the closer the coefficient of correlation is to ±1, the stronger the linear relationship between X and Y. On the other hand, as the coefficient of correlation approaches 0, the linear relationship between X and Y becomes weaker and weaker—regardless of whether it is positive or negative. If, as in Panel B of Figure 13.5, there is no apparent relationship between X and Y, then the coefficient of correlation is 0. (The slope b_1 will be 0, and X is not useful for predicting Y.) Panel D of Figure 13.5 shows a relationship that is perfect (Y can be predicted exactly from knowledge of X), but $r = 0$ because the relationship is nonlinear. Other forms of nonlinear relationships will have other values of r, but r will never equal 1 (or −1) for nonlinear relationships.

In the example of interest to the placement officer, we compute the coefficient of correlation as follows:

$$r = +\sqrt{.7484} = +.865$$

That is, since the slope of the linear regression line is positive, the correlation is positive. The closeness of the coefficient of correlation to +1.0 implies a strong positive association between grade-point index and starting salary of recent graduates from the College of Business.

In the above situation the coefficient of correlation has been computed and interpreted from a regression viewpoint. Nevertheless, as we have previously mentioned, regression analysis and correlation analysis are really two separate techniques. Regression is concerned with prediction, and correlation is concerned with association. In many applications, particularly in the social sciences, the researcher is interested only in measuring association between X and Y, not in using one variable to predict the values of another. In such situations (where the researcher is not interested in the regression equation) the coefficient of correlation can be computed directly by using the following formula:

$$r = \frac{(\sum XY) - \dfrac{(\sum X)(\sum Y)}{n}}{\sqrt{\sum X^2 - \dfrac{(\sum X)^2}{n}}\sqrt{\sum Y^2 - \dfrac{(\sum Y)^2}{n}}} \tag{13.13}$$

This can be represented more simply as

$$r = \frac{SS_{XY}}{\sqrt{SS_X\, SS_Y}} \qquad \textbf{(13.13a)}$$

If the linear association between X and Y is positive, the numerator of Equation (13.13) will be positive, so that r is positive. On the other hand, if the linear association is negative, the numerator of Equation (13.13) will be negative, so that r is negative. Finally, if there is no linear association between X and Y in the data, the numerator of Equation (13.13) will be zero, so that r is zero.

Problems

• 13.33 A group of ice cream lovers tried several brands of ice cream. Each member of the group rated the flavor of each brand; the ratings for a brand were averaged, so that each brand has an overall flavor rating. To see whether these subjective flavor judgments were related to any objective characteristics of the ice cream, the group gathered data on (among other things) the fat content of each ice cream. The data are presented below:

Flavor rating	Fat content
6.3	8.0
8.9	12.5
5.5	7.9
4.3	6.8
7.2	10.0
8.0	13.8
9.6	14.1

(a) Plot the flavor rating (Y) by the fat content (X). Does the plot seem linear?
(b) Find the correlation between flavor rating and fat content.

13.34 A gym teacher in an elementary school timed her students in the 50-yard dash. Each student was timed twice; their times (in seconds) are reported below:

First try	Second try
8.4	8.1
7.9	7.9
7.5	7.3
6.3	6.0
7.2	7.3
8.1	8.0
6.8	6.2
7.3	7.1
6.9	7.0

(a) Construct a scatter diagram of the times, using the second try as the Y variable and the first try as the X variable. Does the relationship seem to be linear?
(b) Find the correlation between first-try time and second-try time.

13.35 A psychologist gave students in a statistics class a questionnaire from which she calculated a measure of how much test anxiety each student had (a higher score indicates more anxiety about being tested). She later collected data on how each student did on the final examination in the course. These data are presented below:

Anxiety scale score	Final exam score
35	85
73	79
52	82
20	91
68	74
89	70
55	82
41	88

(a) Draw a scatter diagram of final examination score (Y) by anxiety score (X). Does the line appear approximately straight?
(b) Calculate the correlation between exam score and anxiety level.

13.36 A manufacturer of metal parts wants to find a way of testing their strength without breaking them. A salesman tells the manufacturer that his company's new machine, a magneto-spectral-electro-mechanical densitometer, will take measurements of a metal part and calculate its strength without damaging it in any way. The manufacturer decides to test the machine and measures several parts, which he then tests in the traditional way (which breaks them) to find their actual strength. The reading of the machine and the actual strength for the parts are reported below:

Machine reading	Actual strength
4.1	4.3
5.1	5.0
9.8	9.6
7.5	7.6
8.1	8.0
6.5	6.4
7.1	7.1
9.1	9.4
9.6	9.5
8.4	8.5

(a) Plot the actual strength (Y) versus the machine reading (X). Does the plot appear straight?
(b) Find the correlation between actual strength and the machine reading.

13.37 Calculate the correlation coefficient for the data on software quality and price in Problem 13.2 on page 373.

• 13.38 Calculate the correlation coefficient for the data on car weight and mileage in Problem 13.6 on page 375.

13.39 Calculate the correlation coefficient for the data on education and income in Problem 13.8 on page 375.

13.40 Calculate the correlation coefficient for the data on weight and height in Problem 13.10 on page 376.

13.8 INFERENTIAL METHODS IN REGRESSION AND CORRELATION: AN OVERVIEW

To this point in the chapter our emphasis has been both *exploratory* and *descriptive*. That is, we have been attempting to find the underlying structure in our data and then express this relationship between X and Y through methods of descriptive statistics.

- We have developed the least-squares linear regression equation.
- We have measured the scatter about the sample regression line.
- We have described the strength of the relationship between X and Y.

For the remainder of this chapter, however, our emphasis shifts toward the *confirmatory* or *inferential*. That is, we must determine whether all our endeavors have been fruitful.

- We must determine whether a *significant* linear relationship exists.
- We must evaluate the aptness of the linear fit.
- We must demonstrate the use of confidence intervals in regression and correlation models.

To these ends we focus on the fundamental aspects of model development and evaluation through methods of statistical inference.

13.9 TESTING THE SIGNIFICANCE OF THE LINEAR RELATIONSHIP

To test the significance of the simple linear relationship, we may use any one of three methods:

1. A t test for the true slope, β_1 (see Section 13.9.1).
2. A t test for the true coefficient of correlation, ρ (called "rho") (see Section 13.9.2).
3. An ANOVA approach (see Section 13.9.3).

Each of these methods results in the same conclusion regarding the significance (or lack thereof) of the simple linear relationship.[4]

[4] If we were to use a statistical package (see Chapter 15) for studying the relationship between X and Y, an understanding of all three methods would assist in the interpretation of the computer output.

13.9.1 A t Test for the True Slope, β_1

We can determine whether a significant simple linear relationship between X and Y exists by testing whether β_1 (the true population slope) is equal to zero. If this hypothesis is rejected, we can conclude that there is evidence of a simple linear relationship.

The null and alternate hypotheses can be stated as follows:

$$H_0\colon\ \beta_1 = 0 \qquad \text{(the true slope is zero)}$$
$$H_1\colon\ \beta_1 \neq 0 \qquad \text{(the true slope is not zero)}$$

The test of the null hypothesis is given by the t ratio,

$$t = \frac{b_1 - \beta_1}{S_{b_1}} = \frac{b_1}{S_{b_1}} \qquad \textbf{(13.14)}$$

where the standard error of the sample slope, denoted by S_{b_1}, is computed from

$$S_{b_1} = \frac{S_{Y|X}}{\sqrt{SS_X}} \qquad \textbf{(13.15)}$$

Using an α level of significance, the null hypothesis is rejected if $t > t_{n-2}$ (the upper tail critical value) or if $t < -t_{n-2}$ (the lower tail critical value), where the critical values of the t distribution with $n - 2$ degrees of freedom are obtained from Table C.3.

For the example of interest to the placement officer, the following information is obtained from Table 13.2 and the results from Sections 13.4 and 13.5:

$$b_1 = +3.9282, \qquad S_{Y|X} = .835, \qquad n = 30$$

$$\Sigma X = 87.0, \qquad \Sigma X^2 = 256.06, \qquad SS_X = 256.06 - \frac{(87)^2}{30} = 3.76$$

Therefore, from Equation (13.15)

$$S_{b_1} = \frac{.835}{\sqrt{3.76}} = .4305$$

To test the null hypothesis we have

$$t = \frac{b_1 - \beta_1}{S_{b_1}} = \frac{+3.9282 - 0}{.4305} = +9.12$$

If we use a .05 level of significance, we see from Table C.3 that the critical values of the t distribution (with 28 degrees of freedom) are ± 2.048. This is depicted in Figure 13.6.

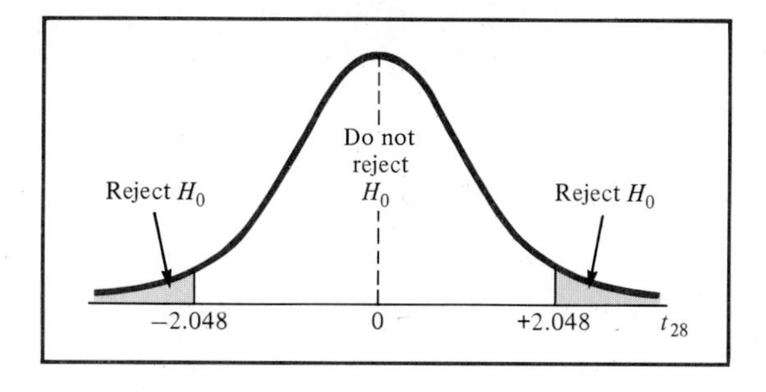

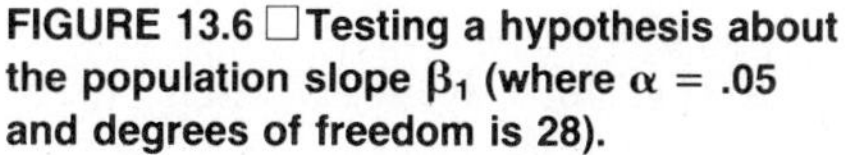

FIGURE 13.6 □ Testing a hypothesis about the population slope β_1 (where $\alpha = .05$ and degrees of freedom is 28).

Since $t = +9.12 > +2.048$, the null hypothesis is rejected. It may be concluded that a highly significant simple linear relationship exists. Therefore we have determined that the true slope β_1 is not zero. That is, the fitted sample regression equation is significant and can be used for prediction purposes.

13.9.2 A t Test for the True Coefficient of Correlation, ρ

We can also determine whether a significant simple linear relationship between X and Y exists by testing whether ρ (the true population coefficient of correlation) is equal to zero. If this hypothesis is rejected, we can conclude that there is evidence of a simple linear relationship.

The null and alternative hypotheses can be stated as follows:

H_0: $\rho = 0$ (the true correlation is zero)

H_1: $\rho \neq 0$ (the true correlation is not zero)

The test of the null hypothesis is given by the t ratio,

$$t = \frac{r - \rho}{S_r} = \frac{r}{S_r} \qquad \textbf{(13.16)}$$

where the standard error of the coefficient of correlation, denoted by S_r, is computed from

$$S_r = \sqrt{\frac{1 - r^2}{n - 2}} \qquad \textbf{(13.17)}$$

Using an α level of significance, the null hypothesis is rejected if $t > t_{n-2}$ (the upper tail critical value) or if $t < -t_{n-2}$ (the lower tail critical value), where the critical values on the t distribution with $n - 2$ degrees of freedom are obtained from Table C.3.

For the example of interest to the placement officer, the following results were obtained in Sections 13.6 and 13.7:

$$r^2 = .7484, \qquad 1 - r^2 = .2516, \qquad r = +.865$$

Therefore, from Equation (13.17)

$$S_r = \sqrt{\frac{.2516}{28}} = .0948$$

To test the null hypothesis we have

$$t = \frac{r - \rho}{S_r} = \frac{+.865 - 0}{.0948} = +9.12$$

If we use a .05 level of significance, we see from Table C.3 that the critical values on the t distribution (with 28 degrees of freedom) are ±2.048. This is displayed in Figure 13.7.

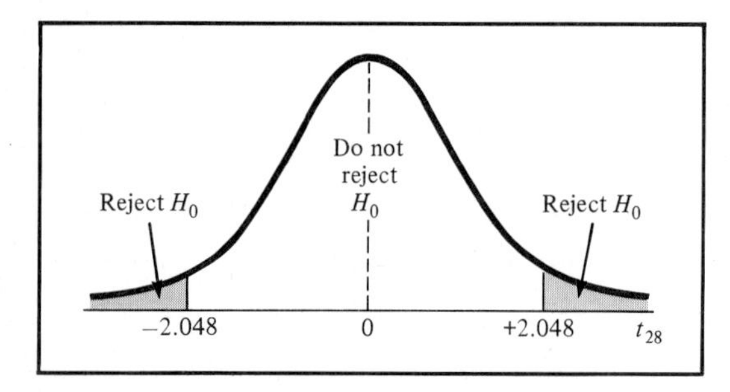

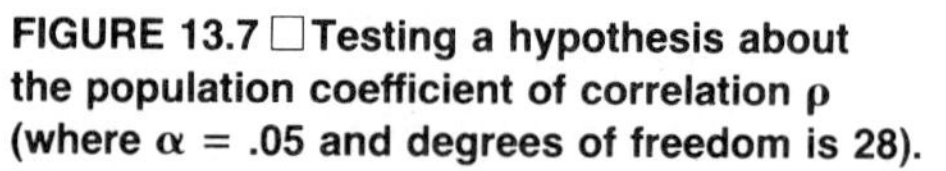

FIGURE 13.7 □ Testing a hypothesis about the population coefficient of correlation ρ (where α = .05 and degrees of freedom is 28).

Since $t = +9.12 > +2.048$, the null hypothesis is rejected. It may be concluded that a highly significant simple linear relationship exists. That is, we have determined that the true coefficient of correlation ρ is not zero.[5]

[5] Except for rounding errors, the t test for the slope and the t test for the coefficient of correlation [Equations (13.14) and (13.16)] give identical results.

TABLE 13.3 ☐ Testing the linear relationship by ANOVA

Source of variation	Degrees of freedom	Sum of squares	Mean square	F
Due to linear regression (i.e., explained)	1	$SSR = \Sigma(\hat{Y} - \bar{Y})^2 = b_0(\Sigma Y) + b_1(\Sigma XY) - \frac{(\Sigma Y)^2}{n}$	$MSR = \frac{SSR}{1}$	$F = \frac{MSR}{MSE}$
Error (i.e., unexplained)	$n - 2$	$SSE = \Sigma(Y - \hat{Y})^2 = \Sigma Y^2 - b_0(\Sigma Y) - b_1(\Sigma XY)$	$MSE = \frac{SSE}{n-2} = S_{Y\mid X}^2$	
Total	$n - 1$	$SST = \Sigma(Y - \bar{Y})^2 = \Sigma Y^2 - \frac{(\Sigma Y)^2}{n}$	$MST = \frac{SST}{n-1} = S_Y^2$	

13.9.3 An ANOVA Approach

Since the variation in Y has already been partitioned into two components [see Equations (13.7)–(13.10)], we may set up the ANOVA table as in Table 13.3.

We can determine whether or not there is a significant linear relationship between X and Y by testing whether the true slope β_1 is zero. The null and alternative hypotheses are:

H_0: $\beta_1 = 0$ (i.e., a simple linear relationship is not present)
H_1: $\beta_1 \neq 0$ (i.e., a simple linear relationship is present)

The null hypothesis is tested by obtaining the F ratio (see Table 13.3):

$$F = \frac{\text{MSR}}{\text{MSE}} \tag{13.18}$$

and, using an α level of significance, the null hypothesis is rejected if $F > F_{1,n-2}$, the critical value on the F distribution with 1 and $n - 2$ degrees of freedom as in Table C.5.[6]

By using Equations (13.7)–(13.10) for the example of interest to the placement officer, the three measures of variation were already computed on page 381:

explained variation:	SSR = 58.0199
unexplained variation:	SSE = 19.5071
total variation:	SST = 77.5270

[6] Recall from Chapter 12 that the F test is used when testing the ratio of two variances. When testing for the significance of the linear relationship, the measure of random error is called the error variance, so that the F test is the ratio of the variance due to the relationship between X and Y divided by the error variance as shown in Equation (13.18). Interestingly, the error variance is also the square of the standard error of estimate ($S_{Y\mid X}$), which was described in Section 13.5.

Hence the ANOVA table shown as Table 13.4 is obtained.

If we use a .05 level of significance, we see from Table C.5 that the critical value of the F distribution (with 1 and 28 degrees of freedom) is 4.20. This is depicted in Figure 13.8.

From Table 13.4, since

$$F = \frac{\text{MSR}}{\text{MSE}} = 83.28 > F_{1,28} = 4.20$$

the null hypothesis is rejected, and it may be concluded that a highly significant simple linear relationship exists.

TABLE 13.4 □ Testing the linear relationship

Source of variation	Degrees of freedom	Sum of squares	Mean square	F
Due to linear regression (i.e., explained)	1	58.0199	58.0199	83.28
Error (i.e., unexplained)	28	19.5071	.69668	
Total	29	77.5270		

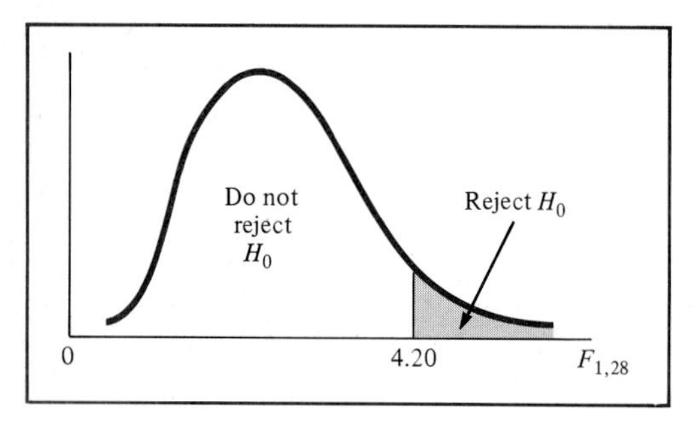

FIGURE 13.8 □ Testing for the significance of the linear relationship using ANOVA (where α = .05 and degrees of freedom are 1 and 28).

Problems

13.41 Using the data on software price and quality in Problem 13.2 on page 373 (see also Problems 13.14, 13.24, and 13.37), test for the significance of the simple linear relationship (α = .05) by using:

(a) The t test for the true slope, β_1.

(b) The t test for the true coefficient of correlation, ρ.

(c) An ANOVA approach.

• **13.42** Using the data on car weight and mileage in Problem 13.6 on page 375 (see also Problems 13.18, 13.28, and 13.38), test for the significance of the simple linear relationship (α = .01) by using:
- **(a)** The t test for the true slope, β_1.
- **(b)** The t test for the true coefficient of correlation, ρ.
- **(c)** An ANOVA approach.

13.43 Using the data on education and income in Problem 13.8 on page 375 (see also Problems 13.20, 13.30, and 13.39), test for the significance of the simple linear relationship (α = .10) by using:
- **(a)** The t test for the true slope, β_1.
- **(b)** The t test for the true coefficient of correlation, ρ.
- **(c)** An ANOVA approach.

13.44 Using the data on weight and height in Problem 13.10 on page 376 (see also Problems 13.22, 13.32, and 13.40), test for the significance of the simple linear relationship (α = .05) by using:
- **(a)** The t test for the true slope, β_1.
- **(b)** The t test for the true coefficient of correlation, ρ.
- **(c)** An ANOVA approach.

13.45 Using the data on pollution and mortality in Problem 13.3 on page 373 (see also Problems 13.15 and 13.25), test for the significance of the simple linear relationship (α = .05) by using:
- **(a)** A t test for the true slope, β_1.
- **(b)** The ANOVA approach.

13.46 Using the data on temperature and cricket chirps in Problem 13.4 on page 374 (see also Problems 13.16 and 13.26), test for the significance of the simple linear relationship (α = .01) by using:
- **(a)** A t test for the true slope, β_1.
- **(b)** The ANOVA approach.

13.47 Using the data on executions and homicide in Problem 13.5 on page 374 (see also Problems 13.17 and 13.27), test for the significance of the simple linear relationship (α = .10) by using:
- **(a)** A t test for the true slope, β_1.
- **(b)** The ANOVA approach.

• **13.48** Using the data on alcohol consumption and the reading on a machine to test consumption in Problem 13.7 on page 375 (see also Problems 13.19 and 13.29), test for the significance of the simple linear relationship (α = .05) by using:
- **(a)** A t test for the true slope, β_1.
- **(b)** The ANOVA approach.

13.49 Using the data in Problem 13.9 on page 376 (see also Problems 13.21 and 13.31) on distance between cities and the time needed to drive between the cities, test for the significance of the simple linear relationship (α = .10) by using:
- **(a)** A t test for the true slope, β_1.
- **(b)** The ANOVA approach.

• 13.50 Using the data on ice cream flavor and fat content in Problem 13.33 on page 387, do a t test for the true coefficient of correlation, ρ. ($\alpha = .05$)

13.51 Using the data on test anxiety and final examination score in Problem 13.35 on page 388, do a t test for the true coefficient of correlation, ρ. ($\alpha = .10$)

13.52 Using the data on actual metal strength and the machine estimate of metal strength in Problem 13.36 on page 388, do a t test for the true coefficient of correlation, ρ. ($\alpha = .05$)

13.10 CONFIDENCE-INTERVAL ESTIMATION

Now that a significant linear relationship between X and Y has been found, we may obtain confidence-interval estimates of the true slope β_1 and the true coefficient of correlation ρ. In addition, we may obtain a confidence-interval estimate of the true mean response $\mu_{Y|X}$ and a predicted individual value Y_I in order to use our regression equation for purposes of prediction.

13.10.1 Estimating the True Slope, β_1

A $100(1 - \alpha)\%$ confidence interval estimate of the true slope β_1 is given by

$$b_1 \pm t_{n-2} S_{b_1} \tag{13.19}$$

For the placement officer's data, $b_1 = 3.9282$ and $S_{b_1} = .4305$. When using a 95% confidence coefficient, $t_{28} = 2.048$, and we have

$$3.9282 \pm 2.048(.4305)$$

$$3.9282 \pm .8817$$

so that

$$+3.05 \leq \beta_1 \leq +4.81$$

With 95% confidence the placement officer can interpret this to mean that for each additional achieved unit in the grade-point index, the starting salaries offered to recent graduates majoring in business will increase by \$3,050 to \$4,810.

13.10.2 Estimating the True Coefficient of Correlation, ρ

Since a significant positive correlation between the grade-point index and the starting salary has been found, the placement officer might also be interested in obtaining a confidence-interval estimate of the true correlation coefficient, ρ. Analogous to our other interval estimates, we would expect a $100(1 - \alpha)\%$ confidence-interval estimate of ρ to have the form

$$r \pm t_{n-2} S_r$$

Unfortunately, unless ρ equals zero, the sampling distribution of the sample correlation coefficient r is not normally distributed, and we cannot use the above form for estimation purposes. To adjust for this, we utilize the **Fisher Z transformation** (Reference 5) whose values are provided in Table C.7. For any correlation coefficient, this transformation gives a value which does have a normal distribution, so that we can compute a confidence interval. But the confidence interval is for the transformed values, so we then use Table C.7 in the other direction, translating back into correlations.

The procedure for obtaining the confidence-interval estimate of ρ takes the following six steps:

OBTAINING THE CONFIDENCE-INTERVAL ESTIMATE OF ρ

1. Compute r.
2. Transform r to $\mathcal{Z}_{F_r}$ using Table C.7.
3. Instead of S_r, use $\hat{\sigma}_{\mathcal{Z}_{Fr}}$ (the estimated standard error of the transformed correlation value), where

$$\hat{\sigma}_{\mathcal{Z}_{Fr}} = \frac{1}{\sqrt{n-3}} \tag{13.20}$$

4. Instead of t_{n-2}, use Z, the $100[1 - (\alpha/2)]$ percentile point of the standardized normal distribution (Table C.2).
5. Obtain the $100(1 - \alpha)\%$ confidence-interval estimate of $\mathcal{Z}_{F_\rho}$ from

$$\mathcal{Z}_{Fr} \pm Z\hat{\sigma}_{\mathcal{Z}_{Fr}} \tag{13.21}$$

6. Using Table C.7, reconvert the lower and upper limits of Equation (13.21), the confidence-interval estimate of $\mathcal{Z}_{F_\rho}$ back to units of r and then obtain the $100(1 - \alpha)\%$ confidence-interval estimate of ρ.

For the example of interest to the placement officer, the 95% confidence-interval estimate of the true correlation coefficient ρ is computed as follows:

1. Using Equation (13.12), $r = +.865$.
2. Rounding r to two decimal places, we use Table C.7 and transform $r = +.87$ to $\mathcal{Z}_{Fr} = +1.333$ as follows:

r	Z_{F_r}	r	Z_{F_r}	r	Z_{F_r}	r	Z_{F_r}	r	Z_{F_r}
.00	.000	.20	.203	.40	.424	.60	.693	.80	1.099
.01	.010	.21	.213	.41	.436	.61	.709	.81	1.127
.02	.020	.22	.224	.42	.448	.62	.725	.82	1.157
.03	.030	.23	.234	.43	.460	.63	.741	.83	1.188
.04	.040	.24	.245	.44	.472	.64	.758	.84	1.221
.05	.050	.25	.255	.45	.485	.65	.775	.85	1.256
.06	.060	.26	.266	.46	.497	.66	.793	.86	1.293
.07	.070	.27	.277	.47	.510	.67	.811	.87 →	1.333

3. Using Equation (13.20), $\hat{\sigma}_{Z_{Fr}} = 1/\sqrt{27} = .19245$.
4. For a 95% confidence coefficient, $Z = \pm 1.96$ (see Table C.2).
5. From Equation (13.21) the 95% confidence-interval estimate of Z_{F_ρ} is

$$Z_{F_r} \pm Z\hat{\sigma}_{Z_{Fr}}$$
$$+1.333 \pm 1.96(.19245)$$
$$+1.333 \pm .377$$

so that

$$+.956 \leq Z_{F_\rho} \leq +1.710$$

6. Using the closest Z_{F_r} values in Table C.7 for the lower and upper limits of the confidence interval, we reconvert back to the scale of r as follows:

r	Z_{F_r}	r	Z_{F_r}	r	Z_{F_r}	r	Z_{F_r}	r	Z_{F_r}
.00	.000	.20	.203	.40	.424	.60	.693	.80	1.099
⋮	⋮	⋮	⋮	⋮	⋮	⋮	⋮	⋮	⋮
.10	.100	.30	.310	.50	.549	.70	.867	.90	1.472
.11	.110	.31	.320	.51	.563	.71	.887	.91	1.528
.12	.121	.32	.332	.52	.576	.72	.908	.92	1.589
.13	.131	.33	.343	.53	.590	.73	.929	.93	1.658
.14	.141	.34	.354	.54	.604	.74 ←	.950	.94 ←	1.738

Thus, the 95% confidence-interval estimate of the true correlation coefficient ρ is given by

$$+.74 \leq \rho \leq +.94$$

In the population there is evidence of a strong positive correlation between the grade-point index and the starting salary.

★ 13.10.3 Confidence Interval for the True Mean Response, $\mu_{Y|X}$ (Optional Subsection)

The primary purpose of developing the regression equation is for prediction. That is, we want to be able to take advantage of the relationship between the predictor

★ Note that this is an optional subsection that may be skipped without any loss of continuity.

variable X and the response variable Y so that by using the former we can obtain a better estimate of the *mean response* than if we had merely evaluated the same Y values alone. However, as we mentioned on page 369, to use the sample regression equation for prediction purposes it is important that we consider only the *relevant* range of the predictor variable X. That is, we should not extrapolate beyond the relevant range of X values. Hence, for the data in Table 13.1 the placement officer can utilize the sample regression equation to predict expected starting salaries based on any grade-point index from 2.1 through 3.7. Although in the population there may have been recent graduates whose grade-point indexes either exceeded 3.7 or were at the minimum of 2.0 (for graduation purposes), such individuals were not observed in the particular sample listed in Table 13.1. Such indexes, then, are inappropriate for the sample regression equation developed here. A prediction based on a grade-point index of 0.0 is, of course, not only outside the relevant range but also is a grade-point level impossible for a college graduate to obtain.

THE TRUE MEAN RESPONSE

In Section 13.4 we predicted the starting salary for a recent graduate who had achieved a 3.0 grade-point index. Such a result, however, is a *point estimate* (as was described in Section 9.1) of $\mu_{Y|X}$, the true mean starting salary for all recent graduates who have achieved a 3.0 grade-point index while majoring in business. Suppose now the placement officer wishes to estimate with 95% confidence the true mean starting salary for all recent graduates of the College of Business who had achieved a grade-point index of 3.0. In general, a $100(1 - \alpha)\%$ **confidence-interval estimate of the true mean response** $\mu_{Y|X}$ at a particular value of X (say X_i) is given by

$$\hat{Y}_i \pm t_{n-2} S_{Y|X} \sqrt{\frac{1}{n} + \frac{(X_i - \bar{X})^2}{\sum X^2 - \frac{(\sum X)^2}{n}}} \qquad \textbf{(13.22)}$$

where $\hat{Y}_i = b_0 + b_1 X_i$, the predicted value of Y when X is fixed at X_i.

To obtain the confidence-interval estimate we first use Equation (13.2) to compute the point estimate of $\mu_{Y|X}$:

$$\hat{Y}_i = 6.42 + 3.93(3.0) = 18.19$$

Thus the point estimate of $\mu_{Y|X}$ is 18.19 thousands of dollars. To form the 95% confidence-interval estimate we have

$$t_{28} = 2.048, \quad S_{Y|X} = .835, \quad n = 30$$
$$\bar{X} = 2.90, \quad \sum X = 87.0, \quad \sum X^2 = 256.06$$

Thus, from Equation (13.22) we obtain

$$18.19 \pm 2.048(.835)\sqrt{\frac{1}{30} + \frac{(3.00 - 2.90)^2}{256.06 - \frac{(87.0)^2}{30}}}$$

$$18.19 \pm 2.048(.835)\sqrt{\frac{1}{30} + \frac{(0.10)^2}{3.76}}$$

$$18.19 \pm .324$$

so that

$$17.87 \leq \mu_{Y|X} \leq 18.51$$

Therefore, the placement officer estimates that for all recent graduates majoring in business who have achieved a 3.0 grade-point index, the true mean starting salary is between \$17,870 and \$18,510. She has 95% confidence that this interval correctly estimates the true mean starting salary for such individuals.

If the placement officer also wishes to obtain other 95% confidence-interval estimates of the true mean starting salaries for various grade-point indexes in the relevant range and plots those lower and upper limits for each of the obtained confidence intervals, the scatter diagram will demonstrate the presence of a *band* effect around the fitted regression line.

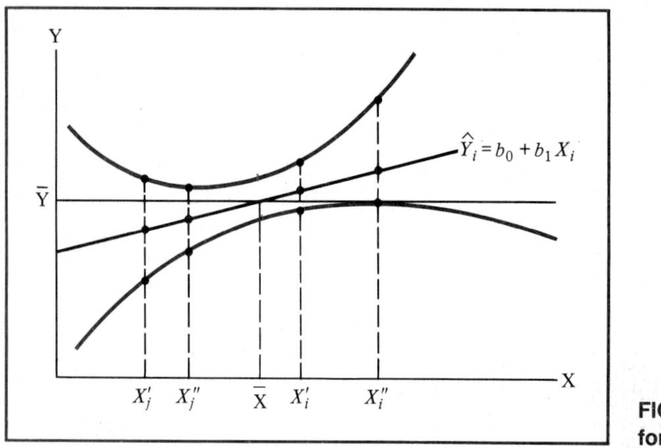

FIGURE 13.9 □ Interval estimates for $\mu_{Y|X}$ for different values of X.

As depicted in Figure 13.9, the width of the confidence interval changes for different values of X. In addition, an examination of Equation (13.22) indicates that the width of the confidence interval depends upon several other factors:

- The sample size n.
- The desired amount of confidence, $1 - \alpha$, which determines t_{n-2}.
- The sample mean $\bar{X}$.
- The total variation in the X values (that is, $\Sigma X^2 - [(\Sigma X)^2/n]$).
- The variation around the regression line, $S_{Y|X}$.

If these other factors are held constant, then the width of the interval depends on the selected value of X. As observed from Figure 13.9, the closer the selected value of X is to $\overline{X}$, the narrower the confidence interval will be. The *minimum width* occurs at values of X identical to the mean $\overline{X}$. On the other hand, for selected values of X which greatly differ from $\overline{X}$, the interval estimate of the true mean response $\mu_{Y|X}$ becomes much wider. These much wider (and less useful) intervals are the tradeoff for predicting at such values of X.

★ 13.10.4 Prediction Interval for the Individual Response (Optional Subsection)

By obtaining such interval estimates from Equation (13.22), the placement officer would be able to provide current seniors majoring in business with information pertaining to *average* starting salary based on achieved grade-point index.

Much more important to an individual student, however, is information pertaining to the starting salary that he or she could expect based on an achieved grade-point index. Thus, a 100(1 − α)% **prediction interval of the individual response**, Y_I, at a particular value of X, say X_i, is given by

$$\hat{Y}_i \pm t_{n-2} S_{Y|X} \sqrt{1 + \frac{1}{n} + \frac{(X_i - \overline{X})^2}{\sum X^2 - \frac{(\sum X)^2}{n}}} \qquad \textbf{(13.23)}$$

where $\hat{Y}_i = b_0 + b_1 X_i$, the predicted value of Y when X is fixed at X_i.

Notice that this formula is the same as that for the confidence interval for the mean response, except that the quantity under the square-root sign has 1 added to it here. This shows, as we would expect, that the interval for predicting an individual response is wider than that for predicting the mean response; it is harder to predict an individual's behavior than the behavior of the average person.

As an example, suppose the placement officer wishes to inform a particular individual with an achieved 3.0 grade-point index as to his or her expected starting salary. Modifying the calculations in the previous section, we add the 1 under the square-root sign to get:

★ Note that this is an optional subsection that may be skipped without any loss of continuity.

$$18.19 \pm 2.048(.835)\sqrt{1 + \frac{1}{30} + \frac{(3.00 - 2.90)^2}{256.06 - \frac{(87.0)^2}{30}}}$$

$$= 18.19 \pm 2.048(.835)\sqrt{1 + .03333 + .00266}$$

$$= 18.19 \pm 2.048(.835)\sqrt{1.03599}$$

$$= 18.19 \pm 1.74$$

so that

$$16.45 < Y_I < 19.93$$

Therefore, our 95% prediction interval estimate for starting salary for an individual with a grade-point index of 3.0 is from $16,450 to $19,930.

Problems

Note: A star indicates a problem or part of a problem that requires a study of an optional section or subsection of this chapter.

13.53 Using the data on software price and quality in Problem 13.2 on page 373 (see also Problems 13.37 and 13.41):

- **(a)** Calculate a 95% confidence interval estimate for the true slope.
- **(b)** Calculate a 95% confidence interval estimate of the true correlation coefficient.
- ★**(c)** Calculate a 95% confidence interval for the mean for software priced at $500.
- ★**(d)** Calculate a 95% prediction interval for an individual software program priced at $500.
- **(e)** Write a brief description of your findings.

• 13.54 Using the data on car weight and mileage in Problem 13.6 on page 375 (see also Problems 13.18, 13.28, 13.38, and 13.42):

- **(a)** Calculate a 95% confidence interval estimate for the true slope.
- **(b)** Calculate a 95% confidence interval estimate of the true correlation coefficient.
- ★**(c)** Calculate a 95% confidence interval for the mean for cars weighing 2,000 pounds.
- ★**(d)** Calculate a 95% prediction interval for an individual car weighing 2,000 pounds.
- **(e)** Write a brief description of your findings.

13.55 Using the data on education and income in Problem 13.8 on page 375 (see also Problems 13.20, 13.30, 13.39, and 13.43):

- **(a)** Calculate a 95% confidence interval estimate for the true slope.
- **(b)** Calculate a 95% confidence interval estimate of the true correlation coefficient.
- ★**(c)** Calculate a 95% confidence interval for the mean for people with 12 years of education.
- ★**(d)** Calculate a 95% prediction interval for an individual with 12 years of education.
- **(e)** Write a brief description of your findings.

13.56 Using the data on weight and height in Problem 13.10 on page 376 (see also Problems 13.22, 13.32, 13.40, and 13.44):

(a) Calculate a 95% confidence interval estimate for the true slope.

(b) Calculate a 95% confidence interval estimate of the true correlation coefficient.

★**(c)** Calculate a 95% confidence interval for the mean for men who are 70 inches tall.

★**(d)** Calculate a 95% prediction interval for an individual male who is 70 inches tall.

(e) Write a brief description of your findings.

13.57 Based on the results from Example 13.1 on pages 370–372 and 382–383, set up 95% interval estimates of:

(a) The time needed to complete a task by a *particular* college senior who has consumed 2.0 ounces of alcohol.

(b) The *average* time needed to complete a task by college seniors who have consumed 2.0 ounces of alcohol.

(c) Interpret your findings.

13.58 Calculate a 95% confidence interval for the slope of the data on pollution and mortality in Problem 13.3 on page 373 (see also Problems 13.15, 13.25, and 13.45).

13.59 Calculate a 95% confidence interval for the slope of the data on distance between cities and the time needed to drive between the cities (Problem 13.9 on page 376; see also Problems 13.21, 13.31, and 13.49).

• 13.60 Calculate a 95% confidence interval for the correlation between ice cream flavor and fat content (Problem 13.33 on page 387; see also Problem 13.50).

13.61 Calculate a 95% confidence interval for the correlation between test anxiety and final examination score (Problem 13.35 on page 388; see also Problem 13.51).

13.62 Calculate a 95% confidence interval for the correlation between actual metal strength and the machine estimate of strength (Problem 13.36 on page 388; see also Problem 13.52).

13.11 ASSUMPTIONS OF SIMPLE LINEAR REGRESSION AND CORRELATION

13.11.1 Introduction: The Need for Studying the Aptness of the Fitted Model

The next step is to study the appropriateness of our significant simple linear model. Perhaps a **curvilinear model** would actually fit the data better. On the other hand, had the hypothesis test for the simple linear relationship proved insignificant, there would have been no point in using the fitted linear regression model further. Nevertheless, even in such cases we would still want to examine the appropriateness of the linear model. The results of such an examination would either assist us in searching for a more appropriate (*curvilinear*) relationship between X and Y or suggest that it would be better to seek other predictor variables and develop a **multiple regression model** (see References 1, 2, 6). If none of these could be achieved, we would then make inferential statements using only the observed set of responses (Y)—thereby not utilizing the insignificant relationship with X.

In order to examine the aptness of our fitted model, it is important to understand the assumptions upon which it was based.

13.11.2 Assumptions

In our investigations into hypothesis testing and the analysis of variance (ANOVA), we noted that the appropriate application of a particular statistical procedure depends on how well a set of assumptions for that procedure are met. The assumptions necessary for regression and correlation analysis are analogous to those of the ANOVA, since they fall under the general heading of **linear models** (see References 1 and 2). Although there are some differences in the assumptions made by the regression and correlation models (see References 2 or 6), this topic is beyond the scope of this text. We will consider here only the assumptions necessary for regression analysis. *These assumptions apply only to inference, not to descriptive statistics.* We can still calculate correlations and regression lines when they are violated, but we cannot do valid significance tests using the procedures discussed in this chapter.

The four major assumptions of the simple linear regression model are:

1. Normality.
2. Linearity.
3. Independence.
4. Homoscedasticity.

NORMALITY

The first assumption, **normality,** requires that the value of Y be normally distributed at each value of X (see Figure 13.10). Like the t test and the analysis-of-variance F test, regression analysis is ''robust'' against departures from the normality assumption. That is, as long as the distribution of Y values around each X is not extremely

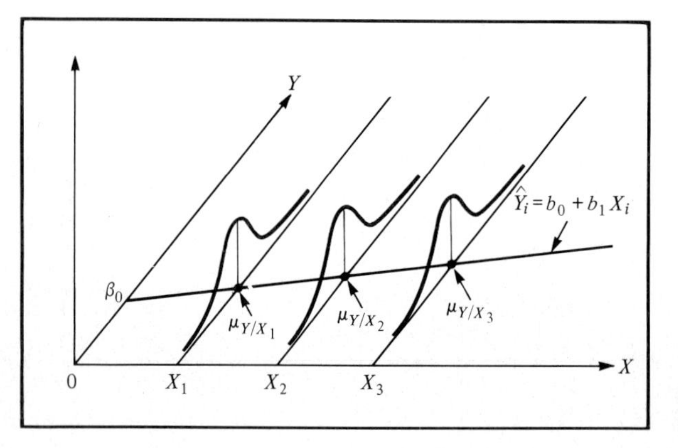

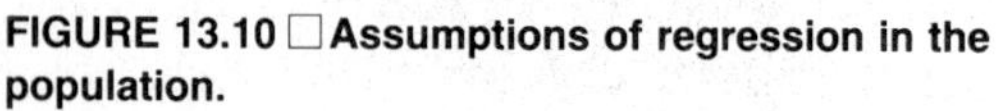
FIGURE 13.10 □ Assumptions of regression in the population.

different from a normal distribution, inferences concerning the regression coefficients will not be seriously affected.

The normality assumption will be examined further in Section 13.12.3.

LINEARITY

The second assumption, **linearity,** specifies the functional relationship between X and Y. From Figure 13.10, we observe that at each fixed X the corresponding (population) mean of the Y values $\mu_{Y|X}$ is a straight-line function of X (that is, $\hat{Y}_i = \mu_{Y|X} = \beta_0 + \beta_1 X_i$).

The linearity assumption will be examined further in Section 13.12.2.

INDEPENDENCE

The third assumption, **independence,** requires that the observed Y values be independent of one another for each value of X.[7] Referring to Figure 13.10, if X is fixed at X_1, the observed Y values from the corresponding population are assumed to be independent of each other. Similarly, if X is fixed at X_2, the observed Y values from the corresponding population are also assumed to be independent, and so on. In addition, the population of possible Y values for one level of X is independent of the population of Y values for any other level of X.

As we noted when discussing the analysis of variance, the independence assumption often is violated by data that are collected over a period of time. For example, in economic data the observed Y values (say sales revenues or stock market prices) for one particular time period may be correlated with the previous and subsequent time periods. These types of models fall under the general heading of **time-series analysis** (see, for example, Reference 3) and are beyond the scope of this text.

The independence assumption will be investigated further in Section 13.12.3.

HOMOSCEDASTICITY

The fourth assumption, **homoscedasticity,** requires that the variation in the Y values around the line of regression be constant for all values of X. As depicted in Figure 13.10, this means that Y varies the same amount when X is fixed at a low value as when X is fixed at a high value.

Uniformity in spread (i.e., homoscedasticity) about the regression line is quite an important assumption for using the least-squares method. If there are serious departures from this assumption, either data transformations or **weighted least-squares** methods can be applied (see References 2 or 6).

The homoscedasticity assumption will be examined further in Section 13.12.3.

---

[7] More formally we may speak of this as the independence-of-errors assumption. The independence-of-errors assumption requires that the errors (that is, the residual differences between observed and predicted values of Y) should be independent for each value of X.

13.11.3 Interpreting the Assumptions

To summarize for the model thus far developed, we note that for any fixed level of X the population of Y values has a *normal distribution* whose particular *mean* $\mu_{Y|X}$ *changes linearly* with changes in X but whose *variance* $\sigma^2_{Y|X}$ is the same at each value of X.

THE PLACEMENT OFFICER'S PROBLEM REVISITED

In terms of the example of interest to the placement officer, it should be assumed that at each possible grade-point index there exists an entire population of *independent* starting-salary values which are *normally distributed* with a mean $\mu_{Y|X}$ (which changes linearly with changes in the grade-point index) and with a variance $\sigma^2_{Y|X}$ (which remains constant with changes in the grade-point index). That is, starting salaries are a linear function of the grade-point index. Moreover, for recent graduates with a particular grade-point index the possible starting-salary values are independent and normally distributed. Furthermore, variation in starting salaries obtained by recent graduates with low grade-point indexes is the same as the variation in starting salaries obtained by recent graduates having high grade-point indexes.

★ 13.12 RESIDUAL ANALYSIS (OPTIONAL SECTION)

13.12.1 Introduction

Throughout this chapter we have relied upon a simple regression model in which the dependent variable was predicted based on a straight-line relationship with a single independent variable. In this section we shall introduce a graphical approach called **residual analysis** to assist us in evaluating the worthiness of our model. By using residual analysis, not only are we able to evaluate the aptness of a fitted model directly, but we also can evaluate its appropriateness through a study of potential violations in the assumptions.

13.12.2 Evaluating the Aptness of the Fitted Model Directly

The residual or error values ϵ_i are defined as the difference between the observed and predicted values of the dependent variable for each observation.

★ Note that this is an optional section that may be skipped without any loss of continuity.

That is,

$$\epsilon_i = Y_i - \hat{Y}_i \qquad \textbf{(13.24)}$$

where Y_i is the *observed* value of Y for a given observation X_i
$\hat{Y}_i$ is the *predicted* value of Y for a given observation X_i

The individual residual values ϵ_i can be obtained as part of a computer output (see Chapter 15) or can be computed manually at each value of X using the sample regression line [Equation 13.2)].

We may evaluate the aptness of the fitted regression model by plotting the residuals on the vertical axis against the corresponding values of the independent variable X on the horizontal axis for all n pairs of observations. If the fitted model is appropriate for the data, there will be no apparent pattern in this plot of the residuals versus X. However, if the fitted model is not appropriate, there will be a relationship between ϵ and X. Such a pattern can be observed in Figure 13.11. Panel A of the figure depicts a situation in which there is a significant simple *linear* relationship between X and Y; however, it would seem possible that a *curvilinear* model between the two variables might be even more appropriate. This effect appears to be highlighted in Panel B of the figure, which displays the residual plot of ϵ versus X. An obvious curvilinear effect between X and ϵ is indicated.

By plotting the residuals we have essentially "filtered out" or removed the *linear* trend of X with Y, thereby exposing the "lack of fit" in the simple linear model.

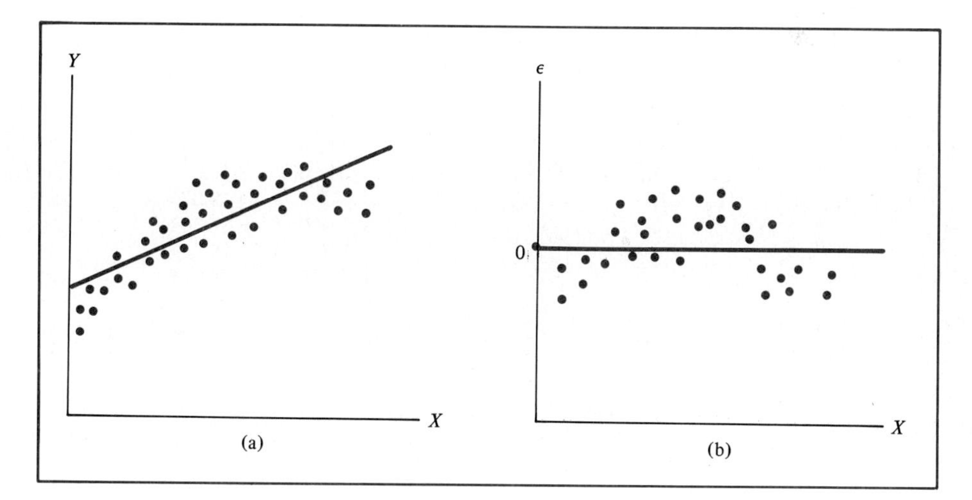

FIGURE 13.11 □ Studying the appropriateness of the simple linear regression model.

Thus from Panels A and B we may conclude that the curvilinear model may be a better fit and should be evaluated in place of the simple linear model. This topic, however, is beyond the scope of this text. (See References 1, 2, and 6).

For the problem of interest to the placement officer, Table 13.5 presents the observed, predicted, and residual values of the response variable (starting salary) in the simple linear model we have fitted.

The residuals have been plotted against the independent variable (grade-point index) in the scatter diagram depicted in Figure 13.12. From this we observe that there is no apparent pattern in this residual plot. That is, the residuals seem to be spread uniformly above and below $\epsilon = 0$ for all values of X. Thus we may conclude that our fitted model appears to be appropriate.

TABLE 13.5 □ Observed, predicted, and residual values for the placement officer's data

Student	X	Y	$\hat{Y}$	$\epsilon = Y - \hat{Y}$
1	2.7	17.0	17.021	− .021
2	3.1	17.7	18.593	− .893
3	3.0	18.6	18.200	.400
4	3.3	20.5	19.379	1.121
5	3.1	19.1	18.593	.507
6	2.4	16.4	15.842	.558
7	2.9	19.3	17.807	1.493
8	2.1	14.5	14.663	− .163
9	2.6	15.7	16.628	− .928
10	3.2	18.6	18.986	− .386
11	3.0	19.5	18.200	1.300
12	2.2	15.0	15.056	− .056
13	2.8	18.0	17.414	.586
14	3.2	20.0	18.986	1.014
15	2.9	19.0	17.807	1.193
16	3.0	17.4	18.200	− .800
17	2.6	17.3	16.628	.672
18	3.3	18.1	19.379	−1.279
19	2.9	18.0	17.807	.193
20	2.4	16.2	15.842	.358
21	2.8	17.5	17.414	.086
22	3.7	21.3	20.951	.349
23	3.1	17.2	18.593	−1.393
24	2.8	17.0	17.414	− .414
25	3.5	19.6	20.165	− .565
26	2.7	16.6	17.021	− .421
27	2.6	15.0	16.628	−1.628
28	3.2	18.4	18.986	− .586
29	2.9	17.3	17.807	− .507
30	3.0	18.5	18.200	.300

SOURCE: Table 13.1.

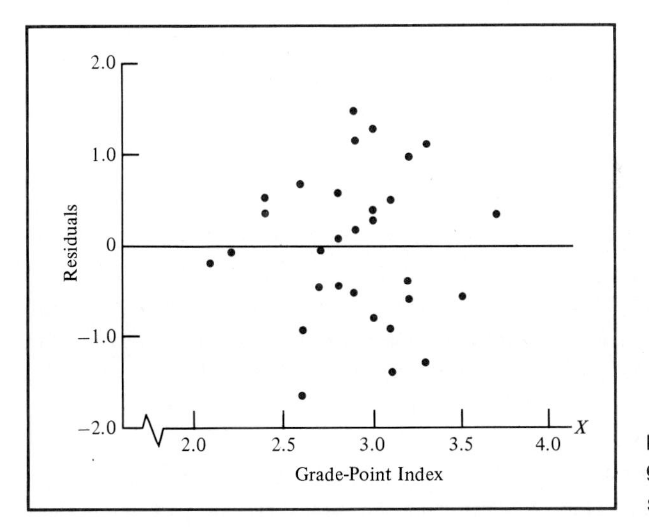

FIGURE 13.12 □ Plotting the residuals versus grade-point index.

SOURCE: Table 13.5.

13.12.3 Evaluating the Aptness of the Fitted Model by Examining Its Assumptions

HOMOSCEDASTICITY

The assumption of homoscedasticity can be evaluated from the plot of ϵ with X. For the problem of interest to the placement officer, no apparent violation in this assumption is observed (see Figure 13.12). The residuals appear to fluctuate uniformly about zero for all values of X.

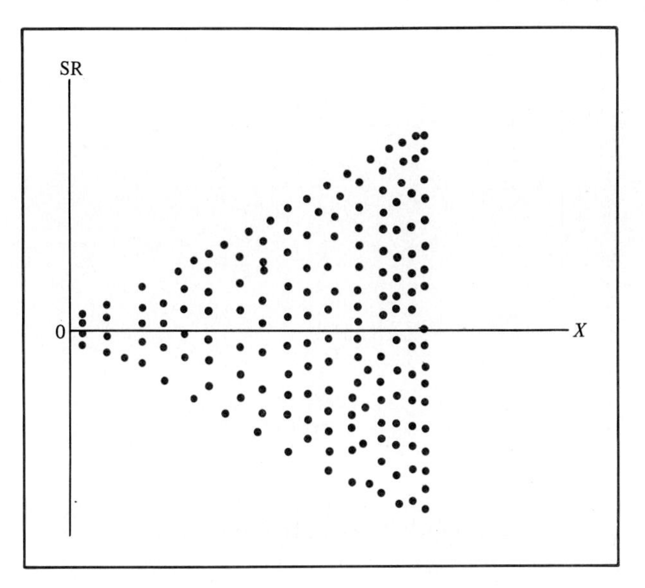

FIGURE 13.13 □ Violations of homoscedasticity.

On the other hand, we can observe a **fanning effect** for the *hypothetical* plot of ϵ with X in Figure 13.13 on page 409. The residuals seem to "fan out" as X increases—thereby demonstrating a lack of homogeneity in the variances of the Y values at each level of X. (Methods for treating this problem are presented in Reference 6.)

NORMALITY

The assumption of normality can be evaluated from a residual analysis by tallying the residuals into a frequency distribution and displaying the results in a histogram (see Section 4.4). In addition, if the sample size is large enough, the normality assumption can also be examined by using a **goodness-of-fit test** (see Reference 1).

For the problem of interest to the placement officer, the 30 residual values are tallied into a frequency distribution as shown in Table 13.6, with the results then plotted in a histogram as depicted in Figure 13.14. Since the plot of these data appears approximately bell-shaped, there would be no reason to suspect that the normality assumption has been violated.

TABLE 13.6 □ Frequency distribution of 30 residual values

Residual values	No. of observations
−2.5 but less than −1.5	2
−1.5 but less than −0.5	7
−0.5 but less than +0.5	12
+0.5 but less than +1.5	7
+1.5 but less than +2.5	2
Total	30

SOURCE: Table 13.5.

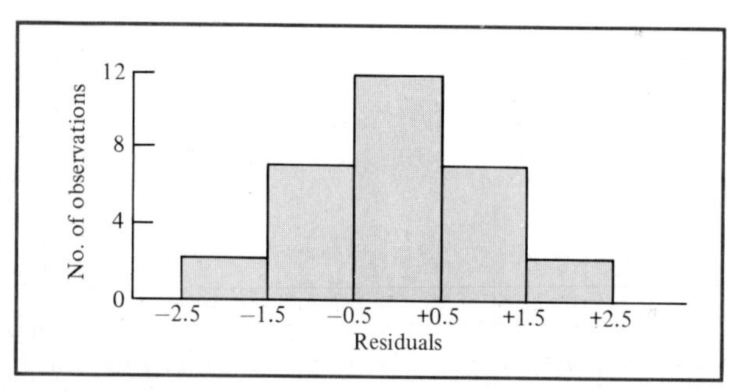

FIGURE 13.14 □ Histogram of 30 residual values.

SOURCE: Table 13.6.

INDEPENDENCE

The assumption of independence can be evaluated by plotting the residuals in the order in which the observed data were obtained. Any resulting patterns which visually emerge through such a scatter plot would indicate a potential violation in the independence assumption.[8] Thus, for the problem of interest to the placement officer, the randomness and independence of our observed sequence of 30 observations can be evaluated from Figure 13.15.

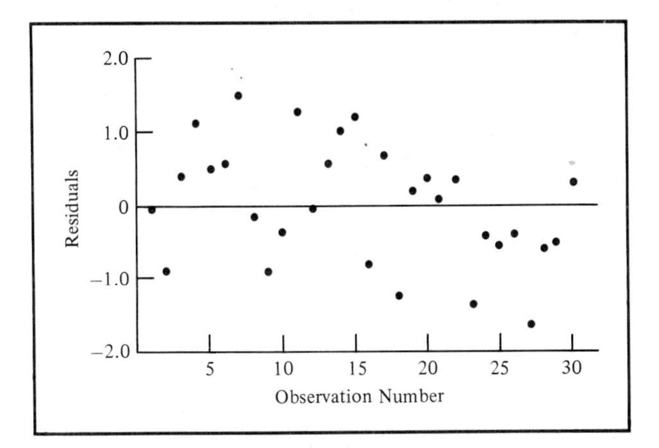

FIGURE 13.15 □ Studying the residuals in an observed sequence.

A careful examination of Figure 13.15 reveals a possible pattern in these residual plots. In the first half of the sequence most of the residuals are positive, while in the second half the majority are negative. Thus the researcher might seriously be concerned that the independence assumption has been violated. Fortunately, however, this anomaly can be explained by the fact that the first 15 individuals listed in Table 13.1 had majored in accountancy while the others had majored in marketing/management. Within these two groups, the residual plot in Figure 13.15 does not indicate any pattern, and hence the assumption of independence no longer appears to be violated. (Nevertheless, what we have just observed is that the categorical variable "business major" may have an influence on starting salary. See References 1, 2 and 6 for a discussion of how these categorical variables can be considered in regression analysis.)

Problems

13.63 Describe the assumptions underlying statistical inference in regression and correlation. Show—using plots, charts, etc.—cases where these assumptions are violated, and cases in which they are satisfied.

---

[8] For example, when data are collected over periods of time, there is often an autocorrelation effect among successive observations. That is, there exists a correlation between a particular observation and those values which precede and succeed it—thereby violating the assumption of independence. See Reference 2 for a discussion of methods that may be used to remove such effects.

13.64 What is a residual? What are residuals used for in regression?

Note: The following exercises use data sets with fewer data values than should actually be used in a residual analysis, because it would be too tedious to do the calculations for a realistic data set without using a computer.

13.65 For the data on software ratings and prices in Problem 13.2 on page 373:
(a) Produce a table with the predicted values and residuals for each point in the data set.
(b) Construct a histogram of the residuals, and comment on whether there is anything unusual about any values or the distribution as a whole.
(c) Plot the residuals (on the Y axis) versus the X values.
(d) Interpret your findings.

13.66 For the data on temperature and cricket chirps in Problem 13.4 on page 374:
(a) Produce a table with the predicted values and residuals for each point in the data set.
(b) Construct a histogram of the residuals, and comment on whether there is anything unusual about any values or the distribution as a whole.
(c) Plot the residuals (on the Y axis) versus the X values.
(d) Interpret your findings.

• 13.67 For the data on beer consumption and the reading of the breath-testing machine (Problem 13.7 on page 375):
(a) Produce a table with the predicted values and residuals for each point in the data set.
(b) Construct a histogram of the residuals, and comment on whether there is anything unusual about any values or the distribution as a whole.
(c) Plot the residuals (on the Y axis) versus the X values.
(d) Interpret your findings.

13.68 For the data on education and income in Problem 13.8 on page 375:
(a) Produce a table with the predicted values and residuals for each point in the data set.
(b) Construct a histogram of the residuals, and comment on whether there is anything unusual about any values or the distribution as a whole.
(c) Plot the residuals (on the Y axis) versus the X values.
(d) Interpret your findings.

13.13 SUMMARY

In this chapter we have developed the concepts of regression and correlation—both descriptively and inferentially. In particular, we have learned to fit the simple linear regression model (by using the method of least squares) as well as measure the strength of the linear relationship between X and Y (by calculating the coefficient of correlation). Moreover, we have learned how to test for the significance of the simple linear relationship and, in its presence, use the regression model for prediction. Finally, we explored ways to evaluate the aptness of the fitted model.

While we have accomplished much, it must again be pointed out that the subject of regression and correlation is very broad. Extensions to curvilinear analysis, categorical variables, and multiple regression and correlation analysis are well beyond the scope of an introductory text. If your data requires these techniques, we encourage you to examine the references.

Supplementary Problems

Note: A star indicates a problem or part of a problem that requires a study of an optional section or subsection of this chapter.

13.69 A statistician for a large American automobile manufacturer would like to develop a statistical model for predicting delivery time (the days between the ordering of the car and the actual delivery of the car) of custom-ordered new automobiles. The statistician believes that there is a *linear* relationship between the number of options ordered on the car and delivery time. A random sample of 16 cars is selected with the results given below:

Relating delivery time with options ordered

Car	Number of options ordered, X	Delivery time, Y (in days)
1	3	25
2	4	32
3	4	26
4	7	38
5	7	34
6	8	41
7	9	39
8	11	46
9	12	44
10	12	51
11	14	53
12	16	58
13	17	61
14	20	64
15	23	66
16	25	70

(a) Set up a scatter diagram.
(b) Use the least-squares method to find the regression coefficients b_0 and b_1.
(c) Interpret the meaning of the Y intercept b_0 and the slope b_1 in this problem.
(d) If a car was ordered that had 16 options, how many days would you predict it would take to be delivered?
(e) Compute the standard error of the estimate.
(f) Compute the coefficient of determination r^2 and interpret its meaning in this problem.
(g) Compute the coefficient of correlation r.
★**(h)** Set up a 95% confidence-interval estimate of the average delivery time for all cars ordered with 16 options.

★**(i)** Set up a 95% prediction-interval estimate of the delivery time for an individual car that was ordered with 16 options.
(j) At the .05 level of significance, is there evidence of a linear relationship between number of options and delivery time?
(k) Set up a 95% confidence-interval estimate of the true slope.
(l) Set up a 95% confidence-interval estimate of the true correlation coefficient.
★**(m)** Perform a residual analysis on your results and determine the adequacy of the fit of the model.

13.70 An official of a local racetrack would like to develop a model to forecast the amount of money bet (in millions of dollars) based on attendance. A random sample of 10 days is selected with the results given in the table below.

Relating betting with attendance

Day	Attendance (thousands)	Amount bet (millions of dollars)
1	14.5	.70
2	21.2	.83
3	11.6	.62
4	31.7	1.10
5	46.8	1.27
6	31.4	1.02
7	40.0	1.15
8	21.0	.80
9	16.3	.71
10	32.1	1.04

[*Hint*: Determine which is the independent and which is the dependent variable.]
(a) Set up a scatter diagram.
(b) Assuming a linear relationship, use the least-squares method to find the regression coefficients b_0 and b_1.
(c) Interpret the meaning of the slope b_1 in this problem.
(d) Predict the amount bet for a day on which attendance is 20,000.
(e) Compute the standard error of the estimate.
(f) Compute the coefficient of determination r^2 and interpret its meaning in this problem.
(g) Compute the coefficient of correlation r.
★**(h)** Set up a 99% confidence-interval estimate of the average amount of money bet when attendance is 20,000.
★**(i)** Set up a 99% prediction interval for the amount of money bet on a day in which attendance is 20,000.
(j) At the .01 level of significance, is there evidence of a linear relationship between the amount of money bet and attendance?
(k) Set up a 99% confidence-interval estimate of the true slope.
(l) Set up a 99% confidence interval estimate of the true correlation coefficient.
(m) Discuss why you should not predict the amount bet on a day on which the attendance exceeded 46,800 or was below 11,600.

★**(n)** Perform a residual analysis on your results and determine the adequacy of the fit of the model.

13.71 The controller of a large department store chain would like to predict the account balance at the end of a billing period based upon the number of transactions made during the billing period. A random sample of 12 accounts was selected with the results given below.

Relating account balance with transactions

Account	Number of transactions	Account balance
1	1	$ 15
2	2	36
3	3	40
4	3	63
5	4	69
6	5	78
7	6	84
8	7	100
9	10	175
10	10	120
11	12	150
12	15	198

[*Hint*: Determine which is the independent and which is the dependent variable.]

(a) Set up a scatter diagram.
(b) Assuming a linear relationship, use the least-squares method to find the regression coefficients b_0 and b_1.
(c) Interpret the meaning of the slope b_1 in this problem.
(d) Predict the account balance for an account which has had five transactions in the last billing period.
(e) Compute the standard error of the estimate.
(f) Compute the coefficient of determination r^2 and interpret its meaning in this problem.
(g) Compute the coefficient of correlation r.
★**(h)** Set up a 95% confidence-interval estimate of the average account balance for all accounts in which there have been five transactions in the last billing period.
★**(i)** Set up a 95% prediction interval of the account balance for an individual account in which there has been five transactions in the last billing period.
(j) At the .05 level of significance, is there evidence of a linear relationship between the number of transactions and account balance?
(k) Set up a 95% confidence-interval estimate of the true slope.
(l) Set up a 95% confidence-interval estimate of the true correlation coefficient.
★**(m)** Perform a residual analysis on your results and determine the adequacy of the fit of the model.

• 13.72 The Director of Graduate Studies at a large college of business would like to be able to predict the grade-point index (GPI) of students in an MBA program based upon Graduate Management Aptitude Test (GMAT) score. A sample of 20 students who had completed two years in the program was selected; the results are presented on page 416.

Relating GPI to GMAT score

Observation	GMAT score	GPI
1	688	3.72
2	647	3.44
3	652	3.21
4	608	3.29
5	680	3.91
6	617	3.28
7	557	3.02
8	599	3.13
9	616	3.45
10	594	3.33
11	567	3.07
12	542	2.86
13	551	2.91
14	573	2.79
15	536	3.00
16	639	3.55
17	619	3.47
18	694	3.60
19	718	3.88
20	759	3.76

[*Hint*: First determine which is the independent variable and which is the dependent variable.]

(a) Assuming a linear relationship, use the least-squares method to find the regression coefficients b_0 and b_1.
(b) Interpret the meaning of the Y intercept b_0 and the slope b_1 in this problem.
(c) Use the regression model developed in (a) to predict the grade-point index for a student with a GMAT score of 600.
(d) Compute the standard error of the estimate.
(e) Compute the coefficient of determination r^2 and interpret its meaning in this problem.
(f) Compute the coefficient of correlation r.
(g) At the .05 level of significance, is there evidence of a linear relationship between GMAT score and grade-point index?
★**(h)** Set up a 95% confidence-interval estimate for the average grade-point index of students with a GMAT score of 600.
★**(i)** Set up a 95% prediction-interval estimate of the grade-point index for a particular student with a GMAT score of 600.
(j) Set up a 95% confidence-interval estimate of the population slope.
(k) Set up a 95% confidence-interval estimate of the true correlation coefficient.
★**(l)** Perform a residual analysis on your results and determine the adequacy of the fit of the model.

13.73 The following data represents the price per eight ounces of club soda and seltzer for eleven brands at a supermarket in New York City:

Brand	Club soda	Seltzer
1	15	13
2	17	20
3	14	14
4	14	14
5	14	10
6	20	17
7	14	14
8	9	11
9	14	14
10	10	10
11	17	17

(a) Compute the coefficient of correlation r between the price of club soda and seltzer.
(b) At the .01 level of significance, is there evidence of a significant correlation between the price of club soda and seltzer?
(c) Set up a 99% confidence-interval estimate of ρ.

• 13.74 The following data represents the price per pound and number of grams of protein per 100 grams of different types of meat sold at a local supermarket:

Meat	Price per pound ($)	Protein (grams)
Beef chuck (lean and fat)	1.82	24.9
Beef chuck (lean only)	1.89	30.7
Beef bottom (lean and fat)	4.06	29.5
Beef bottom (lean only)	4.49	31.8
Beef rib (lean and fat)	5.55	20.9
Beef rib (lean only)	7.89	25.3
Beef roast (lean and fat)	2.36	26.2
Beef roast (lean only)	2.79	28.3
Sirloin steak (lean and fat)	3.76	26.8
Sirloin steak (lean only)	4.12	30.1

(a) Compute the coefficient of correlation r between price and amount of protein.
(b) At the .10 level of significance, is there evidence of a significant correlation between price and amount of protein?
(c) Set up a 90% confidence-interval estimate of ρ.

Database Exercises

The following problems refer to the sample data obtained from the questionnaire of Figure 2.5 and presented in Figure 2.10. They should be solved with the aid of a computer package. Use α = .05 throughout.

□ 13.75 Construct a scatter diagram of salary (question 7) versus undergraduate grade-point average (question 15). Does the relationship appear approximately linear? Find the correlation between these variables. Use regression to predict salary from grade-point average and interpret the coefficients.

□ 13.76 Construct a scatter diagram of the number of areas in which the alumnus has worked (question 12) by the number of years since graduation (question 1). Does this look as you would expect? Find the correlation between the two variables. Use regression to predict number of areas based on the number of years since graduation.

□ 13.77 Construct a scatter diagram of grade-point average (question 15) by height (question 19) for females (question 18). Is there a significant correlation between the variables? If so, use regression to predict grade-point average based on height and interpret the regression coefficients.

□ 13.78 Construct a scatter diagram of salary (question 7) versus height (question 19) for males (question 18). Is there a significant correlation between the variables? If so, use regression to predict salary from height and interpret the regression coefficients.

References

1. BERENSON, M. L., AND D. M. LEVINE, *Basic Business Statistics: Concepts and Applications,* 3d ed. (Englewood Cliffs, N.J.: Prentice-Hall, 1986).
2. BERENSON, M. L., D. M. LEVINE, AND M. GOLDSTEIN, *Intermediate Statistical Methods and Applications: A Computer Package Approach* (Englewood Cliffs, N.J.: Prentice-Hall, 1983).
3. BOWERMAN, B. L., AND R. T. O'CONNELL, *Forecasting and Time-Series* (North Scituate, Mass.: Duxbury Press, 1979).
4. DRAPER, N. R., AND H. SMITH, *Applied Regression Analysis,* 2d ed. (New York: John Wiley, 1981).
5. FISHER, R. A., *Statistical Methods and Scientific Inference*, 2d ed. (New York: Hafner Press, 1959).
6. NETER, J., W. WASSERMAN, AND M. H. KUTNER, *Applied Linear Statistical Models,* 2d ed. (Homewood, Ill.: Richard D. Irwin, 1985).

Chapter 14

Nonparametric Methods

14.1 INTRODUCTION: WHAT'S AHEAD

In the preceding four chapters a variety of hypothesis-testing situations were described. In most of these situations the so-called "classical" or **parametric methods** of hypothesis testing were employed. Nevertheless, often the researcher in the social sciences or in business must decide what kinds of testing procedures to choose if:

1. The measurements attained on the data are only qualitative (i.e., *nominally* scaled) or in ranks (i.e., *ordinally* scaled).
2. The assumptions underlying the use of the classical methods cannot be met.
3. The situation requires a study of such features as *randomness, trend, independence, symmetry,* or *goodness-of-fit* rather than the testing of hypotheses about particular population parameters.

For such circumstances as these, **nonparametric methods** of hypothesis testing have been devised. In fact, two such methods have already been discussed—the McNemar test for changes in proportions (see Section 11.7) and the chi-square test for independence (see Sections 11.4–11.6).[1] This chapter will focus on the development and use of six additional nonparametric methods:

1. The Wald-Wolfowitz one-sample runs test for randomness.
2. The Wilcoxon one-sample signed-ranks test.
3. The Wilcoxon paired-sample signed-ranks test.
4. The Wilcoxon rank-sum test for two independent samples.
5. The Kruskal-Wallis test for *c* independent samples.
6. The Spearman rank correlation procedure.

The goals of this chapter are:

1. To show the need for nonparametric methods.
2. To demonstrate the usefulness of nonparametric methods by comparing a selected set against their classical counterparts.

14.2 CLASSICAL VERSUS NONPARAMETRIC PROCEDURES

14.2.1 Classical Procedures

Classical or **parametric procedures** have three distinguishing characteristics. First, they require that the level of measurement attained on the collected data be in the form of an *interval* scale or *ratio* scale (as described in Chapter 2). Second, they involve hypothesis testing of specified parameters (such as $\mu_X = 30$ or $\beta_1 = 0$ or $\rho = 0$). Third, classical procedures require very stringent assumptions and are valid only if these assumptions hold. Among these assumptions are:

1. That the sample data be randomly drawn from a population which is normally distributed.
2. That the observations be independent of each other.
3. For situations concerning central tendency for which two or more samples have been drawn, that they be drawn from normal populations having equal variances (so that any differences between the populations will be in central tendency).

The sensitivity of classical procedures to violations in the assumptions has been considered in the statistical literature [References 1 and 2]. Some test procedures are said to be "robust" because they are relatively insensitive to slight violations in the assumptions. However, with gross violations in the assumptions both the

[1] Both of these methods deal with qualitative data and are not (necessarily) concerned with parameters. The McNemar procedure may be viewed as a test of symmetry, while the chi-square technique seeks to determine whether or not a relationship between two qualitative variables is present.

true level of significance and the power of a test may differ sharply from what otherwise would be expected.

14.2.2 Nonparametric Procedures

When classical methods of hypothesis testing are not applicable, an appropriate **nonparametric procedure** (References 2 and 3) can be selected.

> **Nonparametric test procedures may be broadly defined as either: (1) Those whose test statistic does not depend upon the form of the underlying population distribution from which the sample data were drawn. (2) Those which are not concerned with the parameters of a population. (3) Those for which the data are of "insufficient strength" to warrant meaningful arithmetic operations.**

The one-, two-, and *c*-sample tests for proportions, the chi-square test for independence, and the McNemar test for changes (all of which were discussed in Chapter 11) fit such a definition. Indeed, when testing a hypothesis about a single proportion, we have previously employed what is called (in the nonparametric literature) the **binomial test.** For example, in Case Study H (see page 308) involving a survey of attendees at a microcomputer show, the binomial test may be broadly categorized as nonparametric for two reasons. First, each observation (that is, attendee) was "grossly measured" (that is, classified as presently owning or not owning a personal computer). Second, the test statistic did not depend upon the underlying population from which the sample data were drawn. For example, regardless of whether the variable "amount spent on a personal computer system" actually follows a normal (or any other shaped) distribution, the variable "ownership" is obtained from the categorization of each person as having spent either zero dollars or more than zero dollars buying a computer. The binomial test statistic is independent of the form of the particular underlying distribution for the quantitative random variable of interest, amount spent.

On the other hand, aside from the fact that the data gathered are qualitative (nominally scaled), the chi-square test for independence may be broadly classified as nonparametric for a different reason—it is not concerned with parameters.[2] In Section 11.4 the researcher sought to ascertain whether or not one's intention to return to the microcomputer show next year was independent of one's purchasing behavior at the present show. Other kinds of nonparametric procedures that are in this category are tests for randomness (see Section 14.4), tests for trend, tests for symmetry (see Section 11.7), and tests for goodness-of-fit.

---

[2] That is, traditionally this test has not been considered as parametric. The development of new techniques for analyzing qualitative data—specifically log-linear models—have changed that situation.

14.3 ADVANTAGES AND DISADVANTAGES OF USING NONPARAMETRIC METHODS

The use of nonparametric methods offers numerous advantages. Six are summarized below.

1. Nonparametric methods may be used on all types of data—qualitative data (nominal scaling), data in rank form (ordinal scaling), as well as data that have been measured more precisely (interval or ratio scaling).
2. Nonparametric methods are generally easy to apply and quick to compute when the sample sizes are small. Sometimes they are as simple as just counting how often some feature appears in the data. Thus they are often used for pilot or preliminary studies and/or in situations in which quick answers are desired.
3. Nonparametric methods make fewer, less stringent assumptions (which are more easily met) than do the classical procedures. Hence they enjoy wider applicability and yield a more general, broad-based set of conclusions.
4. Nonparametric methods permit the solution of problems that do not involve the testing of population parameters.
5. Nonparametric methods may be more economical than classical procedures, since the researcher may increase power and yet save money, time, and labor by collecting larger samples of data which are more grossly measured (that is, qualitative data or data in rank form), thereby solving the problem more quickly.
6. Depending on the particular procedure selected, nonparametric methods may be equally (or almost) as powerful as the classical procedure when the assumptions of the latter are met, and when they are not met may be quite a bit more powerful.

It should now be apparent that nonparametric methods may be advantageously employed in a variety of situations. On the other hand, nonparametric procedures have some shortcomings.

1. It is disadvantageous to use nonparametric methods when all the assumptions of the classical procedures can be met and the data are measured on either an interval or ratio scale. Unless classical procedures are employed in these instances, the researcher is not taking full advantage of the data. Information is lost when we convert such collected data (from an interval or ratio scale) to either ranks (ordinal scale) or categories (nominal scale). In particular, in such circumstances, some very quick and simple nonparametric tests have much less power than the classical procedures, and should usually be avoided.
2. As the sample size gets larger, data manipulations required for nonparametric procedures are sometimes laborious unless appropriate computer software is available.
3. Often special tables of critical values are needed, and these are not as readily available as are the tables of the normal, t, χ^2, and F critical values.

Now that we have discussed some of the advantages and disadvantages of nonparametric methods, it is apparent that such procedures do not provide a panacea for all problems of statistical data analysis. Nevertheless, they comprise an important and useful body of statistical techniques. In the sections that follow we will consider the development of six simple but important nonparametric techniques that may prove highly useful under differing circumstances. To help motivate the discussion concerning these techniques let us consider an illustrative example.

Case K—The Advertising Account Executive's Problem

An account executive for a large advertising agency has decided to initiate a study of the effect of advertising on product perception for one of the agency's clients—a well-known pen manufacturer.

Five different examples of advertising copy were to be compared in the marketing of a ballpoint pen. Advertisement A tended to *greatly undersell* the pen's characteristics. Advertisement B tended to *slightly undersell* the pen's characteristics. Advertisement C tended to *slightly oversell* the pen's characteristics. Advertisement D tended to *greatly oversell* the pen's characteristics. Advertisement E tended to *correctly state* the pen's characteristics.

A sample of 30 adult participants were tested one at a time in individual cubicles in the testing studio. The five advertisements were randomly assigned to the 30 participants so that there would be six individuals responding to each advertisement. After reading the advertisement and developing a sense of "product expectation," each participant unknowingly received the same type of pen to evaluate. The participants were permitted to test their pen; their evaluations would presumably be influenced by the plausibility of the advertising copy. The participants were then asked to rate the pen from 1 to 7 on four product-characteristic scales as shown below:

	Scale						
Product characteristic	**Extremely poor** ↓			**Neutral** ↓			**Extremely good** ↓
Appearance	1	2	3	4	5	6	7
Durability	1	2	3	4	5	6	7
General feel	1	2	3	4	5	6	7
Writing performance	1	2	3	4	5	6	7

In addition, each participant was asked to determine the selling price based on what he or she thought the pen was actually worth.

The *combined* scores of the four ratings (appearance, durability, feel, and writing performance) along with the stated selling price are recorded in Table 14.1 in the order in which the participants were tested.

The account executive also wanted to evaluate the reactions of a "younger audience" (for example, high-school students)—particularly when such potential users of the product were subjected to advertisement E (which attempted to correctly state the pen's characteristics). Table 14.2 presents the combined ratings data for a sample of eight high-school students subjected to advertisement E.

At a meeting with the agency's consulting statistician it was suggested that appropriate nonparametric methods should be considered when analyzing the "combined ratings" data displayed in Tables 14.1 and 14.2. The statistician stated that many

TABLE 14.1 □ Combined ratings and stated selling prices based on five advertisements

Participant	Advertisement	Combined rating	Stated selling price
1	B	19	1.00
2	E	15	.49
3	B	28	2.49
4	E	14	.49
5	A	23	1.69
6	D	7	.29
7	C	12	.50
8	A	17	.85
9	C	9	.35
10	C	17	.59
11	D	13	.45
12	C	17	.65
13	C	8	.25
14	D	15	.69
15	E	23	1.50
16	E	16	.79
17	D	10	.39
18	E	20	1.19
19	D	11	.35
20	E	23	1.49
21	B	22	1.25
22	A	24	1.59
23	A	23	1.39
24	D	6	.19
25	C	18	.99
26	B	20	1.15
27	B	21	1.29
28	A	27	2.29
29	A	25	1.99
30	B	26	1.89

TABLE 14.2 □ Combined ratings of eight high-school students subjected to advertisement E

Student	Combined rating
a	24
b	19
c	21
d	13
e	17
f	19
g	22
h	18

researchers would argue that the "product-characteristic scales" used do not truly satisfy the criteria of interval or ratio scaling and, therefore, that nonparametric methods are more appropriate.

Based on this meeting with the consulting statistician, the account executive desired that the research efforts address the following questions:

1. With respect to the combined ratings given by the 30 adults, is there evidence of some *pattern* in the data or can the collected sequence of ratings be considered *random*?
2. Is there evidence that the median stated selling price for adults subjected to advertisements A or B (which *undersell* the pen's characteristics) exceeds $1.09?
3. Is there evidence of a difference in the combined ratings of adult versus high-school student participants subjected to advertisement E (which attempted to correctly state the pen's characteristics)?
4. Is there evidence of a difference in the combined ratings given by the adults based on exposure to different advertisements? That is, do product perceptions differ for various levels of expectation induced through advertising?
5. Is there evidence of a significant positive correlation between combined product-characteristic ratings and stated selling price?

Each of these questions will be addressed, in turn, in the five sections of this chapter that follow.

14.4 WALD-WOLFOWITZ ONE-SAMPLE RUNS TEST FOR RANDOMNESS

In statistical inference it is usually assumed that the collected data constitute a random sample. Such an assumption, however, may be tested by the employment of a nonparametric procedure called the **Wald-Wolfowitz one-sample runs test for randomness**.

The null hypothesis of randomness may be tested by observing the *order* or *sequence* in which the items are obtained. If each item is assigned one of two symbols, such as S and F (for success and failure), depending either on whether the item possesses a particular property or on the amount or magnitude in which the property is possessed, the randomness of the sequence may be investigated. If the sequence is randomly generated, the items will be independent and identically distributed. This means that the value of an item will be independent both of its position in the sequence and of the values of the items which precede it and follow it. On the other hand, if an item in the sequence is affected by the items which precede it or succeed it so that the probability of its occurrence varies from one position to another, the process is not considered random. In such cases either similar items would tend to cluster together (such as when a trend in the data is present) or the similar items would alternatingly mix so that some systematic periodic effect would exist.

To study whether or not an observed sequence is random, we will consider as the test statistic the **number of runs** present in the data.

A run is defined as a consecutive series of similar items which are bounded by items of a different type (or by the beginning or ending of the sequence).

For instance, suppose that the following is the observed occurrence of an experiment of flipping a coin 20 times:

HHHHHHHHHHTTTTTTTTTT

In this sequence there are two runs—a run of 10 heads followed by a run of 10 tails. With similar items tending to cluster, such a sequence would not be considered random even though, as would theoretically be expected, 10 of the 20 outcomes are heads and 10 are tails.

At the other extreme, suppose that the following sequence is obtained when flipping a coin 20 times:

HTHTHTHTHTHTHTHTHTHT

In this sequence there are 20 runs—10 runs of one head each and 10 runs of one tail each. With such an alternating pattern, this sequence could not be considered random because there are too many runs.

On the other hand, if, as shown below, the sequence of responses to the 20 coin tosses is thoroughly mixed, the number of runs will neither be too few nor too many, and the process may then be considered random:

HHTTHHHHTTTTTHTHTTHH

Therefore, in testing for randomness, what is essential is the ordering or positioning of the items in the sequence, not the frequency of items of each type.

14.4.1 Procedure

To perform the test of the null hypothesis of randomness, let the total sample size n be decomposed into two parts, n_1 successes and n_2 failures. The test statistic U, the total number of runs, is then obtained by counting. For a two-tailed test, if U is either too large or too small, we may reject the null hypothesis of randomness in favor of the alternative that the sequence is not random. If both n_1 and n_2 are less than or equal to 20, Table C.8, Parts 1 and 2, presents the critical values for the test statistic U at the .05 level of significance (two-tailed). If, for a given combination of n_1 and n_2, U is either greater than or equal to the upper critical value or less than or equal to the lower critical value, the null hypothesis of randomness may be rejected at the .05 level. However, if U lies between these limits, the null hypothesis of randomness cannot be rejected.

On the other hand, tests for randomness are not always two-tailed. If we are interested in testing for randomness against the specific alternative of a **trend effect**—that there is a tendency for like items to cluster together—a one-tailed test is needed. Here we reject the null hypothesis only if too few runs occur—if the observed value of U is less than or equal to the critical value presented in Table C.8, Part 1, at the .025 level of significance. At the other extreme, if we are interested in testing for randomness against a **systematic** or **periodic effect,** we use a one-tailed test that rejects only if too many runs occur—if the observed value of U is greater than or equal to the critical value given in Table C.8, Part 2, at the .025 level of significance.

Regardless of whether the test is one-tailed or two-tailed, however, for a sample size n greater than 40 (or when either n_1 or n_2 exceeds 20) the test statistic U is approximately normally distributed. Therefore, the following large-sample approximation formula may be used to test the hypothesis of randomness:

$$Z = \frac{U - \mu_U}{\sigma_U} \quad \textbf{(14.1)}$$

where U is the total number of runs

μ_U is the mean value of U; $\mu_U = \frac{2n_1n_2}{n} + 1$

σ_U is the standard deviation of U; $\sigma_U = \sqrt{\frac{2n_1n_2(2n_1n_2 - n)}{n^2(n-1)}}$

n_1 is the number of successes in the sample

n_2 is the number of failures in the sample

n is the sample size; $n = n_1 + n_2$

That is,

$$Z = \frac{U - \left(\frac{2n_1n_2}{n} + 1\right)}{\sqrt{\frac{2n_1n_2(2n_1n_2 - n)}{n^2(n-1)}}} \quad \textbf{(14.2)}$$

and, based on the level of significance selected, the null hypothesis may be rejected if the computed Z value falls in the appropriate region of rejection, depending on whether a two-tailed test or a one-tailed test is used (see Figure 14.1).

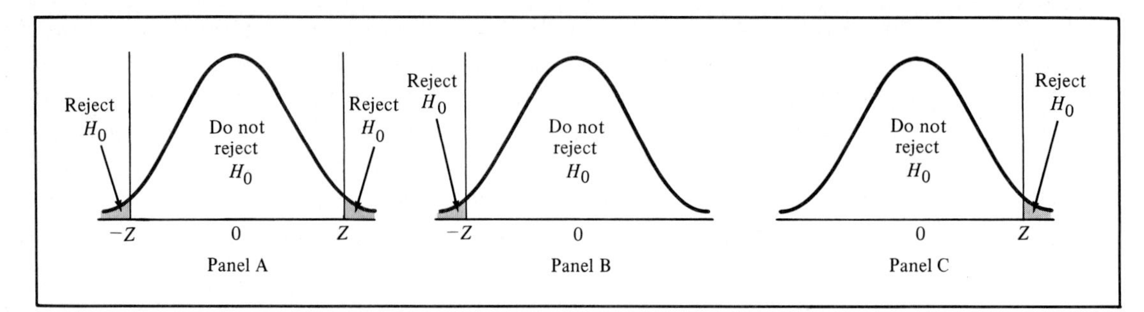

FIGURE 14.1 □ Determining the rejection region. Panel A, two-tailed test. Panel B, one-tailed test, trend effect. Panel C, one-tailed test, periodic effect.

14.4.2 Application

As an example employing the Wald-Wolfowitz one-sample runs test, suppose we examine the first question posed by the account executive: Can the collected sequence of product quality ratings be considered random?

A distinguishing feature of the Wald-Wolfowitz one-sample runs test for randomness is that it may be used not only on data which constitute a nominal scale in which each of the items is classified as success or failure, but also on data measured on an ordinal, interval, or ratio scale. In these cases, each item is classified according to its position with respect to the median of the sequence. For example, from Table 14.1 we may wish to test the null hypothesis that the collected sequence of product quality ratings are randomly distributed with respect to the median rating against the alternative that these ratings are not randomly distributed with respect to the median—that is,

H_0: The collected sequence of ratings is random

H_1: The collected sequence of ratings is not random (*two-tailed*)

To perform the runs test we assign the symbol A to each rating that equals or exceeds the median rating and the symbol B to each rating that is below the median rating. From Table 14.1, the median rating is 17.5. Thus, for the 30 observations we note that 15 are above the median rating ($n_1 = 15$) and 15 are below the median rating ($n_2 = 15$) and we have:

Collection sequence:	1	2	3	4	5	6	7	8	9	10	11	12	13	14	15	16	17	18	19	20	21	22	23	24	25	26	27	28	29	30
Combined rating:	19	15	28	14	23	7	12	17	9	17	13	17	8	15	23	16	10	20	11	23	22	24	23	6	18	20	21	27	25	26
Relationship to median of 17.5:	A	B	A	B	A	B	B	B	B	B	B	B	B	B	A	B	B	A	B	A	A	A	A	B	A	A	A	A	A	A
Runs:	1	2	3	4	5					6					7	8		9	10			11		12			13			

Table C.8, Parts 1 and 2, presents the critical values of the runs test at the .05 level of significance; a facsimile is displayed as Table 14.3 on page 430.

Since $n_1 = 15$ and $n_2 = 15$, we would reject the null hypothesis at the .05 level if $U \geq 22$ or if $U \leq 10$ for this two-tailed test. Here, however, the observed number of runs is 13, so we cannot reject the null hypothesis of randomness. The account executive should be informed that there is no evidence of any relationship between the magnitude of the combined ratings and the order in which they were collected.

EXAMPLE 14.1

An educational researcher wishes to construct a true-false examination containing 50 questions. She desires the sequence of correct responses to be random so that nobody taking the examination can merely guess at the answers by following a particular pattern. The following sequence is obtained:

Questions: 1 2 3 4 5 6 7 8 9 10 11 12 13 14 15 16 17 18 19 20 21 22 23 24 25
Answers: F T T T F T F F F T T F T F F F T T T T F F T T T
Runs: 1 2 3 4 5 6 7 8 9 10 11 12

Questions: 26 27 28 29 30 31 32 33 34 35 36 37 38 39 40 41 42 43 44 45 46 47 48 49 50
Answers: F F T F F F T T F T T F T T F F F F T T T T F F F
Runs: 13 14 15 16 17 18 19 20 21 22 23

Can this sequence be considered random?

Solution:

H_0: Sequence is random
H_1: Sequence is not random (*two-tailed*)
$\alpha = .05$

$$n = 50, \quad n_1 = 25, \quad n_2 = 25, \quad U = 23$$

$$\mu_U = \frac{2n_1n_2}{n} + 1 = \frac{2(25)(25)}{50} + 1 = 26$$

$$\sigma_U = \sqrt{\frac{2n_1n_2(2n_1n_2 - n)}{n^2(n-1)}} = \sqrt{\frac{2(25)(25)[2(25)(25) - 50]}{(50^2)(49)}} = 3.5$$

$$Z = \frac{U - \mu_U}{\sigma_U} = \frac{23 - 26}{3.5} = -.86$$

At the $\alpha = .05$ level of significance the critical Z values are ± 1.96. Since $Z = -.86$ falls between ± 1.96, we do not reject H_0. There is no evidence that the sequence is not random.

TABLE 14.3 □ Obtaining the lower and upper tail critical values U for the runs test where $n_1 = 15$, $n_2 = 15$, and $\alpha = .05$

Part 1. Lower tail ($\alpha = .025$)

n_1 \ n_2	2	3	4	5	6	7	8	9	10	11	12	13	14	15	16	17	18	19	20
2											2	2	2	2	2	2	2	2	2
3					2	2	2	2	2	2	2	2	2	3	3	3	3	3	3
4				2	2	2	3	3	3	3	3	3	3	3	4	4	4	4	4
5			2	2	3	3	3	3	3	4	4	4	4	4	4	4	5	5	5
6		2	2	3	3	3	3	4	4	4	4	5	5	5	5	5	5	6	6
7		2	2	3	3	3	4	4	5	5	5	5	5	6	6	6	6	6	6
8		2	3	3	3	4	4	5	5	5	6	6	6	6	6	7	7	7	7
9		2	3	3	4	4	5	5	5	6	6	6	7	7	7	7	8	8	8
10		2	3	3	4	5	5	5	6	6	7	7	7	7	8	8	8	8	9
11		2	3	4	4	5	5	6	6	7	7	7	8	8	8	9	9	9	9
12	2	2	3	4	4	5	6	6	7	7	7	8	8	8	9	9	9	10	10
13	2	2	3	4	5	5	6	6	7	7	8	8	9	9	9	10	10	10	10
14	2	2	3	4	5	5	6	7	7	8	8	9	9	9	10	10	10	11	11
15	2	3	3	4	5	6	6	7	7	8	8	9	9	10	10	11	11	11	12
16	2	3	4	4	5	6	6	7	8	8	9	9	10	10	11	11	11	12	12
17	2	3	4	4	5	6	7	7	8	9	9	10	10	11	11	11	12	12	13
18	2	3	4	5	5	6	7	8	8	9	9	10	10	11	11	12	12	13	13
19	2	3	4	5	6	6	7	8	8	9	10	10	11	11	12	12	13	13	13
20	2	3	4	5	6	6	7	8	9	9	10	10	11	12	12	13	13	13	14

Part 2. Upper tail ($\alpha = .025$)

n_1 \ n_2	2	3	4	5	6	7	8	9	10	11	12	13	14	15	16	17	18	19	20
2																			
3																			
4				9	9														
5			9	10	10	11	11												
6			9	10	11	12	12	13	13	13	13								
7				11	12	13	13	14	14	14	14	15	15	15					
8				11	12	13	14	14	15	15	16	16	16	16	17	17	17	17	17
9					13	14	14	15	16	16	16	17	17	18	18	18	18	18	18
10					13	14	15	16	16	17	17	18	18	18	19	19	19	20	20
11					13	14	15	16	17	17	18	19	19	19	20	20	20	21	21
12					13	14	16	16	17	18	19	19	20	20	21	21	21	22	22
13						15	16	17	18	19	19	20	20	21	21	22	22	23	23
14						15	16	17	18	19	20	20	21	22	22	23	23	23	24
15						15	16	18	18	19	20	21	22	22	23	23	24	24	25
16							17	18	19	20	21	21	22	23	23	24	25	25	25
17							17	18	19	20	21	22	23	23	24	25	25	26	26
18							17	18	19	20	21	22	23	24	25	25	26	26	27
19							17	18	20	21	22	23	23	24	25	26	26	27	27
20							17	18	20	21	22	23	24	25	25	26	27	27	28

SOURCE: Extracted from Table C.8, Parts 1 and 2.

Problems

14.1 What three general conditions define nonparametric tests?

14.2 What are the advantages and disadvantages of nonparametric tests?

• 14.3 The following table presents the unemployment rates of clerical workers in the United States from 1958 through 1982

Year	Unemployment rates (per thousand)
1958	4.4
1959	3.7
1960	3.8
1961	4.6
1962	4.0
1963	4.0
1964	3.7
1965	3.3
1966	2.9
1967	3.1
1968	3.0
1969	3.0
1970	4.0
1971	4.8
1972	4.7
1973	4.2
1974	4.6
1975	6.6
1976	6.4
1977	5.9
1978	4.9
1979	4.6
1980	5.3
1981	5.7
1982	7.0

SOURCE: Data are extracted from Table 28, *Handbook of Labor Statistics Bulletin 2175*, U.S. Department of Labor, Bureau of Labor Statistics, December 1984.

Using an $\alpha = .025$ level of significance, determine whether the above sequence may be considered random over time or whether there is evidence of some systematic or periodic effect present. [*Hint*: Compare each yearly value against the median of the series of data.]

✗14.4 Give an example of a sequence of results from tossing a coin which (a) has a clear pattern (and is therefore unlikely to be random) *but* (b) is not detected by the Wald-Wolfowitz test as being nonrandom. This demonstrates that the Wald-Wolfowitz test is sensitive to certain departures from randomness but not to others.

✗14.5 You will find that you cannot do a Wald-Wolfowitz test on the data from the high-school students in the example in the previous section as was done for the adults. Why is this so? [*Hint*: Discuss this in the context of the power to detect nonrandom outcomes.]

14.6 For the data on the measurement of the mass of the riton (Problem 3.11 on page 46), see whether there is evidence that the measurements follow a random pattern rather than either steadily increasing or decreasing, or oscillating back and forth. ($\alpha = .05$.)

14.7 For the data on the vitamin C content of oranges (Problem 3.23 on page 49), see whether there is evidence against the oranges having been selected in random order for examination. ($\alpha = .05$.)

• 14.8 Suppose we walk down a street on a Sunday and notice which stores are open (Y) and which are not (N). The results are:

YYYNNYNNNYNNYYYYNYNYYY

Is there evidence that the tendency of stores to be open or closed is not random? ($\alpha = .05$.)

14.5 WILCOXON SIGNED-RANKS TEST

In the preceding section the nonparametric procedure was not concerned with the testing of any particular parameters and as such has no classical counterpart. However, the procedure presented in this section may be used when it is desired to test a hypothesis regarding a parameter reflecting central tendency. This procedure, known as the **Wilcoxon signed-ranks test,** may be chosen over its respective classical counterpart—the one-sample t test (Section 10.6) or the paired-difference t test (Section 10.9)—when the researcher is able to obtain data measured at a higher level than an ordinal scale but does not believe that the assumptions of the classical procedure are sufficiently met. When the assumptions of the t test are violated, the Wilcoxon procedure (which makes fewer and less stringent assumptions than does the t test) is likely to be more powerful in detecting the existence of significant differences than its corresponding classical counterpart. Moreover, even under conditions appropriate to the classical t test, the Wilcoxon signed-ranks test has proven to be almost as powerful.

14.5.1 Procedure

To perform the Wilcoxon signed-ranks test the following six-step procedure may be used:

1. For each item in a sample of n' items we obtain a **difference score** D (to be described in Sections 14.5.3 and 14.5.5 for the one-sample and paired-sample tests, respectively).
2. We then neglect the + and − signs and obtain a set of n' absolute differences $|D|$.
3. We omit from further analysis any absolute difference score of zero, thereby yielding a set of n nonzero absolute difference scores where $n \leq n'$.
4. We then assign ranks R from 1 to n to each of the $|D|$ such that the smallest absolute difference score gets rank 1 and the largest gets rank n. Owing to a lack of precision

in the measuring process, if two or more $|D|$ are equal, they are each assigned the "average rank" of the ranks they otherwise would have individually been assigned had ties in the data not occurred.

5. We now reassign the symbol + or − to each of the n nonzero absolute difference scores $|D|$, depending on whether D was originally positive or negative.
6. The Wilcoxon test statistic W is obtained as the sum of the + ranks.

$$W = \sum R^{(+)} \tag{14.3}$$

Since the sum of the first n integers $(1, 2, \ldots, n)$ is given by $n(n + 1)/2$, the Wilcoxon test statistic W may range from a minimum of 0 (where all the observed difference scores are negative) to a maximum of $n(n + 1)/2$ (where all the observed difference scores are positive). If the null hypothesis were true, we would expect the test statistic W to take on a value close to its mean, $\mu_W = n(n + 1)/4$, while if the null hypothesis were false, we would expect the observed value of the test statistic to be close to one of the extremes.

For samples of $n \leq 20$, Table C.9 may be used for obtaining the critical values of the test statistic W for both one- and two-tailed tests at various levels of significance. For a two-tailed test and for a particular level of significance, if the observed value of W equals or exceeds the upper critical value or is equal to or less than the lower critical value, the null hypothesis may be rejected. For a one-tailed test in the positive direction the decision rule is to reject the null hypothesis if the observed value of W equals or exceeds the upper critical value, while for a one-tailed test in the negative direction the decision rule is to reject the null hypothesis if the observed value of W is less than or equal to the lower critical value.

For samples of $n > 20$, the test statistic W is approximately normally distributed, and the following large-sample approximation formula may be used for testing the null hypothesis:

$$Z = \frac{W - \mu_W}{\sigma_W} \tag{14.4}$$

where W is the sum of the positive ranks; $W = \sum R^{(+)}$

μ_W is the mean value of W; $\mu_W = \dfrac{n(n + 1)}{4}$

σ_W is the standard deviation of W; $\sigma_W = \sqrt{\dfrac{n(n + 1)(2n + 1)}{24}}$

n is the number of nonzero absolute difference scores in the sample

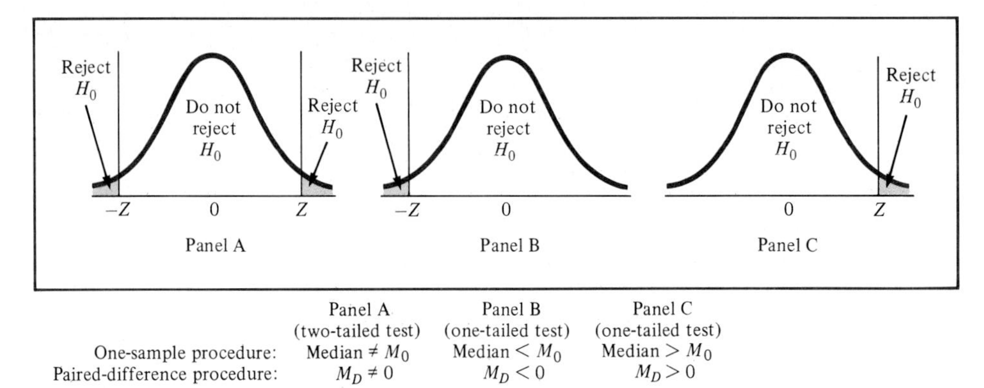

FIGURE 14.2 □ Determining the rejection region using the Wilcoxon signed-ranks test.

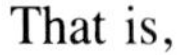
That is,

$$Z = \frac{W - \dfrac{n(n+1)}{4}}{\sqrt{\dfrac{n(n+1)(2n+1)}{24}}} \qquad \textbf{(14.5)}$$

and, based on the level of significance selected, the null hypothesis may be rejected if the computed Z value falls in the appropriate region of rejection depending on whether a two-tailed or one-tailed test is used (see Figure 14.2).

The *only* difference between the one-sample and the paired-sample tests described below is in how the ranks are computed. After that step is done, the computational procedures are identical.

14.5.2 One-Sample Test

If the researcher is interested in testing a hypothesis regarding a specified population median M_0 based on data from a single sample, the Wilcoxon (one-sample) signed-ranks test may be used. The test of the null hypothesis may be one-tailed or two-tailed:

Two-tailed test	One-tailed test	One-tailed test
H_0: Median $= M_0$	H_0: Median $\geq M_0$	H_0: Median $\leq M_0$
H_1: Median $\neq M_0$	H_1: Median $< M_0$	H_1: Median $> M_0$

The assumptions necessary for performing the one-sample test are

1. That the observed data $(X_1, X_2, \ldots, X_{n'})$ constitute a random sample of n' independent values from a population with unknown median.

2. That the underlying random phenomenon of interest be continuous.
3. That the observed data be measured at a higher level than the ordinal scale.
4. That the underlying population be (approximately) symmetrical.[3]

14.5.3 Application

To demonstrate the use of the Wilcoxon (one-sample) signed-ranks test, suppose we examine the second question posed by the account executive: Is there evidence that the median stated selling price for adults subjected to advertisements A or B (which *undersell* the pen's characteristics) exceeds $1.09?

The pertinent data extracted from Table 14.1 follow:

Advertisement:	B	B	A	A	B	A	A	B	B	A	A	B
Stated selling price ($):	1.00	2.49	1.69	.85	1.25	1.59	1.39	1.15	1.29	2.29	1.99	1.89

The following null and alternative hypotheses are established:

$$H_0: \text{ Median} \leq \$1.09$$
$$H_1: \text{ Median} > \$1.09$$

and the test is one-tailed.

To perform the one-sample test the first step is to obtain a set of difference scores D between each of the observed values X and the hypothesized median M_0—that is

$$D = X - M_0$$

The remaining steps of the six-step procedure are developed in Table 14.4 on page 436. The test statistic W is obtained as the sum of the positive ranks.

$$W = \Sigma R^{(+)} = 12 + 8 + 3 + 7 + 6 + 1 + 4 + 11 + 10 + 9 = 71$$

From Table 14.4 we note that 10 of the 12 nonzero absolute difference scores exceed the claimed median of $1.09.

To test for significance, we compare the observed value of the test statistic, $W = 71$, to the upper tail critical value presented in Table C.9 for $n = 12$ and for an α level selected at .05. As shown in Table 14.5, this critical value is 61. Since $W > 61$, the null hypothesis may be rejected at the .05 level of significance. There is evidence that the true median stated selling price for adults exposed to advertisements A or B exceeds $1.09. This information should be very useful to the account executive when assisting the client (i.e., the pen manufacturer) in establishing the pen's retail price.

---

[3] Such an assumption as symmetry, however, is not as stringent as an assumption of normality. Recall that not all symmetrical distributions are normal, although all normal distributions are symmetrical.

TABLE 14.4 □ Setting up the Wilcoxon (one-sample) signed-ranks test

Stated selling price, X	$D = X - 1.09$	$\|D\|$	R	Sign of D
1.00	− .09	.09	2.0	−
2.49	+1.40	1.40	12.0	+
1.69	+ .60	.60	8.0	+
.85	− .24	.24	5.0	−
1.25	+ .16	.16	3.0	+
1.59	+ .50	.50	7.0	+
1.39	+ .30	.30	6.0	+
1.15	+ .06	.06	1.0	+
1.29	+ .20	.20	4.0	+
2.29	+1.20	1.20	11.0	+
1.99	+ .90	.90	10.0	+
1.89	+ .80	.80	9.0	+

TABLE 14.5 □ Obtaining the upper tail critical value W for Wilcoxon one-sample signed-ranks test where $n = 12$ and $\alpha = .05$

	One-tailed: $\alpha = .05$	$\alpha = .025$	$\alpha = .01$	$\alpha = .005$
	Two-tailed: $\alpha = .10$	$\alpha = .05$	$\alpha = .02$	$\alpha = .01$
n		(Lower, Upper)		
5	0,15			
6	2,19	0,21		
7	3,25	2,26	0,28	
8	5,31	3,33	1,35	0,36
9	8,37	5,40	3,42	1,44
10	10,45	8,47	5,50	3,52
11	13,53	10,56	7,59	5,61
12	17,61	13,65	10,68	7,71
13	21,70	17,74	12,79	10,81
14	25,80	21,84	16,89	13,92
15	30,90	25,95	19,101	16,104

SOURCE: Extracted from Table C.9.

It is interesting that the large-sample approximation formula [Equation (14.5)] for the test statistic W yields excellent results for samples as small as 8. With the data above, for a sample of $n = 12$,

$$Z = \frac{W - \frac{n(n+1)}{4}}{\sqrt{\frac{n(n+1)(2n+1)}{24}}} = \frac{71 - \frac{12(13)}{4}}{\sqrt{\frac{12(13)(25)}{24}}} = \frac{71 - 39}{\sqrt{162.5}} = \frac{32}{12.75} = 2.51$$

Since $Z = +2.51$ is greater than the critical Z value of $+1.645$, the null hypothesis would also be rejected. However, since Table C.9 is available for $n \le 20$, it is both simpler and more accurate to merely look up the critical value in the table and avoid these computations when possible.

Problems

14.9 A manufacturer of batteries claims that the median capacity of a certain type of battery the company produces is 140 ampere-hours. An independent consumer protection agency wishes to test the credibility of the manufacturer's claim and measures the capacity of a random sample of $n' = 20$ batteries from a recently produced batch. The results are as follows:

137.0	140.0	138.3	139.0	144.3	139.1	141.7	137.3	133.5	138.2
141.1	139.2	136.5	136.5	135.6	138.0	140.9	140.6	136.3	134.1

The consumer protection agency is interested in whether or not the manufacturer's claim is being overstated. At the $\alpha = .05$ level of significance, is there evidence that the (median) capacity is significantly below 140 ampere-hours?

14.10 For the data on creativity test scores (Problem 3.12 on page 46) test whether there is evidence that the median score is different from 15. ($\alpha = .05$)

14.11 Test whether there is evidence that the median death rate in industrialized countries (Problem 3.13 on page 46) is less than 1,200 per 100,000 population. ($\alpha = .05$)

• 14.12 Is there evidence that the median price of dot matrix computer printers (Problem 3.15 on page 47) is less than $700? ($\alpha = .05$)

• 14.13 For the data on tax revenues in Problem 3.62 on page 69, is there evidence that the median revenue is different from $20,000? ($\alpha = .05$)

14.5.4 Paired-Sample Test

In the social sciences and in marketing research it is frequently of interest to examine differences between two *related* groups. We discussed this situation in Chapter 10. For example, in test-marketing a product under two different advertising conditions, a sample of test markets can be *matched* (*paired*) based on the test-market population size and/or other socioeconomic and demographic variables. Moreover, when performing a taste-testing experiment, each subject in the sample could be used as his or her own control so that *repeated measurements* on the same individual are obtained. For such situations, involving either matched (paired) items or repeated measurements of the same item, the Wilcoxon (paired-sample) signed-ranks test may be used when the t test is not appropriate. The test of the null hypothesis that the population median difference M_D is zero may be one-tailed or two-tailed:

Two-tailed test	One-tailed test	One-tailed test
H_0: $M_D = 0$	H_0: $M_D \geq 0$	H_0: $M_D \leq 0$
H_1: $M_D \neq 0$	H_1: $M_D < 0$	H_1: $M_D > 0$

The assumptions necessary for performing the paired-sample test are that

1. The observed data constitute a random sample of n' independent items or individuals, each with two measurements $(X_{11}, X_{21}), (X_{12}, X_{22}), \ldots, (X_{1n'}, X_{2n'})$, or the observed data constitute a random sample of n' independent pairs of items or individuals so that (X_{1i}, X_{2i}) represents the observed values for each member of the matched pair $(i = 1, 2, \ldots, n')$.
2. The underlying variable of interest be continuous.
3. The observed data be measured at a higher level than the ordinal scale—i.e., at the interval or ratio level.
4. The distribution of the population of difference scores between repeated measurements or between matched (paired) items or individuals be (approximately) symmetric.[4]

14.5.5 Application

Influenza and its effects are considered dangerous for adults over 60, and many doctors recommend that their patients in this age group be inoculated with an anti-influenza vaccine each fall. To demonstrate the use of the Wilcoxon (paired-sample) signed-ranks test, suppose that a pharmaceutical manufacturer is producing a new vaccine for a particular strain of influenza. In testing the effects of this vaccine it is desired to study the change in body temperature taken immediately before and one hour after the inoculation. If the vaccine is effective, symptoms of a mild case of influenza—a slightly stuffy nose, sore throat, chills, and a rise in bodily temperature—are noted within one hour after the vaccine is administered.

The results which follow are for a sample of $n' = 10$ male patients (aged 60 and over) who volunteered for the experiment.

	Volunteer									
	a	*b*	*c*	*d*	*e*	*f*	*g*	*h*	*i*	*j*
Temperature before (°F)	98.8	98.6	97.5	98.0	98.7	98.4	98.7	98.6	98.3	98.6
Temperature after (°F)	99.8	98.8	98.4	99.9	99.4	98.4	98.6	101.2	99.0	99.1

The question that must be answered is whether or not this new vaccine is effective. That is, for males aged 60 and over, is there any evidence of a significant increase in the body temperature one hour after the vaccination?

The following null and alternative hypotheses are established:

$$H_0: \ M_D \leq 0$$
$$H_1: \ M_D > 0$$

and the test is one-tailed.

[4] See footnote 3 on page 435.

To perform the paired-sample test, the first step of the six-step procedure is to obtain a set of difference scores D between each of the n' paired observations:

$$D = X_1 - X_2$$

In our example, we obtain a set of n' difference scores from

$$D = X_{\text{after}} - X_{\text{before}}$$

If the vaccine is effective, bodily temperature is expected to rise, so that the difference scores will tend to be *positive* values (and H_0 will be rejected). On the other hand, if the vaccine is not effective, we can expect some D values to be positive, others to be negative, and some to show no change (that is, $D = 0$). If this is the case, the difference scores will average near zero (that is, $\overline{D} \cong 0$) and H_0 will not be rejected.

The remaining steps of the six-step procedure are developed in Table 14.6. Notice that these are exactly the same steps as for the one-sample test. From this table we note that volunteer f is discarded from the study (because his difference score is zero) and that eight of the remaining $n = 9$ difference scores have a positive sign. The test statistic W is obtained as the sum of the positive ranks:

$$W = \sum R^{(+)} = 7 + 2 + 6 + 8 + 4.5 + 9 + 4.5 + 3 = 44$$

Since $n = 9$, we use Table C.9 to determine the upper tail critical value for this one-tailed test for an α level selected at .05. The upper tail critical value is 37. Since $W > 37$, the null hypothesis may be rejected at the .05 level of significance. There is evidence to support the contention that bodily temperature significantly increases among males over 60 years of age within one hour of the vaccination.

TABLE 14.6 ☐ Setting up the Wilcoxon (paired-sample) signed-ranks test

	Temperature (°F)					
Volunteer	**After, X_1**	**Before, X_2**	**$D = X_i - X_2$**	**$\|D\|$**	**R**	**Sign of D**
a	99.8	98.8	+1.0	1.0	7.0	+
b	98.8	98.6	+0.2	0.2	2.0	+
c	98.4	97.5	+0.9	0.9	6.0	+
d	99.9	98.0	+1.9	1.9	8.0	+
e	99.4	98.7	+0.7	0.7	4.5	+
f	98.4	98.4	0.0	0.0	. . .	Discard
g	98.6	98.7	−0.1	0.1	1.0	−
h	101.2	98.6	+2.6	2.6	9.0	+
i	99.0	98.3	+0.7	0.7	4.5	+
j	99.1	98.6	+0.5	0.5	3.0	+

Problems

14.14 An educational researcher claims that the maturation of a student, regardless of other factors, leads, on average, to a significant point increase in the combined verbal and mathematics SAT scores. The following table presents the combined verbal and mathematical SAT scores for a random sample of $n' = 20$ high-school students throughout the United States who participated in the testing program once during their junior year and once during their senior year without taking any special test-preparatory programs in between:

Student	Combined verbal and mathematics SAT scores	
	Senior year	**Junior year**
Al	1,050	1,010
Barbara	1,330	1,250
Claire	1,210	1,050
Daisy	980	920
Ethel	790	810
Fred	1,120	1,080
Gale	780	720
Harry	1,050	960
Ida	1,540	1,450
Jeri	1,070	1,060
Karen	1,030	1,000
Lou	1,130	1,070
Mort	890	890
Nathan	1,080	1,110
Oliver	1,560	1,510
Penny	1,240	1,170
Quinn	990	900
Richard	1,280	1,280
Sally	1,140	1,120
Trudy	790	720

Using an α level of .05, is there evidence of a significant point increase in the combined verbal and mathematics SAT scores for such students?

14.15 A swimming instructor did an experiment to see whether hyperventilation increased the amount of time a swimmer could spend underwater. She tested several swimmers who tried to swim as far as possible underwater. Each swimmer made two attempts; one was done normally, but before the other attempt the swimmer spent two minutes taking fast deep breaths (hyperventilating) to try to increase the amount of oxygen in the bloodstream. The distance each swimmer went underwater under each condition is reported below:

Condition	Swimmer						
	A	**B**	**C**	**D**	**E**	**F**	**G**
Normal	53	29	78	35	47	40	24
Hyperventilated	71	34	90	51	42	40	31

Is there any evidence that the median distance is lower for the normal condition than for the hyperventilated condition? ($\alpha = .05$)

• 14.16 A federal agency tested the relative effectiveness of two brands of disinfectants designed for use in hospitals to prevent infections. Nine hospitals used the two disinfectants; each used one disinfectant in one ward, the other in another (similar) ward. The percentage of patients getting an infection during the six-month test period was:

	Hospital								
Disinfectant	**1**	**2**	**3**	**4**	**5**	**6**	**7**	**8**	**9**
Brand X	12	5	3	9	14	4	2	5	6
Brand Y	10	6	6	10	14	5	8	4	9

Is there any evidence of a difference between Brand X and Brand Y in the median percentage of patients getting an infection? ($\alpha = .10$)

• 14.17 The T & A Blech Tax Preparation Company Inc. claimed that taxpayers would save money by having their firm prepare the individual's tax return. A consumer agency had people who had already prepared their return go to Blech's offices to get their taxes done by Blech's preparers. The taxes each person would pay if they paid what they calculated and if they paid what Blech calculated are presented below:

	Tax return preparer	
Taxpayer	**Blech**	**Self**
Jose	1,459	1,910
Marcia	3,250	2,900
Alexis	1,190	1,200
Harry	8,100	7,650
Jean	13,200	15,390
Marc	9,210	9,100
JR	255,970	33,120
Billy	210	140
Richard	1,290	1,320
Ted	130	0
Bruce	5,190	6,123

Is there evidence that the claim is not valid? ($\alpha = .05$)

14.18 The ratings of the television shows of two networks are shown below; the ratings are for shows which compete in the same time slots:

Time slot:	1	2	3	4	5	6	7	8	9
Network A:	14.5	21.4	9.9	12.8	19.2	28.1	14.2	23.6	13.2
Network B:	12.3	21.9	8.8	11.1	18.1	26.4	11.0	20.1	11.5

(The ratings are the percentages of homes with a television which are watching the show.) Decide whether there is evidence that one network is stronger than the other in these time slots. ($\alpha = .05$)

14.6 WILCOXON RANK-SUM TEST

In Section 10.8, when the researcher was interested in testing for differences in the means of two independent groups of data, the two-sample t test [Equation (10.4)] was selected. To use the two-sample t test, however, it is necessary to make a set of stringent assumptions. In particular, it is necessary that the two independent samples be randomly drawn from normal populations having equal variances and that the data be measured on at least an interval scale. However, it is frequently the case in studies of consumer behavior, marketing research, and experimental psychology that only ordinal type data can be obtained. Since the classical two-sample t test could not be used in such situations, an appropriate nonparametric technique is needed. On the other hand, even if data possessing the properties of an interval or ratio scale are obtained, the researcher may feel that the assumptions of the t test are unrealistic for the set of data. In such a circumstance a very simple but powerful nonparametric procedure known as the **Wilcoxon rank-sum test** may be used. Like the Wilcoxon signed-ranks test of the previous section, the Wilcoxon rank-sum test has proven to be almost as powerful as its classical counterpart under conditions appropriate to the latter and is likely to be more powerful when the assumptions of the t test are not met.

14.6.1 Procedure

To perform the Wilcoxon rank-sum test we must replace the observations in the two samples of size n_1 and n_2 with their combined ranks (unless, of course, the obtained data contained the ranks initially). The ranks are assigned in such manner that rank 1 is given to the smallest of the $n = n_1 + n_2$ combined observations, rank 2 is given to the second smallest, and so on until rank n is given to the largest. If several values are tied, we assign each the average of the ranks that would otherwise have been assigned.

For convenience we must establish that whenever the two sample sizes are unequal, we will let n_1 represent the *smaller*-sized sample and n_2 the *larger*-sized sample. The Wilcoxon rank-sum test statistic T_1 is merely the sum of the ranks assigned to the n_1 observations in the smaller sample. (For equal-sized samples, either group may be selected for determining T_1.)

For any integer value n, the sum of the first n consecutive integers may easily be calculated as $n(n + 1)/2$. The test statistic T_1 plus the sum of the ranks assigned to the n_2 items in the second sample, T_2, must therefore be equal to this value; that is,

$$T_1 + T_2 = \frac{n(n + 1)}{2} \qquad \textbf{(14.6)}$$

so that Equation (14.6) can serve as a check on the ranking procedure.

The test of the null hypothesis may be one-tailed or two-tailed:

Two-tailed test	One-tailed test	One-tailed test
H_0: $M_1 = M_2$ H_1: $M_1 \neq M_2$	H_0: $M_1 \geq M_2$ H_1: $M_1 < M_2$	H_0: $M_1 \leq M_2$ H_1: $M_1 > M_2$

where M_1 = population median for the first group, having n_1 sample observations
M_2 = population median for the second group, having n_2 sample observations

When both samples n_1 and n_2 are $\leq$ 10, Table C.10 may be used to obtain the critical values of the test statistic T_1 for both one- and two-tailed tests at various levels of significance. For a two-tailed test and for a particular level of significance, if the computed value of T_1 equals or exceeds the upper critical value or is less than or equal to the lower critical value, the null hypothesis may be rejected. For one-tailed tests having the alternative H_1: $M_1 > M_2$, the decision rule is to reject the null hypothesis if the observed value of T_1 equals or exceeds the upper critical value, while for one-tailed tests having the alternative H_1: $M_1 < M_2$ the decision rule is to reject the null hypothesis if the observed value of T_1 is less than or equal to the lower critical value.

For large sample sizes the test statistic T_1 is approximately normally distributed. The following large-sample approximation formula may be used for testing the null hypothesis when sample sizes are outside the range of Table C.10:

$$Z = \frac{T_1 - \mu_{T_1}}{\sigma_{T_1}} \tag{14.7}$$

where T_1 is the sum of the ranks assigned to the n_1 observations in the first sample
μ_{T_1} is the mean value of T_1; $\mu_{T_1} = n_1(n + 1)/2$
σ_{T_1} is the standard deviation of T_1; $\sigma_{T_1} = \sqrt{n_1 n_2(n + 1)/12}$
and $n_1 \leq n_2$

That is,

$$Z = \frac{T_1 - \dfrac{n_1(n + 1)}{2}}{\sqrt{\dfrac{n_1 n_2(n + 1)}{12}}} \tag{14.8}$$

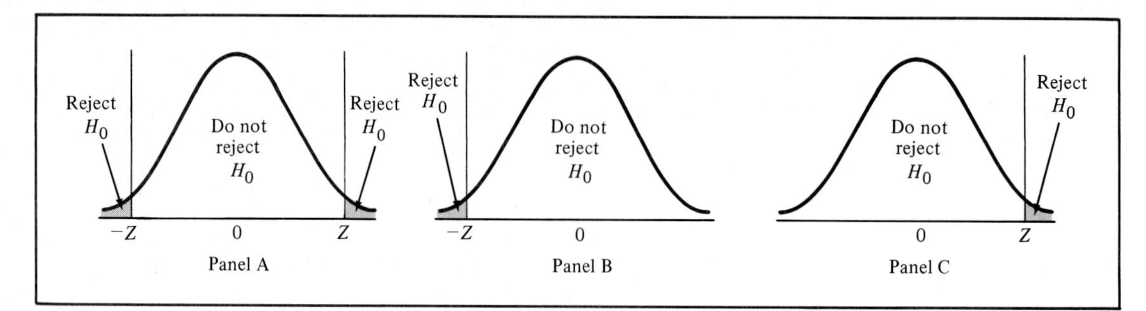

FIGURE 14.3 □ Determining the rejection region. Panel A, two-tailed test, $M_1 \neq M_2$. Panel B, one-tailed test, $M_1 < M_2$. Panel C, one-tailed test, $M_1 > M_2$.

Based on the level of significance selected, the null hypothesis may be rejected if the computed Z value falls in the appropriate region of rejection, depending on whether a two-tailed or a one-tailed test is used (see Figure 14.3).

14.6.2 Application

As an example using the Wilcoxon rank-sum test, suppose we examine the third question posed by the account executive: Is there evidence of a difference in the combined product-characteristic ratings of adult versus high-school-student participants subjected to advertisement E? That is, do adults and high-school students perceive the product's quality differently?

The pertinent data, extracted from Tables 14.1 and 14.2, are displayed in Table 14.7.

Since the account executive is not specifying which of the two groups is likely

TABLE 14.7 □ A comparison of combined product-characteristic ratings for 6 adults and 8 high-school students subjected to advertisement E

Adults	Students
15	24
14	19
23	21
16	13
20	17
23	19
	22
	18

SOURCE: Data are extracted from Tables 14.1 and 14.2.

to give higher product-characteristic ratings, the test is two-tailed and the following null and alternative hypotheses are established:

H_0: $M_1 = M_2$ (the median product characteristic ratings are equal)
H_1: $M_1 \neq M_2$ (the median product characteristic ratings are different)

To perform the Wilcoxon rank-sum test we form the combined ranking of the ratings obtained from the $n_1 = 6$ adults with the ratings obtained from the $n_2 = 8$ high-school students as in Table 14.8.

TABLE 14.8 □ Forming the combined ranks

Adults		High-school students	
Rating	**Rank**	**Rating**	**Rank**
15	3.0	24	14.0
14	2.0	19	7.5
23	12.5	21	10.0
16	4.0	13	1.0
20	9.0	17	5.0
23	12.5	19	7.5
		22	11.0
		18	6.0

SOURCE: Data are extracted from Tables 14.1 and 14.2.

We then obtain the test statistic T_1, the sum of the ranks assigned to the smaller sample:

$$T_1 = 3 + 2 + 12.5 + 4 + 9 + 12.5 = 43$$

As a check on the ranking procedure we also obtain T_2 and use Equation (14.6) to show that the sum of the first $n = 14$ integers is equal to $T_1 + T_2$:

$$T_1 + T_2 = \frac{n(n+1)}{2}$$

$$43 + 62 = \frac{14(15)}{2}$$

$$105 = 105$$

Had one or both of the samples been greater than 10, the large-sample approximation formula [Equation (14.8)] would have been used for testing. Here, since both n_1 and $n_2 \leq 10$, Table C.10 is employed to obtain the critical values for the statistic T_1. A copy of this table is presented as Table 14.9.

TABLE 14.9 □ Obtaining the lower and upper tail critical values T_1 for the Wilcoxon rank-sum test where $n_1 = 6$, $n_2 = 8$, and $\alpha = .05$

	α		n_1						
n_2	One-tailed	Two-tailed	4	5	6	7	8	9	10
4	.05	.10	11,25						
	.025	.05	10,26						
	.01	.02							
	.005	.01							
5	.05	.10	12,28	19,36					
	.025	.05	11,29	17,38					
	.01	.02	10,30	16,39					
	.005	.01		15,40					
6	.05	.10	13,31	20,40	28,50				
	.025	.05	12,32	18,42	26,52				
	.01	.02	11,33	17,43	24,54				
	.005	.01	10,34	16,44	23,55				
7	.05	.10	14,34	21,44	29,55	39,66			
	.025	.05	13,35	20,45	27,57	36,69			
	.01	.02	11,37	18,47	25,59	34,71			
	.005	.01	10,38	16,49	24,60	32,73			
8	.05	.10	15,37	23,47	31,59	41,71	51,85		
	.025	.05	14,38	21,49	29,61	38,74	49,87		
	.01	.02	12,40	19,51	27,63	35,77	45,91		
	.005	.01	11,41	17,53	25,65	34,78	43,93		

SOURCE: Extracted from Table C.10.

Since $n_1 = 6$ and $n_2 = 8$, it is observed that (at the .05 level of significance) the lower and upper critical values for the two-tailed test are, respectively, 29 and 61. Since the observed value of the test statistic $T_1 = 43$ falls between these critical values, the null hypothesis cannot be rejected. It may therefore be said that there is no evidence of any difference between the median product-characteristic ratings given by the two groups exposed to advertisement E.

Problems

• 14.19 An epidemiologist at a large hospital wished to determine whether cholesterol levels were significantly higher among women who had used birth control pills versus those who did not. A random sample of $n = 12$ women under age 30 who had used birth control pills for at least two consecutive years was selected from the hospital records along with a random sample of $n = 12$ women under age 30 who had not used birth control pills. The following data were obtained:

Users:	245 260 287 241 220 203 240 230 317 285 265 208
Nonusers:	220 253 263 200 190 210 274 217 185 230 215 222

Using α = .05, is there evidence that cholesterol levels are significantly higher among the users of birth control pills?

14.20 A security analyst wishes to compare the dividend yields of stocks traded on the American Stock Exchange with those traded on the New York Stock Exchange. (Yield is the dividend per share divided by the price per share.) Random samples of eight issues from the American Stock Exchange and ten issues from the New York Stock Exchange are selected with results as follows:

American Stock Exchange:	5.1 1.4 1.6 5.7 9.7 9.1 11.2 8.2
New York Stock Exchange:	2.8 7.3 9.8 7.0 9.5 5.5 5.6 10.8 4.7 6.5

Using an α level of .05, is there evidence of a significant difference in the median dividend yields for stocks traded on the American Stock Exchange versus those traded on the New York Stock Exchange?

• 14.21 Use the techniques of this chapter to see whether the acceleration of German and Japanese cars (Problem 3.75 on page 73) differ. How do the results compare to those for Problem 10.32 on page 291? (α = .10)

14.22 A New York television station decided to do a story comparing two commuter railroads in the area—the Long Island Railroad (LIRR) and New Jersey Transit (NJT). The researchers at the station sampled the performance of several scheduled train runs of each railroad. The data on how many minutes early (negative numbers) or late (positive numbers) each train was are presented below:

LIRR:	5	−1	39	9	12	21	15	52	18	23		
NJT:	8	4	10	4	12	5	4	9	15	33	14	7

Is there evidence that the railroads differ in their median tendencies to be late? (α = .01.) What conclusions about the lateness of the two railroads can be made?

14.23 For the data on television show ratings in Problem 14.18 on page 441, pretend that there was no pairing, and analyze the data using the methods of this section. Discuss any differences in the outcomes of the two analyses. (α = .05)

14.24 To investigate the optimal meal to eat before running a marathon, marathoners were randomly divided into two groups which ate either spaghetti or steak and eggs on the morning of the marathon. The times (in hours and minutes) to run the marathon for the members of each group were:

Spaghetti:	3:45 5:12 3:21 2:41 4:49 4:01 3:28 3:12 4:51
Steak and Eggs:	4:32 2:53 3:44 3:12 4:44 3:18 4:59 8:43 3:52 4:28 4:01

Is there any evidence that the meal eaten before running makes a difference in the time it takes to run a marathon? (α = .10)

14.7 KRUSKAL-WALLIS TEST FOR c INDEPENDENT SAMPLES

The **Kruskal-Wallis test for *c* independent samples** (where $c > 2$) may be considered an extension of the Wilcoxon rank sum test for two independent samples discussed in Section 14.6. Thus the Kruskal-Wallis test enjoys the same power properties relative to the analysis-of-variance F test (Chapter 12) as does the Wilcoxon rank-sum test relative to the t test for two independent samples (Section 10.8). That is, the Kruskal-Wallis procedure has proven to be almost as powerful as the F test under conditions appropriate to the latter and even more powerful than the classical procedure when its assumptions (see Section 12.6) are violated.

14.7.1 Procedure

The Kruskal-Wallis procedure is most often used to test whether c independent sample groups have been drawn from populations possessing equal medians. That is,

$$H_0\text{: } M_1 = M_2 = M_3 = \cdots = M_c$$
$$H_1\text{: Not all } M_j\text{'s are equal (where } j = 1, 2, \ldots, c)$$

For such situations, it is necessary to assume that:

1. The c samples are randomly and independently drawn from their respective populations.
2. The underlying random phenomenon of interest is continuous (to avoid ties).
3. The observed data constitute at least an ordinal scale of measurement, both within and between the c samples.
4. The c populations have the same *spread*.
5. The c populations have the same *shape*.

Interestingly, the Kruskal-Wallis procedure still makes less stringent assumptions than does the F test. To employ the Kruskal-Wallis procedure, the measurements need only be ordinal over all sample groups, and the common population distributions need only be continuous—their common shapes are irrelevant. On the other hand, to utilize the classical F test the level of measurement must be more sophisticated, and we must assume that the c samples are coming from underlying normal populations having equal variances.

The Kruskal-Wallis test is very simple to use. To perform the test we must first (if necessary) replace the observations in the c samples with their combined ranks such that rank 1 is given to the smallest of the combined observations and rank n to the largest of the combined observations (where $n = n_1 + n_2 + \cdots + n_c$). If any values are tied, they are assigned the average of the ranks they would otherwise have been assigned if ties had not been present in the data.

The Kruskal-Wallis test statistic H may be computed from

$$H = \left[\frac{12}{n(n+1)} \sum_j \frac{T_j^2}{n_j}\right] - 3(n+1) \qquad \textbf{(14.9)}$$

where n is the total number of observations over the combined samples, i.e., $n = n_1 + n_2 + \cdots + n_c$
n_j is the number of observations in the jth sample; $j = 1, 2, \ldots, c$
T_j^2 is the square of the sum of the ranks assigned to the jth sample

Part of this formula may look familiar: $\sum_j (T_j^2/n_j)$ is very much like part of the formula for computing the between-groups sum of squares in the analysis of variance. In fact, if we were to compute the between sum of squares using the ranks instead of the raw data, and then multiply by the constant $12/\{n(n+1)\}$, we would get H. H is, therefore, a measure of discrepancy among the mean ranks of the groups, rather than a measure of discrepancy among the mean scores of the groups as in ANOVA.

As the sample sizes in each group get large (greater than 5), the test statistic H may be approximated by the χ^2 distribution with $c - 1$ degrees of freedom. Thus for any selected level of significance α, the decision rule would be to reject the null hypothesis if the computed value of H exceeds the critical χ^2 value and not to reject the null hypothesis if H is less than or equal to the critical χ^2 value (see Figure 14.4). The critical χ^2 values are given in Table C.4.

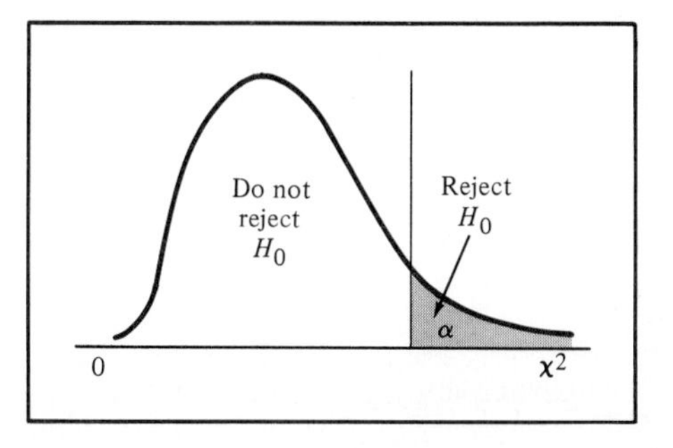

FIGURE 14.4 □ Determining the rejection region.

14.7.2 Application

As an application of the Kruskal-Wallis procedure, suppose that we refer to the fourth (and perhaps most important) question posed by the account executive: Is there evidence of a difference in the (median) combined ratings given by the adults based on exposure to different advertisements? That is, do product perceptions differ for various levels of expectations induced through advertising?

TABLE 14.10 □ Combined product-quality ratings based on five advertisements

Advertisements				
A	B	C	D	E
23	19	12	7	15
17	28	9	13	14
24	22	17	15	23
23	20	17	10	16
27	21	8	11	20
25	26	18	6	23

SOURCE: Table 14.1.

The answer to this question will have a direct influence on the selection of the particular advertisement to be used for the product. Recall that advertisements A and B tended to undersell the pen's characteristics, advertisements C and D tended to oversell or exaggerate the pen's characteristics, and advertisement E attempted to correctly state the pen's characteristics. The client (i.e., the pen manufacturer) has a real fear that if an advertisement is too implausible and creates an unrealistic product image, the public will be disappointed and not repurchase the item after seeing it and "trying it out." The pertinent data, extracted from Table 14.1, are displayed in Table 14.10.

The null hypothesis to be tested is that the median combined product-quality ratings are equal regardless of the advertisement read. The alternative is that at least two of the advertisements lead to median product-quality ratings which differ. Thus,

$$H_0\text{: } M_A = M_B = M_C = M_D = M_E$$
$$H_1\text{: Not all the medians are equal}$$

Converting the 30 ratings to ranks, we obtain Table 14.11.

We note that in the combined ranking, the last participant exposed to advertisement

TABLE 14.11 □ Converting data to ranks

Advertisements				
A	B	C	D	E
23.5	17.0	7.0	2.0	10.5
14.0	30.0	4.0	8.0	9.0
26.0	21.0	14.0	10.5	23.5
23.5	18.5	14.0	5.0	12.0
29.0	20.0	3.0	6.0	18.5
27.0	28.0	16.0	1.0	23.5

D had the lowest score, 6, and received a rank of 1; the first participant in this group had the second lowest score, 7, and received a rank of 2; and so on. We also note, for example, that the third participant exposed to advertisement D and the first participant exposed to advertisement E both had scores of 15 and received a rank of 10.5, since they were tied for the tenth and eleventh lowest scores.

After all the ranks are assigned, we then obtain the sum of the ranks for each group:

$$T_A = 143.0, \quad T_B = 134.5, \quad T_C = 58.0, \quad T_D = 32.5, \quad T_E = 97.0$$

As a check on the rankings we have:

$$T_A + T_B + T_C + T_D + T_E = \frac{n(n+1)}{2}$$

$$143.0 + 134.5 + 58.0 + 32.5 + 97.0 = \frac{30(31)}{2}$$

$$465 = 465$$

Now Equation (14.9) is employed to test the null hypothesis:

$$H = \left[\frac{12}{n(n+1)} \sum_j \frac{T_j^2}{n_j}\right] - 3(n+1)$$

$$= \frac{12}{30(31)}\left[\frac{(143.0)^2}{6} + \frac{(134.5)^2}{6} + \frac{(58.0)^2}{6} + \frac{(32.5)^2}{6} + \frac{(97.0)^2}{6}\right] - 3(31)$$

$$= \frac{12}{930}\left[\frac{52{,}368.5}{6}\right] - 93$$

$$= 112.62 - 93 = 19.62$$

Using Table C.4, the critical χ^2 value having $c - 1 = 4$ degrees of freedom and corresponding to a .05 level of significance is 9.488. Since the computed value of the test statistic H exceeds this critical value, we may reject the null hypothesis and conclude that there is evidence of a difference in the median product-quality ratings induced through exposure to different advertisements. That is, product perceptions significantly differ for various levels of expectations induced through advertising.

Although further analyses are outside the scope of this text (see Reference 4), it is clear from the rank totals T_j in each of the five groups that the highest combined product-quality ratings were given by participants exposed to advertisements A and B—both of which tended to *undersell* the pen's true characteristics. The lowest ratings were given by participants exposed to advertisements C and D—both of which tended to *exaggerate* the pen's characteristics. Hence the account executive may recommend to the pen manufacturer that advertisements A and B be considered for actual use when marketing the product.

Problems

• 14.25 To examine the effects of the work environment on attitude toward work, an industrial psychologist randomly assigned a group of 18 recently hired sales trainees to three "home rooms"—six trainees per room. The rooms were identical except for wall color. One was light green, another light blue, and the third deep red.

During the week-long training program, the trainees stayed mainly in their respective "home rooms." At the end of the program, an attitude scale was used to measure each trainee's attitude toward work. The following data were obtained:

Room color		
Light green	**Light blue**	**Deep red**
46	59	34
51	54	29
48	47	43
42	55	40
58	49	45
50	44	34

A low score indicates a poor attitude, a high score a good attitude. Using a level of significance of .05, is there evidence that work environment has an effect on attitude toward work? If so, which environments seem best and worst?

14.26 The following data represent the percentage of fat content for samples of eight 1-pound packages of frankfurters for each of five brands tested by a consumers' interest group:

Brand of frankfurters				
A	**B**	**C**	**D**	**E**
26	25	31	33	21
19	23	20	36	29
31	28	25	30	20
28	18	31	35	30
22	27	23	32	28
20	26	30	39	26
27	20	22	34	25
24	24	29	37	23

Using a level of significance of .05, is there evidence of a difference in the percentage of fat content for these five brands? If so, which brands would you avoid? Which would you buy?

14.27 Suppose that in the study described in Problem 14.24 on page 447 to determine the best breakfast to eat before running a marathon, a third group was included which ate no solid food, but drank a quart of a specially prepared formula which contained electrolytes normally lost during heavy exercise. The times run by this group are presented below:

2:54 4:12 3:19 3:42 4:01 3:33 2:59

Test the hypothesis that the times of the three groups do not differ significantly. (α = .10)

14.28 The percentage increase in the price of stocks of companies in four different industry groups over a ten-year period were:

Industry	Percentage increase in stock prices						
A	213	1040	321	112	421		
B	155	240	121	290	174	226	
C	490	381	402	648			
D	280	178	310	249	281	353	399

Is there evidence that the prices of stocks in the industries have increased at different rates? (α = .05)

• 14.29 Farmers were randomly assigned to test one of five varieties of cotton plant. The yields for each farmer (in pounds per acre) were:

Variety	Yield (pounds per acre)						
I	421	501	458	521	489	533	
J	486	452	429	438	461		
K	504	568	533	489	522		
L	432	467	401	455	463	503	499
M	487	512	469	531	514		

Is there evidence that the variety of plant made a difference in yield? (α = .01) If so, which varieties seem best and worst?

14.8 SPEARMAN'S RANK CORRELATION PROCEDURE

Among the various statistical methods based on ranks, the **Spearman rank correlation** procedure was the earliest to be developed. For more than three-quarters of a century this procedure has continued to be widely used for studying the association between two variables—primarily because of its simplicity and its power. That is, we shall observe that the Spearman rank correlation procedure is simple to use and quick to apply. Moreover, it has proven to be almost as powerful as its classical counterpart—the **Pearson (product-moment) correlation** method described in Chapter 13—under conditions favorable to the latter and even more powerful than the parametric method when its assumptions are violated.

14.8.1 Procedure

Let us suppose that we obtain a sample of n subjects, each measured on two variables, X and Y. To investigate the degree of possible association between X and Y we must focus on how the values of X "relate" to the values of Y. A *direct* or *positive association* is said to exist if subjects possessing high values of X also possess high values of Y while subjects having low values of X also contain low values of Y. At the other extreme, an *indirect* or *negative association* is said to exist if subjects possessing high values of X also contain low values of Y, and vice versa.

To study the amount or degree of association between the two variables X and Y, we may compute r_S, the **Spearman coefficient of rank correlation,** provided that in our data we have attained at least an ordinal level of measurement for each variable.

The Spearman coefficient of rank correlation may be obtained using the following five steps:

1. Replace the n values of X by their ranks R_X by giving the rank of 1 to the smallest X and the rank of n to the largest. If two or more X values are tied, they are each assigned the average rank of the rank positions they otherwise would have been assigned individually had ties not occurred.
2. Replace the n values of Y by their ranks R_Y as in step 1.
3. For each of the n subjects, obtain a set of **rank difference scores**

$$d = R_X - R_Y$$

4. Obtain $\Sigma\, d^2$, the sum of each of the squared rank difference scores.
5. The Spearman coefficient of rank correlation, r_S, is given by the following formula:

$$r_S = 1 - \frac{6\,\Sigma\, d^2}{n(n^2 - 1)} \qquad \textbf{(14.10)}$$

The Spearman rank correlation is equivalent to computing the usual Pearson correlation on the ranks instead of the raw data. Equation (14.10) is a simplified computational formula, but Equation (13.13) on page 386 would give the same result.

The rank correlation coefficient is a relative measure which varies from -1 (a perfect negative relationship between X and Y) to $+1$ (a perfect positive relationship between X and Y). The closer r_S is to these extremes, the *stronger* is the relationship between X and Y. The closer r_S is to 0, the *weaker* is the relationship between X and Y. If X and Y are independent of each other, there is no relationship and thus the coefficient of rank correlation is 0.

To evaluate whether or not a significant relationship between X and Y exists, we may test the null hypothesis of independence against the alternative that there is a relationship. The test may either be one-tailed or two-tailed. For a two-tailed test, the null and alternative hypotheses are:

H_0: The two variables X and Y are independent ($\rho_S = 0$)

H_1: There is either a positive or negative association between X and Y ($\rho_S \neq 0$)

On the other hand, if the researcher is specifically interested in a particular type of relationship, a one-tailed test would be needed. For a test of positive correlation between X and Y the hypotheses are:

H_0: There is no positive association between X and Y ($\rho_S \leq 0$)

H_1: There is a positive association between X and Y ($\rho_S > 0$)

while for a test of negative correlation between X and Y the hypotheses are:

H_0: There is no negative association between X and Y ($\rho_S \geq 0$)

H_1: There is a negative association between X and Y ($\rho_S < 0$)

To test the null hypothesis the following large-sample approximation formula may be used (provided that the sample size n is not very small):

$$Z = r_s \sqrt{n - 1} \tag{14.11}$$

and, based on the level of significance selected, the null hypothesis may be rejected if the computed Z value falls in the appropriate region of rejection, depending on whether a two-tailed or one-tailed test is employed (see Figure 14.5).

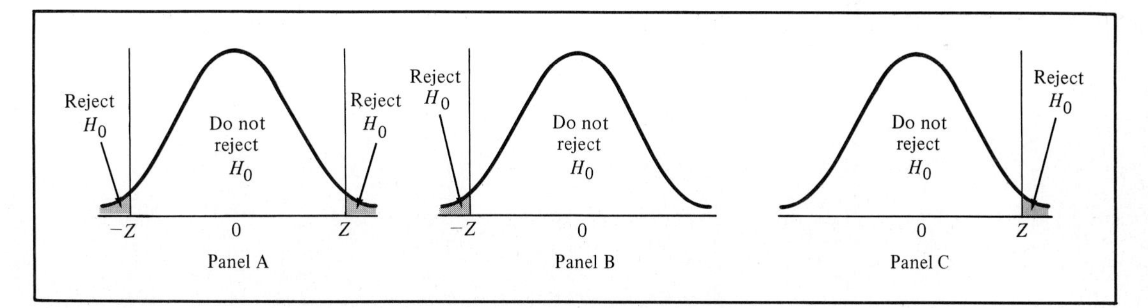

FIGURE 14.5 □ Determining the rejection region. Panel A, two-tailed test. Panel B, one-tailed test, negative association. Panel C, one-tailed test, positive association.

14.8.2 Application

To demonstrate the use of the Spearman rank correlation procedure, suppose we refer to the fifth and final question posed by the account executive: Is there evidence of a significant positive relationship between combined product-characteristic ratings and stated selling price? That is, does the public perceive a direct correlation between the quality and price of such a product as ballpoint pens?

The pertinent data were displayed in Table 14.1 (page 424). Since the direction of the alternative hypothesis has been specified (that is, the account executive is interested in determining whether there is evidence of a significant positive association), the test is one-tailed.

The following null and alternative hypotheses may be stated:

H_0: $\rho_S \leq 0$ [There is no positive association]

H_1: $\rho_S > 0$ [There is a positive association (*one-tailed*)]

The five-step procedure to obtain the Spearman coefficient of rank correlation r_S is developed in Table 14.12 using the data extracted from Table 14.1.

TABLE 14.12 □ Determining Spearman's coefficient of rank correlation

Subject	Combined rating, X	Rank, R_X	Stated selling price, $, Y	Rank, R_Y	$d = R_X - R_Y$	d^2
1	19	17.0	1.00	17.0	0.0	0.00
2	15	10.5	.49	8.5	2.0	4.00
3	28	30.0	2.49	30.0	0.0	0.00
4	14	9.0	.49	8.5	0.5	0.25
5	23	23.5	1.69	26.0	−2.5	6.25
6	7	2.0	.29	3.0	−1.0	1.00
7	12	7.0	.50	10.0	−3.0	9.00
8	17	14.0	.85	15.0	−1.0	1.00
9	9	4.0	.35	4.5	−0.5	0.25
10	17	14.0	.59	11.0	3.0	9.00
11	13	8.0	.45	7.0	1.0	1.00
12	17	14.0	.65	12.0	2.0	4.00
13	8	3.0	.25	2.0	1.0	1.00
14	15	10.5	.69	13.0	−2.5	6.25
15	23	23.5	1.50	24.0	−0.5	0.25
16	16	12.0	.79	14.0	−2.0	4.00
17	10	5.0	.39	6.0	−1.0	1.00
18	20	18.5	1.19	19.0	−0.5	0.25
19	11	6.0	.35	4.5	1.5	2.25
20	23	23.5	1.49	23.0	0.5	0.25
21	22	21.0	1.25	20.0	1.0	1.00
22	24	26.0	1.59	25.0	1.0	1.00
23	23	23.5	1.39	22.0	1.5	2.25
24	6	1.0	.19	1.0	0.0	0.00
25	18	16.0	.99	16.0	0.0	0.00
26	20	18.5	1.15	18.0	0.5	0.25
27	21	20.0	1.29	21.0	−1.0	1.00
28	27	29.0	2.29	29.0	0.0	0.00
29	25	27.0	1.99	28.0	−1.0	1.00
30	26	28.0	1.89	27.0	1.0	1.00
						58.50

From this table we observe that $\Sigma\, d^2 = 58.5$ and, using Equation (14.10),

$$
\begin{aligned}
r_S &= 1 - \frac{6 \Sigma d^2}{n(n^2 - 1)} \\
&= 1 - \frac{6(58.5)}{30(900 - 1)} \\
&= 1 - .013 \\
&= .987
\end{aligned}
$$

To test the null hypothesis we have:

$$
\begin{aligned}
Z &= r_S \sqrt{n - 1} \\
&= (.987)\sqrt{30 - 1} \\
&= 5.32
\end{aligned}
$$

Using a level of significance of .05, the one-tailed test has a critical Z value of $+1.645$. Since Z exceeds this critical value, the null hypothesis may be rejected. The account executive may conclude that there is evidence of a highly significant positive correlation between the perceived quality of the product and what one would expect to pay for it.

Problems

14.30 An educational psychologist wishes to study whether or not a significant positive relationship exists between verbal facility (as measured by the Peabody test) and geometrical intuition (as measured by the Raven test). A sample of 15 ten-year-old girls yields the following results:

Subject	Verbal, X	Geometrical, Y
Alice	67	24
Clare	84	37
Dora	77	21
Ethel	85	38
Kathy	89	40
Kiki	72	32
Laurie	93	36
Lila	65	26
Lola	78	25
Lori	88	39
Merri	80	31
Nina	79	27
Patsy	75	29
Sandi	81	22
Terri	82	30

At the .05 level of significance, is there evidence that a positive relationship between the two variables exists?

• 14.31 Given the following set of "social-status striving scores" (X) and "conformity ratings" (Y) for a sample of $n = 12$ male college sophomores pledging for various fraternities on campus:

	Student											
	1	2	3	4	5	6	7	8	9	10	11	12
Social status (X)	49	65	38	50	72	83	79	95	71	66	68	62
Conformity (Y)	13	16	15	21	22	25	24	23	19	18	20	17

Using $\alpha = .05$, is there evidence of a significant positive association between social status striving and conformity among male college sophomores pledging for fraternities?

14.32 Compute the Spearman correlation coefficient and the ordinary Pearson correlation coefficient (see Section 13.7) for the following data, and tell why they give different results:

X:	1	2	3	4	5	6
Y:	1	8	27	64	125	216

✓14.33 For the data on the ratings of television shows (Problem 14.18 on page 441) compute the Spearman rank order correlation. Use this result to explain why there can be different answers to Problems such as 14.18 and 14.23.

14.34 A chemical reaction is timed at several different temperatures; the results are:

Temperature (°F):	100	150	200	250	300	350
Time needed (seconds):	843	211	164	69	22	17

Without doing any calculations, tell what the Spearman correlation is between the variables.

• 14.35 For the data on quality and price of computer software (Problem 13.2 on page 373) compute the rank-order correlation. How similar is this to the Pearson correlation you computed for those data?

14.9 SUMMARY

In this chapter we have provided but a brief introduction to the very broad subject of nonparametric methods of hypothesis testing. The need for such methods arose for the following three reasons:

1. The measurements attained on behavioral science data were often only qualitative (nominal scale) or in ranks (ordinal scale), and procedures were required to make appropriate tests.

2. The researcher often had no knowledge of the form of the population from which the sample data were drawn and believed that the assumptions for the classical procedure were unrealistic for a given endeavor.
3. The researcher was not always interested in testing hypotheses about particular population parameters and required methods for treating problems regarding randomness, trend, symmetry, and goodness of fit.

For these reasons a set of nonparametric procedures was devised. Six such nonparametric methods were presented here. Such methods may be broadly described as those which: make fewer and less stringent assumptions than do their parametric counterparts; are generally easier to apply and quicker to compute than their classical counterparts; and yield a more general set of conclusions than do their parametric counterparts. In addition, nonparametric methods are applicable to very small sample sizes (from which exact probabilities may be computed) and also lend themselves readily to preliminary or pilot studies from which more detailed and sophisticated analyses can later be undertaken. Finally, for those situations in which the researcher has a choice, it has been found that some of the nonparametric techniques are either as powerful or almost as powerful in their ability to detect statistically significant differences (when they do exist) as their classical counterparts under conditions appropriate to the latter, and may actually be much more powerful than the classical procedures when their assumptions do not hold.

On the other hand, especially with larger sample sizes, when the conditions of the classical methods are met it would be considered wasteful of information to merely convert sophisticated levels of measurement into ranks (ordinal scale) or into categorized groupings (nominal scale) just so that a nonparametric procedure could be employed. Thus, in such circumstances as these, the classical procedures should be utilized.

Supplementary Problems

14.36 Starting at the beginning of the table of random numbers (Table C.1), record the sequence of high digits (5, 6, 7, 8, 9) and low digits (0, 1, 2, 3, 4) for the first 100 digits observed. Can this resulting sequence of high and low digits be considered random? ($\alpha = .05$.)

14.37 During the period from 1960 to 1985 there was an increase in the federal budget outlays for veterans' benefits and services. During this period, however, total federal outlays for all functions also increased. The data in the accompanying table present the percentage of total federal outlays for veterans' benefits and services during the 26-year period from 1960 to 1985. With respect to fluctuations above and below the median, is there any evidence of trend over the 26-year period? ($\alpha = .025$.)

Percentage of total federal outlays to veterans' benefits and services

Year	Percentage
1960	5.9
1961	5.8
1962	5.3
1963	5.0
1964	4.8
1965	4.8
1966	4.4
1967	4.4
1968	3.9
1969	4.1
1970	4.4
1971	4.6
1972	4.6
1973	4.9
1974	5.0
1975	5.1
1976	5.0
1977	4.5
1978	4.2
1979	4.0
1980	3.7
1981	3.5
1982	3.3
1983	3.1
1984	3.0
1985 (estimated)	2.8

SOURCE: Data are extracted from Tables 491 and 540, *Statistical Abstract of the United States*, U.S. Department of Commerce, 1986.

14.38 A cigarette manufacturer claims that the tar content of a new brand of cigarettes is 17 milligrams. A random sample of 24 cigarettes is selected and the tar content measured. The results are shown below in milligrams:

16.9	16.6	17.3	17.5	17.0	17.2	16.1	16.4	17.3	15.9	17.7	18.3
15.6	16.8	17.1	17.2	16.4	18.1	17.4	16.7	16.9	16.0	16.5	17.8

Is there evidence that the (median) tar content of this new brand is different from 17 milligrams? (α = .01.)

• 14.39 An actuary of a particular insurance company wants to examine the records of larceny claims filed by persons insured under a household goods policy. In the past the median claim was for $85. A random sample of 18 claims is taken and the results are as follows:

$140	$92	$35	$202	$80	$87	$80	$100	$47
$25	$160	$68	$50	$65	$310	$90	$75	$120

Is there evidence that the median claim significantly increased? (α = .05.)

14.40 A statistics professor taught two special sections of a basic course in which the 10 students in each section were considered outstanding. She used a ''traditional'' method of instruction (T) in one section and an ''experimental'' method (E) in the other. At the end of the semester she ranked the students based on their performance from 1 (worst) to 20 (best).

T:	1	2	3	5	9	10	12	13	14	15
E:	4	6	7	8	11	16	17	18	19	20

For this instructor was there any evidence of a difference in performance based on the two methods? ($\alpha = .05$.)

14.41 A consumer protection agency wishes to perform a ''life test'' to investigate the differences in the median time to failure of electronic calculators supplied with AAA alkaline batteries operated under normal-temperature versus high-temperature conditions. A total of 16 electronic calculators of the same type are randomly assigned to the two groups so that 8 are examined under normal temperature and 8 are examined under high temperature. All the calculators are started simultaneously, and the following data are the times to failure (in hours):

Normal temperature:	9.3	12.7	12.9	14.9	16.1	16.3	17.9	23.1
High temperature:	6.0	8.2	8.8	11.9	12.3	14.7	14.8	15.6

Is there evidence that calculators operating under high temperature have a significantly shorter life than those operating under normal temperature? ($\alpha = .05$.)

14.42 An industrial psychologist desires to test whether the reaction times of assembly-line workers are equivalent under three different learning methods. From a group of 25 new employees, nine are randomly assigned to method A, eight to method B, and eight to method C. The data below represent the rankings from 1 (fastest) to 25 (slowest) of the reaction times to complete a task given by the industrial psychologist after the learning period.

Method		
A	**B**	**C**
2	1	5
3	6	7
4	8	11
9	15	12
10	16	13
14	17	18
19	21	24
20	22	25
23		

Is there evidence of a difference in the reaction times for these three learning methods? ($\alpha = .01$.)

• 14.43 The director of personnel of a large company would like to determine whether two personnel interviewers have been evaluating job applicants in a similar manner. It is important for the company to have consistency (high correlation) between its interviewers, so that job placement is independent of the interviewer. Two interviewers were asked to rank a set of 14 applicants in order of preference, with rank 1 given to the most preferred applicant and rank 14 to the least preferred applicant. The results were as follows:

	Interviewer	
Applicant	***X***	**Y**
A	3	2
B	5	4
C	1	3
D	2	1
E	7	5
F	4	7
G	6	8
H	8	6
I	11	11
J	12	12
K	10	9
L	9	10
M	13	13
N	14	14

Measure the rank correlation between the two interviewers. Is there evidence of a significant *positive* relationship? ($\alpha = .05$.)

14.44 The Quantitative Methods Society of a business school wanted to study the preferences of statistics majors and computer majors for different academic subjects. A ranking by the students in each of the two majors is presented in the accompanying table. Measure the correlation in the preferences of statistics and computer majors. Is there evidence of a significant *positive* relationship? ($\alpha = .05$.)

Subject	Statistics majors	Computer majors
Accounting	6	8
Mathematics	2	5
Statistics	1	3
Computers	3	1
Psychology	8	6
Finance	5	7
Marketing	9	10
Management	10	4
Law	11	11
English	13	12
Political science	12	13
History	14	14
Biology	7	9
Economics	4	2

14.45 A psychologist wishes to construct a true-false examination containing 30 questions. She desires the sequence of correct responses to be random so that nobody taking the examination can merely guess at the answers by following a particular pattern. The following sequence is obtained:

F T T T F T F F F T T F T F F F T T T T F F T T T F F T F F

Can this sequence be considered random? ($\alpha = .05$.)

14.46 A machine being used for packaging cereals has been set so that on the average 13 ounces of cereal will be packaged per box. The quality-control engineer wishes to test the machine setting and selects a sample of 30 consecutive cereal packages filled during the production process. Their weights are recorded below in row sequence (from left to right):

13.2	13.3	13.1	13.7	13.3	13.0	13.1	12.3	12.6	12.5
13.0	13.2	13.4	13.6	13.7	13.4	13.3	12.9	12.8	12.6
12.3	12.4	13.5	13.4	13.2	13.5	13.6	13.1	13.3	13.1

Do these data indicate a lack of randomness in the sequence of underfills and overfills, or can the production process be considered as "in control"? ($\alpha = .05$.)

14.47 In Problem 14.46 a machine used for packaging cereals has been set so that on the average 13 ounces of cereal will be packaged per box. Using the data of Problem 14.46, the weights from a sample of 30 cereal packages are obtained. Is there evidence that the (median) weight per box is different from 13 ounces? ($\alpha = .05$.) [*Note*: To test this claim we must assume that the observed sequence is random. Therefore, let us assume that the data were collected in column sequence (from top to bottom) rather than in row sequence (from left to right).]

14.48 A women's organization wishes to know whether starting salaries of men and women entering business differ. A random sample of 10 male seniors with B+ indexes and a random sample of 9 female seniors with B+ indexes are selected, and each person is permitted a total of three interviews with companies in marketing, public relations, or advertising. The following data are the reported "best offers" made to the individuals:

Male		**Female**	
$21,000	$22,200	$21,200	$17,000
$18,800	$21,500	$17,250	$19,750
$24,000	$17,500	$20,100	$18,250
$20,000	$19,500	$16,000	$19,000
$21,000	$22,000	$18,500	

Is there any evidence that women are not as well paid as men when their qualifications are similar? ($\alpha = .05$.)

A batch of 20 AAA carbon-zinc batteries are randomly assigned to four groups (so that there are five batteries per group.) Each group of batteries is then subjected to a particular temperature level—low, normal, high, and very high. The batteries are simultaneously tested under these temperatures, and the times to failure (in hours) are recorded below:

Low	**Normal**	**High**	**Very high**
8.0	7.6	6.0	5.1
8.1	8.2	6.3	5.6
9.2	9.8	7.1	5.9
9.4	10.9	7.7	6.7
11.7	12.3	8.9	7.8

Do the four temperature levels yield the same median battery lives? ($\alpha = .05$.)

14.50 Using the data on lawnmower sales (Problem 10.30 on page 286), is there evidence that the median is greater than 5? (Use $\alpha = .05$.) Compare your results to those obtained previously with the t test.

14.51 Using the data on travel times (Problem 10.70 on page 304), is there evidence that the median travel time of employees is less than 60 minutes? (Use $\alpha = .01$.) Compare your results to those obtained previously with the t test.

14.52 Using the data on processing times (Problem 10.44 on page 298), is there evidence that the old computer system has a greater median processing time than the new computer system? (Use $\alpha = .01$.) Compare your results to those obtained previously with the t test.

14.53 Using the data on calculator prices (Problem 10.76 on page 305), is there evidence that the unknown brand has a lower median price? (Use $\alpha = .01$.) Compare your results to those obtained previously with the t test.

14.54 Using the data on monthly rental prices of unfurnished studio apartments (Problem 3.68 on page 70), is there evidence that the median rent is higher in Manhattan than in Brooklyn? (Use $\alpha = .05$.) Compare your results to those obtained previously with the t test in Problem 10.31, page 290.

14.55 Using the data on college tuition (Problem 3.76 on page 73, is there any evidence of a difference in the median tuition rates of colleges in the Northeast versus Midwest? (Use $\alpha = .05$.) Compare your results to those obtained previously with the t test in Problem 10.33, page 291.

14.56 Using the data from Problem 12.23 on page 358, is there evidence of a difference in the median sales among the three aisle locations? (Use $\alpha = .01$.) Compare your results to those obtained previously with the F test.

14.57 Using the data from Problem 12.8 on page 350, is there evidence of a difference in the median life among the four brands of transistor batteries? (Use $\alpha = .05$.) Compare your results to those obtained previously with the F test.

14.58 Using the data from Problem 13.73 on page 416, is there evidence of a significant rank correlation between the price of club soda and seltzer? (Use $\alpha = .01$.) Compare your results to those obtained previously with the t test.

14.59 Using the data from Problem 13.74 on page 417, is there evidence of a significant rank correlation between price and amount of protein? (Use $\alpha = .10$,) Compare your results to those obtained previously with the t test.

Database Exercises

The following problems refer to the sample data obtained from the questionnaire of Figure 2.5 and presented in Figure 2.10. They should be solved with the aid of a computer package. Use $\alpha = .05$ throughout.

□ **14.60** Suppose that the questionnaires were returned by the sample of alumni in the order displayed in the database. Using the following variables, see whether the order appears to be random with respect to these variables:

(a) Years since graduation (question 1).
(b) Whether graduate school was attended (question 5).
(c) Salary (question 7).

(**d**) Grade point average (question 15).
(**e**) Gender (question 18).
(**f**) Age (question 20).

□ **14.61** Is there evidence that the median salary of alumni association members is greater than $40,000? (See question 7.)

□ **14.62** Is there evidence that the median age of alumni association members is greater than 35? (See question 20.)

□ **14.63** Is there evidence of a difference in the median salary of male versus female alumni association members? (See questions 7 and 18.)

□ **14.64** Is there evidence of a difference in the median number of years since graduation for male versus female alumni association members? (See questions 1 and 18.)

□ **14.65** Is there evidence of a difference in the median salary of alumni association members based on undergraduate major? (See questions 7 and 2.)

□ **14.66** Is there evidence of a difference in the median grade-point average of alumni association members based on undergraduate major? (See questions 15 and 2.)

□ **14.67** Is there evidence of a difference in the median age of alumni association members based on undergraduate major? (See questions 20 and 2.)

□ **14.68** Is there evidence of a difference in the median salary among alumni who work in different types of organizations? (See questions 7 and 8.)

□ **14.69** Is there evidence of a difference in the median age of alumni association members based on primary reason for attending the college? (See questions 20 and 3.)

□ **14.70** Is there evidence of a relationship between height and salary among the male members of the alumni association? (See questions 19, 7, and 18.)

□ **14.71** Is there evidence of a relationship between undergraduate grade-point average and salary for females? (See questions 15, 7, and 18.)

References

1. BRADLEY, J. V., *Distribution-Free Statistical Tests* (Englewood Cliffs, N.J.: Prentice-Hall, 1968).
2. CONOVER, W. J., *Practical Nonparametric Statistics*, 2d ed. (New York: Wiley, 1980).
3. DANIEL, W. W., *Applied Nonparametric Statistics* (Boston: Houghton Mifflin, 1978).
4. MARASCUILO, L., AND M. MCSWEENEY, *Nonparametric and Distribution-Free Methods for the Social Sciences* (Belmont, Calif.: Wadsworth, 1977).

Chapter 15

Using Computers for Statistical Analysis

15.1 INTRODUCTION: WHAT'S AHEAD

Statistical analysis can be made much simpler by the use of calculators and computers to do the arithmetic calculations involved. Calculators, of course, only do the calculations which we specify; computers can do much more. In this chapter we describe the general principles behind the operation of computers and the development of statistical software which assists us in data analysis and statistical problem solving.

15.2 THE ROLE OF THE COMPUTER

Most tools and machines are constructed to do only one task, and to do that task in only one way. A can opener opens cans, a hammer drives nails, and a radio translates electronic signals into sound. Some devices can be adjusted to the problem at hand: for example, the temperature of an oven can be set at any of a number of levels. But the oven is still meant to perform only one task: heat food.

A computer, on the other hand, is much more versatile. It can do a statistical analysis one minute, help predict the state of the economy the next, and then format a book (such as this one) and print it. The computer does this because its actions can be controlled by giving it a different set of instructions (called a **program**) to execute at different times. One person would have the computer run a statistics program, another a forecasting program, and a third a text formatting program.

All programs are written in one of a number of **languages** which the computer can "understand." These languages have been developed because human languages are (so far) too complicated for computers to interpret easily. Computer languages have a relatively small "vocabulary" and a simple, unambiguous **syntax** (set of rules for constructing statements in the language). To write a program, we must learn these aspects of the language.

Fortunately, we do not always need to learn how to program in order to make effective use of a computer. Once a program has been written to do a task (such as to compute the mean and variance of a set of numbers), anyone can use that program to perform the task without having to understand exactly how the program was written or what the individual statements in it mean. A **package** is a set of such programs, written for a common purpose (such as statistical analysis), and constructed to make it easy for people to use (even if they don't know how to write such a program). In later sections of this chapter we will see how three widely used computer packages (SPSS^X, Minitab, and STATGRAPHICS) are used to do statistical analyses. In the section that follows, however, we shall focus on how a simple computer program is developed. Similar (but more complex) programs form components of statistical computer packages.

★ 15.3 WRITING A COMPUTER PROGRAM (OPTIONAL SECTION)

Although we do not need to know a computer language in order to do simple analyses using computer packages, for illustrative purposes we will demonstrate a simple program written in one version of a language called FORTRAN (which stands for FORmula TRANslation). This language, written primarily for scientific work, is today the most widely used language for this purpose. Examining this program will show (1) what a program written in a common programming language looks like, and (2) that it is much harder to write a program than to use a program which someone else has written.

The program[1] displayed in Figure 15.1 reads a series of numbers, counts how

★ Note that this is an optional section that may be skipped without any loss of continuity.

[1] In order to keep the exposition as simple as possible we have left off a few instructions, and have added one which doesn't really exist.

```
$JOB
       N = 0
       S = 0
    1  READ,X
       N = N + 1
       S = S + X
       IF(END-OF-DATA) GO TO 2
       GO TO 1
    2  AVG = S/N
       PRINT,N,AVG
       STOP
       END
$ENTRY
    5
   12
    8
    3.5
    3
    9
$STOP
```

FIGURE 15.1 □ A FORTRAN Program to Compute the Mean

many numbers have been read, calculates the mean, and prints the mean and number of observations.

Every program is read from top to bottom, unless an instruction in the program says to do otherwise. Here, the first instruction creates a variable called N and sets its value to zero. The second instruction creates a variable called S, also started (**initialized** in computer terminology) at zero.

Next, the computer is told to read a number, and to call that number X. Where is the number read from? All data to be read are placed after the $ENTRY statement. Every time the computer performs a READ instruction, it reads from the next line in the list after the $ENTRY statement. So, the first time this instruction is performed (**executed**), the variable X is given the value 5, since that is the first number in the list.

The following statement adds 1 to the variable N. Here, N is used to keep track of how many numbers we have read; each time we read a number, we add 1 to the total number read. Note that this is not interpreted in the usual sense of an algebraic equation, but that the right-hand side describes a calculation which is to be performed, and the left-hand side names that variable which will contain the result of the calculation.

The statement $S = S + X$ adds X to the number called S. The purpose here is to obtain the sum of all the numbers we read. Since one number at a time is read, the total eventually gets accumulated and is called S.

The next instruction checks to see if we have finished reading all the data in our list. If we have, then we go on to the statement which has the number 2 in front of it. If not, then we execute the next statement in the program, which says to go to the statement with the number 1 in front of it (the READ statement). By using the IF statement, we can keep going back to read new data as long as we need to, but we can then go on afterward. The set of statements which keep being repeated as long as we have data is called a **loop.**

After we have read all the data in our list, we have the total number of observations stored as a variable called N, and the sum of all the observations stored as a variable called S. Now we calculate the average, called AVG, by dividing the sum S by the number of observations N. Last, we print the results.

Many beginners find it difficult to follow programs. It will be easier if, every time a variable is first mentioned, we draw a box with the variable name on top of the box. We then put in the box whatever number is indicated. For example, the first two statements in the above program would lead us to do the following:

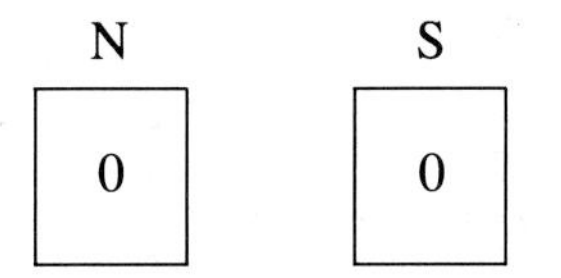

Each time the value of a variable is changed by a statement in the program, we do the calculation indicated and put the new value in the correct box. For example, when the READ statement is executed the first time, a variable X is created, and we should now have

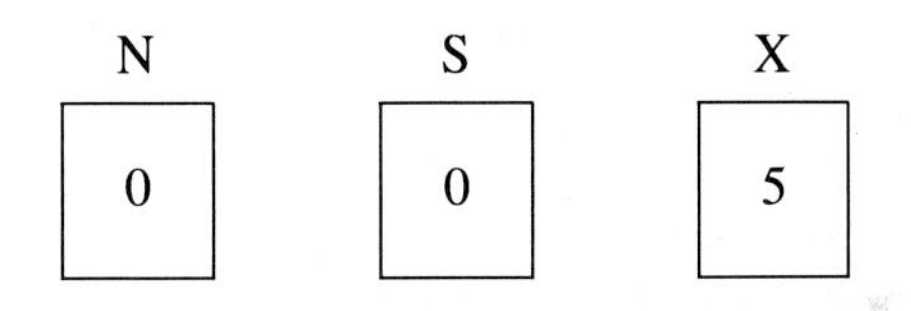

When the statement N = N + 1 is executed, it says to take whatever was in the box marked N, add 1 to it, and place the result in the box marked N. So now we have

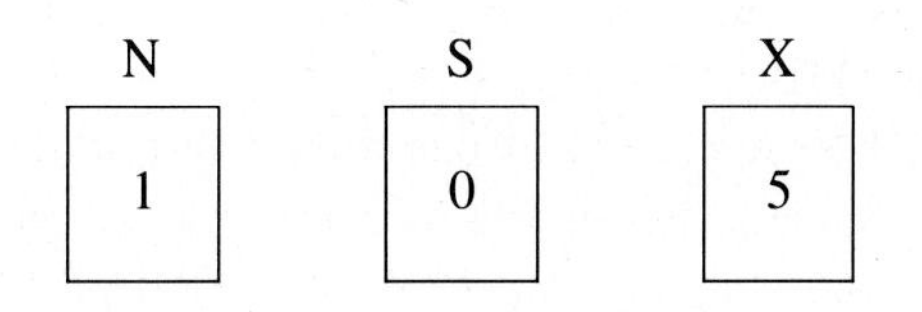

The statement S = S + X says to take the value in S, add to it the value in X, and put the result in the box marked S:

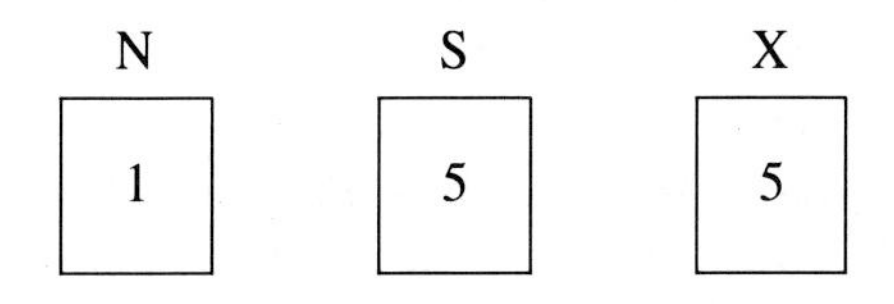

After the last data entry is made, our boxes appear as follows:

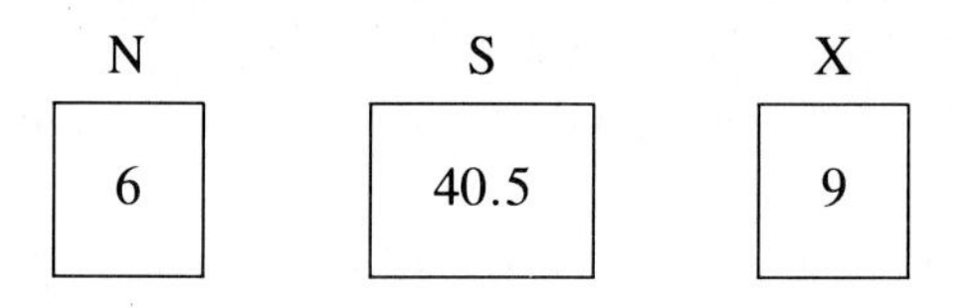

By following the whole program through in this way, it will be easier to understand the logic of the program.

All of this may seem like a lot of work to calculate the average. Writing one or two such programs makes us appreciate statistical packages, where programming is generally unnecessary. With such packages, we merely describe the statistics we want computed, and we do not have to tell the computer how to do the calculations.

15.4 INTRODUCTION TO STATISTICAL PACKAGES

The increasing use of computers to assist in the management and analysis of information in all fields has stimulated the demand for various types of software. Statistical packages allow relatively unskilled users to benefit from utilizing computers as a tool in data analysis and statistical problem solving. The standardized nature of a statistical package insures that the results of the computerized analyses will be reliable and that the time spent on implementing the statistical programs will be minimized.

While some of the more broad-based statistical packages have been used for more than twenty years on mainframe and minicomputer systems, over the past decade numerous statistical packages have been developed in conjunction with the rapid growth in the use of personal computers. Some of these statistical packages were designed specifically for special research interests such as medicine and the health sciences, reliability and quality-control engineering, and economic/financial forecasting. Others were designed to accommodate broader or more general aspects of data analysis and statistical problem solving. In this chapter, three of these more broad-based statistical packages will be introduced—SPSS^X, Minitab, and STATGRAPHICS.

The main thrust of the remaining sections of this chapter is to briefly familiarize the reader with the use of statistical packages as aids in data analysis and statistical problem solving. The three statistical packages selected for this demonstration are among the more widely used commercially available packages that have been developed to date. Among others, SPSS^X, Minitab, and STATGRAPHICS are packages used by researchers in colleges and universities, in scientific laboratories, in medical facilities, in governmental institutions, and in business organizations.

Depending on the type of computer system available and statistical package we select, our data can be processed in either **batch** or **interactive mode.**

Batch processing originated with the use of large mainframe computer systems. It refers to gathering one or more computer jobs for processing by the computer system at the same time. In our case, a **job** is a program containing one or more instructions to be used for analyzing a given set of data. With batch processing, we cannot review the output from the job until it is entirely finished executing, and we cannot modify either any instructions or the data while the whole job is executing.

In contrast, with **interactive processing** we can give instructions one at a time, see the results, and then give further instructions. In this way, we can decide what the next instruction should be on the basis of the results we have already seen. Interactive processing may be performed either on large mainframe computer systems or on personal microcomputer systems.

Statistical packages differ in the method used to give the instructions about the analyses to be performed. In a **command-driven** package, statements are constructed by the user to indicate what action should be taken. In a **menu-driven** package, the user is presented with a list of actions from which he or she can choose. While a command-driven package requires a knowledge of its particular syntax in order to run a program, a menu-driven package requires only a knowledge of what keystrokes to make (i.e., ''buttons to push''). Therefore, the menu-driven packages are generally much easier to use. Nevertheless, using either type of package is far more efficient than writing the entire computer program itself.

Among the three statistical packages to be demonstrated here, SPSS^{X} and Minitab are command-driven, Minitab may be employed in either the batch or interactive processing of data, while STATGRAPHICS in menu-driven and is used interactively.

THE ALUMNI ASSOCIATION PRESIDENT'S STUDY REVISITED

In the sections that follow we shall use the data from the results of the alumni association president's study described in Chapter 2 (see Case A, page 13) to demonstrate the value of employing statistical packages to assist in data analysis. In that study the president of the alumni association of a well-known college had hired a statistician to help develop a questionnaire and conduct a survey to obtain information on the past achievements, attitudes, and interests as well as current activities of its membership. The questionnaire is depicted in Figure 2.5 (pages 15 and 16), and the survey responses for 200 members of the alumni association are displayed in Figure 2.10 (pages 28 to 31).

In Figure 2.10 each row of data represents the responses by a particular person to each of the questions which comprise the survey. Each set of adjacent columns **(field)** represents the responses to a question (values of the variable) for all respondents. For computer usage, each of the questions (i.e., variables of interest) has been given a name **(code** or **variable label)** containing no more than eight characters. These code labels are also indicated in Figure 2.8, which depicts the responses to the survey by Jackie Daniels (see page 26), the first alumni association member selected in the sample.

In any large-scale data analysis, a variety of statistical procedures will be performed on the same set of data at different times. It is, therefore, convenient to permanently store a set of data as a **data file** so that it may easily be accessed whenever the researcher wishes to use it. Hence, a data file for all 200 alumni association respondents (Figure 2.10) would be created and stored on disk for a mainframe computer system or stored on either a floppy diskette or a hard disk drive for a personal microcomputer system.

In the remaining sections of this chapter we will demonstrate how SPSS^X, Minitab, and STATGRAPHICS can be used to assist in the analysis of several questions of interest to the alumni association president. For each question to be studied, the specific command or instruction needed to invoke the appropriate statistical procedure will be described and the output generated by the command or instruction will be interpreted.

15.5 USING SPSS^X

SPSS stands for Statistical Package for the Social Sciences (see Reference 3). There are several versions of the SPSS package; the one we will demonstrate here, SPSS^X, is a very complete package designed to be used on medium- to large-size computers (i.e., minicomputers and mainframe computers, respectively).[2]

To illustrate some of the capabilities of SPSS^X, we shall seek answers to the following four questions posed by the president of the alumni association:

1. What does the alumni association membership profile look like with respect to gender, choice of major, and most valuable subject taken?
2. What does the alumni association membership profile look like with respect to age, height, grade-point index, and salary?
3. Is there evidence of a relationship between gender and choice of major?
4. Is there evidence of a relationship between grade-point index and salary so that the former could be used for predicting the latter?

To answer these questions we run the SPSS^X program depicted in Figure 15.2.

There are two main segments of any set of instructions for doing a statistical analysis using any package. One defines the data to be analyzed; the other specifies the procedures to be used. Each segment can have one or more parts, with some parts being optional and others depending on the particular analysis to be done. The SPSS^X commands necessary for defining the data set are listed in the top portion of the program in Figure 15.2, whereas the SPSS^X commands needed for specifying particular statistical procedures are listed in the bottom portion. The first part of each command is the name of the command (such as FILE HANDLE, DATA LIST, VARIABLE LABELS, VALUE LABELS, FREQUENCIES, etc.).

[2] Another version, SPSS-PC, is an interactive package designed to be used on personal microcomputers. The syntax of SPSS-PC is very similar to SPSS^X.

```
FILE HANDLE   ALUMNI   NAME='[DMR]ALUMNI.DAT'
DATA LIST     FILE=ALUMNI /
    YEARS 1-2 MAJOR 4  SALARY 14-16  SUBJCAR 34-35
    GPI 40-42 (1)  SEX 48 HEIGHT 50-51 AGE 53-54 RESPNO 56-58
VARIABLE LABELS
    MAJOR      'UNDERGRADUATE MAJOR AREA'
    SALARY     'ANNUAL SALARY IN THOUSANDS'
    SUBJCAR    'SUBJECT MOST VALUABLE TO CAREER'
    GPI        'UNDERGRADUATE GRADE POINT INDEX'
    HEIGHT     'HEIGHT IN INCHES'
    AGE        'AGE IN YEARS'
VALUE LABELS
    MAJOR      1 'BUSINESS' 2 'HUMANITY,SOC.SCI' 3 'SCI.MATH' /
    SUBJCAR
          1 'ACCOUNTING' 2 'ART' 3 'BLACK, HISPANIC' 4 'COMPUTERS'
          5 'ECON, FINANCE' 6 'EDUCATION' 7 'ENGLISH COMP' 8 'ENG LIT'
          9 'FOREIGN LANG' 10 'HISTORY' 11 'LAW' 12 'MANAGEMENT'
         13 'MARKETING' 14 'MATH' 15 'MUSIC' 16 'NATURAL SCIENCE'
         17 'PHIL, RELIGION' 18 'PHYS ED' 19 'POL SCIENCE'
         20 'PSYCHOLOGY' 21 'PUBL ADMIN' 22 'SECRETARIAL'
         23 'SOC-ANTHRO' 24 'SPEECH' 25 'STATISTICS' 26 'OTHER' /
    SEX    1 'MALE' 2 'FEMALE'   /
 /*  TO ANSWER QUESTION 1:
FREQUENCIES  VARIABLES = SEX, MAJOR, SUBJCAR / HISTOGRAM /
 /*  TO ANSWER QUESTION 2:
CONDESCRIPTIVE  AGE HEIGHT GPI SALARY
STATISTICS   ALL
 /*  TO ANSWER QUESTION 3:
CROSSTABS    TABLES = SEX BY MAJOR
OPTIONS  3  14           /* PRINTS ROW PERCENTAGES, EXPECTED FREQS
STATISTICS  1            /* DO A CHI-SQUARE TEST OF INDEPENDENCE
 /*  TO ANSWER QUESTION 4:
REGRESSION   DESCRIPTIVES = DEFAULTS /
             VARIABLES = SALARY GPI /
             DEPENDENT = SALARY /
             ENTER GPI /
```

FIGURE 15.2 □ SPSS^X Commands for Alumni Association Study

Following the command name are the **parameters** and **keywords** which give the necessary details pertaining to the command.

The first command, FILE HANDLE, tells SPSS^X that we will be reading the data from a file stored permanently on the computer. The "handle" of the file (ALUMNI) is a short name by which we will refer to the file within our command statements. The name by which the computer knows the file is "[DMR]ALUMNI. DAT".

The DATA LIST command tells SPSS^X that it will find the data in a file with a name ("handle") of ALUMNI. In addition, the values of a variable called YEARS will be found in columns 1 through 2, the values of MAJOR will be found in column 4, and so on. (For simplicity, we do not include the specifications for every variable in the data set, but only the ones we analyze here.)

The DATA LIST command is necessary to specify exactly which data are to be used, and the names of the variables. The next two commands are not strictly necessary, but are valuable for *documenting* the nature of the data. The VARIABLE LABELS command allows us to attach longer labels to the name of each variable, which makes it easier for us to remember (and for others who might look at our program to figure out) what we meant by the name. The VALUE LABELS command is used for qualitative variables to indicate what each category represents. For example, if we use the numbers 1 and 2 to represent the gender of participants in the survey, does a 1 mean male or female?

This completes the specification of the characteristics of the data which will be analyzed. Next is the specification of the analyses to be performed. First we want to obtain a summary table for some of the qualitative variables (here, SEX, MAJOR, and SUBJCAR). We have also asked to see bar charts (HISTOGRAMs) for each of these variables. Before specifying each analysis, we have included a comment (line starting with /*) to remind ourselves of the purpose of each analysis.

The CONDESCRIPTIVE command is meant to obtain certain DESCRIPTIVE statistics of CONtinuous variables. For these variables, we would want information on the mean, variance, standard deviation, and perhaps other characteristics of our quantitative variables. The next subcommand (STATISTICS) tells which of these should be calculated and printed. By using the keyword ALL, we have asked for all of the descriptive statistics available to be printed.

The CROSSTABS command requests a cross tabulation of the variables SEX and MAJOR. The OPTIONS and STATISTICS subcommands which follow indicate that the (1) row percentages (2) expected (theoretical) frequencies, and (3) chi-square test of independence should be reported. The comments after the /* in the commands are not necessary but help remind us what we have done when we don't have a manual to check. Unless we use these procedures quite frequently, we would be unlikely to remember that for the CROSSTABS command statistic 1 is the chi-square test, or that option 14 requests that the expected frequencies be printed.

The last procedure requested is a regression. We have specified that the **default** descriptive statistics be printed. The default is what will happen if we don't specify otherwise. By using the default in a computer program, we are letting the person who wrote the program decide what will happen. While this often may be what we would choose to do, we should always check the documentation of computer programs we use to make sure that we agree with the choices which are being made for us.

The variables included in this analysis are SALARY and GPI. We would like to see if grade-point index in school predicts salary later in life. We specify that SALARY is the DEPENDENT variable in the next line. Then we ENTER GPI into the regression equation. This means that GPI will be put into the equation as a predictor (independent) variable. This completes the specification of the regression.

Excerpts from the output from SPSS^X are displayed in Figures 15.3 through 15.6. First, in Figure 15.3, the summary tables and bar charts are presented. We can see that 143 of the 200 people in the study were male, and that this is 71.5%

of the sample. For major areas, 118 people (almost 3/5) chose business, about 30% chose humanities and social sciences, and about 1/9 chose sciences and mathematics. About 2/3 of the sample believe that accounting courses were the most useful in their career. A large percentage of the remaining people indicated business-related courses as most important (finance, economics, management, marketing).

A potential danger of using defaults is seen in the bar chart of SUBJCAR. Because of the large number of categories (26) of this variable, SPSS^X has combined some of them together. Fortunately, because not all categories were used by respondents,

```
FREQUENCIES  VARIABLES = SEX, MAJOR, SUBJCAR / HISTOGRAM /

SEX
                                                               VALID       CUM
  VALUE LABEL                  VALUE  FREQUENCY  PERCENT  PERCENT  PERCENT

MALE                               1        143     71.5      71.5      71.5
FEMALE                             2         57     28.5      28.5     100.0
                                        -------  -------   -------
                               TOTAL        200    100.0     100.0

    COUNT      VALUE    ONE SYMBOL EQUALS APPROXIMATELY  4.00 OCCURRENCES

      143      1.00   ************************************
       57      2.00   **************
                     I.........I.........I.........I.........I.........I
                     0        40        80       120       160       200
                                   HISTOGRAM FREQUENCY

VALID CASES      200       MISSING CASES      0

- - - - - - - - - - - - - - - - - - - - - - - - - - - - - - - - - - - -

MAJOR      UNDERGRADUATE MAJOR AREA
                                                               VALID       CUM
  VALUE LABEL                  VALUE  FREQUENCY  PERCENT  PERCENT  PERCENT

BUSINESS                           1        118     59.0      59.0      59.0
HUMANITY,SOC.SCI                   2         59     29.5      29.5      88.5
SCI.MATH                           3         23     11.5      11.5     100.0
                                        -------  -------   -------
                               TOTAL        200    100.0     100.0

    COUNT      VALUE    ONE SYMBOL EQUALS APPROXIMATELY  4.00 OCCURRENCES

      118      1.00   ******************************
       59      2.00   ***************
       23      3.00   ******
                     I.........I.........I.........I.........I.........I
                     0        40        80       120       160       200
                                   HISTOGRAM FREQUENCY

VALID CASES      200       MISSING CASES      0
```

FIGURE 15.3 □ **SPSS^X Output for FREQUENCIES Command**

SUBJCAR SUBJECT MOST VALUABLE TO CAREER

VALUE LABEL	VALUE	FREQUENCY	PERCENT	VALID PERCENT	CUM PERCENT
ACCOUNTING	1	134	67.0	67.0	67.0
ECON, FINANCE	5	12	6.0	6.0	73.0
ENGLISH COMP	7	6	3.0	3.0	76.0
FOREIGN LANG	9	1	.5	.5	76.5
HISTORY	10	1	.5	.5	77.0
LAW	11	1	.5	.5	77.5
MANAGEMENT	12	13	6.5	6.5	84.0
MARKETING	13	16	8.0	8.0	92.0
PSYCHOLOGY	20	5	2.5	2.5	94.5
PUBL ADMIN	21	1	.5	.5	95.0
SECRETARIAL	22	1	.5	.5	95.5
STATISTICS	25	5	2.5	2.5	98.0
OTHER	26	4	2.0	2.0	100.0
	TOTAL	200	100.0	100.0	

```
   COUNT   MIDPOINT   ONE SYMBOL EQUALS APPROXIMATELY  4.00 OCCURRENCES

       0      -1.5
       0        .0
     134       1.5   **********************************
       0       3.0
      12       4.5   ***
       0       6.0
       6       7.5   **
       1       9.0
       2      10.5   *
      13      12.0   ***
      16      13.5   ****
       0      15.0
       0      16.5
       0      18.0
       5      19.5   *
       1      21.0
       1      22.5
       0      24.0
       9      25.5   **
       0      27.0
       0      28.5
                    I....+....I....+....I....+....I....+....I....+....I
                    0        40        80       120       160       200
                              HISTOGRAM FREQUENCY

VALID CASES      200        MISSING CASES      0
```

FIGURE 15.3 □ ***(continued)***

the effect is not too bad here. Nevertheless, for all three of the qualitative variables studied, the tables seem to be much easier to read and interpret than the corresponding bar charts.

Next, in Figure 15.4, are the descriptive statistics summarizing the continuous variables AGE, HEIGHT, GPI, and SALARY. The mean age in the sample is

```
CONDESCRIPTIVE  AGE HEIGHT GPI SALARY
STATISTICS   ALL

NUMBER OF VALID OBSERVATIONS (LISTWISE) =       200.0C

VARIABLE  AGE        AGE IN YEARS

MEAN              48.115              S.E. MEAN          .917
STD DEV           12.969              VARIANCE        168.203
KURTOSIS          -1.223              S.E. KURT          .342
SKEWNESS            .020              S.E. SKEW          .172
RANGE             52.000              MINIMUM              25
MAXIMUM               77              SUM            9623.000

VALID OBSERVATIONS -       200       MISSING OBSERVATIONS -        0

- - - - - - - - - - - - - - - - - - - - - - - - - - - - - - - - - - -
VARIABLE  HEIGHT     HEIGHT IN INCHES

MEAN              68.210              S.E. MEAN          .262
STD DEV            3.709              VARIANCE         13.755
KURTOSIS           -.438              S.E. KURT          .342
SKEWNESS           -.330              S.E. SKEW          .172
RANGE             19.000              MINIMUM              58
MAXIMUM               77              SUM           13642.000

VALID OBSERVATIONS -       200       MISSING OBSERVATIONS -        0

- - - - - - - - - - - - - - - - - - - - - - - - - - - - - - - - - - -
VARIABLE  GPI        UNDERGRADUATE GRADE POINT INDEX

MEAN               2.838              S.E. MEAN          .027
STD DEV             .385              VARIANCE           .149
KURTOSIS           -.121              S.E. KURT          .342
SKEWNESS            .240              S.E. SKEW          .172
RANGE              1.900              MINIMUM             2.0
MAXIMUM              3.9              SUM             567.600

VALID OBSERVATIONS -       200       MISSING OBSERVATIONS -        0

- - - - - - - - - - - - - - - - - - - - - - - - - - - - - - - - - - -
VARIABLE  SALARY     ANNUAL SALARY IN THOUSANDS

MEAN              54.770              S.E. MEAN         1.910
STD DEV           27.015              VARIANCE        729.826
KURTOSIS           4.541              S.E. KURT          .342
SKEWNESS           1.458              S.E. SKEW          .172
RANGE            175.000              MINIMUM              10
MAXIMUM              185              SUM           10954.000

VALID OBSERVATIONS -       200       MISSING OBSERVATIONS -
```

FIGURE 15.4 □ SPSS^X Output for CONDESCRIPTIVE Command

about 48 years, with a standard deviation of almost 13. The youngest person in the sample is 25, the oldest is 77. We also got some useless information (such as the sum of the ages) by asking for all available statistics. In addition, we got information on some statistics which were not discussed in this book (such as kurtosis).

```
CROSSTABS     TABLES = SEX BY MAJOR
OPTIONS  3   14           /* PRINTS ROW PERCENTAGES, EXPECTED FREQS
STATISTICS   1            /* DO A CHI-SQUARE TEST OF INDEPENDENCE

 - - - - - - C R O S S T A B U L A T I O N   O F  - - - - - -
   SEX             BY  MAJOR     UNDERGRADUATE MAJOR AREA
- - - - - - - - - - - - - - - - - - - - - - - - - - - - - - - -

                    MAJOR
          COUNT   |
          EXP VAL |BUSINESS HUMANITY SCI.MATH   ROW
          ROW PCT |         ,SOC.SCI            TOTAL
                  |        1|       2|       3|
SEX       --------+--------+--------+--------+
                1 |   92   |   37   |   14   |   143
  MALE            |  84.4  |  42.2  |  16.4  |  71.5%
                  |  64.3% |  25.9% |   9.8% |
                  +--------+--------+--------+
                2 |   26   |   22   |    9   |    57
  FEMALE          |  33.6  |  16.8  |   6.6  |  28.5%
                  |  45.6% |  38.6% |  15.8% |
                  +--------+--------+--------+
           COLUMN    118       59       23       200
            TOTAL   59.0%    29.5%    11.5%    100.0%

 CHI-SQUARE    D.F.       SIGNIFICANCE       MIN E.F.      CELLS WITH E.F.< 5
 ----------    ----       ------------       --------      ------------------

   5.93273       2          0.0515            6.555           NONE

NUMBER OF MISSING OBSERVATIONS =       0
```

FIGURE 15.5 □ SPSS^X Output for CROSSTABS Command

From Figure 15.5, in the table cross-tabulating SEX and MAJOR, there are three numbers in each cell. The first is the number of people in that cell (for example, 92 males majoring in business); the second is the expected value (i.e., theoretical frequency) if the two variables are independent; and the third is the row proportion. For example, the table shows that almost 2/3 of the males (64.3%) majored in business, but less than half of the females (about 46%) did so. A larger proportion of females in this sample majored in the social sciences and humanities, and in the sciences and mathematics, than did males. (But notice that more males than females majored in these areas; we shouldn't confuse proportions with the actual numbers who majored in these areas.) The chi-square test for independence shows, however, that the relationship between the variables is not significant at the .05 level (but just barely—the p value is .0515). We therefore have no statistical evidence of a relationship between gender and choice of major in the population from which this sample was drawn.

The final analysis we requested was a regression of salary on grade-point index. The output in Figure 15.6 lists the descriptive statistics we requested (the defaults;

```
REGRESSION    DESCRIPTIVES = DEFAULTS /
              VARIABLES = SALARY GPI /
              DEPENDENT = SALARY /
              ENTER GPI /

      * * * *   M U L T I P L E   R E G R E S S I O N   * * * *

              Mean   Std Dev   Label

SALARY      54.770    27.015   ANNUAL SALARY IN THOUSANDS
GPI          2.838      .385   UNDERGRADUATE GRADE POINT INDEX

N of Cases =    200

Correlation:

             SALARY        GPI

SALARY        1.000       -.061
GPI           -.061       1.000

        * * * *   M U L T I P L E   R E G R E S S I O N   * * * *

Equation Number 1   Dependent Variable.. SALARY   ANNUAL SALARY IN THOUSANDS
Variable(s) Entered on Step Number  1..  GPI  UNDERGRADUATE GRADE POINT INDEX
Multiple R            .06111          R Square               .00373
Adjusted R Square   -.00130           Standard Error       27.03281

Analysis of Variance
                       DF       Sum of Squares       Mean Square
Regression              1           542.44431          542.44431
Residual              198        144692.97569          730.77260

F =     .74229      Signif F =   .3900

------------------ Variables in the Equation ------------------

Variable              B          SE B         Beta          T   Sig T

GPI           -4.282951      4.971148    -.061114       -.862   .3900
(Constant)    66.925015     14.237023                   4.701   .0000
```

FIGURE 15.6 □ SPSS^X Output for REGRESSION Command

here the mean and standard deviation). Next we see that the correlation between the two variables is $-.061$. This is not only very low, but if significant it would be in the counterintuitive direction: lower grade-point indices associated with higher salaries! The analysis-of-variance summary table shows that, indeed, knowledge of GPI does not help predict salary: the F value is .74, $t = -.862$, and $p = .39$.

15.6 USING Minitab

Minitab, a command-driven statistical package, was originally developed in 1972 as a student-oriented adaptation of the National Bureau of Standards' "Omnitab" system (Reference 2). Minitab was intended as a simple, multipurpose statistical package which would be useful to students majoring in various disciplines. It is available on a wide variety of computers, including microcomputers, minicomputers, and mainframes. Moreover, it can be used in either the batch or interactive modes.

To demonstrate the use of Minitab we shall seek answers to the following questions posed by the president of the alumni association:

1. What is the estimate of the average current annual salary for all full-time employed members of the alumni association?
2. Is there evidence that the true mean reported grade-point index of all the alumni association members exceeds 2.5?
3. Is there evidence of a difference in the true mean reported grade-point index for male versus female alumni association members?
4. Is there evidence of a difference in the average age of alumni association members based on their undergraduate major—business, humanities and social sciences, or science and mathematics?
5. Can the collected sample sequence of 200 survey responses (see the printout from Figure 2.10 on pages 28 to 31) be considered random with respect to the median age of 48.5 years?

To answer these questions, the Minitab program displayed in Figure 15.7 describes the commands needed to access the stored data file (here named BLRDATA), gives computer code labels (names) to the particular survey questions (i.e., variables) of interest, and uses the appropriate statistical procedures.

Minitab is a command-driven statistical package. We observe from Figure 15.7 that the commands are typed in free format across the length of a line. Descriptive information pertaining to the command (including commas and blank spaces) may be part of the command line. In addition, all Minitab commands have a brief form

```
READ DATA FROM FILE 'BLRDATA' INTO C1-C22
NOTE: NAMES ARE GIVEN TO VARIABLES OF INTEREST
NAME C2='MAJOR' C7='SALARY' C16='GPI' C19='SEX' C21='AGE'
NOTE: CATEGORICAL LABELS ARE GIVEN TO QUALITATIVE VARIABLES
NOTE: FOR 'MAJOR'  1=BUSINESS  2=HUMANITY,SOC.SCI  3=SCI.MATH
NOTE: FOR 'SEX'  1=MALE   2=FEMALE
NOTE: TO ANSWER QUESTION 1
TINTERVAL WITH 95 PERCENT CONFIDENCE FOR DATA IN 'SALARY'
NOTE: TO ANSWER QUESTION 2
TTEST OF MU = 2.5, DATA IN 'GPI';
ALTERNATIVE 1.
NOTE: TO ANSWER QUESTION 3
TWOT 95 PERCENT CONFIDENCE, DATA IN 'GPI' AND GROUPS IN 'SEX';
POOLED;
ALTERNATIVE 0.
NOTE: TO ANSWER QUESTION 4
ONEWAY, DATA IN 'AGE' AND LEVELS IN 'MAJOR'
NOTE: TO ANSWER QUESTION 5
RUNS ABOVE AND BELOW 48.5 FOR 'AGE'
STOP
```

FIGURE 15.7 □ Minitab Commands for Alumni Association Study

in which the first word of the command phrase may be abbreviated to four letters and all auxiliary words such as OF, THE, INTO, AND, WITH, and VERSUS can be deleted.

As examples, from the first line of Figure 15.7 the command

READ DATA FROM FILE 'BLRDATA' INTO C1-C22

could have been written more succinctly as

READ 'BLRDATA' C1-C22

while from the eighth line the command

TINTERVAL WITH 95 PERCENT CONFIDENCE FOR DATA IN 'SALARY'

could have been given as

TINT 95 'SALARY'

For clarity, however, this practice is not recommended.

Many of the Minitab commands have optional subcommands that may be applied in specific situations. As observed in lines 13, 14, and 15 in Figure 15.7, when a subcommand is to be used, the Minitab command line ends with a semicolon (;) and each subcommand follows on a separate line. A period (.) is included on the line that contains the last subcommand. If several subcommands are to be used, each subcommand line except the last ends with a semicolon.

Let us now take a look at the various Minitab commands. The READ command on the first line accesses the data file BLRDATA comprising the 200 responses to the alumni association membership survey (Figure 2.10 on pages 28–31). The questionnaire contained 21 parts and, in addition, each respondent was given a code number. Hence, the data file BLRDATA contained 22 fields of information (C1-C22).

Next, the NAME command on line 3 is used to provide names for the variables of interest. From the questionnaire (Figure 2.5 on pages 15–16) we see that the variable MAJOR was in the second of our 22 fields; thus C2='MAJOR' makes MAJOR the name of variable C2, and so on. We also observe that when using Minitab, an apostrophe (') is always placed around the name given to a variable (or to a data file).

To answer the president's first question requires a confidence-interval estimate for the mean. The TINTERVAL command is displayed in line eight. Following the command the only important **parameters** and **keywords** are 95 (the percent confidence desired) and SALARY (the name of the variable of interest). As previously mentioned, all other words are superflous to Minitab; they are given for sake of clarity.

A one-sample *t* test is needed to answer the president's second question. Hence, the TTEST command is presented on line 10. The important parameter here is the 2.5, the value for the hypothesized mean. The important keyword is GPI, the name of the variable of interest. Line 11 contains the subcommand ALTERNATIVE for this TTEST command. The value 1 specifies a one-sided test with the rejection region in the right tail.

Under the assumptions of normality and homoscedasticity, the pooled t test would be appropriate for answering the president's third question. Thus, the TWOT command is displayed on line 13. The 95 represents the desired percent confidence (so that 5% would represent the desired level of significance). Following this value, the name of the quantitative variable of interest is given along with the qualitative variable for which two categories are to be compared. The order here is important. In our case, the grade-point indices of male alumni are to be compared against those of female alumni. Two subcommands follow the TWOT command. The POOLED subcommand refers to the type of t test employed, while the value of 0 for the ALTERNATIVE subcommand specifies a two-tailed hypothesis test.

Provided that certain assumptions hold, a one-way analysis of variance would be appropriate for answering the president's fourth question. Hence, the ONEWAY command is presented on line 17. The structure of this command is similar to that of the previous TWOT command. The name of the quantitative variable of interest is given along with the qualitative variable for which several categories are to be compared. Again, the order here is important. In our case, the president wished to compare the ages of alumni based on their chosen major area of study. No subcommands are available for the ONEWAY command.

To answer the president's fifth question the nonparametric Wald-Wolfowitz one-sample runs test for randomness would be appropriate. The RUNS command is shown on line 19. Following the command, the only important parameters and keywords are the 48.5 (the median value above and below which we want to compare consecutive responses for randomness) and AGE (the particular variable of interest).

The STOP command is used to end a Minitab program.

For clarity in programming, it is always advisable to use comment statements. Such documentation permits the researcher to more easily review and interpret programs written at an earlier date. In Minitab, comment statements may begin with the word NOTE. All comments following the word NOTE on a particular line are not processed by Minitab. However, these comments are printed as part of the Minitab output and serve to enhance the clarity of the program. In Figure 15.7 some comment statements were employed to introduce the commands (and where necessary, subcommands) needed to answer the five questions posed in this section by the president of the alumni association.

Figures 15.8 to 15.12 on pages 483–484 represent the Minitab output which will assist in answering these five questions.

From Figure 15.8 we note that with 95% confidence the estimate for the true average annual salary of all full-time employed members of the alumni association is between \$51,000 and \$58,540.

From Figure 15.9 we observe that there is evidence that the true mean reported grade-point index of all the alumni association members exceeds 2.5. Using the t test with 199 degrees of freedom, the null hypothesis H_0: $\mu_X \leq 2.5$ is rejected in

```
MTB > NOTE: TO ANSWER QUESTION 1
MTB > TINTERVAL WITH 95 PERCENT CONFIDENCE FOR DATA IN 'SALARY'

                 N       MEAN     STDEV   SE MEAN   95.0 PERCENT C.I.
SALARY         200      54.77     27.02      1.91   (  51.00,    58.54)
```

FIGURE 15.8 □ Using the TINTERVAL Command in Minitab

```
MTB > NOTE: TO ANSWER QUESTION 2
MTB > TTEST OF MU = 2.5, DATA IN 'GPI';
SUBC> ALTERNATIVE 1.

TEST OF MU = 2.5000 VS MU G.T. 2.5000

                N      MEAN     STDEV   SE MEAN        T    P VALUE
GPI           200     2.838     0.385     0.027    12.40     0.0000
```

FIGURE 15.9 □ Using the TTEST Command in Minitab

favor of the alternative H_1: $\mu_X > 2.5$ because the computed t statistic is 12.40 and its corresponding one-tailed p value (which is rounded to .000 in the output) is less than .05.

From Figure 15.10 we see that there is no evidence of any real difference between the true mean reported grade-point indexes of male versus female alumni association members. Assuming normality and equal variability in the grade-point indexes of the male and female groups, the t test of the null hypothesis H_0: $\mu_F = \mu_M$ results in a t statistic of $-.59$ with 198 degrees of freedom. Since the corresponding two-tailed p value of .55 is greater than .05, the null hypothesis is not rejected.

From Figure 15.11 we note that there is evidence of a difference in the average age of alumni association members based on undergraduate major. That is, using a .05 level of significance, the null hypothesis of no differences in the average age

```
         MTB > NOTE: TO ANSWER QUESTION 3
         MTB > TWOT 95 PERCENT CONFIDENCE, DATA IN 'GPI' AND GROUPS IN 'SEX';
         SUBC> POOLED;
         SUBC> ALTERNATIVE 0.

         TWOSAMPLE T FOR GPI
         SEX     N      MEAN     STDEV   SE MEAN
Female = 2      57     2.812     0.400     0.053
  Male = 1     143     2.848     0.380     0.032

         95 PCT CI FOR MU 2 - MU 1: (-0.155, 0.083)
         TTEST MU 2 = MU 1 (VS NE): T=-0.59 P=0.55 DF=198.0
```

FIGURE 15.10 □ Using the TWOT Command in Minitab

```
MTB > NOTE: TO ANSWER QUESTION 4
MTB > ONEWAY, DATA IN 'AGE' AND LEVELS IN 'MAJOR'

ANALYSIS OF VARIANCE ON AGE
SOURCE      DF        SS        MS         F
MAJOR        2      2201      1101      6.93
ERROR      197     31271       159
TOTAL      199     33472
                                      INDIVIDUAL 95 PCT CI'S FOR MEAN
                                      BASED ON POOLED STDEV
      LEVEL        N      MEAN     STDEV  ----------+---------+---------+------
  Business = 1   118     50.69     13.17                      (-----*----)
Hum. & S.S. = 2   59     43.24     12.43  (-------*-------)
Sci. & Math = 3   23     47.39      9.56       (------------*------------)
                                          ----------+---------+---------+------
  POOLED STDEV =         12.60                    44.0      48.0      52.0   Age
```

FIGURE 15.11 □ Using the ONEWAY Command in Minitab

(H_0: $\mu_1 = \mu_2 = \mu_3$) based on the three types of undergraduate majors is rejected. $F = 6.93 > F_{2,197} \cong 3.00$—obtained from Table C.5. The attained significance level (i.e., the p value) is less than .005. Although Minitab does not permit a direct pairwise comparison between the groups (see the Scheffé procedure in Section 12.7), it is clear from the lack of overlap in the 95% confidence-interval estimates of the individual group means that alumni who had majored in business (level 1) are, on average, significantly older than alumni who had majored in the humanities and social sciences (level 2).

From Figure 15.12 we see that at the .05 level of significance there is no reason to reject the null hypothesis of randomness in the sequence of collected responses with respect to the median age of 48.5 years. The observed number of runs U is 107, the expected number of runs under the null hypothesis of randomness is 101.0, and the p value is .3952.

```
MTB > NOTE: TO ANSWER QUESTION 5
MTB > RUNS ABOVE AND BELOW 48.5 FOR 'AGE'

    AGE

    K =     48.5000


    THE OBSERVED NO. OF RUNS = 107 = U
    THE EXPECTED NO. OF RUNS = 101.0000 = μ_U
   100 OBSERVATIONS ABOVE K   100 BELOW
              THE TEST IS SIGNIFICANT AT  0.3952 = p value
              CANNOT REJECT AT ALPHA = 0.05
```

FIGURE 15.12 □ Using the RUNS Command in Minitab

15.7 USING STATGRAPHICS

In this section we shall discuss STATGRAPHICS (see Reference 4), a menu-driven[3] statistical package developed expressly for use on a microcomputer. STATGRAPHICS can be considered to be both a statistical package and a graphics package, since it includes a wide variety of statistical procedures and high-resolution graphic plots.

The user accesses the STATGRAPHICS package by entering the expression

STATGRAF

Once the system is accessed, the user is asked to provide information about his or her computer system. The system then creates a STATGRAPHICS data directory, which consists of all the variables that can be analyzed.

The first STATGRAPHICS menu, displayed in Figure 15.13 on page 486, is the "Main Menu" of 22 Sections, labeled A through V, that represent the various statistical procedures and data management features included in STATGRAPHICS. When selected, each choice on this Main Menu presents its own menu of selections. How do we choose one of the alternatives from the Main Menu?

The first time the Main Menu is displayed during a session the **cursor** (a highlighted character on the screen) is located at the first selection, Data Management. The cursor may be moved to other items on the menu by pressing various keys. Once the cursor is positioned at the desired choice, the Enter key (like a return key on a typewriter) is pressed, and STATGRAPHICS presents the menu corresponding to that choice.[4]

From Figure 15.13 (the Main Menu), we observe that some of the choices relate either to procedures that are beyond the scope of this text or to data management utilities that refer to how data may be entered, edited, and imported from or exported to other software packages. Since our purpose in this chapter is merely to illustrate the use of STATGRAPHICS rather than to cover all its aspects, we shall focus on obtaining results pertaining to the following questions posed by the president of the alumni association:

1. What are some characteristics which describe the alumni membership with respect to number of years since graduation, annual salary, and the number of different occupational areas in which the alumni have worked?
2. How can we graphically portray, through exploratory data analysis techniques, the number of years since graduation and the annual salary?
3. Is there evidence of a relationship in the cross-tabulation of the undergraduate major with graduate school attendance?
4. Is it useful to use number of years since graduation to predict annual salary?

[3] For more sophisticated users, STATGRAPHICS can also be used in a command mode by merely pressing the F8 function key on a standard IBM keyboard.

[4] We may return to the previous step by pressing the Escape key.

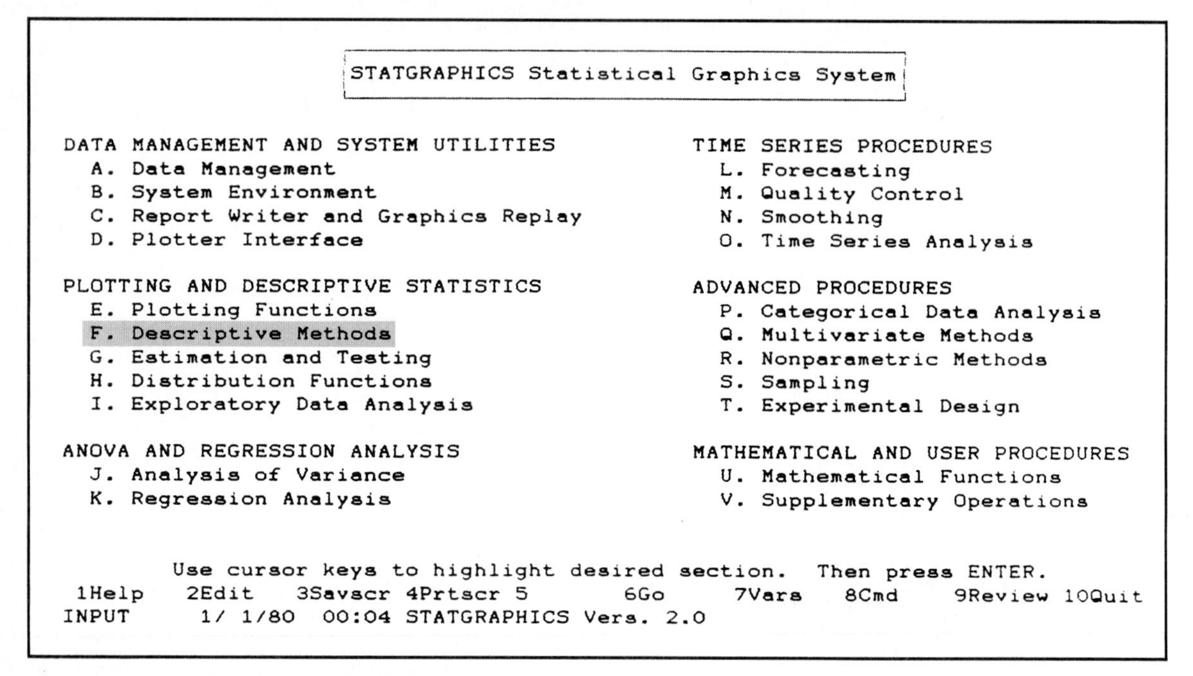

FIGURE 15.13 □ STATGRAPHICS Main menu.

In order to obtain descriptive statistics for the desired variables we need to choose Descriptive Methods from the Main Menu. After pressing Enter, we will be at the Descriptive Methods Menu (see Figure 15.14). The first selection on this menu is Summary Statistics. If the Enter key is pressed with the cursor at this selection, we will reach the Summary Statistics panel illustrated in Figure 15.15. We observe from this panel that two fields (or sets of information) need to be provided, the Data Vectors and the Statistics. The Data Vectors merely refer to the variables for which we would like to obtain descriptive statistics. If we choose, either we can

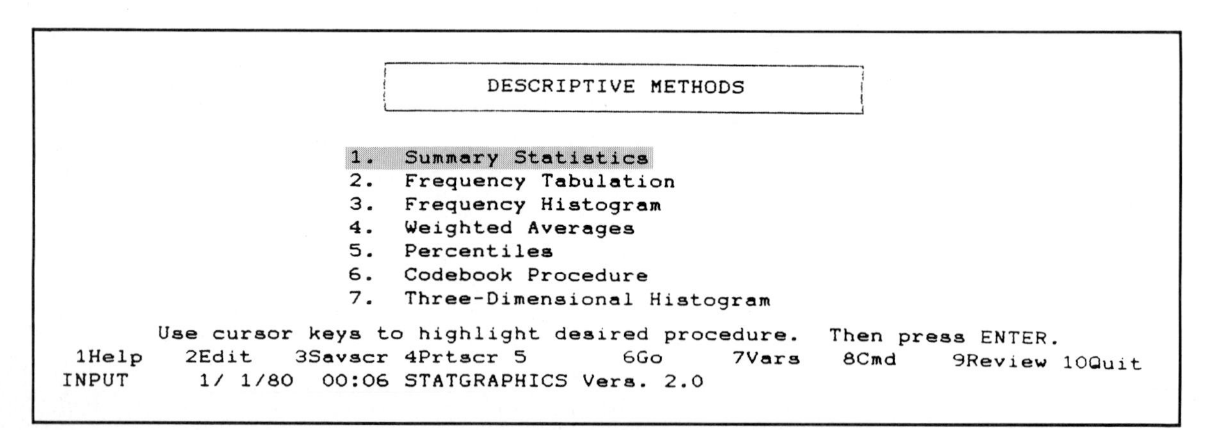

FIGURE 15.14 □ STATGRAPHICS Descriptive Methods menu.

```
                              Summary Statistics
-------------------------------------------------------------------------------
Data vectors: BLRDATA.YEARS
              BLRDATA.SALARY
              BLRDATA.AREAS









Statistics: ABEFHIJKLM
(a) Average         (f) Std. deviation   (k) Lower quartile  (p) Kurtosis
(b) Median          (g) Std. error       (l) Upper quartile  (q) Std. kurtosis
(c) Mode            (h) Minimum          (m) Interquar. range
(d) Geometric mean  (i) Maximum          (n) Skewness
(e) Variance        (j) Range            (o) Std. skewness

                    Complete input fields and press F6.
 1Help   2Edit   3Savscr 4Prtscr 5      6Go      7Vars    8Cmd     9Review 10Quit
INPUT     1/ 1/80  00:08 STATGRAPHICS Vers. 2.0                          STATS
```

FIGURE 15.15 □ STATGRAPHICS Summary Statistics panel.

enter numeric data or we can indicate the name of a file that contains the variables that we wish to analyze.

From Figure 15.15, we can see that we wish to use the variables BLRDATA. YEARS, BLRDATA.SALARY, and BLRDATA.AREAS. Each variable appears on a separate line. The first part of the name (BLRDATA), which must be in capital letters, refers to the file name that contains the data from the alumni survey.[5] The mnemonic name BLRDATA was chosen to represent the fact that the data was from the Berenson, Levine, and Rindskopf text. The names immediately following the period represent the variables for which we desire descriptive statistics.

The Statistics panel provides for the computation of any or all of the statistics listed below the panel. If we wish to obtain all the statistics available, we would leave the panel blank. Since we wished to obtain the sample size, average, median, variance, standard deviation, minimum, maximum, range, lower quartile, upper quartile, and interquartile range, we merely entered the letters ABEFHIJKLM on the display that corresponded to these statistics. We obtain printed output by pressing the F4 function key followed by the Enter key.

The results obtained are illustrated in Figure 15.16. We observe that the names

[5] The data could have been entered line by line as in Figure 2.10 by accessing the File Operations choice from the Data Management menu. However since a file had already been set up for use with SPSS^X and Minitab, the Import Files choice from the Data Management menu was selected and a STATGRAPHICS file was created.

Variable:	YEARS	SALARY	AREAS
Sample size	200	200	200
Average	23.255	54.77	2.14
Median	22	52	2
Variance	187.467	729.826	2.83457
Standard deviation	13.6919	27.0153	1.68362
Minimum	5	10	1
Maximum	61	185	12
Range	56	175	11
Lower quartile	11	34	1
Upper quartile	35	71	3
Interquartile range	24	37	2

```
Use Esc to Quit, Cursor keys or Page Number:                      Page 1.1 of 1.1
 1Help    2Edit    3Savscr 4Prtscr 5Prtopt 6Go      7Vars    8Cmd      9Review 10Quit
INPUT      1/ 1/80  00:12 STATGRAPHICS Vers. 2.0                              STATS
```

FIGURE 15.16 □ Output obtained from Summary Statistics panel.

of the statistics computed correspond to the rows, while each column represents a different variable.

Now that the various descriptive statistics have been obtained, we turn to the next part of our analysis, providing box-and-whisker plots for the number of years since graduation and the annual salary of alumni. We may develop these plots by selecting the Exploratory Data Analysis choice from the Main Menu. After pressing Enter, we will be at the Exploratory Data Analysis menu (see Figure 15.17).

We note that the first selection on this menu is the box-and-whisker plot. If the Enter key is pressed with the cursor at this selection, we reach the box-and-whisker plot panel which presents the words "Data Vector." At this point we type in the name of a variable for which a box-and-whisker plot is desired. Only one variable name may be entered. Once the variable name is entered in this field, the F6 function key is pressed and the box-and-whisker plot is displayed.[6] The same procedure would be repeated for the BLRDATA.SALARY variable. Figure 15.18 illustrates the box-and-whisker plots for these two variables placed side by side. The plot for SALARY looks slightly different from the plots we constructed in Chapter 5, because several outliers are plotted as separate points.

The third question of interest concerned the possible relationship between undergraduate major and graduate-school attendance. In this situation we need to have STATGRAPHICS develop a two-way cross-tabulation and compute the chi-square test statistic. This may be accomplished by selecting Categorical Data Analysis from the Main Menu. After pressing Enter, we will be at the Categorical Data Analysis menu (see Figure 15.19 on page 490). The first selection on this menu is Crosstabulation. If the Enter key is pressed with the cursor at this selection, we

[6] To obtain a printed copy we would press the F4 function key and then the Enter key.

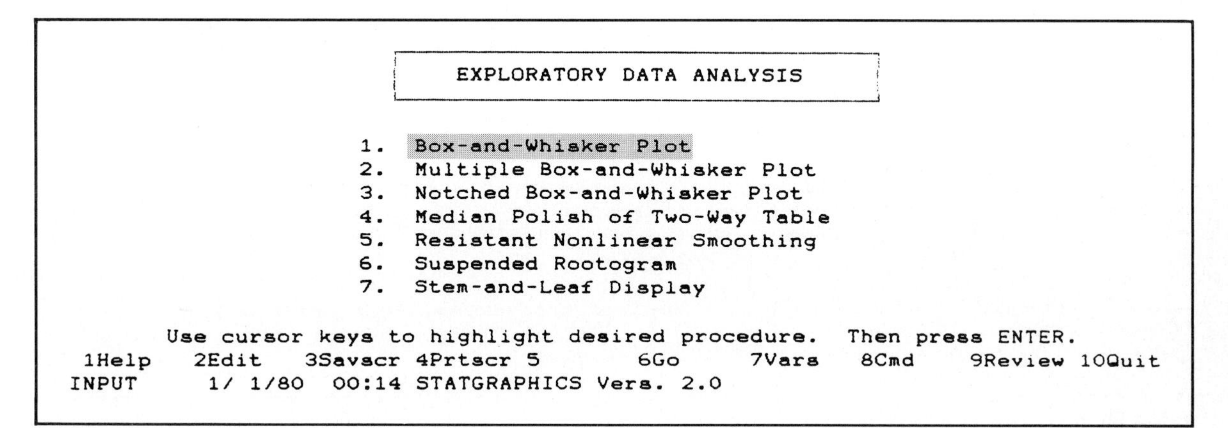

FIGURE 15.17 □ **Exploratory Data Analysis menu.**

have the Crosstabulation panel (see Figure 15.20 on page 490). In this panel are three sets of information that we need to provide—the Data Vectors, the Labels, and the Percentages. The Data Vectors refer to the variables that we wish to crosstabulate. From Figure 15.20 we observe that we are using BLRDATA.MAJOR

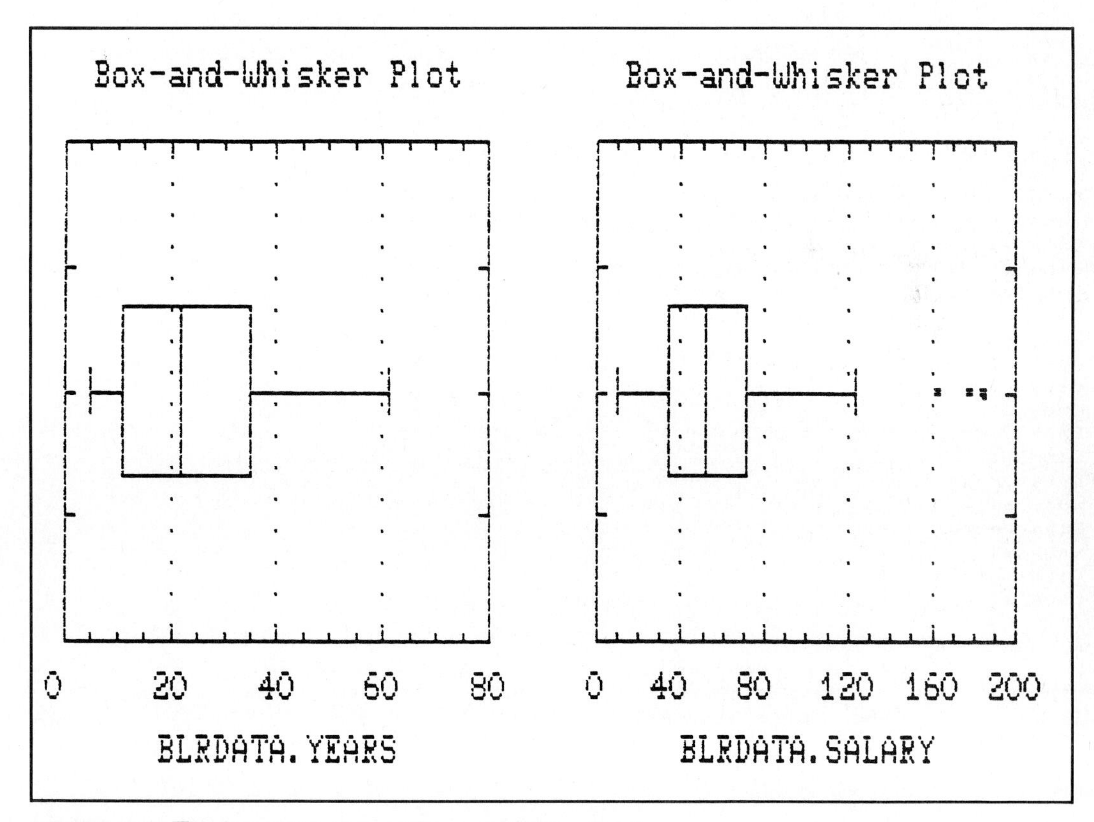

FIGURE 15.18 □ **Output obtained from Box-and-Whisker panel.**

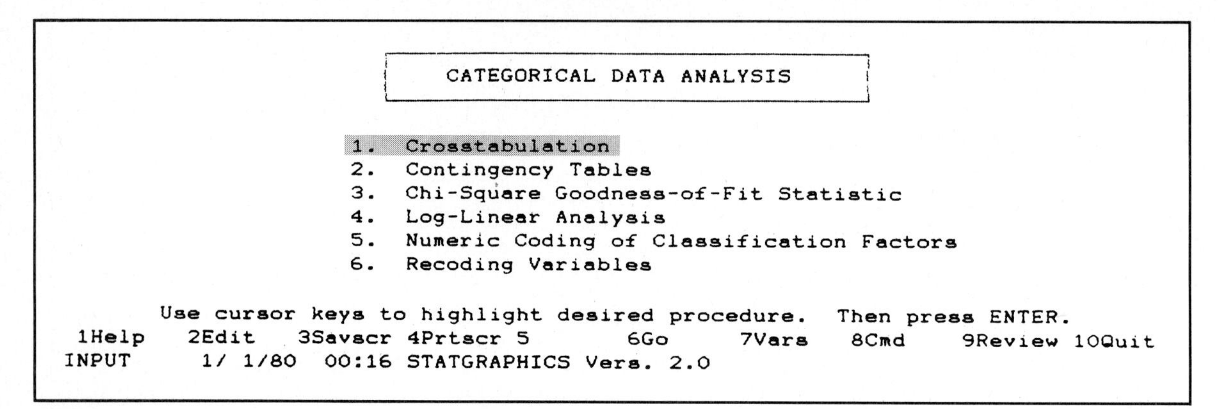

```
                    CATEGORICAL DATA ANALYSIS

                 1.  Crosstabulation
                 2.  Contingency Tables
                 3.  Chi-Square Goodness-of-Fit Statistic
                 4.  Log-Linear Analysis
                 5.  Numeric Coding of Classification Factors
                 6.  Recoding Variables

     Use cursor keys to highlight desired procedure.  Then press ENTER.
 1Help   2Edit   3Savscr 4Prtscr 5       6Go     7Vars   8Cmd    9Review 10Quit
INPUT     1/ 1/80  00:16 STATGRAPHICS Vers. 2.0
```

FIGURE 15.19 □ Categorical Data Analysis menu.

and BLRDATA.GRAD. Each variable appears on a separate line. Once each variable has been entered, the cursor is moved to the Labels area in order to provide a category label. This is accomplished through the use of the RESHAPE utility. The use of RESHAPE is complicated; we will not illustrate its use (but we have used it here and the labels will appear on the output).

Once the Data Vectors and Category Labels have been completed, the cursor can be positioned at the Percentage field. Three types of percentages can be ob-

```
                              Crosstabulation
---------------------------------------------------------------------------
Data vector A: BLRDATA.MAJOR
Labels: 3 11 RESHAPE 'Business   Humanities Science/Mt '
Data vector B: BLRDATA.GRAD
Labels: 2 3 RESHAPE 'YesNo '
Data vector C:
Labels:
Data vector D:
Labels:
Data vector E:
Labels:
Data vector F:
Labels:
Data vector G:
Labels:
Data vector H:
Labels:
Data vector I:
Labels:

Percentages: Tablewise
                     Complete input fields and press F6.
 1Help   2Edit   3Savscr 4Prtscr 5       6Go     7Vars   8Cmd    9Review 10Quit
INPUT     1/ 1/80  00:18 STATGRAPHICS Vers. 2.0                        CTAB
```

FIGURE 15.20 □ Crosstabulation panel.

tained—Columnwise, Rowwise, or Tablewise. Columnwise (the default value) calculates percentages based upon the column total frequencies, Rowwise calculates percentages based on the row total frequencies, and Tablewise calculates percentages based on the total frequency in the entire table. The Percentage type is selected by pressing the space bar until the desired type is displayed.

When the Crosstabulation panel is complete, the F6 function key is pressed and the cross-tabulation table is developed. A separate menu "pops up" to ask whether you wish to display the table, display the statistics, save the counts, or save the labels. The results obtained are illustrated in Figure 15.21. The summary statistics displayed include the chi-square statistic as well as other measures not discussed in

Crosstabulation of BLRDAT

BLRDATA.GR / BLRDATA.MAJ	Yes	No	Row Total
Business	58 29.0	60 30.0	118 59.0
Humanities	25 12.5	34 17.0	59 29.5
Science/Mt	16 8.0	7 3.5	23 11.5
Column Total	99 49.5	101 50.5	200 100.0

```
Use Esc to Quit, Cursor keys or Page Number:                        Page 1.1 of 1.1
 1Help   2Edit   3Savscr 4Prtscr 5Prtopt 6Go     7Vars     8Cmd     9Review 10Quit
INPUT      1/ 1/80   00:21 STATGRAPHICS Vers. 2.0                          CTAB
```

Summary Statistics for Contingency Tables (Page 1)

Chi-square	D.F.	Significance
4.90901	2	0.0859057

Statistic	Symmetric	With rows dependent	With columns dependent
Lambda	0.04972	0.00000	0.09091
Uncertainty Coeff.	0.01553	0.01361	0.01807
Somer's D	-0.04289	-0.04510	-0.04088
Eta	0.07258	0.15667	0.07258

```
                          Press ENTER to continue.
 1Help    2Edit    3Savscr 4Prtscr 5        6Go      7Vars     8Cmd      9Review 10Quit
INPUT      1/ 1/80   00:23 STATGRAPHICS Vers. 2.0                              CTAB
```

FIGURE 15.21 □ **Crosstabulation output.**

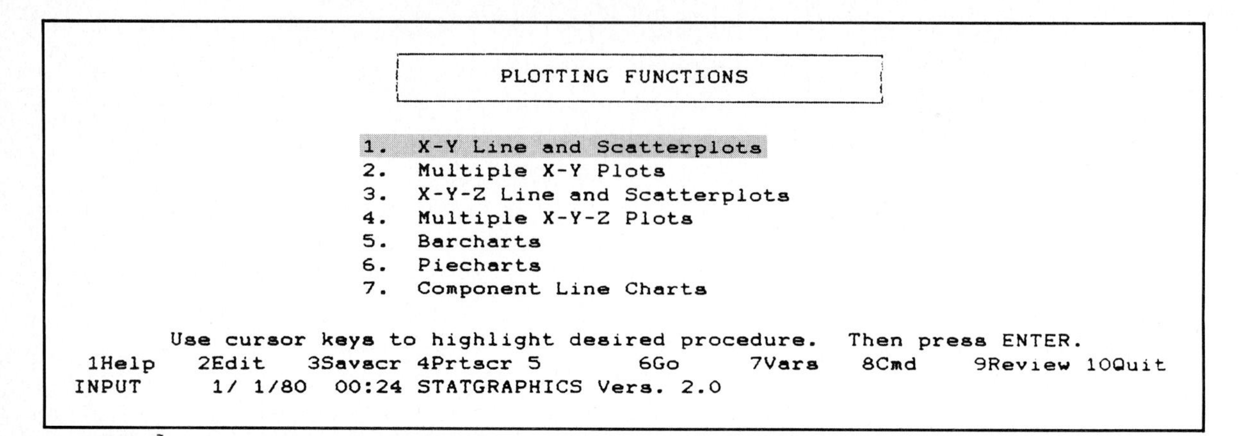

FIGURE 15.22 □ Plotting Functions menu.

this text. Since the computed chi-square statistic of 4.90901 with 2 degrees of freedom is less than the critical value of 5.991 at the .05 level of significance (and the p value is .0859), we would conclude that there is no evidence of a relationship between undergraduate major and subsequent attendance at graduate school.

For the last question of interest we wish to develop a simple linear regression model to predict salary based upon the number of years since graduation. We may begin this analysis by obtaining a scatter plot of the two variables by selecting the Plotting Functions choice from the Main menu. After pressing Enter, we will be at the Plotting Functions menu (see Figure 15.22). The first selection, X-Y Line and

```
                          X-Y Line and Scatterplots
--------------------------------------------------------------------------------
                                                                          Log
X: BLRDATA.YEARS                                                          No
Y: BLRDATA.SALARY                                                         No


Point labels:

Point codes:
Point colors:
Std. err. X:
Std. err. Y:


Points: Yes
Lines:  No

                   Complete input fields and press F6.
 1Help   2Edit   3Savscr 4Prtscr 5      6Go     7Vars    8Cmd     9Review 10Quit
INPUT      1/ 1/80  00:26 STATGRAPHICS Vers. 2.0                         PLOT
```

FIGURE 15.23 □ X-Y Line and Scatterplots panel.

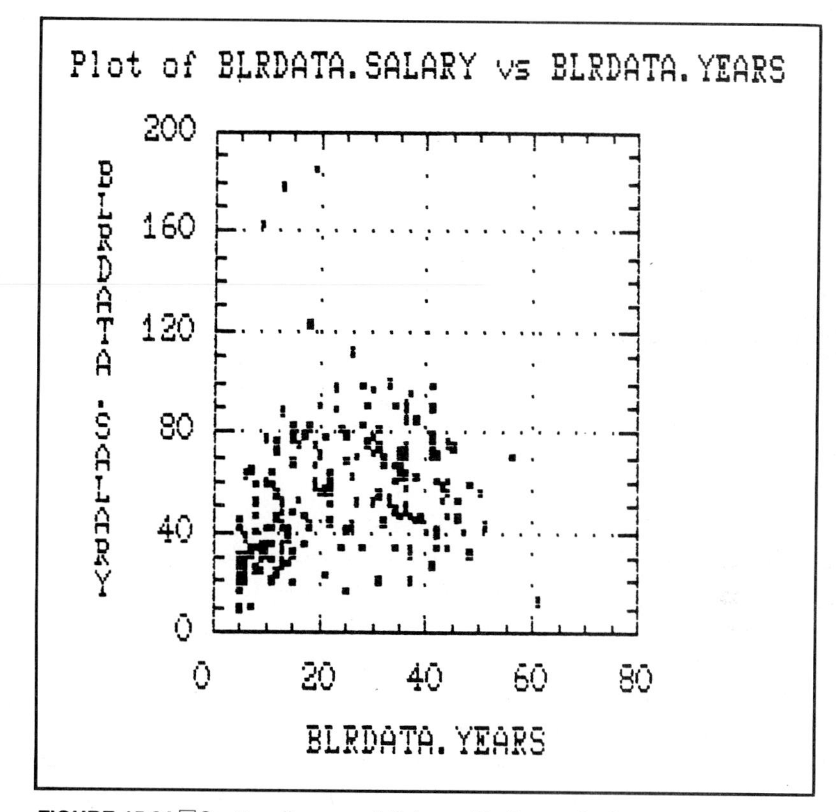

FIGURE 15.24 ☐ **Scatter diagram of Salary with Years obtained from STATGRAPHICS.**

Scatterplots, is our desired choice on this menu. Pressing the Enter key brings us to the X-Y Line and Scatterplots panel (see Figure 15.23). We see that in this panel there are four sets of information that must be provided—X, Y, Points, and Lines. X, of course, refers to the independent variable (BLRDATA.YEARS), Y refers to the dependent variable (BLRDATA.SALARY), ''points'' refers to whether we want each point plotted (we do), and ''lines'' refers to whether we want lines to connect all the points (we do not). Once this panel is completed, we press the F6 function key to get the scatter plot. Figure 15.24 represents the scatter diagram for our variables of interest.

Once the scatter diagram has been examined, we can proceed by selecting the Regression Analysis choice from the Main Menu. After pressing Enter, we will be at the Regression Analysis menu (see Figure 15.25 on page 494). To perform Simple Regression analysis we position the cursor at this line and press the Enter key.

Figure 15.26 on page 494 displays the Regression Analysis panel that we have now reached. We note that six sets of information must be provided—the Dependent variable, the Independent variable, the Model, Confidence limits, Prediction limits,

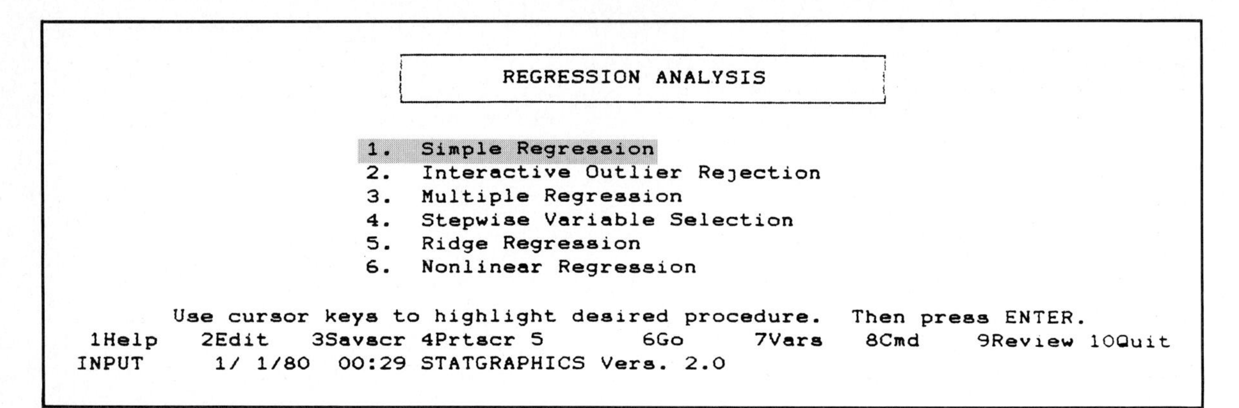

FIGURE 15.25 □ Regression Analysis menu.

and Point labels. For our example, the dependent variable is BLRDATA.SALARY, while the independent variable is BLRDATA.YEARS. There are four models that can be selected: linear (our choice), multiplicative, exponential, and reciprocal (see References 1 and 4). Confidence limits represent interval estimates of the mean response. Confidence limits appear as bands on an X-Y plot that includes the fitted line of regression. If limits are not desired, zeros can be entered in this field. Prediction limits represent estimates for individual observations. Information would be entered in the same manner as confidence limits. Finally, Point labels provide an interactive labeling capability for individual observations.

```
                              Simple Regression
-------------------------------------------------------------------------
Dependent variable: BLRDATA.SALARY

Independent variable: BLRDATA.YEARS

Model: Linear

Confidence limits: 000.00

Prediction limits: 000.00

Point labels:

                    Complete input fields and press F6.
 1Help   2Edit   3Savscr 4Prtscr 5        6Go     7Vars   8Cmd    9Review 10Quit
INPUT     1/ 1/80  00:30 STATGRAPHICS Vers. 2.0                        REG
```

FIGURE 15.26 □ Regression Analysis panel.

```
Regression Analysis - Linear model: Y = a+bX

Dependent variable: BLRDATA.SALARY          Independent variable: BLRDATA.YEARS

                               Standard           T            Prob.
Parameter        Estimate        Error          Value          Level

Intercept         43.1807       3.65901        11.8012     2.66454E-15
Slope            0.498358      0.135675        3.67317      3.08241E-4

                         Analysis of Variance

Source              Sum of Squares     Df   Mean Square     F-Ratio   Prob. Level
Model                    9265.3319      1     9265.3319     13.4922        .00031
Error                    135970.09    198        686.72

Total (Corr.)            145235.42    199

Correlation Coefficient = 0.252577         R-squared =     6.38 percent
Stnd. Error of Est. = 26.2053

                        Press ENTER to continue.
 1Help   2Edit   3Savscr 4Prtscr 5       6Go      7Vars    8Cmd    9Review 10Quit
INPUT      1/ 1/80  00:36 STATGRAPHICS Vers. 2.0                           REG
```

FIGURE 15.27 ☐ Regression analysis output from STATGRAPHICS.

Once the panel is completed, the F6 function key is pressed to perform the regression analysis. Figure 15.27 represents the regression analysis output for our example. We note that $b_0 = 43.1807$ and $b_1 = .498358$, and the t statistic for the significance of β_1 is 3.67317 (the p value is 3.08241E-4, which is scientific notation for .000308241). Thus, there is highly significant evidence of a linear relationship between years since graduation and salary. The second panel of Figure 15.27 presents the Analysis of Variance approach toward determining the significance of β_1. The F value of 13.4922 (with a p value of .00031) is equivalent to the previously mentioned t test. We may also note that r^2 is equal to only .0638 or 6.38%. Therefore, although the results are significantly different from what would be expected if there were no linear relationship, only 6.38% of the variation in salary can be explained by variation in the number of years since graduation.

If the F6 function key is pressed, a pop-up menu will appear that provides four options—Plot fitted line, Plot residuals, Save residuals, and Save predictions. If Plot fitted line is selected, the output displayed in Figure 15.28 will be obtained. After pressing Escape, and then positioning the cursor on Plot residuals and following a similar sequence of keystrokes, the residual plot illustrated in Figure 15.29 may be obtained.

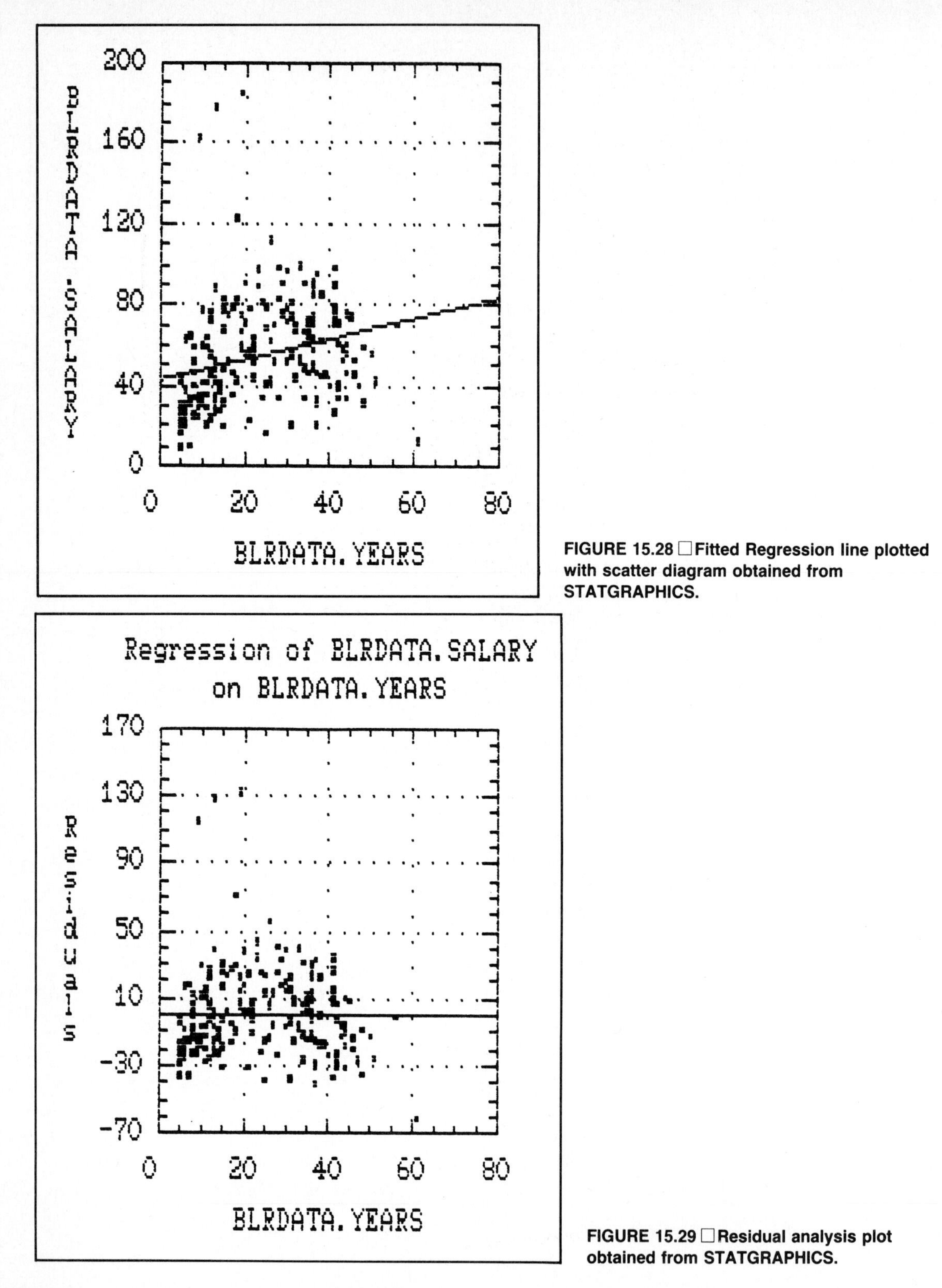

FIGURE 15.28 □ **Fitted Regression line plotted with scatter diagram obtained from STATGRAPHICS.**

FIGURE 15.29 □ **Residual analysis plot obtained from STATGRAPHICS.**

15.8 SUMMARY

In this chapter we have discussed three different statistical packages that are currently available for data analysis—SPSS^X, Minitab, and STATGRAPHICS. These packages represent only a small selection from the myriad of statistical packages that are currently available (see References 5 and 6). However, they do represent a wide variety in terms of style of usage (command versus menu driven, and batch versus interactive use). There is much overlap in the statistical procedures they have available, but even so they may implement the same procedure in different ways. Of foremost concern is that the user look for a package that contains all the statistical procedures that he or she might want to use, rather than select procedures that merely happen to be available. If several packages satisfy this requirement, then the choice can be made on the basis of preferences among various styles of usage.

References

1. Berenson, M. L., D. M. Levine, and M. Goldstein, *Intermediate Statistical Methods: A Computer Package Approach* (Englewood Cliffs, N.J.: Prentice-Hall, 1983).
2. Ryan, T. A., B. L. Joiner, and B. F. Ryan, *Minitab Student Handbook,* 2d ed. (North Scituate, Mass.: Duxbury Press, 1986).
3. *SPSS^X User's Guide* (New York: McGraw-Hill, 1983).
4. *STATGRAPHICS User's Guide* (Rockville, Md.: STSC Inc., 1986).
5. Stoll, M., "Statistical Software," *PC Week*, September 23, 1986, pp. 77–94.
6. Woodward, W. A., A. C. Elliott, and H. L. Gray, *Directory of Statistical Microcomputer Software, 1985 Edition* (New York: Marcel Dekker, 1985).

Appendix A

Review of Arithmetic, Algebra, and Summation Notation

A.1 RULES FOR ARITHMETIC OPERATIONS

The following is a summary of various rules for arithmetic operations. Each rule is illustrated by a numerical example.

RULE	EXAMPLE
1. $a + b = b + a$	$2 + 1 = 1 + 2 = 3$
2. $a + (b + c) = (a + b) + c$	$5 + (7 + 4) = (5 + 7) + 4 = 16$
3. $a - b \neq b - a$ unless $a = b$	$9 - 7 = 2$ but $7 - 9 = -2$
4. $a \times b = b \times a$	$7 \times 6 = 6 \times 7 = 42$
5. $a \times (b + c) = (a \times b) + (a \times c)$	$2 \times (3 + 5) = (2 \times 3) + (2 \times 5) = 16$
6. $a \div b \neq b \div a$ unless $a = b$	$12 \div 3 \neq 3 \div 12$
7. $\frac{a}{c} + \frac{b}{c} = \frac{a + b}{c}$	$\frac{7}{2} + \frac{3}{2} = \frac{7 + 3}{2} = 5$
8. $\frac{a}{b} + \frac{a}{c} \neq \frac{a}{b + c}$	$\frac{3}{4} + \frac{3}{5} \neq \frac{3}{4 + 5}$

RULE	EXAMPLE
9. $\frac{1}{a} + \frac{1}{b} = \frac{b + a}{ab}$	$\frac{1}{3} + \frac{1}{5} = \frac{5 + 3}{(3)(5)} = \frac{8}{15}$
10. $\frac{a}{b} \times \frac{c}{d} = \frac{a \times c}{b \times d}$	$\frac{2}{3} \times \frac{6}{7} = \frac{2 \times 6}{3 \times 7} = \frac{12}{21}$
11. $\frac{a}{b} \div \frac{c}{d} = \frac{a \times d}{b \times c}$	$\frac{5}{8} \div \frac{3}{7} = \frac{5 \times 7}{8 \times 3} = \frac{35}{24}$

A.2 RULES FOR ALGEBRA: EXPONENTS AND SQUARE ROOTS

The following is a summary of various rules for algebraic operations. Each rule is illustrated by a numerical example.

RULE	EXAMPLE
1. $X^a \cdot X^b = X^{a+b}$	$4^2 \cdot 4^3 = 4^5$
2. $(X^a)^b = X^{ab}$	$(2^2)^3 = 2^6$
3. $(X^a/X^b) = X^{a-b}$	$\frac{3^5}{3^3} = 3^2$
4. $\frac{X^a}{X^a} = X^0 = 1$	$\frac{3^4}{3^4} = 3^0 = 1$
5. $\sqrt{XY} = \sqrt{X}\sqrt{Y}$	$\sqrt{(25)(4)} = \sqrt{25}\sqrt{4} = 10$
6. $\sqrt{\frac{X}{Y}} = \frac{\sqrt{X}}{\sqrt{Y}}$	$\sqrt{\frac{16}{100}} = \frac{\sqrt{16}}{\sqrt{100}} = .40$

A.3 SUMMATION NOTATION

Since the operation of addition occurs so frequently in statistics, the special symbol Σ (sigma) is used to denote "take the sum of." Suppose, for example, that we have a set of n values for some variable X. The expression ΣX means that these n values are to be added together. Thus

$$\Sigma X = X_1 + X_2 + X_3 + \cdots + X_n$$

The use of the summation notation can be illustrated in the following problem. Suppose that we have five observations of a variable X: $X_1 = 2$, $X_2 = 0$, $X_3 = -1$, $X_4 = 5$, and $X_5 = 7$. Thus

$$\begin{aligned}\Sigma X &= X_1 + X_2 + X_3 + X_4 + X_5 \\ &= 2 + 0 + (-1) + 5 + 7 = 13\end{aligned}$$

In statistics we are also frequently involved with summing the squared values of a variable. Thus

$$\Sigma X^2 = X_1^2 + X_2^2 + X_3^2 + \cdots + X_n^2$$

and, in our example, we have

$$\begin{aligned}\Sigma X^2 &= X_1^2 + X_2^2 + X_3^2 + X_4^2 + X_5^2 \\ &= 2^2 + 0^2 + (-1)^2 + 5^2 + 7^2 \\ &= 4 + 0 + 1 + 25 + 49 \\ &= 79\end{aligned}$$

We should realize here that ΣX^2, the sum of the squares, is not the same as $(\Sigma X)^2$, the square of the sum; that is,

$$\Sigma X^2 \neq (\Sigma X)^2$$

In our example the sum of squares is equal to 79. This is not equal to the square of the sum, which is $13^2 = 169$.

Another frequently used operation involves the sum of the product. That is, suppose that we have two variables, X and Y, each having n observations. Then,

$$\Sigma XY = X_1Y_1 + X_2Y_2 + X_3Y_3 + \cdots + X_nY_n$$

Continuing with our previous example, suppose that there is also a second variable Y whose five values are $Y_1 = 1$, $Y_2 = 3$, $Y_3 = -2$, $Y_4 = 4$, and $Y_5 = 3$. Then,

$$\begin{aligned}\Sigma XY &= X_1Y_1 + X_2Y_2 + X_3Y_3 + X_4Y_4 + X_5Y_5 \\ &= 2(1) + 0(3) + (-1)(-2) + 5(4) + 7(3) \\ &= 2 + 0 + 2 + 20 + 21 \\ &= 45\end{aligned}$$

In computing ΣXY the first value of X is multiplied by the first value of Y, the second value of X is multiplied by the second value of Y, etc. These products are then summed in order to obtain the desired result.

Note that the sum of products is not equal to the product of the individual sums; that is,

$$\Sigma XY \neq (\Sigma X)(\Sigma Y)$$

In our example, $\Sigma X = 13$ and $\Sigma Y = 1 + 3 + (-2) + 4 + 3 = 9$, so that

$$(\Sigma X)(\Sigma Y) = 13(9) = 117.$$

This is not the same as ΣXY, which equals 45.

Before studying the four basic rules of performing operations with summation notation, it is helpful to present the values for each of the five observations of X and Y in a tabular format.

Observation	X	Y
1	2	1
2	0	3
3	−1	−2
4	5	4
5	7	3
	$\Sigma X = 13$	$\Sigma Y = 9$

RULE 1 The summation of the sum of two variables is equal to the sum of the values of each summed variable.

$$\Sigma (X + Y) = \Sigma X + \Sigma Y$$

Thus, in our example,

$$\begin{aligned}\Sigma (X + Y) &= (2 + 1) + (0 + 3) + (-1 + (-2)) + (5 + 4) + (7 + 3)\\ &= 3 + 3 + (-3) + 9 + 10\\ &= 22 = \Sigma X + \Sigma Y = 13 + 9 = 22\end{aligned}$$

RULE 2 The sum of a difference between the values of two variables is equal to the difference between the summed values of the variables.

$$\Sigma (X - Y) = \Sigma X - \Sigma Y$$

Thus, in our example,

$$\begin{aligned}\Sigma (X - Y) &= (2 - 1) + (0 - 3) + (-1 - (-2)) + (5 - 4) + (7 - 3)\\ &= 1 + (-3) + 1 + 1 + 4\\ &= 4 = \Sigma X - \Sigma Y = 13 - 9 = 4\end{aligned}$$

RULE 3 The sum of a constant times a variable is equal to that constant times the sum of the values of the variable.

$$\Sigma (CX) = C (\Sigma X)$$

where C is a constant.

Thus, in our example, if $C = 2$,

$$\begin{aligned}\Sigma (CX) = \Sigma (2X) &= 2(2) + 2(0) + 2(-1) + 2(5) + 2(7)\\ &= 4 + 0 + (-2) + 10 + 14\\ &= 26 = 2 (\Sigma X) = 2(13) = 26\end{aligned}$$

RULE 4 A constant summed n times will be equal to n times the value of the constant.

$$\Sigma\, C = nC$$

where C is a constant.

Thus, if the constant $C = 2$ is summed five times, we have

$$\begin{aligned}\Sigma\, C &= 2 + 2 + 2 + 2 + 2 \\ &= 10 = 5(2) = 10\end{aligned}$$

To illustrate how these summation rules are used, we may demonstrate one of the mathematical properties pertaining to the average or arithmetic mean (see Section 3.4); that is,

$$\Sigma\,(X - \overline{X}) = 0$$

This property states that the sum of the differences between each observation and the arithmetic mean is zero. This can be proven mathematically in the following manner:

1. Using summation rule 2, we have

$$\Sigma\,(X - \overline{X}) = \Sigma\, X - \Sigma\, \overline{X}$$

2. Since, for any fixed set of data, $\overline{X}$ can be considered a constant, from summation rule 4 we have

$$\Sigma\, \overline{X} = n\overline{X}$$

Therefore,

$$\Sigma\,(X - \overline{X}) = \Sigma\, X - n\overline{X}$$

3. However, from Equation (3.1), since

$$\overline{X} = \frac{\Sigma\, X}{n} \qquad \text{then} \qquad n\overline{X} = \Sigma\, X$$

Therefore,

$$\Sigma\,(X - \overline{X}) = \Sigma\, X - \Sigma\, X = 0.$$

Problem

Suppose that there are six observations for the variables X and Y such that $X_1 = 2$, $X_2 = 1$, $X_3 = 5$, $X_4 = -3$, $X_5 = 1$, and $X_6 = -2$, and $Y_1 = 4$, $Y_2 = 0$, $Y_3 = -1$, $Y_4 = 2$, $Y_5 = 7$, and $Y_6 = -3$. Compute each of the following.

(a) ΣX

(b) ΣY

(c) ΣX^2

(d) ΣY^2

(e) ΣXY

(f) $\Sigma (X + Y)$

(g) $\Sigma (X - Y)$

(h) $\Sigma (X - 3Y + 2X^2)$

(i) $\Sigma (CX)$, where $C = -1$

(j) $\Sigma (X - 3Y + C)$, where $C = +3$

Appendix B

Statistical Symbols and Greek Alphabet

B.1 STATISTICAL SYMBOLS

Symbol	Meaning	Symbol	Meaning
$=$	equals	$\neq$	not equal
$\cong$	approximately equal to		
$>$	greater than	$<$	less than
$\geq$	greater than or equal to	$\leq$	less than or equal to

B.2 GREEK ALPHABET

Greek letter	Greek name	English equivalent	Greek letter	Greek name	English equivalent
Α α	Alpha	a	Ν ν	Nu	n
Β β	Beta	b	Ξ ξ	Xi	x
Γ γ	Gamma	g	Ο ο	Omicron	ŏ
Δ δ	Delta	d	Π π	Pi	p
Ε ϵ	Epsilon	ĕ	Ρ ρ	Rho	r
Ζ ζ	Zeta	z	Σ σ ς	Sigma	s
Η η	Eta	ē	Τ τ	Tau	t
Θ θ ϑ	Theta	th	Υ υ	Upsilon	u
Ι ι	Iota	i	Φ ϕ φ	Phi	ph
Κ κ	Kappa	k	Χ χ	Chi	ch
Λ λ	Lambda	l	Ψ ψ	Psi	ps
Μ μ	Mu	m	Ω ω	Omega	ō

Appendix C

Tables

LIST OF TABLES

TABLE C.1 □ Table of random numbers

	Column							
Row	*00000 12345*	*00001 67890*	*11111 12345*	*11112 67890*	*22222 12345*	*22223 67890*	*33333 12345*	*33334 67890*
01	49280	88924	35779	00283	81163	07275	89863	02348
02	61870	41657	07468	08612	98083	97349	20775	45091
03	43898	65923	25078	86129	78496	97653	91550	08078
04	62993	93912	30454	84598	56095	20664	12872	64647
05	33850	58555	51438	85507	71865	79488	76783	31708
06	97340	23364	88472	04334	63919	36394	11095	92470
07	70543	29776	10087	10072	55980	64688	68239	20461
08	89382	93809	00796	95945	34101	81277	66090	88872
09	37818	72142	67140	50785	22380	16703	53362	44940
10	60430	22834	14130	96593	23298	56203	92671	15925
11	82975	66158	84731	19436	55790	69229	28661	13675
12	39087	71938	40355	54324	08401	26299	49420	59208
13	55700	24586	93247	32596	11865	63397	44251	43189
14	14756	23997	78643	75912	83832	32768	18928	57070
15	32166	53251	70654	92827	63491	04233	33825	69662
16	23236	73751	31888	81718	06546	83246	47651	04877
17	45794	26926	15130	82455	78305	55058	52551	47182
18	09893	20505	14225	68514	46427	56788	96297	78822
19	54382	74598	91499	14523	68479	27686	46162	83554
20	94750	89923	37089	20048	80336	94598	26940	36858
21	70297	34135	53140	33340	42050	82341	44104	82949
22	85157	47954	32979	26575	57600	40881	12250	73742
23	11100	02340	12860	74697	96644	89439	28707	25815
24	36871	50775	30592	57143	17381	68856	25853	35041
25	23913	48357	63308	16090	51690	54607	72407	55538
26	79348	36085	27973	65157	07456	22255	25626	57054
27	92074	54641	53673	54421	18130	60103	69593	49464
28	06873	21440	75593	41373	49502	17972	82578	16364
29	12478	37622	99659	31065	83613	69889	58869	29571
30	57175	55564	65411	42547	70457	03426	72937	83792
31	91616	11075	80103	07831	59309	13276	26710	73000
32	78025	73539	14621	39044	47450	03197	12787	47709
33	27587	67228	80145	10175	12822	86687	65530	49325
34	16690	20427	04251	64477	73709	73945	92396	68263
35	70183	58065	65489	31833	82093	16747	10386	59293
36	90730	35385	15679	99742	50866	78028	75573	67257
37	10934	93242	13431	24590	02770	48582	00906	58595
38	82462	30166	79613	47416	13389	80268	05085	96666
39	27463	10433	07606	16285	93699	60912	94532	95632
40	02979	52997	09079	92709	90110	47506	53693	49892
41	46888	69929	75233	52507	32097	37594	10067	67327
42	53638	83161	08289	12639	08141	12640	28437	09268
43	82433	61427	17239	89160	19666	08814	37841	12847
44	35766	31672	50082	22795	66948	65581	84393	15890
45	10853	42581	08792	13257	61973	24450	52351	16602
46	20341	27398	72906	63955	17276	10646	74692	48438
47	54458	90542	77563	51839	52901	53355	83281	19177
48	26337	66530	16687	35179	46560	00123	44546	79896
49	34314	23729	85264	05575	96855	23820	11091	79821
50	28603	10708	68933	34189	92166	15181	66628	58599

TABLE C.1 *(continued)*

Row	Column 00000 12345	00001 67890	11111 12345	11112 67890	22222 12345	22223 67890	33333 12345	33334 67890
51	66194	28926	99547	16625	45515	67953	12108	57846
52	78240	43195	24837	32511	70880	22070	52622	61881
53	00833	88000	67299	68215	11274	55624	32991	17436
54	12111	86683	61270	58036	64192	90611	15145	01748
55	47189	99951	05755	03834	43782	90599	40282	51417
56	76396	72486	62423	27618	84184	78922	73561	52818
57	46409	17469	32483	09083	76175	19985	26309	91536
58	74626	22111	87286	46772	42243	68046	44250	42439
59	34450	81974	93723	49023	58432	67083	36876	93391
60	36327	72135	33005	28701	34710	49359	50693	89311
61	74185	77536	84825	09934	99103	09325	67389	45869
62	12296	41623	62873	37943	25584	09609	63360	47270
63	90822	60280	88925	99610	42772	60561	76873	04117
64	72121	79152	96591	90305	10189	79778	68016	13747
65	95268	41377	25684	08151	61816	58555	54305	86189
66	92603	09091	75884	93424	72586	88903	30061	14457
67	18813	90291	05275	01223	79607	95426	34900	09778
68	38840	26903	28624	67157	51986	42865	14508	49315
69	05959	33836	53758	16562	41081	38012	41230	20528
70	85141	21155	99212	32685	51403	31926	69813	58781
71	75047	59643	31074	38172	03718	32119	69506	67143
72	30752	95260	68032	62871	58781	34143	68790	69766
73	22986	82575	42187	62295	84295	30634	66562	31442
74	99439	86692	90348	66036	48399	73451	26698	39437
75	20389	93029	11881	71685	65452	89047	63669	02656
76	39249	05173	68256	36359	20250	68686	05947	09335
77	96777	33605	29481	20063	09398	01843	35139	61344
78	04860	32918	10798	50492	52655	33359	94713	28393
79	41613	42375	00403	03656	77580	87772	86877	57085
80	17930	00794	53836	53692	67135	98102	61912	11246
81	24649	31845	25736	75231	83808	98917	93829	99430
82	79899	34061	54308	59358	56462	58166	97302	86828
83	76801	49594	81002	30397	52728	15101	72070	33706
84	36239	63636	38140	65731	39788	06872	38971	53363
85	07392	64449	17886	63632	53995	17574	22247	62607
86	67133	04181	33874	98835	67453	59734	76381	63455
87	77759	31504	32832	70861	15152	29733	75371	39174
88	85992	72268	42920	20810	29361	51423	90306	73574
89	79553	75952	54116	65553	47139	60579	09165	85490
90	41101	17336	48951	53674	17880	45260	08575	49321
91	36191	17095	32123	91576	84221	78902	82010	30847
92	62329	63898	23268	74283	26091	68409	69704	82267
93	14751	13151	93115	01437	56945	89661	67680	79790
94	48462	59278	44185	29616	76537	19589	83139	28454
95	29435	88105	59651	44391	74588	55114	80834	85686
96	28340	29285	12965	14821	80425	16602	44653	70467
97	02167	58940	27149	80242	10587	79786	34959	75339
98	17864	00991	39557	54981	23588	81914	37609	13128
99	79675	80605	60059	35862	00254	36546	21545	78179
00	72335	82037	92003	34100	29879	46613	89720	13274

SOURCE: Reprinted from *A Million Random Digits with 100,000 Normal Deviates* by The RAND Corporation (New York: The Free Press, 1955).

TABLE C.2 □ The standardized normal distribution

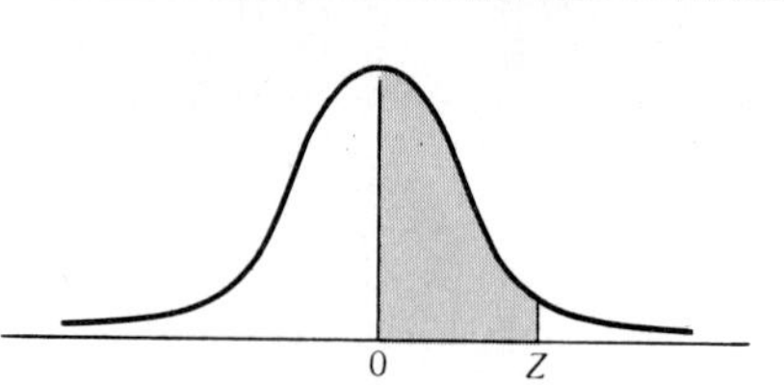

Entry represents area under the standardized normal distribution from the mean to Z

Z	.00	.01	.02	.03	.04	.05	.06	.07	.08	.09
0.0	.0000	.0040	.0080	.0120	.0160	.0199	.0239	.0279	.0319	.0359
0.1	.0398	.0438	.0478	.0517	.0557	.0596	.0636	.0675	.0714	.0753
0.2	.0793	.0832	.0871	.0910	.0948	.0987	.1026	.1064	.1103	.1141
0.3	.1179	.1217	.1255	.1293	.1331	.1368	.1406	.1443	.1480	.1517
0.4	.1554	.1591	.1628	.1664	.1700	.1736	.1772	.1808	.1844	.1879
0.5	.1915	.1950	.1985	.2019	.2054	.2088	.2123	.2157	.2190	.2224
0.6	.2257	.2291	.2324	.2357	.2389	.2422	.2454	.2486	.2518	.2549
0.7	.2580	.2612	.2642	.2673	.2704	.2734	.2764	.2794	.2823	.2852
0.8	.2881	.2910	.2939	.2967	.2995	.3023	.3051	.3078	.3106	.3133
0.9	.3159	.3186	.3212	.3238	.3264	.3289	.3315	.3340	.3365	.3389
1.0	.3413	.3438	.3461	.3485	.3508	.3531	.3554	.3577	.3599	.3621
1.1	.3643	.3665	.3686	.3708	.3729	.3749	.3770	.3790	.3810	.3830
1.2	.3849	.3869	.3888	.3907	.3925	.3944	.3962	.3980	.3997	.4015
1.3	.4032	.4049	.4066	.4082	.4099	.4115	.4131	.4147	.4162	.4177
1.4	.4192	.4207	.4222	.4236	.4251	.4265	.4279	.4292	.4306	.4319
1.5	.4332	.4345	.4357	.4370	.4382	.4394	.4406	.4418	.4429	.4441
1.6	.4452	.4463	.4474	.4484	.4495	.4505	.4515	.4525	.4535	.4545
1.7	.4554	.4564	.4573	.4582	.4591	.4599	.4608	.4616	.4625	.4633
1.8	.4641	.4649	.4656	.4664	.4671	.4678	.4686	.4693	.4699	.4706
1.9	.4713	.4719	.4726	.4732	.4738	.4744	.4750	.4756	.4761	.4767
2.0	.4772	.4778	.4783	.4788	.4793	.4798	.4803	.4808	.4812	.4817
2.1	.4821	.4826	.4830	.4834	.4838	.4842	.4846	.4850	.4854	.4857
2.2	.4861	.4864	.4868	.4871	.4875	.4878	.4881	.4884	.4887	.4890
2.3	.4893	.4896	.4898	.4901	.4904	.4906	.4909	.4911	.4913	.4916
2.4	.4918	.4920	.4922	.4925	.4927	.4929	.4931	.4932	.4934	.4936
2.5	.4938	.4940	.4941	.4943	.4945	.4946	.4948	.4949	.4951	.4952
2.6	.4953	.4955	.4956	.4957	.4959	.4960	.4961	.4962	.4963	.4964
2.7	.4965	.4966	.4967	.4968	.4969	.4970	.4971	.4972	.4973	.4974
2.8	.4974	.4975	.4976	.4977	.4977	.4978	.4979	.4979	.4980	.4981
2.9	.4981	.4982	.4982	.4983	.4984	.4984	.4985	.4985	.4986	.4986
3.0	.49865	.49869	.49874	.49878	.49882	.49886	.49889	.49893	.49897	.49900
3.1	.49903	.49906	.49910	.49913	.49916	.49918	.49921	.49924	.49926	.49929
3.2	.49931	.49934	.49936	.49938	.49940	.49942	.49944	.49946	.49948	.49950
3.3	.49952	.49953	.49955	.49957	.49958	.49960	.49961	.49962	.49964	.49965
3.4	.49966	.49968	.49969	.49970	.49971	.49972	.49973	.49974	.49975	.49976
3.5	.49977	.49978	.49978	.49979	.49980	.49981	.49981	.49982	.49983	.49983
3.6	.49984	.49985	.49985	.49986	.49986	.49987	.49987	.49988	.49988	.49989
3.7	.49989	.49990	.49990	.49990	.49991	.49991	.49992	.49992	.49992	.49992
3.8	.49993	.49993	.49993	.49994	.49994	.49994	.49994	.49995	.49995	.49995
3.9	.49995	.49995	.49996	.49996	.49996	.49996	.49996	.49996	.49997	.49997

TABLE C.3 □ Critical values of t

For a particular number of degrees of freedom, entry represents the critical value of t corresponding to a specified upper tail area α

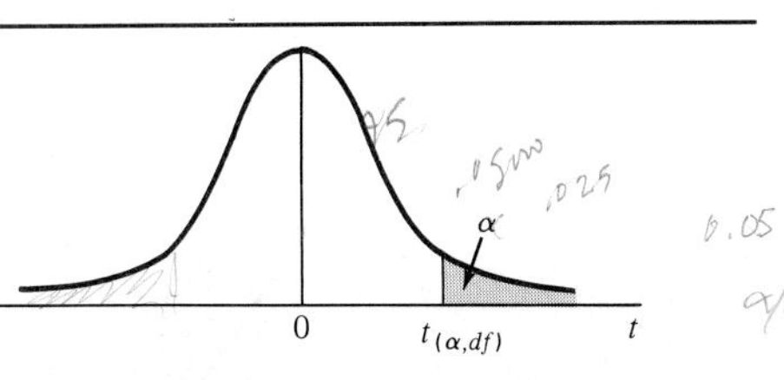

Degrees of Freedom	Upper Tail Areas .25	.10	.05	.025	.01	.005
1	1.0000	3.0777	6.3138	12.7062	31.8207	63.6574
2	0.8165	1.8856	2.9200	4.3027	6.9646	9.9248
3	0.7649	1.6377	2.3534	3.1824	4.5407	5.8409
4	0.7407	1.5332	2.1318	2.7764	3.7469	4.6041
5	0.7267	1.4759	2.0150	2.5706	3.3649	4.0322
6	0.7176	1.4398	1.9432	2.4469	3.1427	3.7074
7	0.7111	1.4149	1.8946	2.3646	2.9980	3.4995
8	0.7064	1.3968	1.8595	2.3060	2.8965	3.3554
9	0.7027	1.3830	1.8331	2.2622	2.8214	3.2498
10	0.6998	1.3722	1.8125	2.2281	2.7638	3.1693
11	0.6974	1.3634	1.7959	2.2010	2.7181	3.1058
12	0.6955	1.3562	1.7823	2.1788	2.6810	3.0545
13	0.6938	1.3502	1.7709	2.1604	2.6503	3.0123
14	0.6924	1.3450	1.7613	2.1448	2.6245	2.9768
15	0.6912	1.3406	1.7531	2.1315	2.6025	2.9467
16	0.6901	1.3368	1.7459	2.1199	2.5835	2.9208
17	0.6892	1.3334	1.7396	2.1098	2.5669	2.8982
18	0.6884	1.3304	1.7341	2.1009	2.5524	2.8784
19	0.6876	1.3277	1.7291	2.0930	2.5395	2.8609
20	0.6870	1.3253	1.7247	2.0860	2.5280	2.8453
21	0.6864	1.3232	1.7207	2.0796	2.5177	2.8314
22	0.6858	1.3212	1.7171	2.0739	2.5083	2.8188
23	0.6853	1.3195	1.7139	2.0687	2.4999	2.8073
24	0.6848	1.3178	1.7109	2.0639	2.4922	2.7969
25	0.6844	1.3163	1.7081	2.0595	2.4851	2.7874
26	0.6840	1.3150	1.7056	2.0555	2.4786	2.7787
27	0.6837	1.3137	1.7033	2.0518	2.4727	2.7707
28	0.6834	1.3125	1.7011	2.0484	2.4671	2.7633
29	0.6830	1.3114	1.6991	2.0452	2.4620	2.7564
30	0.6828	1.3104	1.6973	2.0423	2.4573	2.7500
31	0.6825	1.3095	1.6955	2.0395	2.4528	2.7440
32	0.6822	1.3086	1.6939	2.0369	2.4487	2.7385
33	0.6820	1.3077	1.6924	2.0345	2.4448	2.7333
34	0.6818	1.3070	1.6909	2.0322	2.4411	2.7284
35	0.6816	1.3062	1.6896	2.0301	2.4377	2.7238
36	0.6814	1.3055	1.6883	2.0281	2.4345	2.7195
37	0.6812	1.3049	1.6871	2.0262	2.4314	2.7154
38	0.6810	1.3042	1.6860	2.0244	2.4286	2.7116
39	0.6808	1.3036	1.6849	2.0227	2.4258	2.7079
40	0.6807	1.3031	1.6839	2.0211	2.4233	2.7045
41	0.6805	1.3025	1.6829	2.0195	2.4208	2.7012
42	0.6804	1.3020	1.6820	2.0181	2.4185	2.6981
43	0.6802	1.3016	1.6811	2.0167	2.4163	2.6951
44	0.6801	1.3011	1.6802	2.0154	2.4141	2.6923
45	0.6800	1.3006	1.6794	2.0141	2.4121	2.6896

TABLE C.3 *(continued)*

Degrees of Freedom	*Upper Tail Areas*					
	.25	*.10*	*.05*	*.025*	*.01*	*.005*
46	0.6799	1.3002	1.6787	2.0129	2.4102	2.6870
47	0.6797	1.2998	1.6779	2.0117	2.4083	2.6846
48	0.6796	1.2994	1.6772	2.0106	2.4066	2.6822
49	0.6795	1.2991	1.6766	2.0096	2.4049	2.6800
50	0.6794	1.2987	1.6759	2.0086	2.4033	2.6778
51	0.6793	1.2984	1.6753	2.0076	2.4017	2.6757
52	0.6792	1.2980	1.6747	2.0066	2.4002	2.6737
53	0.6791	1.2977	1.6741	2.0057	2.3988	2.6718
54	0.6791	1.2974	1.6736	2.0049	2.3974	2.6700
55	0.6790	1.2971	1.6730	2.0040	2.3961	2.6682
56	0.6789	1.2969	1.6725	2.0032	2.3948	2.6665
57	0.6788	1.2966	1.6720	2.0025	2.3936	2.6649
58	0.6787	1.2963	1.6716	2.0017	2.3924	2.6633
59	0.6787	1.2961	1.6711	2.0010	2.3912	2.6618
60	0.6786	1.2958	1.6706	2.0003	2.3901	2.6603
61	0.6785	1.2956	1.6702	1.9996	2.3890	2.6589
62	0.6785	1.2954	1.6698	1.9990	2.3880	2.6575
63	0.6784	1.2951	1.6694	1.9983	2.3870	2.6561
64	0.6783	1.2949	1.6690	1.9977	2.3860	2.6549
65	0.6783	1.2947	1.6686	1.9971	2.3851	2.6536
66	0.6782	1.2945	1.6683	1.9966	2.3842	2.6524
67	0.6782	1.2943	1.6679	1.9960	2.3833	2.6512
68	0.6781	1.2941	1.6676	1.9955	2.3824	2.6501
69	0.6781	1.2939	1.6672	1.9949	2.3816	2.6490
70	0.6780	1.2938	1.6669	1.9944	2.3808	2.6479
71	0.6780	1.2936	1.6666	1.9939	2.3800	2.6469
72	0.6779	1.2934	1.6663	1.9935	2.3793	2.6459
73	0.6779	1.2933	1.6660	1.9930	2.3785	2.6449
74	0.6778	1.4931	1.6657	1.9925	2.3778	2.6439
75	0.6778	1.2929	1.6654	1.9921	2.3771	2.6430
76	0.6777	1.2928	1.6652	1.9917	2.3764	2.6421
77	0.6777	1.2926	1.6649	1.9913	2.3758	2.6412
78	0.6776	1.2925	1.6646	1.9908	2.3751	2.6403
79	0.6776	1.2924	1.6644	1.9905	2.3745	2.6395
80	0.6776	1.2922	1.6641	1.9901	2.3739	2.6387
81	0.6775	1.2921	1.6639	1.9897	2.3733	2.6379
82	0.6775	1.2920	1.6636	1.9893	2.3727	2.6371
83	0.6775	1.2918	1.6634	1.9890	2.3721	2.6364
84	0.6774	1.2917	1.6632	1.9886	2.3716	2.6356
85	0.6774	1.2916	1.6630	1.9883	2.3710	2.6349
86	0.6774	1.2915	1.6628	1.9879	2.3705	2.6342
87	0.6773	1.2914	1.6626	1.9876	2.3700	2.6335
88	0.6773	1.2912	1.6624	1.9873	2.3695	2.6329
89	0.6773	1.2911	1.6622	1.9870	2.3690	2.6322
90	0.6772	1.2910	1.6620	1.9867	2.3685	2.6316
91	0.6772	1.2909	1.6618	1.9864	2.3680	2.6309
92	0.6772	1.2908	1.6616	1.9861	2.3676	2.6303
93	0.6771	1.2907	1.6614	1.9858	2.3671	2.6297
94	0.6771	1.2906	1.6612	1.9855	2.3667	2.6291
95	0.6771	1.2905	1.6611	1.9853	2.3662	2.6286

TABLE C.3 *(continued)*

Degrees of Freedom	*Upper Tail Areas*					
	.25	*.10*	*.05*	*.025*	*.01*	*.005*
96	0.6771	1.2904	1.6609	1.9850	2.3658	2.6280
97	0.6770	1.2903	1.6607	1.9847	2.3654	2.6275
98	0.6770	1.2902	1.6606	1.9845	2.3650	2.6269
99	0.6770	1.2902	1.6604	1.9842	2.3646	2.6264
100	0.6770	1.2901	1.6602	1.9840	2.3642	2.6259
110	0.6767	1.2893	1.6588	1.9818	2.3607	2.6213
120	0.6765	1.2886	1.6577	1.9799	2.3578	2.6174
∞	0.6745	1.2816	1.6449	1.9600	2.3263	2.5758

TABLE C.4 □ Critical values of χ^2

For a particular number of degrees of freedom, entry represents the critical value of χ^2 corresponding to a specified upper tail area (α)

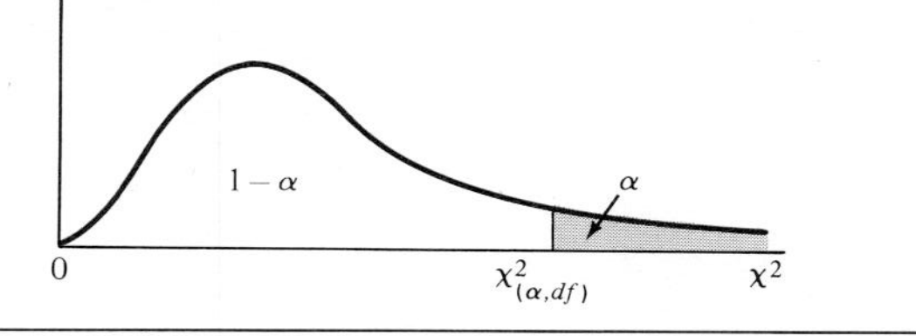

Degrees of Freedom	.995	.99	.975	.95	.90	.75	.25	.10	.05	.025	.01	.005
	Upper Tail Areas (α)											
1			0.001	0.004	0.016	0.102	1.323	2.706	3.841	5.024	6.635	7.879
2	0.010	0.020	0.051	0.103	0.211	0.575	2.773	4.605	5.991	7.378	9.210	10.597
3	0.072	0.115	0.216	0.352	0.584	1.213	4.108	6.251	7.815	9.348	11.345	12.838
4	0.207	0.297	0.484	0.711	1.064	1.923	5.385	7.779	9.488	11.143	13.277	14.860
5	0.412	0.554	0.831	1.145	1.610	2.675	6.626	9.236	11.071	12.833	15.086	16.750
6	0.676	0.872	1.237	1.635	2.204	3.455	7.841	10.645	12.592	14.449	16.812	18.548
7	0.989	1.239	1.690	2.167	2.833	4.255	9.037	12.017	14.067	16.013	18.475	20.278
8	1.344	1.646	2.180	2.733	3.490	5.071	10.219	13.362	15.507	17.535	20.090	21.955
9	1.735	2.088	2.700	3.325	4.168	5.899	11.389	14.684	16.919	19.023	21.666	23.589
10	2.156	2.558	3.247	3.940	4.865	6.737	12.549	15.987	18.307	20.483	23.209	25.188
11	2.603	3.053	3.816	4.575	5.578	7.584	13.701	17.275	19.675	21.920	24.725	26.757
12	3.074	3.571	4.404	5.226	6.304	8.438	14.845	18.549	21.026	23.337	26.217	28.299
13	3.565	4.107	5.009	5.892	7.042	9.299	15.984	19.812	22.362	24.736	27.688	29.819
14	4.075	4.660	5.629	6.571	7.790	10.165	17.117	21.064	23.685	26.119	29.141	31.319
15	4.601	5.229	6.262	7.261	8.547	11.037	18.245	22.307	24.996	27.488	30.578	32.801
16	5.142	5.812	6.908	7.962	9.312	11.912	19.369	23.542	26.296	28.845	32.000	34.267
17	5.697	6.408	7.564	8.672	10.085	12.792	20.489	24.769	27.587	30.191	33.409	35.718
18	6.265	7.015	8.231	9.390	10.865	13.675	21.605	25.989	28.869	31.526	34.805	37.156
19	6.844	7.633	8.907	10.117	11.651	14.562	22.718	27.204	30.144	32.852	36.191	38.582
20	7.434	8.260	9.591	10.851	12.443	15.452	23.828	28.412	31.410	34.170	37.566	39.997
21	8.034	8.897	10.283	11.591	13.240	16.344	24.935	29.615	32.671	35.479	38.932	41.401
22	8.643	9.542	10.982	12.338	14.042	17.240	26.039	30.813	33.924	36.781	40.289	42.796
23	9.260	10.196	11.689	13.091	14.848	18.137	27.141	32.007	35.172	38.076	41.638	44.181
24	9.886	10.856	12.401	13.848	15.659	19.037	28.241	33.196	36.415	39.364	42.980	45.559
25	10.520	11.524	13.120	14.611	16.473	19.939	29.339	34.382	37.652	40.646	44.314	46.928
26	11.160	12.198	13.844	15.379	17.292	20.843	30.435	35.563	38.885	41.923	45.642	48.290
27	11.808	12.879	14.573	16.151	18.114	21.749	31.528	36.741	40.113	43.194	46.963	49.645
28	12.461	13.565	15.308	16.928	18.939	22.657	32.620	37.916	41.337	44.461	48.278	50.993
29	13.121	14.257	16.047	17.708	19.768	23.567	33.711	39.087	42.557	45.722	49.588	52.336
30	13.787	14.954	16.791	18.493	20.599	24.478	34.800	40.256	43.773	46.979	50.892	53.672

TABLE C.4 *(continued)*

Degrees of Freedom	Upper Tail Areas (α)											
	.995	.99	.975	.95	.90	.75	.25	.10	.05	.025	.01	.005
31	14.458	15.655	17.539	19.281	21.434	25.390	35.887	41.422	44.985	48.232	52.191	55.003
32	15.134	16.362	18.291	20.072	22.271	26.304	36.973	42.585	46.194	49.480	53.486	56.328
33	15.815	17.074	19.047	20.867	23.110	27.219	38.058	43.745	47.400	50.725	54.776	57.648
34	16.501	17.789	19.806	21.664	23.952	28.136	39.141	44.903	48.602	51.966	56.061	58.964
35	17.192	18.509	20.569	22.465	24.797	29.054	40.223	46.059	49.802	53.203	57.342	60.275
36	17.887	19.233	21.336	23.269	25.643	29.973	41.304	47.212	50.998	54.437	58.619	61.581
37	18.586	19.960	22.106	24.075	26.492	30.893	42.383	48.363	52.192	55.668	59.892	62.883
38	19.289	20.691	22.878	24.884	27.343	31.815	43.462	49.513	53.384	56.896	61.162	64.181
39	19.996	21.426	23.654	25.695	28.196	32.737	44.539	50.660	54.572	58.120	62.428	65.476
40	20.707	22.164	24.433	26.509	29.051	33.660	45.616	51.805	55.758	59.342	63.691	66.766
41	21.421	22.906	25.215	27.326	29.907	34.585	46.692	52.949	56.942	60.561	64.950	68.053
42	22.138	23.650	25.999	28.144	30.765	35.510	47.766	54.090	58.124	61.777	66.206	69.336
43	22.859	24.398	26.785	28.965	31.625	36.436	48.840	55.230	59.304	62.990	67.459	70.616
44	23.584	25.148	27.575	29.787	32.487	37.363	49.913	56.369	60.481	64.201	68.710	71.893
45	24.311	25.901	28.366	30.612	33.350	38.291	50.985	57.505	61.656	65.410	69.957	73.166
46	25.041	26.657	29.160	31.439	34.215	39.220	52.056	58.641	62.830	66.617	71.201	74.437
47	25.775	27.416	29.956	32.268	35.081	40.149	53.127	59.774	64.001	67.821	72.443	75.704
48	26.511	28.177	30.755	33.098	35.949	41.079	54.196	60.907	65.171	69.023	73.683	76.969
49	27.249	28.941	31.555	33.930	36.818	42.010	55.265	62.038	66.339	70.222	74.919	78.231
50	27.991	29.707	32.357	34.764	37.689	42.942	56.334	63.167	67.505	71.420	76.154	79.490
51	28.735	30.475	33.162	35.600	38.560	43.874	57.401	64.295	68.669	72.616	77.386	80.747
52	29.481	31.246	33.968	36.437	39.433	44.808	58.468	65.422	69.832	73.810	78.616	82.001
53	30.230	32.018	34.776	37.276	40.308	45.741	59.534	66.548	70.993	75.002	79.843	83.253
54	30.981	32.793	35.586	38.116	41.183	46.676	60.600	67.673	72.153	76.192	81.069	84.502
55	31.735	33.570	36.398	38.958	42.060	47.610	61.665	68.796	73.311	77.380	82.292	85.749
56	32.490	34.350	37.212	39.801	42.937	48.546	62.729	69.919	74.468	78.567	83.513	86.994
57	33.248	35.131	38.027	40.646	43.816	49.482	63.793	71.040	75.624	79.752	84.733	88.236
58	34.008	35.913	38.844	41.492	44.696	50.419	64.857	72.160	76.778	80.936	85.950	89.477
59	34.770	36.698	39.662	42.339	45.577	51.356	65.919	73.279	77.931	82.117	87.166	90.715
60	35.534	37.485	40.482	43.188	46.459	52.294	66.981	74.397	79.082	83.298	88.379	91.952

For larger values of degrees of freedom (DF) the expression $z = \sqrt{2\chi^2} - \sqrt{2(DF) - 1}$ may be used and the resulting upper tail area can be obtained from the table of the standardized normal distribution (Table E.2).

TABLE C.5 □ Critical values of *F*

For a particular combination of numerator and denominator degrees of freedom, entry represents the critical values of *F* corresponding to a specified upper tail area (α).

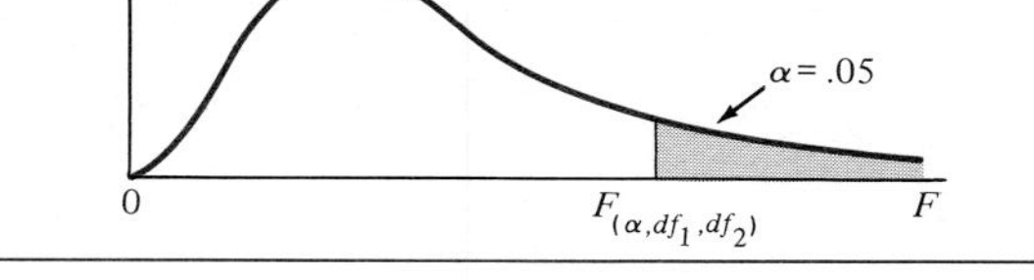

Denominator df_2	Numerator df_1: 1	2	3	4	5	6	7	8	9	10	12	15	20	24	30	40	60	120	∞
1	161.4	199.5	215.7	224.6	230.2	234.0	236.8	238.9	240.5	241.9	243.9	245.9	248.0	249.1	250.1	251.1	252.2	253.3	254.3
2	18.51	19.00	19.16	19.25	19.30	19.33	19.35	19.37	19.38	19.40	19.41	19.43	19.45	19.45	19.46	19.47	19.48	19.49	19.50
3	10.13	9.55	9.28	9.12	9.01	8.94	8.89	8.85	8.81	8.79	8.74	8.70	8.66	8.64	8.62	8.59	8.57	8.55	8.53
4	7.71	6.94	6.59	6.39	6.26	6.16	6.09	6.04	6.00	5.96	5.91	5.86	5.80	5.77	5.75	5.72	5.69	5.66	5.63
5	6.61	5.79	5.41	5.19	5.05	4.95	4.88	4.82	4.77	4.74	4.68	4.62	4.56	4.53	4.50	4.46	4.43	4.40	4.36
6	5.99	5.14	4.76	4.53	4.39	4.28	4.21	4.15	4.10	4.06	4.00	3.94	3.87	3.84	3.81	3.77	3.74	3.70	3.67
7	5.59	4.74	4.35	4.12	3.97	3.87	3.79	3.73	3.68	3.64	3.57	3.51	3.44	3.41	3.38	3.34	3.30	3.27	3.23
8	5.32	4.46	4.07	3.84	3.69	3.58	3.50	3.44	3.39	3.35	3.28	3.22	3.15	3.12	3.08	3.04	3.01	2.97	2.93
9	5.12	4.26	3.86	3.63	3.48	3.37	3.29	3.23	3.18	3.14	3.07	3.01	2.94	2.90	2.86	2.83	2.79	2.75	2.71
10	4.96	4.10	3.71	3.48	3.33	3.22	3.14	3.07	3.02	2.98	2.91	2.85	2.77	2.74	2.70	2.66	2.62	2.58	2.54
11	4.84	3.98	3.59	3.36	3.20	3.09	3.01	2.95	2.90	2.85	2.79	2.72	2.65	2.61	2.57	2.53	2.49	2.45	2.40
12	4.75	3.89	3.49	3.26	3.11	3.00	2.91	2.85	2.80	2.75	2.69	2.62	2.54	2.51	2.47	2.43	2.38	2.34	2.30
13	4.67	3.81	3.41	3.18	3.03	2.92	2.83	2.77	2.71	2.67	2.60	2.53	2.46	2.42	2.38	2.34	2.30	2.25	2.21
14	4.60	3.74	3.34	3.11	2.96	2.85	2.76	2.70	2.65	2.60	2.53	2.46	2.39	2.35	2.31	2.27	2.22	2.18	2.13
15	4.54	3.68	3.29	3.06	2.90	2.79	2.71	2.64	2.59	2.54	2.48	2.40	2.33	2.29	2.25	2.20	2.16	2.11	2.07
16	4.49	3.63	3.24	3.01	2.85	2.74	2.66	2.59	2.54	2.49	2.42	2.35	2.28	2.24	2.19	2.15	2.11	2.06	2.01
17	4.45	3.59	3.20	2.96	2.81	2.70	2.61	2.55	2.49	2.45	2.38	2.31	2.23	2.19	2.15	2.10	2.06	2.01	1.96
18	4.41	3.55	3.16	2.93	2.77	2.66	2.58	2.51	2.46	2.41	2.34	2.27	2.19	2.15	2.11	2.06	2.02	1.97	1.92
19	4.38	3.52	3.13	2.90	2.74	2.63	2.54	2.48	2.42	2.38	2.31	2.23	2.16	2.11	2.07	2.03	1.98	1.93	1.88
20	4.35	3.49	3.10	2.87	2.71	2.60	2.51	2.45	2.39	2.35	2.28	2.20	2.12	2.08	2.04	1.99	1.95	1.90	1.84
21	4.32	3.47	3.07	2.84	2.68	2.57	2.49	2.42	2.37	2.32	2.25	2.18	2.10	2.05	2.01	1.96	1.92	1.87	1.81
22	4.30	3.44	3.05	2.82	2.66	2.55	2.46	2.40	2.34	2.30	2.23	2.15	2.07	2.03	1.98	1.94	1.89	1.84	1.78
23	4.28	3.42	3.03	2.80	2.64	2.53	2.44	2.37	2.32	2.27	2.20	2.13	2.05	2.01	1.96	1.91	1.86	1.81	1.76
24	4.26	3.40	3.01	2.78	2.62	2.51	2.42	2.36	2.30	2.25	2.18	2.11	2.03	1.98	1.94	1.89	1.84	1.79	1.73
25	4.24	3.39	2.99	2.76	2.60	2.49	2.40	2.34	2.28	2.24	2.16	2.09	2.01	1.96	1.92	1.87	1.82	1.77	1.71
26	4.23	3.37	2.98	2.74	2.59	2.47	2.39	2.32	2.27	2.22	2.15	2.07	1.99	1.95	1.90	1.85	1.80	1.75	1.69
27	4.21	3.35	2.96	2.73	2.57	2.46	2.37	2.31	2.25	2.20	2.13	2.06	1.97	1.93	1.88	1.84	1.79	1.73	1.67
28	4.20	3.34	2.95	2.71	2.56	2.45	2.36	2.29	2.24	2.19	2.12	2.04	1.96	1.91	1.87	1.82	1.77	1.71	1.65
29	4.18	3.33	2.93	2.70	2.55	2.43	2.35	2.28	2.22	2.18	2.10	2.03	1.94	1.90	1.85	1.81	1.75	1.70	1.64
30	4.17	3.32	2.92	2.69	2.53	2.42	2.33	2.27	2.21	2.16	2.09	2.01	1.93	1.89	1.84	1.79	1.74	1.68	1.62
40	4.08	3.23	2.84	2.61	2.45	2.34	2.25	2.18	2.12	2.08	2.00	1.92	1.84	1.79	1.74	1.69	1.64	1.58	1.51
60	4.00	3.15	2.76	2.53	2.37	2.25	2.17	2.10	2.04	1.99	1.92	1.84	1.75	1.70	1.65	1.59	1.53	1.47	1.39
120	3.92	3.07	2.68	2.45	2.29	2.17	2.09	2.02	1.96	1.91	1.83	1.75	1.66	1.61	1.55	1.50	1.43	1.35	1.25
∞	3.84	3.00	2.60	2.37	2.21	2.10	2.01	1.94	1.88	1.83	1.75	1.67	1.57	1.52	1.46	1.39	1.32	1.22	1.00

TABLE C.5 *(continued)*

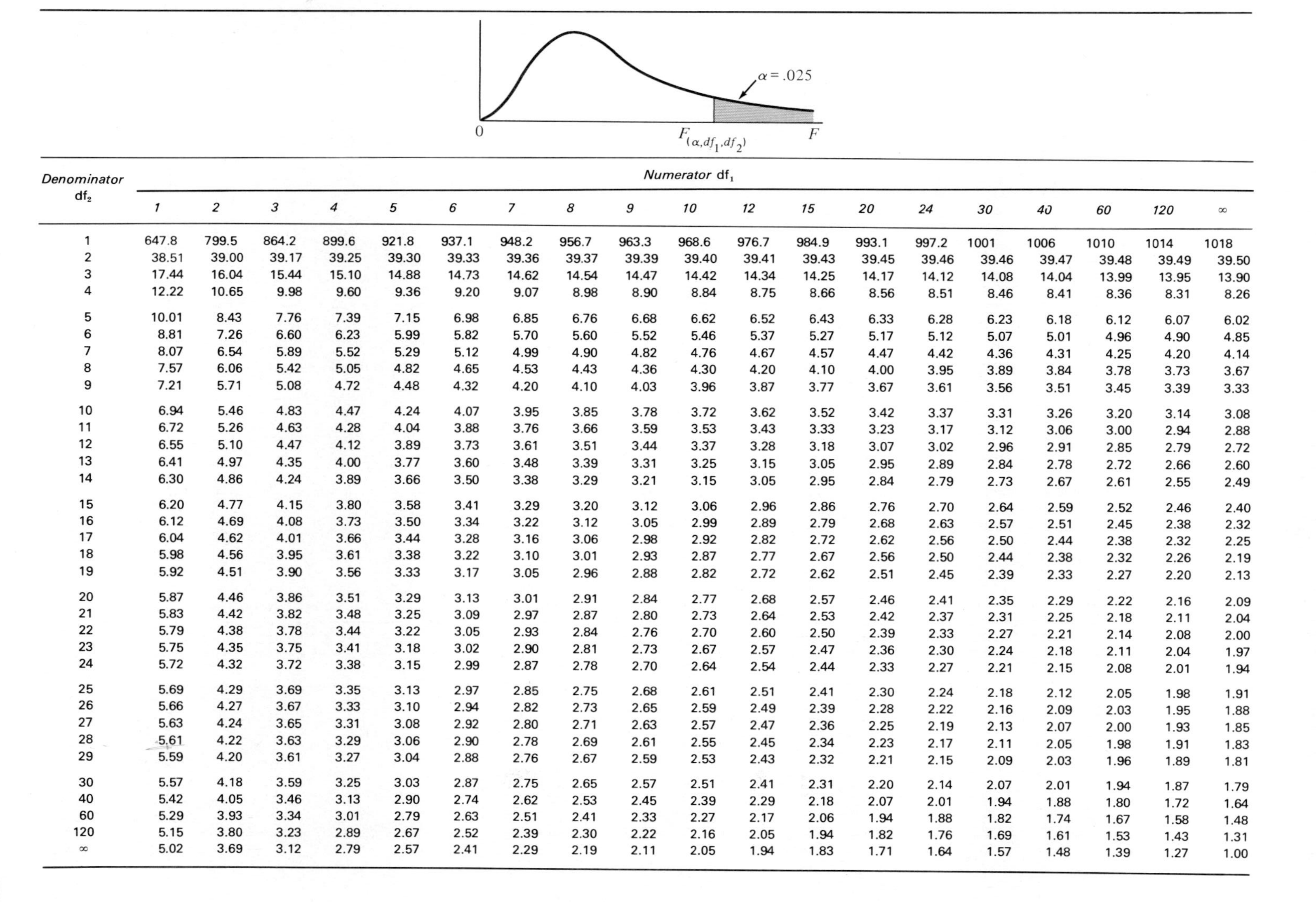

Denominator df_2	Numerator df_1: 1	2	3	4	5	6	7	8	9	10	12	15	20	24	30	40	60	120	∞
1	647.8	799.5	864.2	899.6	921.8	937.1	948.2	956.7	963.3	968.6	976.7	984.9	993.1	997.2	1001	1006	1010	1014	1018
2	38.51	39.00	39.17	39.25	39.30	39.33	39.36	39.37	39.39	39.40	39.41	39.43	39.45	39.46	39.46	39.47	39.48	39.49	39.50
3	17.44	16.04	15.44	15.10	14.88	14.73	14.62	14.54	14.47	14.42	14.34	14.25	14.17	14.12	14.08	14.04	13.99	13.95	13.90
4	12.22	10.65	9.98	9.60	9.36	9.20	9.07	8.98	8.90	8.84	8.75	8.66	8.56	8.51	8.46	8.41	8.36	8.31	8.26
5	10.01	8.43	7.76	7.39	7.15	6.98	6.85	6.76	6.68	6.62	6.52	6.43	6.33	6.28	6.23	6.18	6.12	6.07	6.02
6	8.81	7.26	6.60	6.23	5.99	5.82	5.70	5.60	5.52	5.46	5.37	5.27	5.17	5.12	5.07	5.01	4.96	4.90	4.85
7	8.07	6.54	5.89	5.52	5.29	5.12	4.99	4.90	4.82	4.76	4.67	4.57	4.47	4.42	4.36	4.31	4.25	4.20	4.14
8	7.57	6.06	5.42	5.05	4.82	4.65	4.53	4.43	4.36	4.30	4.20	4.10	4.00	3.95	3.89	3.84	3.78	3.73	3.67
9	7.21	5.71	5.08	4.72	4.48	4.32	4.20	4.10	4.03	3.96	3.87	3.77	3.67	3.61	3.56	3.51	3.45	3.39	3.33
10	6.94	5.46	4.83	4.47	4.24	4.07	3.95	3.85	3.78	3.72	3.62	3.52	3.42	3.37	3.31	3.26	3.20	3.14	3.08
11	6.72	5.26	4.63	4.28	4.04	3.88	3.76	3.66	3.59	3.53	3.43	3.33	3.23	3.17	3.12	3.06	3.00	2.94	2.88
12	6.55	5.10	4.47	4.12	3.89	3.73	3.61	3.51	3.44	3.37	3.28	3.18	3.07	3.02	2.96	2.91	2.85	2.79	2.72
13	6.41	4.97	4.35	4.00	3.77	3.60	3.48	3.39	3.31	3.25	3.15	3.05	2.95	2.89	2.84	2.78	2.72	2.66	2.60
14	6.30	4.86	4.24	3.89	3.66	3.50	3.38	3.29	3.21	3.15	3.05	2.95	2.84	2.79	2.73	2.67	2.61	2.55	2.49
15	6.20	4.77	4.15	3.80	3.58	3.41	3.29	3.20	3.12	3.06	2.96	2.86	2.76	2.70	2.64	2.59	2.52	2.46	2.40
16	6.12	4.69	4.08	3.73	3.50	3.34	3.22	3.12	3.05	2.99	2.89	2.79	2.68	2.63	2.57	2.51	2.45	2.38	2.32
17	6.04	4.62	4.01	3.66	3.44	3.28	3.16	3.06	2.98	2.92	2.82	2.72	2.62	2.56	2.50	2.44	2.38	2.32	2.25
18	5.98	4.56	3.95	3.61	3.38	3.22	3.10	3.01	2.93	2.87	2.77	2.67	2.56	2.50	2.44	2.38	2.32	2.26	2.19
19	5.92	4.51	3.90	3.56	3.33	3.17	3.05	2.96	2.88	2.82	2.72	2.62	2.51	2.45	2.39	2.33	2.27	2.20	2.13
20	5.87	4.46	3.86	3.51	3.29	3.13	3.01	2.91	2.84	2.77	2.68	2.57	2.46	2.41	2.35	2.29	2.22	2.16	2.09
21	5.83	4.42	3.82	3.48	3.25	3.09	2.97	2.87	2.80	2.73	2.64	2.53	2.42	2.37	2.31	2.25	2.18	2.11	2.04
22	5.79	4.38	3.78	3.44	3.22	3.05	2.93	2.84	2.76	2.70	2.60	2.50	2.39	2.33	2.27	2.21	2.14	2.08	2.00
23	5.75	4.35	3.75	3.41	3.18	3.02	2.90	2.81	2.73	2.67	2.57	2.47	2.36	2.30	2.24	2.18	2.11	2.04	1.97
24	5.72	4.32	3.72	3.38	3.15	2.99	2.87	2.78	2.70	2.64	2.54	2.44	2.33	2.27	2.21	2.15	2.08	2.01	1.94
25	5.69	4.29	3.69	3.35	3.13	2.97	2.85	2.75	2.68	2.61	2.51	2.41	2.30	2.24	2.18	2.12	2.05	1.98	1.91
26	5.66	4.27	3.67	3.33	3.10	2.94	2.82	2.73	2.65	2.59	2.49	2.39	2.28	2.22	2.16	2.09	2.03	1.95	1.88
27	5.63	4.24	3.65	3.31	3.08	2.92	2.80	2.71	2.63	2.57	2.47	2.36	2.25	2.19	2.13	2.07	2.00	1.93	1.85
28	5.61	4.22	3.63	3.29	3.06	2.90	2.78	2.69	2.61	2.55	2.45	2.34	2.23	2.17	2.11	2.05	1.98	1.91	1.83
29	5.59	4.20	3.61	3.27	3.04	2.88	2.76	2.67	2.59	2.53	2.43	2.32	2.21	2.15	2.09	2.03	1.96	1.89	1.81
30	5.57	4.18	3.59	3.25	3.03	2.87	2.75	2.65	2.57	2.51	2.41	2.31	2.20	2.14	2.07	2.01	1.94	1.87	1.79
40	5.42	4.05	3.46	3.13	2.90	2.74	2.62	2.53	2.45	2.39	2.29	2.18	2.07	2.01	1.94	1.88	1.80	1.72	1.64
60	5.29	3.93	3.34	3.01	2.79	2.63	2.51	2.41	2.33	2.27	2.17	2.06	1.94	1.88	1.82	1.74	1.67	1.58	1.48
120	5.15	3.80	3.23	2.89	2.67	2.52	2.39	2.30	2.22	2.16	2.05	1.94	1.82	1.76	1.69	1.61	1.53	1.43	1.31
∞	5.02	3.69	3.12	2.79	2.57	2.41	2.29	2.19	2.11	2.05	1.94	1.83	1.71	1.64	1.57	1.48	1.39	1.27	1.00

TABLE C.5 *(continued)*

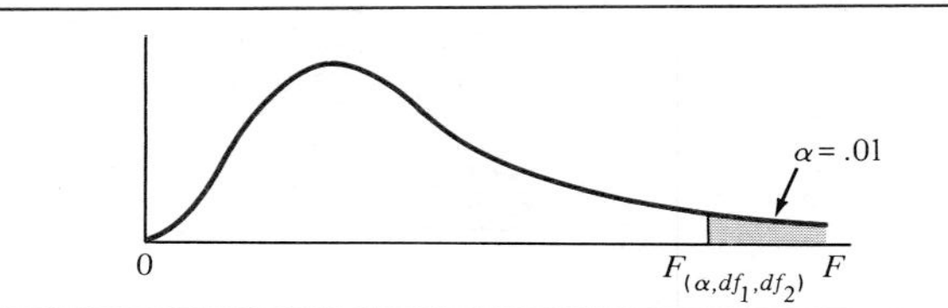

Denominator df_2	Numerator df_1																		
	1	2	3	4	5	6	7	8	9	10	12	15	20	24	30	40	60	120	∞
1	4052	4999.5	5403	5625	5764	5859	5928	5982	6022	6056	6106	6157	6209	6235	6261	6287	6313	6339	6366
2	98.50	99.00	99.17	99.25	99.30	99.33	99.36	99.37	99.39	99.40	99.42	99.43	99.45	99.46	99.47	99.47	99.48	99.49	99.50
3	34.12	30.82	29.46	28.71	28.24	27.91	27.67	27.49	27.35	27.23	27.05	26.87	26.69	26.60	26.50	26.41	26.32	26.22	26.13
4	21.20	18.00	16.69	15.98	15.52	15.21	14.98	14.80	14.66	14.55	14.37	14.20	14.02	13.93	13.84	13.75	13.65	13.56	13.46
5	16.26	13.27	12.06	11.39	10.97	10.67	10.46	10.29	10.16	10.05	9.89	9.72	9.55	9.47	9.38	9.29	9.20	9.11	9.02
6	13.75	10.92	9.78	9.15	8.75	8.47	8.26	8.10	7.98	7.87	7.72	7.56	7.40	7.31	7.23	7.14	7.06	6.97	6.88
7	12.25	9.55	8.45	7.85	7.46	7.19	6.99	6.84	6.72	6.62	6.47	6.31	6.16	6.07	5.99	5.91	5.82	5.74	5.65
8	11.26	8.65	7.59	7.01	6.63	6.37	6.18	6.03	5.91	5.81	5.67	5.52	5.36	5.28	5.20	5.12	5.03	4.95	4.86
9	10.56	8.02	6.99	6.42	6.06	5.80	5.61	5.47	5.35	5.26	5.11	4.96	4.81	4.73	4.65	4.57	4.48	4.40	4.31
10	10.04	7.56	6.55	5.99	5.64	5.39	5.20	5.06	4.94	4.85	4.71	4.56	4.41	4.33	4.25	4.17	4.08	4.00	3.91
11	9.65	7.21	6.22	5.67	5.32	5.07	4.89	4.74	4.63	4.54	4.40	4.25	4.10	4.02	3.94	3.86	3.78	3.69	3.60
12	9.33	6.93	5.95	5.41	5.06	4.82	4.64	4.50	4.39	4.30	4.16	4.01	3.86	3.78	3.70	3.62	3.54	3.45	3.36
13	9.07	6.70	5.74	5.21	4.86	4.62	4.44	4.30	4.19	4.10	3.96	3.82	3.66	3.59	3.51	3.43	3.34	3.25	3.17
14	8.86	6.51	5.56	5.04	4.69	4.46	4.28	4.14	4.03	3.94	3.80	3.66	3.51	3.43	3.35	3.27	3.18	3.09	3.00
15	8.68	6.36	5.42	4.89	4.56	4.32	4.14	4.00	3.89	3.80	3.67	3.52	3.37	3.29	3.21	3.13	3.05	2.96	2.87
16	8.53	6.23	5.29	4.77	4.44	4.20	4.03	3.89	3.78	3.69	3.55	3.41	3.26	3.18	3.10	3.02	2.93	2.84	2.75
17	8.40	6.11	5.18	4.67	4.34	4.10	3.93	3.79	3.68	3.59	3.46	3.31	3.16	3.08	3.00	2.92	2.83	2.75	2.65
18	8.29	6.01	5.09	4.58	4.25	4.01	3.84	3.71	3.60	3.51	3.37	3.23	3.08	3.00	2.92	2.84	2.75	2.66	2.57
19	8.18	5.93	5.01	4.50	4.17	3.94	3.77	3.63	3.52	3.43	3.30	3.15	3.00	2.92	2.84	2.76	2.67	2.58	2.49
20	8.10	5.85	4.94	4.43	4.10	3.87	3.70	3.56	3.46	3.37	3.23	3.09	2.94	2.86	2.78	2.69	2.61	2.52	2.42
21	8.02	5.78	4.87	4.37	4.04	3.81	3.64	3.51	3.40	3.31	3.17	3.03	2.88	2.80	2.72	2.64	2.55	2.46	2.36
22	7.95	5.72	4.82	4.31	3.99	3.76	3.59	3.45	3.35	3.26	3.12	2.98	2.83	2.75	2.67	2.58	2.50	2.40	2.31
23	7.88	5.66	4.76	4.26	3.94	3.71	3.54	3.41	3.30	3.21	3.07	2.93	2.78	2.70	2.62	2.54	2.45	2.35	2.26
24	7.82	5.61	4.72	4.22	3.90	3.67	3.50	3.36	3.26	3.17	3.03	2.89	2.74	2.66	2.58	2.49	2.40	2.31	2.21
25	7.77	5.57	4.68	4.18	3.85	3.63	3.46	3.32	3.22	3.13	2.99	2.85	2.70	2.62	2.54	2.45	2.36	2.27	2.17
26	7.72	5.53	4.64	4.14	3.82	3.59	3.42	3.29	3.18	3.09	2.96	2.81	2.66	2.58	2.50	2.42	2.33	2.23	2.13
27	7.68	5.49	4.60	4.11	3.78	3.56	3.39	3.26	3.15	3.06	2.93	2.78	2.63	2.55	2.47	2.38	2.29	2.20	2.10
28	7.64	5.45	4.57	4.07	3.75	3.53	3.36	3.23	3.12	3.03	2.90	2.75	2.60	2.52	2.44	2.35	2.26	2.17	2.06
29	7.60	5.42	4.54	4.04	3.73	3.50	3.33	3.20	3.09	3.00	2.87	2.73	2.57	2.49	2.41	2.33	2.23	2.14	2.03
30	7.56	5.39	4.51	4.02	3.70	3.47	3.30	3.17	3.07	2.98	2.84	2.70	2.55	2.47	2.39	2.30	2.21	2.11	2.01
40	7.31	5.18	4.31	3.83	3.51	3.29	3.12	2.99	2.89	2.80	2.66	2.52	2.37	2.29	2.20	2.11	2.02	1.92	1.80
60	7.08	4.98	4.13	3.65	3.34	3.12	2.95	2.82	2.72	2.63	2.50	2.35	2.20	2.12	2.03	1.94	1.84	1.73	1.60
120	6.85	4.79	3.95	3.48	3.17	2.96	2.79	2.66	2.56	2.47	2.34	2.19	2.03	1.95	1.86	1.76	1.66	1.53	1.38
∞	6.63	4.61	3.78	3.32	3.02	2.80	2.64	2.51	2.41	2.32	2.18	2.04	1.88	1.79	1.70	1.59	1.47	1.32	1.00

TABLE C.5 (continued)

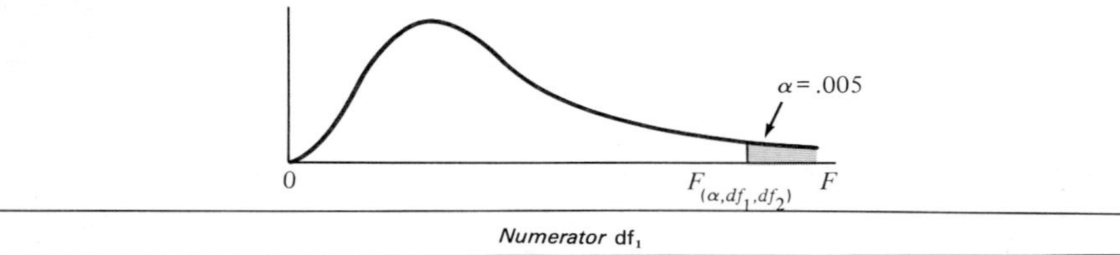

Denominator df_2	Numerator df_1: 1	2	3	4	5	6	7	8	9	10	12	15	20	24	30	40	60	120	∞
1	16211	20000	21615	22500	23056	23437	23715	23925	24091	24224	24426	24630	24836	24940	25044	25148	25253	25359	25465
2	198.5	199.0	199.2	199.2	199.3	199.3	199.4	199.4	199.4	199.4	199.4	199.4	199.4	199.5	199.5	199.5	199.5	199.5	199.5
3	55.55	49.80	47.47	46.19	45.39	44.84	44.43	44.13	43.88	43.69	43.39	43.08	42.78	42.62	42.47	42.31	42.15	41.99	41.83
4	31.33	26.28	24.26	23.15	22.46	21.97	21.62	21.35	21.14	20.97	20.70	20.44	20.17	20.03	19.89	19.75	19.61	19.47	19.32
5	22.78	18.31	16.53	15.56	14.94	14.51	14.20	13.96	13.77	13.62	13.38	13.15	12.90	12.78	12.66	12.53	12.40	12.27	12.14
6	18.63	14.54	12.92	12.03	11.46	11.07	10.79	10.57	10.39	10.25	10.03	9.81	9.59	9.47	9.36	9.24	9.12	9.00	8.88
7	16.24	12.40	10.88	10.05	9.52	9.16	8.89	8.68	8.51	8.38	8.18	7.97	7.75	7.65	7.53	7.42	7.31	7.19	7.08
8	14.69	11.04	9.60	8.81	8.30	7.95	7.69	7.50	7.34	7.21	7.01	6.81	6.61	6.50	6.40	6.29	6.18	6.06	5.95
9	13.61	10.11	8.72	7.96	7.47	7.13	6.88	6.69	6.54	6.42	6.23	6.03	5.83	5.73	5.62	5.52	5.41	5.30	5.19
10	12.83	9.43	8.08	7.34	6.87	6.54	6.30	6.12	5.97	5.85	5.66	5.47	5.27	5.17	5.07	4.97	4.86	4.75	4.64
11	12.23	8.91	7.60	6.88	6.42	6.10	5.86	5.68	5.54	5.42	5.24	5.05	4.86	4.76	4.65	4.55	4.44	4.34	4.23
12	11.75	8.51	7.23	6.52	6.07	5.76	5.52	5.35	5.20	5.09	4.91	4.72	4.53	4.43	4.33	4.23	4.12	4.01	3.90
13	11.37	8.19	6.93	6.23	5.79	5.48	5.25	5.08	4.94	4.82	4.64	4.46	4.27	4.17	4.07	3.97	3.87	3.76	3.65
14	11.06	7.92	6.68	6.00	5.56	5.26	5.03	4.86	4.72	4.60	4.43	4.25	4.06	3.96	3.86	3.76	3.66	3.55	3.44
15	10.80	7.70	6.48	5.80	5.37	5.07	4.85	4.67	4.54	4.42	4.25	4.07	3.88	3.79	3.69	3.58	3.48	3.37	3.26
16	10.58	7.51	6.30	5.64	5.21	4.91	4.69	4.52	4.38	4.27	4.10	3.92	3.73	3.64	3.54	3.44	3.33	3.22	3.11
17	10.38	7.35	6.16	5.50	5.07	4.78	4.56	4.39	4.25	4.14	3.97	3.79	3.61	3.51	3.41	3.31	3.21	3.10	2.98
18	10.22	7.21	6.03	5.37	4.96	4.66	4.44	4.28	4.14	4.03	3.86	3.68	3.50	3.40	3.30	3.20	3.10	2.99	2.87
19	10.07	7.09	5.92	5.27	4.85	4.56	4.34	4.18	4.04	3.93	3.76	3.59	3.40	3.31	3.21	3.11	3.00	2.89	2.78
20	9.94	6.99	5.82	5.17	4.76	4.47	4.26	4.09	3.96	3.85	3.68	3.50	3.32	3.22	3.12	3.02	2.92	2.81	2.69
21	9.83	6.89	5.73	5.09	4.68	4.39	4.18	4.01	3.88	3.77	3.60	3.43	3.24	3.15	3.05	2.95	2.84	2.73	2.61
22	9.73	6.81	5.65	5.02	4.61	4.32	4.11	3.94	3.81	3.70	3.54	3.36	3.18	3.08	2.98	2.88	2.77	2.66	2.55
23	9.63	6.73	5.58	4.95	4.54	4.26	4.05	3.88	3.75	3.64	3.47	3.30	3.12	3.02	2.92	2.82	2.71	2.60	2.48
24	9.55	6.66	5.52	4.89	4.49	4.20	3.99	3.83	3.69	3.59	3.42	3.25	3.06	2.97	2.87	2.77	2.66	2.55	2.43
25	9.48	6.60	5.46	4.84	4.43	4.15	3.94	3.78	3.64	3.54	3.37	3.20	3.01	2.92	2.82	2.72	2.61	2.50	2.38
26	9.41	6.54	5.41	4.79	4.38	4.10	3.89	3.73	3.60	3.49	3.33	3.15	2.97	2.87	2.77	2.67	2.56	2.45	2.33
27	9.34	6.49	5.36	4.74	4.34	4.06	3.85	3.69	3.56	3.45	3.28	3.11	2.93	2.83	2.73	2.63	2.52	2.41	2.29
28	9.28	6.44	5.32	4.70	4.30	4.02	3.81	3.65	3.52	3.41	3.25	3.07	2.89	2.79	2.69	2.59	2.48	2.37	2.25
29	9.23	6.40	5.28	4.66	4.26	3.98	3.77	3.61	3.48	3.38	3.21	3.04	2.86	2.76	2.66	2.56	2.45	2.33	2.21
30	9.18	6.35	5.24	4.62	4.23	3.95	3.74	3.58	3.45	3.34	3.18	3.01	2.82	2.73	2.63	2.52	2.42	2.30	2.18
40	8.83	6.07	4.98	4.37	3.99	3.71	3.51	3.35	3.22	3.12	2.95	2.78	2.60	2.50	2.40	2.30	2.18	2.06	1.93
60	8.49	5.79	4.73	4.14	3.76	3.49	3.29	3.13	3.01	2.90	2.74	2.57	2.39	2.29	2.19	2.08	1.96	1.83	1.69
120	8.18	5.54	4.50	3.92	3.55	3.28	3.09	2.93	2.81	2.71	2.54	2.37	2.19	2.09	1.98	1.87	1.75	1.61	1.43
∞	7.88	5.30	4.28	3.72	3.35	3.09	2.90	2.74	2.62	2.52	2.36	2.19	2.00	1.90	1.79	1.67	1.53	1.36	1.00

SOURCE: Reprinted from E. S. Pearson and H. O. Hartley, eds., *Biometrika Tables for Statisticians*, 3rd ed., 1966, by permission of the *Biometrika* Trustees.

TABLE C.6 □ Table of binomial probabilities

For a given combination of *n* and *p*, entry indicates the probability of obtaining a specified value of *X*. To locate entry: **when *p* ≤ .50,** read *p* across the top heading and both *n* and *X* down the left margin; **when *p* ≥ .50,** read *p* across the bottom heading and both *n* and *X* up the right margin

		p																			
n	*X*	0.01	0.02	0.03	0.04	0.05	0.06	0.07	0.08	0.09	0.10	0.11	0.12	0.13	0.14	0.15	0.16	0.17	0.18	*X*	*n*
2	0	0.9801	0.9604	0.9409	0.9216	0.9025	0.8836	0.8649	0.8464	0.8281	0.8100	0.7921	0.7744	0.7569	0.7396	0.7225	0.7056	0.6889	0.6724	2	
	1	0.0198	0.0392	0.0582	0.0768	0.0950	0.1128	0.1302	0.1472	0.1638	0.1800	0.1958	0.2112	0.2262	0.2408	0.2550	0.2688	0.2822	0.2952	1	
	2	0.0001	0.0004	0.0009	0.0016	0.0025	0.0036	0.0049	0.0064	0.0081	0.0100	0.0121	0.0144	0.0169	0.0196	0.0225	0.0256	0.0289	0.0324	0	2
3	0	0.9703	0.9412	0.9127	0.8847	0.8574	0.8306	0.8044	0.7787	0.7536	0.7290	0.7050	0.6815	0.6585	0.6361	0.6141	0.5927	0.5718	0.5514	3	
	1	0.0294	0.0576	0.0847	0.1106	0.1354	0.1590	0.1816	0.2031	0.2236	0.2430	0.2614	0.2788	0.2952	0.3106	0.3251	0.3387	0.3513	0.3631	2	
	2	0.0003	0.0012	0.0026	0.0046	0.0071	0.0102	0.0137	0.0177	0.0221	0.0270	0.0323	0.0380	0.0441	0.0506	0.0574	0.0645	0.0720	0.0797	1	
	3	0.0000	0.0000	0.0000	0.0001	0.0001	0.0002	0.0003	0.0005	0.0007	0.0010	0.0013	0.0017	0.0022	0.0027	0.0034	0.0041	0.0049	0.0058	0	3
4	0	0.9606	0.9224	0.8853	0.8493	0.8145	0.7807	0.7481	0.7164	0.6857	0.6561	0.6274	0.5997	0.5729	0.5470	0.5220	0.4979	0.4746	0.4521	4	
	1	0.0388	0.0753	0.1095	0.1416	0.1715	0.1993	0.2252	0.2492	0.2713	0.2916	0.3102	0.3271	0.3424	0.3562	0.3685	0.3793	0.3888	0.3970	3	
	2	0.0006	0.0023	0.0051	0.0088	0.0135	0.0191	0.0254	0.0325	0.0402	0.0486	0.0575	0.0669	0.0767	0.0870	0.0975	0.1084	0.1195	0.1307	2	
	3	0.0000	0.0000	0.0001	0.0002	0.0005	0.0008	0.0013	0.0019	0.0027	0.0036	0.0047	0.0061	0.0076	0.0094	0.0115	0.0138	0.0163	0.0191	1	
	4	—	—	0.0000	0.0000	0.0000	0.0000	0.0000	0.0000	0.0001	0.0001	0.0001	0.0002	0.0003	0.0004	0.0005	0.0007	0.0008	0.0010	0	4
5	0	0.9510	0.9039	0.8587	0.8154	0.7738	0.7339	0.6957	0.6591	0.6240	0.5905	0.5584	0.5277	0.4984	0.4704	0.4437	0.4182	0.3939	0.3707	5	
	1	0.0480	0.0922	0.1328	0.1699	0.2036	0.2342	0.2618	0.2866	0.3086	0.3280	0.3451	0.3598	0.3724	0.3829	0.3915	0.3983	0.4034	0.4069	4	
	2	0.0010	0.0038	0.0082	0.0142	0.0214	0.0299	0.0394	0.0498	0.0610	0.0729	0.0853	0.0981	0.1113	0.1247	0.1382	0.1517	0.1652	0.1786	3	
	3	0.0000	0.0001	0.0003	0.0006	0.0011	0.0019	0.0030	0.0043	0.0060	0.0081	0.0105	0.0134	0.0166	0.0203	0.0244	0.0289	0.0338	0.0392	2	
	4	—	0.0000	0.0000	0.0000	0.0000	0.0001	0.0001	0.0002	0.0003	0.0004	0.0007	0.0009	0.0012	0.0017	0.0022	0.0028	0.0035	0.0043	1	
	5	—	—	—	—	—	0.0000	0.0000	0.0000	0.0000	0.0000	0.0000	0.0000	0.0000	0.0001	0.0001	0.0001	0.0001	0.0002	0	5
6	0	0.9415	0.8858	0.8330	0.7828	0.7351	0.6899	0.6470	0.6064	0.5679	0.5314	0.4970	0.4644	0.4336	0.4046	0.3771	0.3513	0.3269	0.3040	6	
	1	0.0571	0.1085	0.1546	0.1957	0.2321	0.2642	0.2922	0.3164	0.3370	0.3543	0.3685	0.3800	0.3888	0.3952	0.3993	0.4015	0.4018	0.4004	5	
	2	0.0014	0.0055	0.0120	0.0204	0.0305	0.0422	0.0550	0.0688	0.0833	0.0984	0.1139	0.1295	0.1452	0.1608	0.1762	0.1912	0.2057	0.2197	4	
	3	0.0000	0.0002	0.0005	0.0011	0.0021	0.0036	0.0055	0.0080	0.0110	0.0146	0.0188	0.0236	0.0289	0.0349	0.0415	0.0486	0.0562	0.0643	3	
	4	—	0.0000	0.0000	0.0000	0.0001	0.0002	0.0003	0.0005	0.0008	0.0012	0.0017	0.0024	0.0032	0.0043	0.0055	0.0069	0.0086	0.0106	2	
	5	—	—	—	—	0.0000	0.0000	0.0000	0.0000	0.0000	0.0001	0.0001	0.0001	0.0002	0.0003	0.0004	0.0005	0.0007	0.0009	1	
	6	—	—	—	—	—	—	—	—	—	0.0000	0.0000	0.0000	0.0000	0.0000	0.0000	0.0000	0.0000	0.0000	0	6

n	X	0.99	0.98	0.97	0.96	0.95	0.94	0.93	0.92	0.91	0.90	0.89	0.88	0.87	0.86	0.85	0.84	0.83	0.82	X	n
7	0	0.9321	0.8681	0.8080	0.7514	0.6983	0.6485	0.6017	0.5578	0.5168	0.4783	0.4423	0.4087	0.3773	0.3479	0.3206	0.2951	0.2714	0.2493	7	
	1	0.0659	0.1240	0.1749	0.2192	0.2573	0.2897	0.3170	0.3396	0.3578	0.3720	0.3827	0.3901	0.3946	0.3965	0.3960	0.3935	0.3891	0.3830	6	
	2	0.0020	0.0076	0.0162	0.0274	0.0406	0.0555	0.0716	0.0886	0.1061	0.1240	0.1419	0.1596	0.1769	0.1936	0.2097	0.2248	0.2391	0.2523	5	
	3	0.0000	0.0003	0.0008	0.0019	0.0036	0.0059	0.0090	0.0128	0.0175	0.0230	0.0292	0.0363	0.0441	0.0525	0.0617	0.0714	0.0816	0.0923	4	
	4	—	0.0000	0.0000	0.0001	0.0002	0.0004	0.0007	0.0011	0.0017	0.0026	0.0036	0.0049	0.0066	0.0086	0.0109	0.0136	0.0167	0.0203	3	
	5	—	—	—	0.0000	0.0000	0.0000	0.0000	0.0001	0.0001	0.0002	0.0003	0.0004	0.0006	0.0008	0.0012	0.0016	0.0021	0.0027	2	
	6	—	—	—	—	—	—	—	0.0000	0.0000	0.0000	0.0000	0.0000	0.0000	0.0000	0.0001	0.0001	0.0001	0.0002	1	
	7	—	—	—	—	—	—	—	—	—	—	—	—	—	—	0.0000	0.0000	0.0000	0.0000	0	7
8	0	0.9227	0.8508	0.7837	0.7214	0.6634	0.6096	0.5596	0.5132	0.4703	0.4305	0.3937	0.3596	0.3282	0.2992	0.2725	0.2479	0.2252	0.2044	8	
	1	0.0746	0.1389	0.1939	0.2405	0.2793	0.3113	0.3370	0.3570	0.3721	0.3826	0.3892	0.3923	0.3923	0.3897	0.3847	0.3777	0.3691	0.3590	7	
	2	0.0026	0.0099	0.0210	0.0351	0.0515	0.0695	0.0888	0.1087	0.1288	0.1488	0.1684	0.1872	0.2052	0.2220	0.2376	0.2518	0.2646	0.2758	6	
	3	0.0001	0.0004	0.0013	0.0029	0.0054	0.0089	0.0134	0.0189	0.0255	0.0331	0.0416	0.0511	0.0613	0.0723	0.0839	0.0959	0.1084	0.1211	5	
	4	0.0000	0.0000	0.0001	0.0002	0.0004	0.0007	0.0013	0.0021	0.0031	0.0046	0.0064	0.0087	0.0115	0.0147	0.0185	0.0228	0.0277	0.0332	4	
	5	—	—	0.0000	0.0000	0.0000	0.0000	0.0001	0.0001	0.0002	0.0004	0.0006	0.0009	0.0014	0.0019	0.0026	0.0035	0.0045	0.0058	3	
	6	—	—	—	—	—	—	0.0000	0.0000	0.0000	0.0000	0.0000	0.0001	0.0001	0.0002	0.0002	0.0003	0.0005	0.0006	2	
	7	—	—	—	—	—	—	—	—	—	—	—	0.0000	0.0000	0.0000	0.0000	0.0000	0.0000	0.0000	1	
	8	—	—	—	—	—	—	—	—	—	—	—	—	—	—	—	—	—	—	0	8
9	0	0.9135	0.8337	0.7602	0.6925	0.6302	0.5730	0.5204	0.4722	0.4279	0.3874	0.3504	0.3165	0.2855	0.2573	0.2316	0.2082	0.1869	0.1676	9	
	1	0.0830	0.1531	0.2116	0.2597	0.2985	0.3292	0.3525	0.3695	0.3809	0.3874	0.3897	0.3884	0.3840	0.3770	0.3679	0.3569	0.3446	0.3312	8	
	2	0.0034	0.0125	0.0262	0.0433	0.0629	0.0840	0.1061	0.1285	0.1507	0.1722	0.1927	0.2119	0.2295	0.2455	0.2597	0.2720	0.2823	0.2908	7	
	3	0.0001	0.0006	0.0019	0.0042	0.0077	0.0125	0.0186	0.0261	0.0348	0.0446	0.0556	0.0674	0.0800	0.0933	0.1069	0.1209	0.1349	0.1489	6	
	4	0.0000	0.0000	0.0001	0.0003	0.0006	0.0012	0.0021	0.0034	0.0052	0.0074	0.0103	0.0138	0.0179	0.0228	0.0283	0.0345	0.0415	0.0490	5	
	5	—	—	0.0000	0.0000	0.0000	0.0001	0.0002	0.0003	0.0005	0.0008	0.0013	0.0019	0.0027	0.0037	0.0050	0.0066	0.0085	0.0108	4	
	6	—	—	—	—	—	0.0000	0.0000	0.0000	0.0000	0.0001	0.0001	0.0002	0.0003	0.0004	0.0006	0.0008	0.0012	0.0016	3	
	7	—	—	—	—	—	—	—	—	—	0.0000	0.0000	0.0000	0.0000	0.0000	0.0000	0.0001	0.0001	0.0001	2	
	8	—	—	—	—	—	—	—	—	—	—	—	—	—	—	—	0.0000	0.0000	0.0000	1	
	9	—	—	—	—	—	—	—	—	—	—	—	—	—	—	—	—	—	—	0	9
10	0	0.9044	0.8171	0.7374	0.6648	0.5987	0.5386	0.4840	0.4344	0.3894	0.3487	0.3118	0.2785	0.2484	0.2213	0.1969	0.1749	0.1552	0.1374	10	
	1	0.0914	0.1667	0.2281	0.2770	0.3151	0.3438	0.3643	0.3777	0.3851	0.3874	0.3854	0.3798	0.3712	0.3603	0.3474	0.3331	0.3178	0.3017	9	
	2	0.0042	0.0153	0.0317	0.0519	0.0746	0.0988	0.1234	0.1478	0.1714	0.1937	0.2143	0.2330	0.2496	0.2639	0.2759	0.2856	0.2929	0.2980	8	
	3	0.0001	0.0008	0.0026	0.0058	0.0105	0.0168	0.0248	0.0343	0.0452	0.0574	0.0706	0.0847	0.0995	0.1146	0.1298	0.1450	0.1600	0.1745	7	
	4	0.0000	0.0000	0.0001	0.0004	0.0010	0.0019	0.0033	0.0052	0.0078	0.0112	0.0153	0.0202	0.0260	0.0326	0.0401	0.0483	0.0573	0.0670	6	
	5	—	—	0.0000	0.0000	0.0001	0.0001	0.0003	0.0005	0.0009	0.0015	0.0023	0.0033	0.0047	0.0064	0.0085	0.0111	0.0141	0.0177	5	
	6	—	—	—	—	0.0000	0.0000	0.0000	0.0000	0.0001	0.0001	0.0002	0.0004	0.0006	0.0009	0.0012	0.0018	0.0024	0.0032	4	
	7	—	—	—	—	—	—	—	—	0.0000	0.0000	0.0000	0.0000	0.0000	0.0001	0.0001	0.0002	0.0003	0.0004	3	
	8	—	—	—	—	—	—	—	—	—	—	—	—	—	0.0000	0.0000	0.0000	0.0000	0.0000	2	
	9	—	—	—	—	—	—	—	—	—	—	—	—	—	—	—	—	—	—	1	
	10	—	—	—	—	—	—	—	—	—	—	—	—	—	—	—	—	—	—	0	10

P

TABLE C.6 *(continued)*

		P																			
n	*X*	0.01	0.02	0.03	0.04	0.05	0.06	0.07	0.08	0.09	0.10	0.11	0.12	0.13	0.14	0.15	0.16	0.17	0.18	*X*	*n*
12	0	0.8864	0.7847	0.6938	0.6127	0.5404	0.4759	0.4186	0.3677	0.3225	0.2824	0.2470	0.2157	0.1880	0.1637	0.1422	0.1234	0.1069	0.0924	12	
	1	0.1074	0.1922	0.2575	0.3064	0.3413	0.3645	0.3781	0.3837	0.3827	0.3766	0.3663	0.3529	0.3372	0.3197	0.3012	0.2821	0.2627	0.2434	11	
	2	0.0060	0.0216	0.0438	0.0702	0.0988	0.1280	0.1565	0.1835	0.2082	0.2301	0.2490	0.2647	0.2771	0.2863	0.2924	0.2955	0.2960	0.2939	10	
	3	0.0002	0.0015	0.0045	0.0098	0.0173	0.0272	0.0393	0.0532	0.0686	0.0852	0.1026	0.1203	0.1380	0.1553	0.1720	0.1876	0.2021	0.2151	9	
	4	0.0000	0.0001	0.0003	0.0009	0.0021	0.0039	0.0067	0.0104	0.0153	0.0213	0.0285	0.0369	0.0464	0.0569	0.0683	0.0804	0.0931	0.1062	8	
	5	—	0.0000	0.0000	0.0001	0.0002	0.0004	0.0008	0.0014	0.0024	0.0038	0.0056	0.0081	0.0111	0.0148	0.0193	0.0245	0.0305	0.0373	7	
	6	—	—	—	0.0000	0.0000	0.0000	0.0001	0.0001	0.0003	0.0005	0.0008	0.0013	0.0019	0.0028	0.0040	0.0054	0.0073	0.0096	6	
	7	—	—	—	—	—	—	0.0000	0.0000	0.0000	0.0000	0.0001	0.0001	0.0002	0.0004	0.0006	0.0009	0.0013	0.0018	5	
	8	—	—	—	—	—	—	—	—	—	—	0.0000	0.0000	0.0000	0.0000	0.0001	0.0001	0.0002	0.0002	4	
	9	—	—	—	—	—	—	—	—	—	—	—	—	—	—	0.0000	0.0000	0.0000	0.0000	3	
	10	—	—	—	—	—	—	—	—	—	—	—	—	—	—	—	—	—	—	2	
	11	—	—	—	—	—	—	—	—	—	—	—	—	—	—	—	—	—	—	1	
	12	—	—	—	—	—	—	—	—	—	—	—	—	—	—	—	—	—	—	0	12
15	0	0.8601	0.7386	0.6333	0.5421	0.4633	0.3953	0.3367	0.2863	0.2430	0.2059	0.1741	0.1470	0.1238	0.1041	0.0874	0.0731	0.0611	0.0510	15	
	1	0.1303	0.2261	0.2938	0.3388	0.3658	0.3785	0.3801	0.3734	0.3605	0.3432	0.3228	0.3006	0.2775	0.2542	0.2312	0.2090	0.1878	0.1678	14	
	2	0.0092	0.0323	0.0636	0.0988	0.1348	0.1691	0.2003	0.2273	0.2496	0.2669	0.2793	0.2870	0.2903	0.2897	0.2856	0.2787	0.2692	0.2578	13	
	3	0.0004	0.0029	0.0085	0.0178	0.0307	0.0468	0.0653	0.0857	0.1070	0.1285	0.1496	0.1696	0.1880	0.2044	0.2184	0.2300	0.2389	0.2452	12	
	4	0.0000	0.0002	0.0008	0.0022	0.0049	0.0090	0.0148	0.0223	0.0317	0.0428	0.0555	0.0694	0.0843	0.0998	0.1156	0.1314	0.1468	0.1615	11	
	5	—	0.0000	0.0001	0.0002	0.0006	0.0013	0.0024	0.0043	0.0069	0.0105	0.0151	0.0208	0.0277	0.0357	0.0449	0.0551	0.0662	0.0780	10	
	6	—	—	0.0000	0.0000	0.0000	0.0001	0.0003	0.0006	0.0011	0.0019	0.0031	0.0047	0.0069	0.0097	0.0132	0.0175	0.0226	0.0285	9	
	7	—	—	—	—	—	0.0000	0.0000	0.0001	0.0001	0.0003	0.0005	0.0008	0.0013	0.0020	0.0030	0.0043	0.0059	0.0081	8	
	8	—	—	—	—	—	—	—	0.0000	0.0000	0.0000	0.0001	0.0001	0.0002	0.0003	0.0005	0.0008	0.0012	0.0018	7	
	9	—	—	—	—	—	—	—	—	—	—	0.0000	0.0000	0.0000	0.0000	0.0001	0.0001	0.0002	0.0003	6	
	10	—	—	—	—	—	—	—	—	—	—	—	—	—	—	0.0000	0.0000	0.0000	0.0000	5	
	11	—	—	—	—	—	—	—	—	—	—	—	—	—	—	—	—	—	—	4	
	12	—	—	—	—	—	—	—	—	—	—	—	—	—	—	—	—	—	—	3	
	13	—	—	—	—	—	—	—	—	—	—	—	—	—	—	—	—	—	—	2	
	14	—	—	—	—	—	—	—	—	—	—	—	—	—	—	—	—	—	—	1	
	15	—	—	—	—	—	—	—	—	—	—	—	—	—	—	—	—	—	—	0	15

n	X	0.99	0.98	0.97	0.96	0.95	0.94	0.93	0.92	0.91	0.90	0.89	0.88	0.87	0.86	0.85	0.84	0.83	0.82	X	n
20	0	0.8179	0.6676	0.5438	0.4420	0.3585	0.2901	0.2342	0.1887	0.1516	0.1216	0.0972	0.0776	0.0617	0.0490	0.0388	0.0306	0.0241	0.0189	20	
	1	0.1652	0.2725	0.3364	0.3683	0.3774	0.3703	0.3526	0.3282	0.3000	0.2702	0.2403	0.2115	0.1844	0.1595	0.1368	0.1165	0.0986	0.0829	19	
	2	0.0159	0.0528	0.0988	0.1458	0.1887	0.2246	0.2521	0.2711	0.2818	0.2852	0.2822	0.2740	0.2618	0.2466	0.2293	0.2109	0.1919	0.1730	18	
	3	0.0010	0.0065	0.0183	0.0364	0.0596	0.0860	0.1139	0.1414	0.1672	0.1901	0.2093	0.2242	0.2347	0.2409	0.2428	0.2410	0.2358	0.2278	17	
	4	0.0000	0.0006	0.0024	0.0065	0.0133	0.0233	0.0364	0.0523	0.0703	0.0898	0.1099	0.1299	0.1491	0.1666	0.1821	0.1951	0.2053	0.2125	16	
	5	—	0.0000	0.0002	0.0009	0.0022	0.0048	0.0088	0.0145	0.0222	0.0319	0.0435	0.0567	0.0713	0.0868	0.1028	0.1189	0.1345	0.1493	15	
	6	—	—	0.0000	0.0001	0.0003	0.0008	0.0017	0.0032	0.0055	0.0089	0.0134	0.0193	0.0266	0.0353	0.0454	0.0566	0.0689	0.0819	14	
	7	—	—	—	0.0000	0.0000	0.0001	0.0002	0.0005	0.0011	0.0020	0.0033	0.0053	0.0080	0.0115	0.0160	0.0216	0.0282	0.0360	13	
	8	—	—	—	—	—	0.0000	0.0000	0.0001	0.0002	0.0004	0.0007	0.0012	0.0019	0.0030	0.0046	0.0067	0.0094	0.0128	12	
	9	—	—	—	—	—	—	—	0.0000	0.0000	0.0001	0.0001	0.0002	0.0004	0.0007	0.0011	0.0017	0.0026	0.0038	11	
	10	—	—	—	—	—	—	—	—	—	0.0000	0.0000	0.0000	0.0001	0.0001	0.0002	0.0004	0.0006	0.0009	10	
	11	—	—	—	—	—	—	—	—	—	—	—	—	0.0000	0.0000	0.0000	0.0001	0.0001	0.0002	9	
	12	—	—	—	—	—	—	—	—	—	—	—	—	—	—	—	0.0000	0.0000	0.0000	8	
	13	—	—	—	—	—	—	—	—	—	—	—	—	—	—	—	—	—	—	7	
	14	—	—	—	—	—	—	—	—	—	—	—	—	—	—	—	—	—	—	6	
	15	—	—	—	—	—	—	—	—	—	—	—	—	—	—	—	—	—	—	5	
	16	—	—	—	—	—	—	—	—	—	—	—	—	—	—	—	—	—	—	4	
	17	—	—	—	—	—	—	—	—	—	—	—	—	—	—	—	—	—	—	3	
	18	—	—	—	—	—	—	—	—	—	—	—	—	—	—	—	—	—	—	2	
	19	—	—	—	—	—	—	—	—	—	—	—	—	—	—	—	—	—	—	1	
	20	—	—	—	—	—	—	—	—	—	—	—	—	—	—	—	—	—	—	0	20
n	X	0.99	0.98	0.97	0.96	0.95	0.94	0.93	0.92	0.91	0.90	0.89	0.88	0.87	0.86	0.85	0.84	0.83	0.82	X	n
		P																			

TABLE C.6 *(continued)*

n	X	0.19	0.20	0.21	0.22	0.23	0.24	0.25	0.26	0.27	0.28	0.29	0.30	0.31	0.32	0.33	0.34	0.35	0.36	X	n
2	0	0.6561	0.6400	0.6241	0.6084	0.5929	0.5776	0.5625	0.5476	0.5329	0.5184	0.5041	0.4900	0.4761	0.4624	0.4489	0.4356	0.4225	0.4096	2	
	1	0.3078	0.3200	0.3318	0.3432	0.3542	0.3648	0.3750	0.3848	0.3942	0.4032	0.4118	0.4200	0.4278	0.4352	0.4422	0.4488	0.4550	0.4608	1	
	2	0.0361	0.0400	0.0441	0.0484	0.0529	0.0576	0.0625	0.0676	0.0729	0.0784	0.0841	0.0900	0.0961	0.1024	0.1089	0.1156	0.1225	0.1296	0	2
3	0	0.5314	0.5120	0.4930	0.4746	0.4565	0.4390	0.4219	0.4052	0.3890	0.3732	0.3579	0.3430	0.3285	0.3144	0.3008	0.2875	0.2746	0.2621	3	
	1	0.3740	0.3840	0.3932	0.4015	0.4091	0.4159	0.4219	0.4271	0.4316	0.4355	0.4386	0.4410	0.4428	0.4439	0.4444	0.4443	0.4436	0.4424	2	
	2	0.0877	0.0960	0.1045	0.1133	0.1222	0.1313	0.1406	0.1501	0.1597	0.1693	0.1791	0.1890	0.1989	0.2089	0.2189	0.2289	0.2389	0.2488	1	
	3	0.0069	0.0080	0.0093	0.0106	0.0122	0.0138	0.0156	0.0176	0.0197	0.0220	0.0244	0.0270	0.0298	0.0328	0.0359	0.0393	0.0429	0.0467	0	3
4	0	0.4305	0.4096	0.3895	0.3702	0.3515	0.3336	0.3164	0.2999	0.2840	0.2687	0.2541	0.2401	0.2267	0.2138	0.2015	0.1897	0.1785	0.1678	4	
	1	0.4039	0.4096	0.4142	0.4176	0.4200	0.4214	0.4219	0.4214	0.4201	0.4180	0.4152	0.4116	0.4074	0.4025	0.3970	0.3910	0.3845	0.3775	3	
	2	0.1421	0.1536	0.1651	0.1767	0.1882	0.1996	0.2109	0.2221	0.2331	0.2439	0.2544	0.2646	0.2745	0.2841	0.2933	0.3021	0.3105	0.3185	2	
	3	0.0222	0.0256	0.0293	0.0332	0.0375	0.0420	0.0469	0.0520	0.0575	0.0632	0.0693	0.0756	0.0822	0.0891	0.0963	0.1038	0.1115	0.1194	1	
	4	0.0013	0.0016	0.0019	0.0023	0.0028	0.0033	0.0039	0.0046	0.0053	0.0061	0.0071	0.0081	0.0092	0.0105	0.0119	0.0134	0.0150	0.0168	0	4
5	0	0.3487	0.3277	0.3077	0.2887	0.2707	0.2536	0.2373	0.2219	0.2073	0.1935	0.1804	0.1681	0.1564	0.1454	0.1350	0.1252	0.1160	0.1074	5	
	1	0.4089	0.4096	0.4090	0.4072	0.4043	0.4003	0.3955	0.3898	0.3834	0.3762	0.3685	0.3601	0.3513	0.3421	0.3325	0.3226	0.3124	0.3020	4	
	2	0.1919	0.2048	0.2174	0.2297	0.2415	0.2529	0.2637	0.2739	0.2836	0.2926	0.3010	0.3087	0.3157	0.3220	0.3275	0.3323	0.3364	0.3397	3	
	3	0.0450	0.0512	0.0578	0.0648	0.0721	0.0798	0.0879	0.0962	0.1049	0.1138	0.1229	0.1323	0.1418	0.1515	0.1613	0.1712	0.1811	0.1911	2	
	4	0.0053	0.0064	0.0077	0.0091	0.0108	0.0126	0.0146	0.0169	0.0194	0.0221	0.0251	0.0283	0.0319	0.0357	0.0397	0.0441	0.0488	0.0537	1	
	5	0.0002	0.0003	0.0004	0.0005	0.0006	0.0008	0.0010	0.0012	0.0014	0.0017	0.0021	0.0024	0.0029	0.0034	0.0039	0.0045	0.0053	0.0060	0	5
6	0	0.2824	0.2621	0.2431	0.2252	0.2084	0.1927	0.1780	0.1642	0.1513	0.1393	0.1281	0.1176	0.1079	0.0989	0.0905	0.0827	0.0754	0.0687	6	
	1	0.3975	0.3932	0.3877	0.3811	0.3735	0.3651	0.3560	0.3462	0.3358	0.3251	0.3139	0.3025	0.2909	0.2792	0.2673	0.2555	0.2437	0.2319	5	
	2	0.2331	0.2458	0.2577	0.2687	0.2789	0.2882	0.2966	0.3041	0.3105	0.3160	0.3206	0.3241	0.3267	0.3284	0.3292	0.3290	0.3280	0.3261	4	
	3	0.0729	0.0819	0.0913	0.1011	0.1111	0.1214	0.1318	0.1424	0.1531	0.1639	0.1746	0.1852	0.1957	0.2061	0.2162	0.2260	0.2355	0.2446	3	
	4	0.0128	0.0154	0.0182	0.0214	0.0249	0.0287	0.0330	0.0375	0.0425	0.0478	0.0535	0.0595	0.0660	0.0727	0.0799	0.0873	0.0951	0.1032	2	
	5	0.0012	0.0015	0.0019	0.0024	0.0030	0.0036	0.0044	0.0053	0.0063	0.0074	0.0087	0.0102	0.0119	0.0137	0.0157	0.0180	0.0205	0.0232	1	
	6	0.0000	0.0001	0.0001	0.0001	0.0001	0.0002	0.0002	0.0003	0.0004	0.0005	0.0006	0.0007	0.0009	0.0011	0.0013	0.0015	0.0018	0.0022	0	6

P

n	X	0.81	0.80	0.79	0.78	0.77	0.76	0.75	0.74	0.73	0.72	0.71	0.70	0.69	0.68	0.67	0.66	0.65	0.64	X	n
7	0	0.2288	0.2097	0.1920	0.1757	0.1605	0.1465	0.1335	0.1215	0.1105	0.1003	0.0910	0.0824	0.0745	0.0672	0.0606	0.0546	0.0490	0.0440	7	
	1	0.3756	0.3670	0.3573	0.3468	0.3356	0.3237	0.3115	0.2989	0.2860	0.2731	0.2600	0.2471	0.2342	0.2215	0.2090	0.1967	0.1848	0.1732	6	
	2	0.2643	0.2753	0.2850	0.2935	0.3007	0.3067	0.3115	0.3150	0.3174	0.3186	0.3186	0.3177	0.3156	0.3127	0.3088	0.3040	0.2985	0.2922	5	
	3	0.1033	0.1147	0.1263	0.1379	0.1497	0.1614	0.1730	0.1845	0.1956	0.2065	0.2169	0.2269	0.2363	0.2452	0.2535	0.2610	0.2679	0.2740	4	
	4	0.0242	0.0287	0.0336	0.0389	0.0447	0.0510	0.0577	0.0648	0.0724	0.0803	0.0886	0.0972	0.1062	0.1154	0.1248	0.1345	0.1442	0.1541	3	
	5	0.0034	0.0043	0.0054	0.0066	0.0080	0.0097	0.0115	0.0137	0.0161	0.0187	0.0217	0.0250	0.0286	0.0326	0.0369	0.0416	0.0466	0.0520	2	
	6	0.0003	0.0004	0.0005	0.0006	0.0008	0.0010	0.0013	0.0016	0.0020	0.0024	0.0030	0.0036	0.0043	0.0051	0.0061	0.0071	0.0084	0.0098	1	
	7	0.0000	0.0000	0.0000	0.0000	0.0000	0.0000	0.0001	0.0001	0.0001	0.0001	0.0002	0.0002	0.0003	0.0003	0.0004	0.0005	0.0006	0.0008	0	7
8	0	0.1853	0.1678	0.1517	0.1370	0.1236	0.1113	0.1001	0.0899	0.0806	0.0722	0.0646	0.0576	0.0514	0.0457	0.0406	0.0360	0.0319	0.0281	8	
	1	0.3477	0.3355	0.3226	0.3092	0.2953	0.2812	0.2670	0.2527	0.2386	0.2247	0.2110	0.1977	0.1847	0.1721	0.1600	0.1484	0.1373	0.1267	7	
	2	0.2855	0.2936	0.3002	0.3052	0.3087	0.3108	0.3115	0.3108	0.3089	0.3058	0.3017	0.2965	0.2904	0.2835	0.2758	0.2675	0.2587	0.2494	6	
	3	0.1339	0.1468	0.1596	0.1722	0.1844	0.1963	0.2076	0.2184	0.2285	0.2379	0.2464	0.2541	0.2609	0.2668	0.2717	0.2756	0.2786	0.2805	5	
	4	0.0393	0.0459	0.0530	0.0607	0.0689	0.0775	0.0865	0.0959	0.1056	0.1156	0.1258	0.1361	0.1465	0.1569	0.1673	0.1775	0.1875	0.1973	4	
	5	0.0074	0.0092	0.0113	0.0137	0.0165	0.0196	0.0231	0.0270	0.0313	0.0360	0.0411	0.0467	0.0527	0.0591	0.0659	0.0732	0.0808	0.0888	3	
	6	0.0009	0.0011	0.0015	0.0019	0.0025	0.0031	0.0038	0.0047	0.0058	0.0070	0.0084	0.0100	0.0118	0.0139	0.0162	0.0188	0.0217	0.0250	2	
	7	0.0001	0.0001	0.0001	0.0002	0.0002	0.0003	0.0004	0.0005	0.0006	0.0008	0.0010	0.0012	0.0015	0.0019	0.0023	0.0028	0.0033	0.0040	1	
	8	0.0000	0.0000	0.0000	0.0000	0.0000	0.0000	0.0000	0.0000	0.0000	0.0000	0.0001	0.0001	0.0001	0.0001	0.0001	0.0002	0.0002	0.0003	0	8
9	0	0.1501	0.1342	0.1199	0.1069	0.0952	0.0846	0.0751	0.0665	0.0589	0.0520	0.0458	0.0404	0.0355	0.0311	0.0272	0.0238	0.0207	0.0180	9	
	1	0.3169	0.3020	0.2867	0.2713	0.2558	0.2404	0.2253	0.2104	0.1960	0.1820	0.1685	0.1556	0.1433	0.1317	0.1206	0.1102	0.1004	0.0912	8	
	2	0.2973	0.3020	0.3049	0.3061	0.3056	0.3037	0.3003	0.2957	0.2899	0.2831	0.2754	0.2668	0.2576	0.2478	0.2376	0.2270	0.2162	0.2052	7	
	3	0.1627	0.1762	0.1891	0.2014	0.2130	0.2238	0.2336	0.2424	0.2502	0.2569	0.2624	0.2668	0.2701	0.2721	0.2731	0.2729	0.2716	0.2693	6	
	4	0.0573	0.0661	0.0754	0.0852	0.0954	0.1060	0.1168	0.1278	0.1388	0.1499	0.1608	0.1715	0.1820	0.1921	0.2017	0.2109	0.2194	0.2272	5	
	5	0.0134	0.0165	0.0200	0.0240	0.0285	0.0335	0.0389	0.0449	0.0513	0.0583	0.0657	0.0735	0.0818	0.0904	0.0994	0.1086	0.1181	0.1278	4	
	6	0.0021	0.0028	0.0036	0.0045	0.0057	0.0070	0.0087	0.0105	0.0127	0.0151	0.0179	0.0210	0.0245	0.0284	0.0326	0.0373	0.0424	0.0479	3	
	7	0.0002	0.0003	0.0004	0.0005	0.0007	0.0010	0.0012	0.0016	0.0020	0.0025	0.0031	0.0039	0.0047	0.0057	0.0069	0.0082	0.0098	0.0116	2	
	8	0.0000	0.0000	0.0000	0.0000	0.0001	0.0001	0.0001	0.0001	0.0002	0.0002	0.0003	0.0004	0.0005	0.0007	0.0008	0.0011	0.0013	0.0016	1	
	9	—	—	—	—	0.0000	0.0000	0.0000	0.0000	0.0000	0.0000	0.0000	0.0000	0.0000	0.0000	0.0000	0.0001	0.0001	0.0001	0	9
10	0	0.1216	0.1074	0.0947	0.0834	0.0733	0.0643	0.0563	0.0492	0.0430	0.0374	0.0326	0.0282	0.0245	0.0211	0.0182	0.0157	0.0135	0.0115	10	
	1	0.2852	0.2684	0.2517	0.2351	0.2188	0.2030	0.1877	0.1730	0.1590	0.1456	0.1330	0.1211	0.1099	0.0995	0.0898	0.0808	0.0725	0.0649	9	
	2	0.3010	0.3020	0.3011	0.2984	0.2942	0.2885	0.2816	0.2735	0.2646	0.2548	0.2444	0.2335	0.2222	0.2107	0.1990	0.1873	0.1757	0.1642	8	
	3	0.1883	0.2013	0.2134	0.2244	0.2343	0.2429	0.2503	0.2563	0.2609	0.2642	0.2662	0.2668	0.2662	0.2644	0.2614	0.2573	0.2522	0.2462	7	
	4	0.0773	0.0881	0.0993	0.1108	0.1225	0.1343	0.1460	0.1576	0.1689	0.1798	0.1903	0.2001	0.2093	0.2177	0.2253	0.2320	0.2377	0.2424	6	
	5	0.0218	0.0264	0.0317	0.0375	0.0439	0.0509	0.0584	0.0664	0.0750	0.0839	0.0933	0.1029	0.1128	0.1229	0.1332	0.1434	0.1536	0.1636	5	
	6	0.0043	0.0055	0.0070	0.0088	0.0109	0.0134	0.0162	0.0195	0.0231	0.0272	0.0317	0.0368	0.0422	0.0482	0.0547	0.0616	0.0689	0.0767	4	
	7	0.0006	0.0008	0.0011	0.0014	0.0019	0.0024	0.0031	0.0039	0.0049	0.0060	0.0074	0.0090	0.0108	0.0130	0.0154	0.0181	0.0212	0.0247	3	
	8	0.0001	0.0001	0.0001	0.0002	0.0002	0.0003	0.0004	0.0005	0.0007	0.0009	0.0011	0.0014	0.0018	0.0023	0.0028	0.0035	0.0043	0.0052	2	
	9	0.0000	0.0000	0.0000	0.0000	0.0000	0.0000	0.0000	0.0000	0.0001	0.0001	0.0001	0.0001	0.0002	0.0002	0.0003	0.0004	0.0005	0.0006	1	
	10	—	—	—	—	—	—	—	—	0.0000	0.0000	0.0000	0.0000	0.0000	0.0000	0.0000	0.0000	0.0000	0.0000	0	10
		P																			

TABLE C.6 *(continued)*

n	X	0.19	0.20	0.21	0.22	0.23	0.24	0.25	0.26	0.27	0.28	0.29	0.30	0.31	0.32	0.33	0.34	0.35	0.36	X	n
		P																			
12	0	0.0798	0.0687	0.0591	0.0507	0.0434	0.0371	0.0317	0.0270	0.0229	0.0194	0.0164	0.0138	0.0116	0.0098	0.0082	0.0068	0.0057	0.0047	12	
	1	0.2245	0.2062	0.1885	0.1717	0.1557	0.1407	0.1267	0.1137	0.1016	0.0906	0.0804	0.0712	0.0628	0.0552	0.0484	0.0422	0.0368	0.0319	11	
	2	0.2897	0.2835	0.2756	0.2663	0.2558	0.2444	0.2323	0.2197	0.2068	0.1937	0.1807	0.1678	0.1552	0.1429	0.1310	0.1197	0.1088	0.0986	10	
	3	0.2265	0.2362	0.2442	0.2503	0.2547	0.2573	0.2581	0.2573	0.2549	0.2511	0.2460	0.2397	0.2324	0.2241	0.2151	0.2055	0.1954	0.1849	9	
	4	0.1195	0.1329	0.1460	0.1589	0.1712	0.1828	0.1936	0.2034	0.2122	0.2197	0.2261	0.2311	0.2349	0.2373	0.2384	0.2382	0.2367	0.2340	8	
	5	0.0449	0.0532	0.0621	0.0717	0.0818	0.0924	0.1032	0.1143	0.1255	0.1367	0.1477	0.1585	0.1688	0.1787	0.1879	0.1963	0.2039	0.2106	7	
	6	0.0123	0.0155	0.0193	0.0236	0.0285	0.0340	0.0401	0.0469	0.0542	0.0620	0.0704	0.0792	0.0885	0.0981	0.1079	0.1180	0.1281	0.1382	6	
	7	0.0025	0.0033	0.0044	0.0057	0.0073	0.0092	0.0115	0.0141	0.0172	0.0207	0.0246	0.0291	0.0341	0.0396	0.0456	0.0521	0.0591	0.0666	5	
	8	0.0004	0.0005	0.0007	0.0010	0.0014	0.0018	0.0024	0.0031	0.0040	0.0050	0.0063	0.0078	0.0096	0.0116	0.0140	0.0168	0.0199	0.0234	4	
	9	0.0000	0.0001	0.0001	0.0001	0.0002	0.0003	0.0004	0.0005	0.0007	0.0009	0.0011	0.0015	0.0019	0.0024	0.0031	0.0038	0.0048	0.0059	3	
	10	—	0.0000	0.0000	0.0000	0.0000	0.0000	0.0000	0.0001	0.0001	0.0001	0.0001	0.0002	0.0003	0.0003	0.0005	0.0006	0.0008	0.0010	2	
	11	—	—	—	—	—	—	—	0.0000	0.0000	0.0000	0.0000	0.0000	0.0000	0.0000	0.0000	0.0001	0.0001	0.0001	1	
	12	—	—	—	—	—	—	—	—	—	—	—	—	—	—	—	0.0000	0.0000	0.0000	0	12
15	0	0.0424	0.0352	0.0291	0.0241	0.0198	0.0163	0.0134	0.0109	0.0089	0.0072	0.0059	0.0047	0.0038	0.0031	0.0025	0.0020	0.0016	0.0012	15	
	1	0.1492	0.1319	0.1162	0.1018	0.0889	0.0772	0.0668	0.0576	0.0494	0.0423	0.0360	0.0305	0.0258	0.0217	0.0182	0.0152	0.0126	0.0104	14	
	2	0.2449	0.2309	0.2162	0.2010	0.1858	0.1707	0.1559	0.1416	0.1280	0.1150	0.1029	0.0916	0.0811	0.0715	0.0627	0.0547	0.0476	0.0411	13	
	3	0.2489	0.2501	0.2490	0.2457	0.2405	0.2336	0.2252	0.2156	0.2051	0.1939	0.1821	0.1700	0.1579	0.1457	0.1338	0.1222	0.1110	0.1002	12	
	4	0.1752	0.1876	0.1986	0.2079	0.2155	0.2213	0.2252	0.2273	0.2276	0.2262	0.2231	0.2186	0.2128	0.2057	0.1977	0.1888	0.1792	0.1692	11	
	5	0.0904	0.1032	0.1161	0.1290	0.1416	0.1537	0.1651	0.1757	0.1852	0.1935	0.2005	0.2061	0.2103	0.2130	0.2142	0.2140	0.2123	0.2093	10	
	6	0.0353	0.0430	0.0514	0.0606	0.0705	0.0809	0.0917	0.1029	0.1142	0.1254	0.1365	0.1472	0.1575	0.1671	0.1759	0.1837	0.1906	0.1963	9	
	7	0.0107	0.0138	0.0176	0.0220	0.0271	0.0329	0.0393	0.0465	0.0543	0.0627	0.0717	0.0811	0.0910	0.1011	0.1114	0.1217	0.1319	0.1419	8	
	8	0.0025	0.0035	0.0047	0.0062	0.0081	0.0104	0.0131	0.0163	0.0201	0.0244	0.0293	0.0348	0.0409	0.0476	0.0549	0.0627	0.0710	0.0798	7	
	9	0.0005	0.0007	0.0010	0.0014	0.0019	0.0025	0.0034	0.0045	0.0058	0.0074	0.0093	0.0116	0.0143	0.0174	0.0210	0.0251	0.0298	0.0349	6	
	10	0.0001	0.0001	0.0002	0.0002	0.0003	0.0005	0.0007	0.0009	0.0013	0.0017	0.0023	0.0030	0.0038	0.0049	0.0062	0.0078	0.0096	0.0118	5	
	11	0.0000	0.0000	0.0000	0.0000	0.0000	0.0001	0.0001	0.0002	0.0002	0.0003	0.0004	0.0006	0.0008	0.0011	0.0014	0.0018	0.0024	0.0030	4	
	12	—	—	—	—	—	0.0000	0.0000	0.0000	0.0000	0.0000	0.0001	0.0001	0.0001	0.0002	0.0002	0.0003	0.0004	0.0006	3	
	13	—	—	—	—	—	—	—	—	—	—	0.0000	0.0000	0.0000	0.0000	0.0000	0.0000	0.0001	0.0001	2	
	14	—	—	—	—	—	—	—	—	—	—	—	—	—	—	—	—	0.0000	0.0000	1	
	15	—	—	—	—	—	—	—	—	—	—	—	—	—	—	—	—	—	—	0	15

n	X	0.81	0.80	0.79	0.78	0.77	0.76	0.75	0.74	0.73	0.72	0.71	0.70	0.69	0.68	0.67	0.66	0.65	0.64	X	n
20	0	0.0148	0.0115	0.0090	0.0069	0.0054	0.0041	0.0032	0.0024	0.0018	0.0014	0.0011	0.0008	0.0006	0.0004	0.0003	0.0002	0.0002	0.0001	20	
	1	0.0693	0.0576	0.0477	0.0392	0.0321	0.0261	0.0211	0.0170	0.0137	0.0109	0.0087	0.0068	0.0054	0.0042	0.0033	0.0025	0.0020	0.0015	19	
	2	0.1545	0.1369	0.1204	0.1050	0.0910	0.0783	0.0669	0.0569	0.0480	0.0403	0.0336	0.0278	0.0229	0.0188	0.0153	0.0124	0.0100	0.0080	18	
	3	0.2175	0.2054	0.1920	0.1777	0.1631	0.1484	0.1339	0.1199	0.1065	0.0940	0.0823	0.0716	0.0619	0.0531	0.0453	0.0383	0.0323	0.0270	17	
	4	0.2168	0.2182	0.2169	0.2131	0.2070	0.1991	0.1897	0.1790	0.1675	0.1553	0.1429	0.1304	0.1181	0.1062	0.0947	0.0839	0.0738	0.0645	16	
	5	0.1627	0.1746	0.1845	0.1923	0.1979	0.2012	0.2023	0.2013	0.1982	0.1933	0.1868	0.1789	0.1698	0.1599	0.1493	0.1384	0.1272	0.1161	15	
	6	0.0954	0.1091	0.1226	0.1356	0.1478	0.1589	0.1686	0.1768	0.1833	0.1879	0.1907	0.1916	0.1907	0.1881	0.1839	0.1782	0.1712	0.1632	14	
	7	0.0448	0.0545	0.0652	0.0765	0.0883	0.1003	0.1124	0.1242	0.1356	0.1462	0.1558	0.1643	0.1714	0.1770	0.1811	0.1836	0.1844	0.1836	13	
	8	0.0171	0.0222	0.0282	0.0351	0.0429	0.0515	0.0609	0.0709	0.0815	0.0924	0.1034	0.1144	0.1251	0.1354	0.1450	0.1537	0.1614	0.1678	12	
	9	0.0053	0.0074	0.0100	0.0132	0.0171	0.0217	0.0271	0.0332	0.0402	0.0479	0.0563	0.0654	0.0750	0.0849	0.0952	0.1056	0.1158	0.1259	11	
	10	0.0014	0.0020	0.0029	0.0041	0.0056	0.0075	0.0099	0.0128	0.0163	0.0205	0.0253	0.0308	0.0370	0.0440	0.0516	0.0598	0.0686	0.0779	10	
	11	0.0003	0.0005	0.0007	0.0010	0.0015	0.0022	0.0030	0.0041	0.0055	0.0072	0.0094	0.0120	0.0151	0.0188	0.0231	0.0280	0.0336	0.0398	9	
	12	0.0001	0.0001	0.0001	0.0002	0.0003	0.0005	0.0008	0.0011	0.0015	0.0021	0.0029	0.0039	0.0051	0.0066	0.0085	0.0108	0.0136	0.0168	8	
	13	0.0000	0.0000	0.0000	0.0000	0.0001	0.0001	0.0002	0.0002	0.0003	0.0005	0.0007	0.0010	0.0014	0.0019	0.0026	0.0034	0.0045	0.0058	7	
	14	—	—	—	—	0.0000	0.0000	0.0000	0.0000	0.0001	0.0001	0.0001	0.0002	0.0003	0.0005	0.0006	0.0009	0.0012	0.0016	6	
	15	—	—	—	—	—	—	—	—	0.0000	0.0000	0.0000	0.0000	0.0001	0.0001	0.0001	0.0002	0.0003	0.0004	5	
	16	—	—	—	—	—	—	—	—	—	—	—	—	0.0000	0.0000	0.0000	0.0000	0.0000	0.0001	4	
	17	—	—	—	—	—	—	—	—	—	—	—	—	—	—	—	—	—	0.0000	3	
	18	—	—	—	—	—	—	—	—	—	—	—	—	—	—	—	—	—	—	2	
	19	—	—	—	—	—	—	—	—	—	—	—	—	—	—	—	—	—	—	1	
	20	—	—	—	—	—	—	—	—	—	—	—	—	—	—	—	—	—	—	0	20
n	X	0.81	0.80	0.79	0.78	0.77	0.76	0.75	0.74	0.73	0.72	0.71	0.70	0.69	0.68	0.67	0.66	0.65	0.64	X	n
										P											

TABLE C.6 *(continued)*

n	X	P 0.37	0.38	0.39	0.40	0.41	0.42	0.43	0.44	0.45	0.46	0.47	0.48	0.49	0.50	X	n
2	0	0.3969	0.3844	0.3721	0.3600	0.3481	0.3364	0.3249	0.3136	0.3025	0.2916	0.2809	0.2704	0.2601	0.2500	2	
	1	0.4662	0.4712	0.4758	0.4800	0.4838	0.4872	0.4902	0.4928	0.4950	0.4968	0.4982	0.4992	0.4998	0.5000	1	
	2	0.1369	0.1444	0.1521	0.1600	0.1681	0.1764	0.1849	0.1936	0.2025	0.2116	0.2209	0.2304	0.2401	0.2500	0	2
3	0	0.2500	0.2383	0.2270	0.2160	0.2054	0.1951	0.1852	0.1756	0.1664	0.1575	0.1489	0.1406	0.1327	0.1250	3	
	1	0.4406	0.4382	0.4354	0.4320	0.4282	0.4239	0.4191	0.4140	0.4084	0.4024	0.3961	0.3894	0.3823	0.3750	2	
	2	0.2587	0.2686	0.2783	0.2880	0.2975	0.3069	0.3162	0.3252	0.3341	0.3428	0.3512	0.3594	0.3674	0.3750	1	
	3	0.0507	0.0549	0.0593	0.0640	0.0689	0.0741	0.0795	0.0852	0.0911	0.0973	0.1038	0.1106	0.1176	0.1250	0	3
4	0	0.1575	0.1478	0.1385	0.1296	0.1212	0.1132	0.1056	0.0983	0.0915	0.0850	0.0789	0.0731	0.0677	0.0625	4	
	1	0.3701	0.3623	0.3541	0.3456	0.3368	0.3278	0.3185	0.3091	0.2995	0.2897	0.2799	0.2700	0.2600	0.2500	3	
	2	0.3260	0.3330	0.3396	0.3456	0.3511	0.3560	0.3604	0.3643	0.3675	0.3702	0.3723	0.3738	0.3747	0.3750	2	
	3	0.1276	0.1361	0.1447	0.1536	0.1627	0.1719	0.1813	0.1908	0.2005	0.2102	0.2201	0.2300	0.2400	0.2500	1	
	4	0.0187	0.0209	0.0231	0.0256	0.0283	0.0311	0.0342	0.0375	0.0410	0.0448	0.0488	0.0531	0.0576	0.0625	0	4
5	0	0.0992	0.0916	0.0845	0.0778	0.0715	0.0656	0.0602	0.0551	0.0503	0.0459	0.0418	0.0380	0.0345	0.0312	5	
	1	0.2914	0.2808	0.2700	0.2592	0.2484	0.2376	0.2270	0.2164	0.2059	0.1956	0.1854	0.1755	0.1657	0.1562	4	
	2	0.3423	0.3441	0.3452	0.3456	0.3452	0.3442	0.3424	0.3400	0.3369	0.3332	0.3289	0.3240	0.3185	0.3125	3	
	3	0.2010	0.2109	0.2207	0.2304	0.2399	0.2492	0.2583	0.2671	0.2757	0.2838	0.2916	0.2990	0.3060	0.3125	2	
	4	0.0590	0.0646	0.0706	0.0768	0.0834	0.0902	0.0974	0.1049	0.1128	0.1209	0.1293	0.1380	0.1470	0.1562	1	
	5	0.0069	0.0079	0.0090	0.0102	0.0116	0.0131	0.0147	0.0165	0.0185	0.0206	0.0229	0.0255	0.0282	0.0312	0	5
6	0	0.0625	0.0568	0.0515	0.0467	0.0422	0.0381	0.0343	0.0308	0.0277	0.0248	0.0222	0.0198	0.0176	0.0156	6	
	1	0.2203	0.2089	0.1976	0.1866	0.1759	0.1654	0.1552	0.1454	0.1359	0.1267	0.1179	0.1095	0.1014	0.0937	5	
	2	0.3235	0.3201	0.3159	0.3110	0.3055	0.2994	0.2928	0.2856	0.2780	0.2699	0.2615	0.2527	0.2436	0.2344	4	
	3	0.2533	0.2616	0.2693	0.2765	0.2831	0.2891	0.2945	0.2992	0.3032	0.3065	0.3091	0.3110	0.3121	0.3125	3	
	4	0.1116	0.1202	0.1291	0.1382	0.1475	0.1570	0.1666	0.1763	0.1861	0.1958	0.2056	0.2153	0.2249	0.2344	2	
	5	0.0262	0.0295	0.0330	0.0369	0.0410	0.0455	0.0503	0.0554	0.0609	0.0667	0.0729	0.0795	0.0864	0.0937	1	
	6	0.0026	0.0030	0.0035	0.0041	0.0048	0.0055	0.0063	0.0073	0.0083	0.0095	0.0108	0.0122	0.0138	0.0156	0	6

n	X	0.63	0.62	0.61	0.60	0.59	0.58	0.57	0.56	0.55	0.54	0.53	0.52	0.51	0.50	X	n
7	0	0.0394	0.0352	0.0314	0.0280	0.0249	0.0221	0.0195	0.0173	0.0152	0.0134	0.0117	0.0103	0.0090	0.0078	7	
	1	0.1619	0.1511	0.1407	0.1306	0.1211	0.1119	0.1032	0.0950	0.0872	0.0798	0.0729	0.0664	0.0604	0.0547	6	
	2	0.2853	0.2778	0.2698	0.2613	0.2524	0.2431	0.2336	0.2239	0.2140	0.2040	0.1940	0.1840	0.1740	0.1641	5	
	3	0.2793	0.2838	0.2875	0.2903	0.2923	0.2934	0.2937	0.2932	0.2918	0.2897	0.2867	0.2830	0.2786	0.2734	4	
	4	0.1640	0.1739	0.1838	0.1935	0.2031	0.2125	0.2216	0.2304	0.2388	0.2468	0.2543	0.2612	0.2676	0.2734	3	
	5	0.0578	0.0640	0.0705	0.0774	0.0847	0.0923	0.1003	0.1086	0.1172	0.1261	0.1353	0.1447	0.1543	0.1641	2	
	6	0.0113	0.0131	0.0150	0.0172	0.0196	0.0223	0.0252	0.0284	0.0320	0.0358	0.0400	0.0445	0.0494	0.0547	1	
	7	0.0009	0.0011	0.0014	0.0016	0.0019	0.0023	0.0027	0.0032	0.0037	0.0044	0.0051	0.0059	0.0068	0.0078	0	7
8	0	0.0248	0.0218	0.0192	0.0168	0.0147	0.0128	0.0111	0.0097	0.0084	0.0072	0.0062	0.0053	0.0046	0.0039	8	
	1	0.1166	0.1071	0.0981	0.0896	0.0816	0.0742	0.0672	0.0608	0.0548	0.0493	0.0442	0.0395	0.0352	0.0312	7	
	2	0.2397	0.2297	0.2194	0.2090	0.1985	0.1880	0.1776	0.1672	0.1569	0.1469	0.1371	0.1275	0.1183	0.1094	6	
	3	0.2815	0.2815	0.2806	0.2787	0.2759	0.2723	0.2679	0.2627	0.2568	0.2503	0.2431	0.2355	0.2273	0.2187	5	
	4	0.2067	0.2157	0.2242	0.2322	0.2397	0.2465	0.2526	0.2580	0.2627	0.2665	0.2695	0.2717	0.2730	0.2734	4	
	5	0.0971	0.1058	0.1147	0.1239	0.1332	0.1428	0.1525	0.1622	0.1719	0.1816	0.1912	0.2006	0.2098	0.2187	3	
	6	0.0285	0.0324	0.0367	0.0413	0.0463	0.0517	0.0575	0.0637	0.0703	0.0774	0.0848	0.0926	0.1008	0.1094	2	
	7	0.0048	0.0057	0.0067	0.0079	0.0092	0.0107	0.0124	0.0143	0.0164	0.0188	0.0215	0.0244	0.0277	0.0312	1	
	8	0.0004	0.0004	0.0005	0.0007	0.0008	0.0010	0.0012	0.0014	0.0017	0.0020	0.0024	0.0028	0.0033	0.0039	0	8
9	0	0.0156	0.0135	0.0117	0.0101	0.0087	0.0074	0.0064	0.0054	0.0046	0.0039	0.0033	0.0028	0.0023	0.0020	9	
	1	0.0826	0.0747	0.0673	0.0605	0.0542	0.0484	0.0431	0.0383	0.0339	0.0299	0.0263	0.0231	0.0202	0.0176	8	
	2	0.1941	0.1831	0.1721	0.1612	0.1506	0.1402	0.1301	0.1204	0.1110	0.1020	0.0934	0.0853	0.0776	0.0703	7	
	3	0.2660	0.2618	0.2567	0.2508	0.2442	0.2369	0.2291	0.2207	0.2119	0.2027	0.1933	0.1837	0.1739	0.1641	6	
	4	0.2344	0.2407	0.2462	0.2508	0.2545	0.2573	0.2592	0.2601	0.2600	0.2590	0.2571	0.2543	0.2506	0.2461	5	
	5	0.1376	0.1475	0.1574	0.1672	0.1769	0.1863	0.1955	0.2044	0.2128	0.2207	0.2280	0.2347	0.2408	0.2461	4	
	6	0.0539	0.0603	0.0671	0.0743	0.0819	0.0900	0.0983	0.1070	0.1160	0.1253	0.1348	0.1445	0.1542	0.1641	3	
	7	0.0136	0.0158	0.0184	0.0212	0.0244	0.0279	0.0318	0.0360	0.0407	0.0458	0.0512	0.0571	0.0635	0.0703	2	
	8	0.0020	0.0024	0.0029	0.0035	0.0042	0.0051	0.0060	0.0071	0.0083	0.0097	0.0114	0.0132	0.0153	0.0176	1	
	9	0.0001	0.0002	0.0002	0.0003	0.0003	0.0004	0.0005	0.0006	0.0008	0.0009	0.0011	0.0014	0.0016	0.0020	0	9
10	0	0.0098	0.0084	0.0071	0.0060	0.0051	0.0043	0.0036	0.0030	0.0025	0.0021	0.0017	0.0014	0.0012	0.0010	10	
	1	0.0578	0.0514	0.0456	0.0403	0.0355	0.0312	0.0273	0.0238	0.0207	0.0180	0.0155	0.0133	0.0114	0.0098	9	
	2	0.1529	0.1419	0.1312	0.1209	0.1111	0.1017	0.0927	0.0843	0.0763	0.0688	0.0619	0.0554	0.0494	0.0439	8	
	3	0.2394	0.2319	0.2237	0.2150	0.2058	0.1963	0.1865	0.1765	0.1665	0.1564	0.1464	0.1364	0.1267	0.1172	7	
	4	0.2461	0.2487	0.2503	0.2508	0.2503	0.2488	0.2462	0.2427	0.2384	0.2331	0.2271	0.2204	0.2130	0.2051	6	
	5	0.1734	0.1829	0.1920	0.2007	0.2087	0.2162	0.2229	0.2289	0.2340	0.2383	0.2417	0.2441	0.2456	0.2461	5	
	6	0.0849	0.0934	0.1023	0.1115	0.1209	0.1304	0.1401	0.1499	0.1596	0.1692	0.1786	0.1878	0.1966	0.2051	4	
	7	0.0285	0.0327	0.0374	0.0425	0.0480	0.0540	0.0604	0.0673	0.0746	0.0824	0.0905	0.0991	0.1080	0.1172	3	
	8	0.0063	0.0075	0.0090	0.0106	0.0125	0.0147	0.0171	0.0198	0.0229	0.0263	0.0301	0.0343	0.0389	0.0439	2	
	9	0.0008	0.0010	0.0013	0.0016	0.0019	0.0024	0.0029	0.0035	0.0042	0.0050	0.0059	0.0070	0.0083	0.0098	1	
	10	0.0000	0.0001	0.0001	0.0001	0.0001	0.0002	0.0002	0.0003	0.0003	0.0004	0.0005	0.0006	0.0008	0.0010	0	10
n	X	0.63	0.62	0.61	0.60	0.59	0.58	0.57	0.56	0.55	0.54	0.53	0.52	0.51	0.50	X	n
								P									

TABLE C.6 *(continued)*

		P															
n	X	0.37	0.38	0.39	0.40	0.41	0.42	0.43	0.44	0.45	0.46	0.47	0.48	0.49	0.50	X	n
12	0	0.0039	0.0032	0.0027	0.0022	0.0018	0.0014	0.0012	0.0010	0.0008	0.0006	0.0005	0.0004	0.0003	0.0002	12	
	1	0.0276	0.0237	0.0204	0.0174	0.0148	0.0126	0.0106	0.0090	0.0075	0.0063	0.0052	0.0043	0.0036	0.0029	11	
	2	0.0890	0.0800	0.0716	0.0639	0.0567	0.0502	0.0442	0.0388	0.0339	0.0294	0.0255	0.0220	0.0189	0.0161	10	
	3	0.1742	0.1634	0.1526	0.1419	0.1314	0.1211	0.1111	0.1015	0.0923	0.0836	0.0754	0.0676	0.0604	0.0537	9	
	4	0.2302	0.2254	0.2195	0.2128	0.2054	0.1973	0.1886	0.1794	0.1700	0.1602	0.1504	0.1405	0.1306	0.1208	8	
	5	0.2163	0.2210	0.2246	0.2270	0.2284	0.2285	0.2276	0.2256	0.2225	0.2184	0.2134	0.2075	0.2008	0.1934	7	
	6	0.1482	0.1580	0.1675	0.1766	0.1851	0.1931	0.2003	0.2068	0.2124	0.2171	0.2208	0.2234	0.2250	0.2256	6	
	7	0.0746	0.0830	0.0918	0.1009	0.1103	0.1198	0.1295	0.1393	0.1489	0.1585	0.1678	0.1768	0.1853	0.1934	5	
	8	0.0274	0.0318	0.0367	0.0420	0.0479	0.0542	0.0611	0.0684	0.0762	0.0844	0.0930	0.1020	0.1113	0.1208	4	
	9	0.0071	0.0087	0.0104	0.0125	0.0148	0.0175	0.0205	0.0239	0.0277	0.0319	0.0367	0.0418	0.0475	0.0537	3	
	10	0.0013	0.0016	0.0020	0.0025	0.0031	0.0038	0.0046	0.0056	0.0068	0.0082	0.0098	0.0116	0.0137	0.0161	2	
	11	0.0001	0.0002	0.0002	0.0003	0.0004	0.0005	0.0006	0.0008	0.0010	0.0013	0.0016	0.0019	0.0024	0.0029	1	
	12	0.0000	0.0000	0.0000	0.0000	0.0000	0.0000	0.0000	0.0001	0.0001	0.0001	0.0001	0.0001	0.0002	0.0002	0	12
15	0	0.0010	0.0008	0.0006	0.0005	0.0004	0.0003	0.0002	0.0002	0.0001	0.0001	0.0001	0.0001	0.0000	0.0000	15	
	1	0.0086	0.0071	0.0058	0.0047	0.0038	0.0031	0.0025	0.0020	0.0016	0.0012	0.0010	0.0008	0.0006	0.0005	14	
	2	0.0354	0.0303	0.0259	0.0219	0.0185	0.0156	0.0130	0.0108	0.0090	0.0074	0.0060	0.0049	0.0040	0.0032	13	
	3	0.0901	0.0805	0.0716	0.0634	0.0558	0.0489	0.0426	0.0369	0.0318	0.0272	0.0232	0.0197	0.0166	0.0139	12	
	4	0.1587	0.1481	0.1374	0.1268	0.1163	0.1061	0.0963	0.0869	0.0780	0.0696	0.0617	0.0545	0.0478	0.0417	11	
	5	0.2051	0.1997	0.1933	0.1859	0.1778	0.1691	0.1598	0.1502	0.1404	0.1304	0.1204	0.1106	0.1010	0.0916	10	
	6	0.2008	0.2040	0.2059	0.2066	0.2060	0.2041	0.2010	0.1967	0.1914	0.1851	0.1780	0.1702	0.1617	0.1527	9	
	7	0.1516	0.1608	0.1693	0.1771	0.1840	0.1900	0.1949	0.1987	0.2013	0.2028	0.2030	0.2020	0.1997	0.1964	8	
	8	0.0890	0.0985	0.1082	0.1181	0.1279	0.1376	0.1470	0.1561	0.1647	0.1727	0.1800	0.1864	0.1919	0.1964	7	
	9	0.0407	0.0470	0.0538	0.0612	0.0691	0.0775	0.0863	0.0954	0.1048	0.1144	0.1241	0.1338	0.1434	0.1527	6	
	10	0.0143	0.0173	0.0206	0.0245	0.0288	0.0337	0.0390	0.0450	0.0515	0.0585	0.0661	0.0741	0.0827	0.0916	5	
	11	0.0038	0.0048	0.0060	0.0074	0.0091	0.0111	0.0134	0.0161	0.0191	0.0226	0.0266	0.0311	0.0361	0.0417	4	
	12	0.0007	0.0010	0.0013	0.0016	0.0021	0.0027	0.0034	0.0042	0.0052	0.0064	0.0079	0.0096	0.0116	0.0139	3	
	13	0.0001	0.0001	0.0002	0.0003	0.0003	0.0004	0.0006	0.0008	0.0010	0.0013	0.0016	0.0020	0.0026	0.0032	2	
	14	0.0000	0.0000	0.0000	0.0000	0.0000	0.0000	0.0001	0.0001	0.0001	0.0002	0.0002	0.0003	0.0004	0.0005	1	
	15	—	—	—	—	—	—	0.0000	0.0000	0.0000	0.0000	0.0000	0.0000	0.0000	0.0000	0	15

n	X	0.63	0.62	0.61	0.60	0.59	0.58	0.57	0.56	0.55	0.54	0.53	0.52	0.51	0.50	X	n
20	0	0.0001	0.0001	0.0001	0.0000	0.0000	0.0000	0.0000	0.0000	0.0000	0.0000	0.0000	—	—	—	20	
	1	0.0011	0.0009	0.0007	0.0005	0.0004	0.0003	0.0002	0.0001	0.0001	0.0001	0.0001	0.0000	0.0000	0.0000	19	
	2	0.0064	0.0050	0.0040	0.0031	0.0024	0.0018	0.0014	0.0011	0.0008	0.0006	0.0005	0.0003	0.0002	0.0002	18	
	3	0.0224	0.0185	0.0152	0.0123	0.0100	0.0080	0.0064	0.0051	0.0040	0.0031	0.0024	0.0019	0.0014	0.0011	17	
	4	0.0559	0.0482	0.0412	0.0350	0.0295	0.0247	0.0206	0.0170	0.0139	0.0113	0.0092	0.0074	0.0059	0.0046	16	
	5	0.1051	0.0945	0.0843	0.0746	0.0656	0.0573	0.0496	0.0427	0.0365	0.0309	0.0260	0.0217	0.0180	0.0148	15	
	6	0.1543	0.1447	0.1347	0.1244	0.1140	0.1037	0.0936	0.0839	0.0746	0.0658	0.0577	0.0501	0.0432	0.0370	14	
	7	0.1812	0.1774	0.1722	0.1659	0.1585	0.1502	0.1413	0.1318	0.1221	0.1122	0.1023	0.0925	0.0830	0.0739	13	
	8	0.1730	0.1767	0.1790	0.1797	0.1790	0.1768	0.1732	0.1683	0.1623	0.1553	0.1474	0.1388	0.1296	0.1201	12	
	9	0.1354	0.1444	0.1526	0.1597	0.1658	0.1707	0.1742	0.1763	0.1771	0.1763	0.1742	0.1708	0.1661	0.1602	11	
	10	0.0875	0.0974	0.1073	0.1171	0.1268	0.1359	0.1446	0.1524	0.1593	0.1652	0.1700	0.1734	0.1755	0.1762	10	
	11	0.0467	0.0542	0.0624	0.0710	0.0801	0.0895	0.0991	0.1089	0.1185	0.1280	0.1370	0.1455	0.1533	0.1602	9	
	12	0.0206	0.0249	0.0299	0.0355	0.0417	0.0486	0.0561	0.0642	0.0727	0.0818	0.0911	0.1007	0.1105	0.1201	8	
	13	0.0074	0.0094	0.0118	0.0146	0.0178	0.0217	0.0260	0.0310	0.0366	0.0429	0.0497	0.0572	0.0653	0.0739	7	
	14	0.0022	0.0029	0.0038	0.0049	0.0062	0.0078	0.0098	0.0122	0.0150	0.0183	0.0221	0.0264	0.0314	0.0370	6	
	15	0.0005	0.0007	0.0010	0.0013	0.0017	0.0023	0.0030	0.0038	0.0049	0.0062	0.0078	0.0098	0.0121	0.0148	5	
	16	0.0001	0.0001	0.0002	0.0003	0.0004	0.0005	0.0007	0.0009	0.0013	0.0017	0.0022	0.0028	0.0036	0.0046	4	
	17	0.0000	0.0000	0.0000	0.0000	0.0001	0.0001	0.0001	0.0002	0.0002	0.0003	0.0005	0.0006	0.0008	0.0011	3	
	18	—	—	—	—	0.0000	0.0000	0.0000	0.0000	0.0000	0.0000	0.0001	0.0001	0.0001	0.0002	2	
	19	—	—	—	—	—	—	—	—	—	—	0.0000	0.0000	0.0000	0.0000	1	
	20	—	—	—	—	—	—	—	—	—	—	—	—	—	—	0	20
		P															

TABLE C.7 □ The Fisher $\mathcal{Z}$ transformation[a]

r	$\mathcal{Z}_r$	r	$\mathcal{Z}_r$	r	$\mathcal{Z}_r$	r	$\mathcal{Z}_r$	r	$\mathcal{Z}_r$
.00	.000	.20	.203	.40	.424	.60	.693	.80	1.099
.01	.010	.21	.213	.41	.436	.61	.709	.81	1.127
.02	.020	.22	.224	.42	.448	.62	.725	.82	1.157
.03	.030	.23	.234	.43	.460	.63	.741	.83	1.188
04	.040	.24	.245	.44	.472	.64	.758	.84	1.221
.05	.050	.25	.255	.45	.485	.65	.775	.85	1.256
.06	.060	.26	.266	.46	.497	.66	.793	.86	1.293
.07	.070	.27	.277	.47	.510	.67	.811	.87	1.333
.08	.080	.28	.288	.48	.523	.68	.829	.88	1.376
.09	.090	.29	.299	.49	.536	.69	.848	.89	1.422
.10	.100	.30	.310	.50	.549	.70	.867	.90	1.472
.11	.110	.31	.320	.51	.563	.71	.887	.91	1.528
.12	.121	.32	.332	.52	.576	.72	.908	.92	1.589
.13	.131	.33	.343	.53	.590	.73	.929	.93	1.658
.14	.141	.34	.354	.54	.604	.74	.950	.94	1.738
.15	.151	.35	.365	.55	.618	.75	.973	.95	1.832
.16	.161	.36	.377	.56	.633	.76	.996	.96	1.946
.17	.172	.37	.388	.57	.648	.77	1.020	.97	2.092
.18	.182	.38	.400	.58	.662	.78	1.045	.98	2.298
.19	.192	.39	.412	.59	.678	.79	1.071	.99	2.647

[a] $\mathcal{Z}_r = \frac{1}{2} \ln [(1 + r)/(1 - r)]$.

TABLE C.8 □ Lower and upper critical values U for the runs test for randomness

Part 1. Lower Tail ($\alpha = .025$)

n_1 \ n_2	2	3	4	5	6	7	8	9	10	11	12	13	14	15	16	17	18	19	20
2											2	2	2	2	2	2	2	2	2
3					2	2	2	2	2	2	2	2	2	3	3	3	3	3	3
4				2	2	2	3	3	3	3	3	3	3	3	4	4	4	4	4
5			2	2	3	3	3	3	3	4	4	4	4	4	4	4	5	5	5
6		2	2	3	3	3	3	4	4	4	4	5	5	5	5	5	5	6	6
7		2	2	3	3	3	4	4	5	5	5	5	5	6	6	6	6	6	6
8		2	3	3	3	4	4	5	5	5	6	6	6	6	6	7	7	7	7
9		2	3	3	4	4	5	5	5	6	6	6	7	7	7	7	8	8	8
10		2	3	3	4	5	5	5	6	6	7	7	7	7	8	8	8	8	9
11		2	3	4	4	5	5	6	6	7	7	7	8	8	8	9	9	9	9
12	2	2	3	4	4	5	6	6	7	7	7	8	8	8	9	9	9	10	10
13	2	2	3	4	5	5	6	6	7	7	8	8	9	9	9	10	10	10	10
14	2	2	3	4	5	5	6	7	7	8	8	9	9	9	10	10	10	11	11
15	2	3	3	4	5	6	6	7	7	8	8	9	9	10	10	11	11	11	12
16	2	3	4	4	5	6	6	7	8	8	9	9	10	10	11	11	11	12	12
17	2	3	4	4	5	6	7	7	8	9	9	10	10	11	11	11	12	12	13
18	2	3	4	5	5	6	7	8	8	9	9	10	10	11	11	12	12	13	13
19	2	3	4	5	6	6	7	8	8	9	10	10	11	11	12	12	13	13	13
20	2	3	4	5	6	6	7	8	9	9	10	10	11	12	12	13	13	13	14

Part 2. Upper Tail ($\alpha = .025$)

n_1 \ n_2	2	3	4	5	6	7	8	9	10	11	12	13	14	15	16	17	18	19	20
2																			
3																			
4				9	9														
5			9	10	10	11	11												
6			9	10	11	12	12	13	13	13	13								
7				11	12	13	13	14	14	14	14	15	15	15					
8				11	12	13	14	14	15	15	16	16	16	16	17	17	17	17	17
9					13	14	14	15	16	16	16	17	17	18	18	18	18	18	18
10					13	14	15	16	16	17	17	18	18	18	19	19	19	20	20
11					13	14	15	16	17	17	18	19	19	19	20	20	20	21	21
12					13	14	16	16	17	18	19	19	20	20	21	21	21	22	22
13						15	16	17	18	19	19	20	20	21	21	22	22	23	23
14						15	16	17	18	19	20	20	21	22	22	23	23	23	24
15						15	16	18	18	19	20	21	22	22	23	23	24	24	25
16							17	18	19	20	21	21	22	23	23	24	25	25	25
17							17	18	19	20	21	22	23	23	24	25	25	26	26
18							17	18	19	20	21	22	23	24	25	25	26	26	27
19							17	18	20	21	22	23	23	24	25	26	26	27	27
20							17	18	20	21	22	23	24	25	25	26	27	27	28

SOURCE: Adapted from F. S. Swed and C. Eisenhart, *Ann. Math. Statist.*, vol. 14, 1943, pp. 83–86.

TABLE C.9 □ Lower and upper critical values *W* of Wilcoxon signed-ranks test

n	*One Tailed:* $\alpha = .05$ *Two-Tailed:* $\alpha = .10$	$\alpha = .025$ $\alpha = .05$	$\alpha = .01$ $\alpha = .02$	$\alpha = .005$ $\alpha = .01$
	(Lower, Upper)			
5	0,15			
6	2,19	0,21		
7	3,25	2,26	0,28	
8	5,31	3,33	1,35	0,36
9	8,37	5,40	3,42	1,44
10	10,45	8,47	5,50	3,52
11	13,53	10,56	7,59	5,61
12	17,61	13,65	10,68	7,71
13	21,70	17,74	12,79	10,81
14	25,80	21,84	16,89	13,92
15	30,90	25,95	19,101	16,104
16	35,101	29,107	23,113	19,117
17	41,112	34,119	27,126	23,130
18	47,124	40,131	32,139	27,144
19	53,137	46,144	37,153	32,158
20	60,150	52,158	43,167	37,173

SOURCE: Material from *Some Rapid Approximate Statistical Procedures*.

TABLE C.10 ☐ Lower and upper critical values T_1 of Wilcoxon rank-sum test

n_2	α		n_1						
	One-Tailed	Two-Tailed	4	5	6	7	8	9	10
4	.05	.10	11,25						
	.025	.05	10,26						
	.01	.02							
	.005	.01							
5	.05	.10	12,28	19,36					
	.025	.05	11,29	17,38					
	.01	.02	10,30	16,39					
	.005	.01		15,40					
6	.05	.10	13,31	20,40	28,50				
	.025	.05	12,32	18,42	26,52				
	.01	.02	11,33	17,43	24,54				
	.005	.01	10,34	16,44	23,55				
7	.05	.10	14,34	21,44	29,55	39,66			
	.025	.05	13,35	20,45	27,57	36,69			
	.01	.02	11,37	18,47	25,59	34,71			
	.005	.01	10,38	16,49	24,60	32,73			
8	.05	.10	15,37	23,47	31,59	41,71	51,85		
	.025	.05	14,38	21,49	29,61	38,74	49,87		
	.01	.02	12,40	19,51	27,63	35,77	45,91		
	.005	.01	11,41	17,53	25,65	34,78	43,93		
9	.05	.10	16,40	24,51	33,63	43,76	54,90	66,105	
	.025	.05	14,42	22,53	31,65	40,79	51,93	62,109	
	.01	.02	13,43	20,55	28,68	37,82	47,97	59,112	
	.005	.01	11,45	18,57	26,70	35,84	45,99	56,115	
10	.05	.10	17,43	26,54	35,67	45,81	56,96	69,111	82,128
	.025	.05	15,45	23,57	32,70	42,84	53,99	65,115	78,132
	.01	.02	13,47	21,59	29,73	39,87	49,103	61,119	74,136
	.005	.01	12,48	19,61	27,75	37,89	47,105	58,122	71,139

SOURCE: Material from *Some Rapid Approximate Statistical Procedures*.

Answers to Selected Problems (·)

CHAPTER 2

2.4 **(a)** Quantitative (discrete); **(b)** qualitative; **(c)** quantitative (discrete); **(d)** quantitative (continuous); **(e)** qualitative; **(f)** quantitative (continuous).

2.37 12 47 83 76 22 65 93 10 61 36 89 58 86 92 71

2.42 Line 1—Column 4 has a 6 for major.
Line 2—Columns 50 and 51 have a 98 for height.
Line 3—Column 6 has a 7 for reason for attending.
Line 4—Columns 40–42 have a 5.5 for GPI.
Line 5—Column 8 has a 4 for current ideology.

CHAPTER 3

3.10 **(a)** $\bar{X} = 32.5$, median = 33.5, midrange = 32; **(b)** total = 13,000.

3.22 $\bar{X} = 701.8$, median = 698, midrange = 732.5.

3.28 Range = 10, $S^2 = 14.29$, $S = 3.78$.

3.39 Range = 389, $S = 108.5$.

3.56 **(a)** 95.8%; **(b)** 95.8%; **(c)** They are more than 75% and 88.9%, respectively.

3.69 **(a)** **(1)** $\bar{X} = .3328$; **(2)** median = .331; **(3)** mode = .362 or .349; **(4)** midrange = .339; **(5)** range = .108; **(6)** $S^2 = .001373$; **(7)** $S = .037$.
(b) Yes, slightly positive (right) skew because the mean is greater than the median.

CHAPTER 4

4.7

	Projected changes	Number
(a)	3.0 – under 3.5	0
	3.5 – under 4.0	4
	4.0 – under 4.5	11
	4.5 – under 5.0	9
	5.0 – under 5.5	2
	5.5 – under 6.0	1
		27

	Projected changes	Number
(b)	3.0 – under 4.0	4
	4.0 – under 5.0	20
	5.0 – under 6.0	3
		27

There are only three classes in part (b), so that not enough distinctions in the data can be observed.

4.12

	Calories	Frequency
(a)	70 – under 80	8
	80 – under 90	13
	90 – under 100	10
	100 – under 110	14
	110 – under 120	7
	120 – under 130	2
		54

	Calories	Frequency
(b)	70 – under 75	1
	75 – under 80	7
	80 – under 85	7
	85 – under 90	6
	90 – under 95	4
	95 – under 100	6
	100 – under 105	3
	105 – under 110	11
	110 – under 115	4
	115 – under 120	3
	120 – under 125	2
		54

With a class interval of 5, the data are more evenly distributed in the classes.

	Calories	First 24, frequency	Last 30, frequency
(c)	70 – under 80	0	8
	80 – under 90	0	13
	90 – under 100	3	7
	100 – under 110	13	1
	110 – under 120	6	1
	120 – under 130	2	0
		24	30

(d) The first 24 observations represent 4% milkfat and the last 30 observations represent 2% milkfat or less.

4.26

Projected changes	Percent
$3.0 < 3.5$	0.0
$3.5 < 4.0$	14.8
$4.0 < 4.5$	40.7
$4.5 < 5.0$	33.3
$5.0 < 5.5$	7.4
$5.5 < 6.0$	3.7
	99.9 (rounding error)

4.32

Calories	4% Milkfat, percentage	2% Milkfat or less, percentage	
70 < 80	0.0	26.7	
80 < 90	0.0	43.3	
90 < 100	12.5	23.3	
100 < 110	54.2	3.3	
110 < 120	25.0	3.3	
120 < 130	8.3	0.0	
	100.0	99.9	(rounding error)

They differ in central tendency and are fairly similar in dispersion.

4.39 **(a)**

(b)

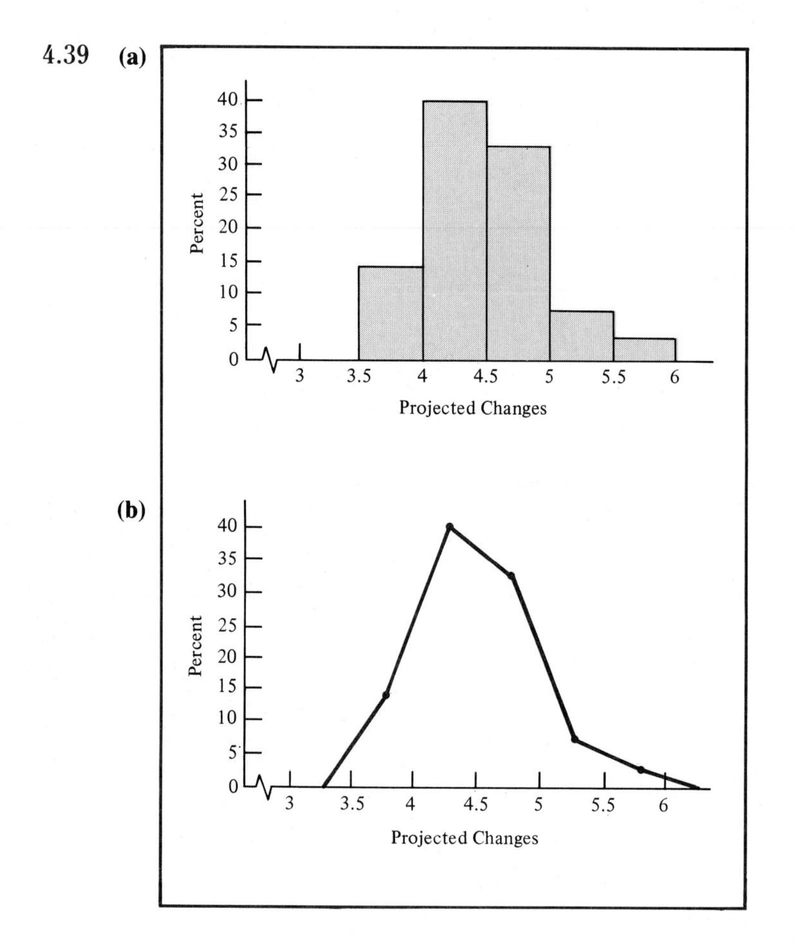

(c) The distribution looks slightly right-skewed; the mean and the median are almost identical, but the midrange is greater than both of these measures.

4.44

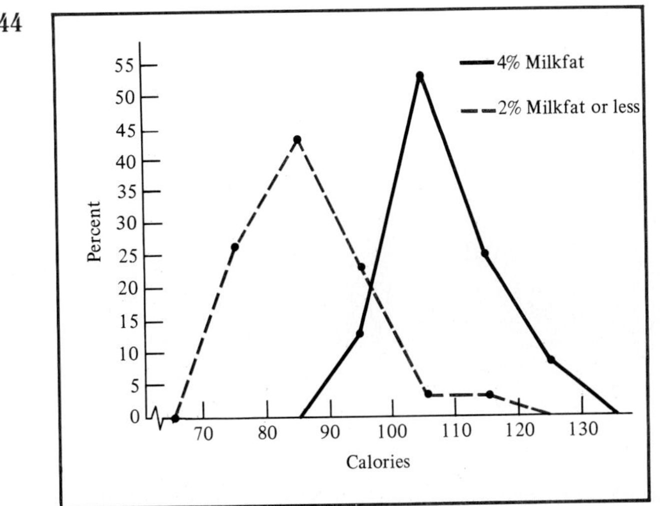

The distributions have approximately the same dispersion and shape, but the calories are much higher for 4% milkfat than for 2% or less milkfat.

4.48 **(a)** Bar chart:

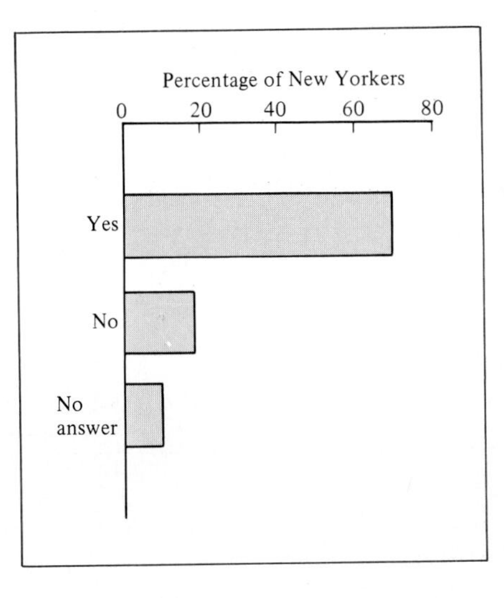

(b) Pie chart:

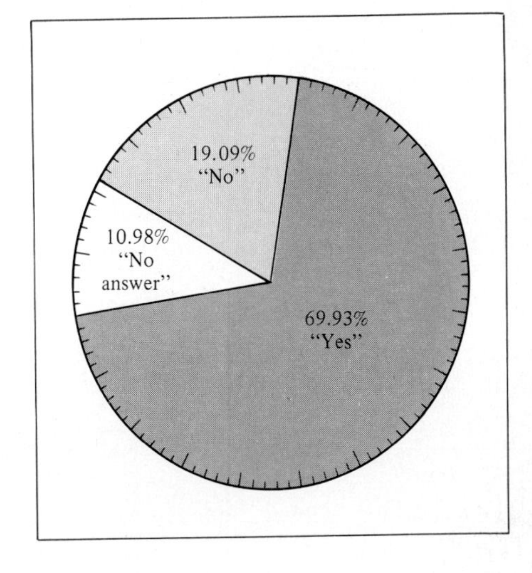

4.56

(a) Sources of funding	Percent
Federal	6.3
State	48.8
Local	44.9
	100.0

(b)

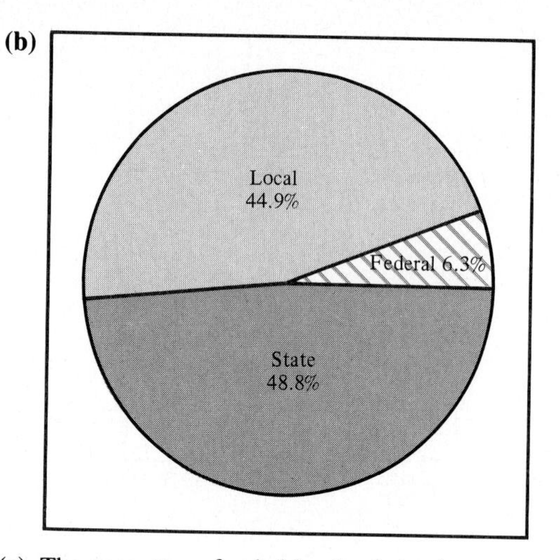

(c) The percentage funded by the federal government is very small.

4.58 **(a)** Providing column percentages:

Management level	Type of college: Ivy league	Other private	Public	Total
High (Sr. V-P or above)	8.5%	6.4%	4.2%	5.6%
Middle	43.6%	58.3%	54.4%	53.8%
Low	47.9%	35.3%	41.4%	40.6%
Total	100.0%	100.0%	100.0%	100.0%

4.71 **(a)**

Per capita alcohol consumption (gallons)	Frequency	Percent
1.0 – under 2.0	8	15.7%
2.0 – under 3.0	31	60.8%
3.0 – under 4.0	9	17.6%
4.0 – under 5.0	1	2.0%
5.0 – under 6.0	2	3.9%
	51	100.0%

(b)

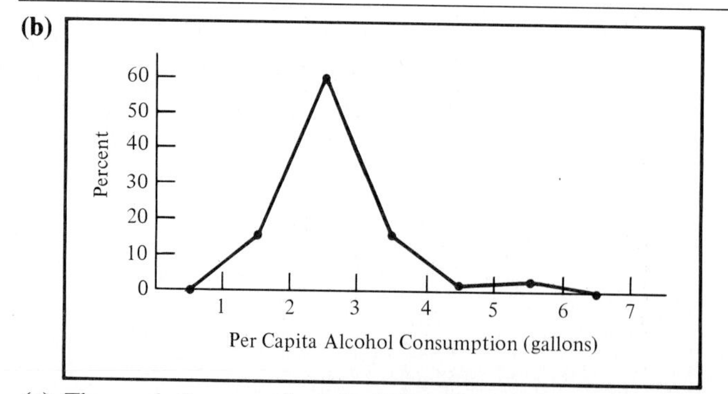

(c) The mode is approximately 2.5 gallons, the range is approximately 5 gallons, and the midrange is approximately 3.5 gallons.

(d) There is no mode, the range is 3.81, and the midrange is 3.435.

4.78

(a) **(b)**

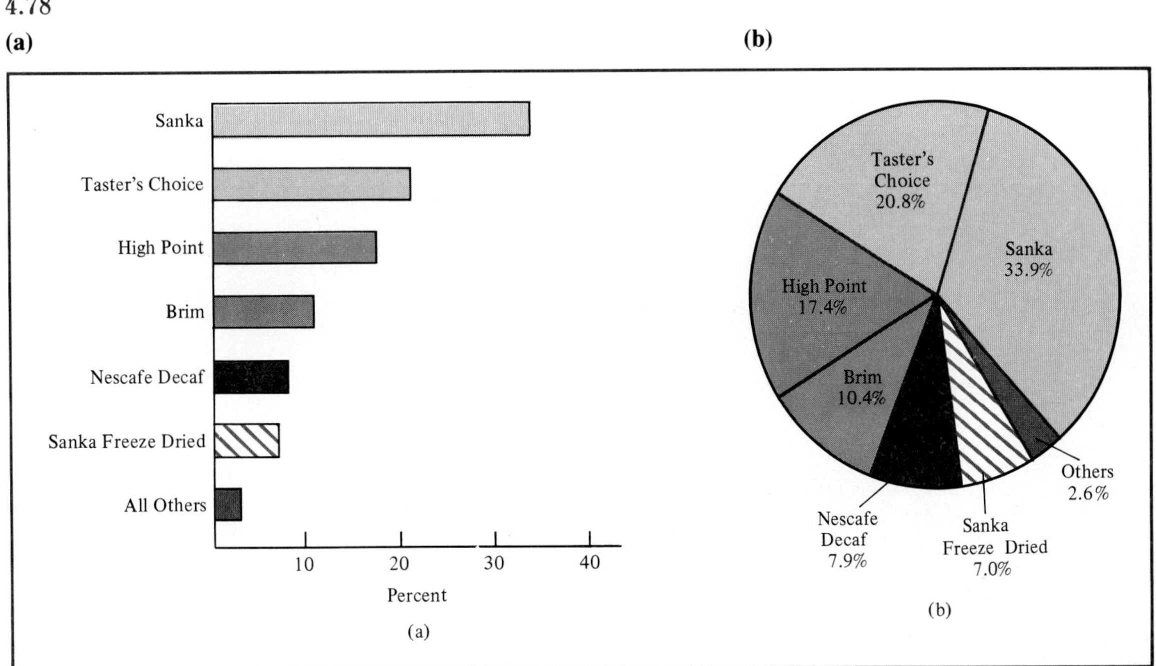

(c) The bar chart seems to display the differences more distinctly.

CHAPTER 5

5.2

0	386679
1	38206359712649
2	149
3	
4	8

0L	3
0H	86679
1L	3203124
1H	8659769
2L	14
2H	9
3L	
3H	
4L	
4H	8

The split stem seems to highlight the data better.

0	366789
1	01223345667899
2	149
3	
4	8

5.9

7	69595693
8	8391912725345
9	8579253901
10	576058097637507
11	308693
12	32

7L	3
7H	6959569
8L	3112234
8H	899755
9L	2301
9H	857959
10L	0030
10H	57658976757
11L	303
11H	869
12L	32

The split stem seems to show the distinction better.

7	35566999
8	1122334557899
9	0123557899
10	000355566777789
11	033689
12	23

5.12 $Q_1 = 9$, $Q_3 = 19$.

5.17 $Q_1 = 83$, $Q_3 = 107$.

5.19 Midhinge = 14, interquartile range = 10.

5.24 Midhinge = 95, interquartile range = 24.

5.26 $X_{\text{smallest}} = 3$, $Q_1 = 9$, median = 13.5, $Q_3 = 19$, $X_{\text{largest}} = 48$. The distribution is right-skewed, since the midrange is much larger than the median; the distance between the median and the largest value is much greater than the distance between the median and the smallest value.

5.31 $X_{\text{smallest}} = 73$, $Q_1 = 83$, median = 96, $Q_3 = 107$, $X_{\text{largest}} = 123$. The distribution is approximately symmetric; the midhinge is slightly less than the median, but the midrange is slightly more than the median.

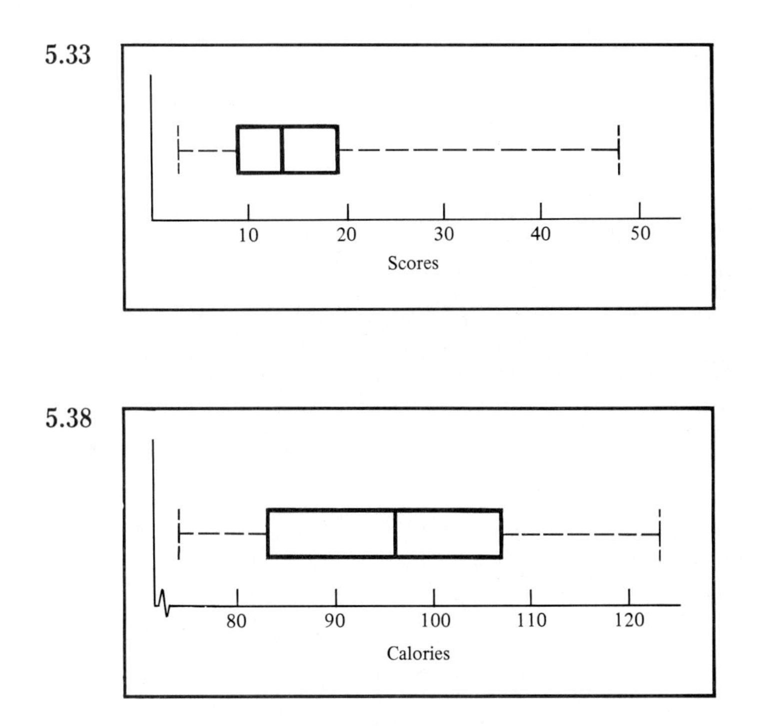

5.43 **(a)** Revised stem-and-leaf displays:

German stem	German leaves	Japanese stem	Japanese leaves
6	2		
7	0	7	9
8	12335	8	9
9		9	277
10	13469	10	5
11	267	11	3558
12		12	8
13	02	13	02
		14	238
		15	01

(b) German: $Q_1 = 8.25$, $Q_3 = 11.4$; Japanese: $Q_1 = 9.7$, $Q_3 = 14.2$.
(c) Midhinge: German = 9.825, Japanese = 11.95;
interquartile range: German = 3.15, Japanese = 4.5.
(d) Five-number summaries:

	$X_{smallest}$	Q_1	Median	Q_3	$X_{largest}$
German:	6.2	8.25	10.3	11.4	13.2
Japanese:	7.9	9.7	11.65	14.2	15.1

(e)

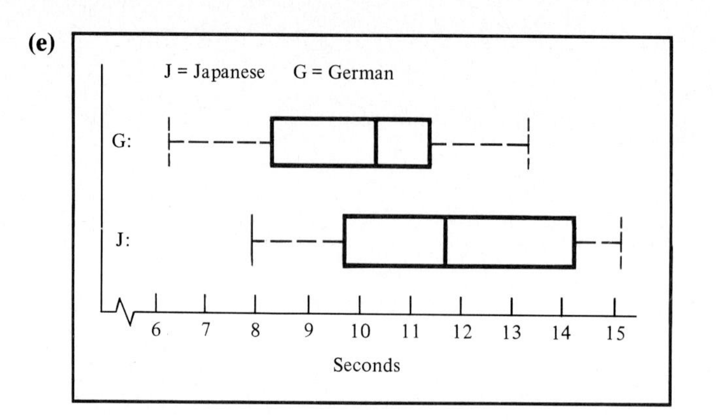

(f) Japanese models required more time than did German models to get from 0 to 60 mph.

CHAPTER 6

6.4 **(a)** P (female) = 141/194 = .727, P(male) = 53/194 = .273;
P(blonde) = 80/194 = .412, P(brunette) = 67/194 = .345;
P(red) = 10/194 = .052, P(black) = 37/194 = .191.
(b) At a location that is more likely to be frequented by women than by men.

6.8 **(a)** Living, dying, getting vaccinated, not getting vaccinated.
(b) P(living) = 2475/3686 = .671, P(dying) = 1211/3686 = .329,
P(getting vaccinated) = 1075/3686 = .292,
P(not getting vaccinated) = 2611/3686 = .708.

6.11 **(a)**

	Apparent hair color				
Gender	**Blonde**	**Brunette**	**Red**	**Black**	**Total**
Female	.351	.222	.036	.119	.728
Male	.062	.124	.015	.072	.273
Total	.413	.346	.051	.191	1.000

(b) The results agree except for rounding errors.

6.15 **(a)**

	Outcome		
Vaccinated	**Live**	**Die**	**Total**
Yes	.226	.066	.292
No	.446	.263	.709
Total	.672	.329	1.000

(b) The results agree except for rounding errors.

6.22 **(a)** 60/194 = .309; **(b)** 155/194 = .799; **(c)** 194/194 = 1.0. Yes, if one occurs, the other cannot occur. Yes, because one of these colors must occur.

6.28 **(a)** $P(\text{blonde}|\text{male}) = 12/53 = .226$, $P(\text{blonde}|\text{female}) = 68/141 = .482$
(b) $P(\text{blonde}) = .412$. The probability is different, depending on gender. This indicates that having blonde hair is not independent of the gender of the individual.
(c) $P(\text{male}|\text{blonde}) = 12/80 = .15$.

6.36 $P(\text{winning nothing}) = .9801$.

6.40 $P(\text{male and blonde}) = 12/194 = .062$, $P(\text{male})P(\text{blonde}) = (53/194)(80/194) = .113$. Since these are not equal, the gender of the individual is not statistically independent of the hair color.

6.46 **(a)** 66; **(b)** 55; **(c)** 121, 110.

6.48 **(a)** 792; **(b)** 120; **(c)** 95,040. The number of permutations is the product of the number of combinations (792) times the number of ways the pictures can be arranged (120).

6.52 **(a)** Average interest in statistics. **(b)** Average interest in statistics and high ability in mathematics. **(c)** Because it satisfies two criteria. **(d)** **(1)** 50/200 = .25; **(2)** 70/200 = .35; **(3)** 80/200 = .40; **(4)** 40/200 = .20; **(e)** **(1)** 60/200 = .30; **(2)** 10/200 = .05; **(3)** 25/200 = .125.

6.54 **(a)** 2/42; **(b)** 10/42.

6.58 **(a)** 130; **(b)** 28.

CHAPTER 7

7.4

	Distribution A	*Distribution B*
(a)	$\mu_X = 1.00$	$\mu_X = 3.00$
(b)	$\sigma_X = 1.22$	$\sigma_X = 1.22$
(c)	Right-skewed	Left-skewed

7.7 **(a)** $\mu_X = \$51.00$. **(b)** $\sigma_X = \$23.11$. **(c)** The expected value of selling beer ($51) exceeds the certain amount ($50) to be made by selling hot dogs. Perhaps the vendor would want to sell hot dogs and receive $50 with certainty.

7.14 **(a)** .6496; **(b)** .1503; **(c)** .1493; **(e)** $\mu_X = 7$, $P(X = 7) = .2668$; **(f)** $\sigma_X = 1.449$. From the Bienayme–Chebyshev inequality $P(4.102 < X < 9.898) \geq .75$. For this distribution $P(4 < X < 10) = .9244$.

7.17 **(a)** .2556; **(b)** .1176; **(c)** .0007.

7.22 **(a)** .0901; **(b)** .8790; **(c)** .3790; **(d)** .1210; **(e)** .8889; **(f)** .1875.

7.24 **(b)** $P(X < 100) = .50$, $P(X < 115) = .8413$, $P(X > 130) = .0228$, $P(X > 145) = .00135$; **(c)** $P(85 \leq X \leq 115) = .6826$, $P(70 \leq X \leq 130) = .9544$; **(d)** 85, 100, 115.

7.30 .0228.

7.32 **(a)** 1.00; **(b)** .9678; **(c)** .409.

7.40 **(a)** Let W represent the results of one field bet:

W	$P(W)$
$-1	20/36
+1	14/36
+2	2/36
	1

(b) $\mu_W = -.056$; **(d)** player loses 5.6 cents per bet; **(d)** house wins 5.6 cents per bet.

7.43 **(a)** **(1)** .7605; **(2)** .6599; **(3)** .1814; **(b)** 6.31 minutes (using $Z = 1.645$).

CHAPTER 8

8.4 **(a)** .1915; **(b)** .0254; **(c)** **(1)** $\mu_{\bar{X}} = 5$, $\sigma_{\bar{X}} = .25$; **(2)** normal; **(3)** .4772; **(4)** .00013.

8.8 **(a)** .9924; **(b)** .0038; **(c)** from 92.65 to 107.35; **(d)** from 70.6 to 129.4.

8.12 **(a)** .4599; **(b)** .0401; **(c)** .4932 and .0068.

8.15 **(1)** $\mu_{\bar{X}} = 5$, $\sigma_{\bar{X}} = .24$; **(2)** normal; **(3)** .4812; **(4)** .00009.

8.20 12.726.

CHAPTER 9

9.2 $[141.55 \leq \mu_X \leq 148.45]$, $[140.065 \leq \mu_X \leq 149.935]$, $[137.26 \leq \mu_X \leq 152.74]$.

9.9 $[.2629 \leq \mu_X \leq .2847]$, $[.2595 \leq \mu_X \leq .2881]$, $[.2497 \leq \mu_X \leq .2979]$.

9.13 $[8.82 \leq \mu_G \leq 10.90]$, $[10.78 \leq \mu_J \leq 13.04]$.

9.19 $[-2.71 \leq \mu_D \leq +0.21]$.

9.23 $[.448 \leq p \leq .752]$.

9.25 $[.279 \leq p \leq .401]$, $[.262 \leq p \leq .418]$, $[.247 \leq p \leq .433]$.

9.30 $n = 36$.

9.31 $n = 12$ for 75% confidence; $n = 25$ for 90% confidence.

9.37 Assuming no prior knowledge of p, $n = 2{,}401$ and $n = 66{,}564$.

9.38 Assuming no prior knowledge of p, $n = 6{,}766$ for 90% confidence and $n = 16{,}641$ for 99% confidence.

9.42 **(a)** $n = 197$; **(b)** $[.082 \leq p \leq .318]$.

9.43 **(a)** $n = 32$; **(b)** $[88.29 \leq \mu_X \leq 101.71]$.

9.55 **(a)** $[14.085 \leq \mu_X \leq 16.515]$; **(b)** $[.601 \leq p \leq .749]$; **(c)** $n = 24$; **(d)** $n = 784$.

CHAPTER 10

10.9 β will increase.

10.14 $Z = +1.67 < 2.58$. Don't reject H_0. There is no evidence that the average yield is different from 140 bushels.

10.19 $t = -0.45 > -3.0545$. Don't reject H_0. There is no evidence that the average life of light bulbs produced is different from 500 hours.

10.21 $t = -6.02 < -2.1009$. Reject H_0. There is evidence that the average amount of vitamin C is different from 120 mg.

10.26 $t = -2.17 < -1.7823$. Reject H_0. There is evidence that the average life of the bulbs is less than 600 hours.

10.27 $t = -6.02 < -2.5524$. Reject H_0. There is evidence that the average amount of vitamin C is less than 120 mg.

10.30 **(a)** $t = 1.313 < 1.8331$. Don't reject H_0. There is no evidence that an average of more than 5 lawnmowers per store are being sold.
(b) That the number of lawnmowers sold per store is normally distributed.

10.32 $t = -2.82 < -1.6924$. Reject H_0. There is evidence of a difference between German and Japanese cars in their average rate of acceleration.

10.35 $t = 9.31 > 1.6759$. Reject H_0. There is evidence that the average number of calories is higher for 4% milkfat cottage cheese than for 2% milkfat cottage cheese.

10.36 **(a)** $t = -6.73 < -1.6644$. Reject H_0. There is evidence of a difference in the average distance that the pens will write.
(b) $Z = -6.62 < -1.645$. Reject H_0. For these data, the conclusions are the same.

10.39 **(a)** $t = 5.00 > 1.7959$. Reject H_0. There is evidence that the average amount of sleep is higher for EZ Sleep than for the placebo.
(b) **(1)** H_0: $\mu_D = 0$, H_1: $\mu_D \neq 0$. **(2)** $t = 5.00 > 2.201$. Reject H_0. In this example, the conclusions are the same.
(c) It makes more sense to do (a) because we want to know whether EZ Sleep is better than a placebo.

10.40 Higher driving skills mean that fewer pylons are knocked over. $t = -2.655 > -2.7181$. Don't reject H_0. There is no evidence that average driving skills are higher (the number of pylons knocked over is lower) for people driving when they are sober.

10.48 The hypothesized value of $\mu_X = 140$ is not included in the 75% and 90% intervals but is included in the 99% confidence interval. Therefore the results of Problem 10.14 at α = .01 are consistent with the 99% confidence interval that indicates no evidence that the average is different from 140 bushels.

10.53 The conclusions are the same because stating the interval includes $\mu_D = 0$ is the same as performing a two-tailed test of hypothesis.

10.57 **(a)** $[-3.29 \leq \mu_G - \mu_J \leq -0.82]$.
(b) $\mu_D = 0$ is not included in the interval.
(c) Since $\mu_D = 0$ is not included in the interval, there is evidence that there is a difference in the average rate of acceleration between German and Japanese cars.
(d) The conclusions are the same, since the confidence-interval estimate is equivalent to the two-tailed t test.

10.62 p value = .3174 for a two-tailed test.

10.64 $t = -0.45$; p value $= P(t < -0.45) + P(t > +0.45)$; p value > 0.50.

10.70 **(a)** $\bar{X} = 47.9167$, $S = 22.71$, $n = 12$, $t = -1.84 > -2.7181$. Don't reject H_0. There is no evidence that the average travel time is less than 60 minutes.
(b) That travel time is normally distributed.

10.76 $t = -5.463 < -2.998$. Reject H_0. There is evidence that the unknown brand sells for a lower price.

CHAPTER 11

11.3 $Z = 1.768 < 2.33$. Don't reject H_0. There is no evidence that more than 20% of the people will buy magazines.

11.4 $Z = -3.55 < -1.96$. Reject H_0. There is evidence that the proportion of voters who will vote Republocrat is different from .50.

11.7 $Z = 1.86 < 1.96$. Don't reject H_0. There is no evidence of a difference in the proportion of males and females who are left-handed.

11.8 $Z = 3.27 > 1.96$. Reject H_0. There is evidence of a difference in the proportion of Republicans and Democrats who favor capital punishment.

11.9 $Z = 1.949 < 2.33$. Don't reject H_0. There is no evidence that a higher proportion of students who received remedial tutoring got degrees.

11.14 **(a)** $\chi^2 = 10.70 > 3.841$; **(b)** $Z^2 = (3.27)^2 \cong 10.70$.

11.15 **(a)** $\chi^2 = 3.80 < 3.841$; **(b)** $Z^2 = (1.949)^2 \cong 3.80$; **(c)** Problem 11.9 is a directional test (H_1: $p_1 > p_2$), whereas the chi-square test is a nondirectional test (H_1: $p_1 \neq p_2$).

11.20 $\chi^2 = .563 < 9.21$. Don't reject H_0. There is no evidence of a relationship between water softness and preference for the new detergent.

11.21 **(a)** $\chi^2 = 5.593 < 5.991$. Don't reject H_0. There is no evidence of a relationship between attitude and results after 10 years.

11.24 **(a)** $\chi^2 = 20.381 > 12.592$. Reject H_0. There is evidence of a relationship between type of number drawn and time of the year.
(b) Thus, it appears that the results of the lottery drawing are different from what would be expected if it were random.

11.25 $\chi^2 = 35.922 > 13.277$. Reject H_0. There is evidence of a relationship between the height of husbands and the height of wives.

11.31 $Z = 2.769 > 1.96$. Reject H_0. There is evidence of a difference in the usage of drugs and natural remedies for colds.

11.33 **(a)** $Z = -3.90 < -2.33$. Reject H_0. There is evidence that the proportion is lower in year 1.
(b) The incentive plan significantly reduced absenteeism.

11.36 $Z = 2.372 > 1.96$. Reject H_0. There is evidence of a difference in the proportion of women in the two groups who eat dinner in a restaurant during the work week.

11.40 $\chi^2 = 9.83 < 13.277$. Don't reject H_0. There is no evidence of a relationship between commuting time and stress.

CHAPTER 12

12.1 **(b)** $\overline{X}_1 = 2.44$, $\overline{X}_2 = 3.18$, $\overline{X}_3 = 3.55$.
(d) $F = 9.06 > 3.81$. There is evidence of a difference in the average life based upon the amount of food intake.

12.7 **(a)** $F = 86.69 > 4.04$. Reject H_0. There is evidence of a difference in the average number of calories between 4% and 2% milkfat cottage cheese.
(b) $t^2 = (9.311)^2 \cong F = 86.69$.

12.8 $F = 4.088 > 3.49$. Reject H_0. There is evidence of a difference in the average life of the four brands.

12.10 $F_C = 2(3.81) = 7.62$. Comparing $\overline{X}_1$ to $\overline{X}_2$, $F = 7.243 < 7.62$; don't reject H_0. Comparing $\overline{X}_1$ to $\overline{X}_3$, $F = 17.778 > 7.62$; reject H_0. Comparing $\overline{X}_2$ to $\overline{X}_3$, $F = 1.976 < 7.62$; don't reject H_0. Comparing $\overline{X}_1$ to $(\overline{X}_2 + \overline{X}_3)/2$, $F = 15.503 > 7.62$. Thus, there is a difference in average life between unlimited food and 80% of normal feeding; there is a difference between unlimited food and the average of 90% and 80% of normal feeding.

12.13 None of the pairwise comparisons are significant. $\overline{X}_1 - \overline{X}_2$: $F = .538 < 10.47$. $\overline{X}_1 - \overline{X}_3$: $F = 8.615 < 10.47$. $\overline{X}_1 - \overline{X}_4$: $F = 6.868 < 10.47$. $\overline{X}_2 - \overline{X}_3$: $F = 4.846 < 10.47$. $\overline{X}_2 - \overline{X}_4$: $F = 3.56 < 10.47$. $\overline{X}_3 - \overline{X}_4$: $F = 0.099 < 10.47$. However, for $(\overline{X}_1 + \overline{X}_2)/2$ versus $(\overline{X}_3 + \overline{X}_4)/2$, $F = 11.626 > 10.47$. Therefore, there is evidence of a difference in the average life of brands 1 and 2 as compared to brands 3 and 4.

12.22 **(b)** $\overline{X}_1 = 3.25$, $\overline{X}_2 = 6$, $\overline{X}_3 = 3.714$.
(c) $F = 10.22 > 3.55$. Reject H_0. There is evidence of a difference in the average number of errors between the three types of dials.
(d) Comparing $\overline{X}_1$ to $\overline{X}_2$, $F = 18.789 > 7.10$; reject H_0. Comparing $\overline{X}_2$ to $\overline{X}_3$, $F = 12.234 > 7.10$; reject H_0. Comparing $(\overline{X}_1 + \overline{X}_3)/2$ to $\overline{X}_2$, $F = 19.662 > 7.10$; reject H_0. There is evidence that the average number of errors is different for the circular analog dial than for each of the other two dials. The average number of errors is different for the circular analog dial than for the average of the digital and vertical analog dials.

12.23 **(a)** $F = 14.10 > 6.36$. Reject H_0. There is evidence of a difference in the average sales between the various aisle locations.
(b) $\overline{X}_1 = 60.67$, $\overline{X}_2 = 20.67$, $\overline{X}_3 = 37.33$. Comparing $\overline{X}_1$ to $\overline{X}_2$, $F = 27.95 > 12.72$; reject H_0. Comparing $\overline{X}_1$ to $\overline{X}_3$, $F = 9.508 < 12.72$; don't reject H_0. Comparing $\overline{X}_2$ to $\overline{X}_3$, $F = 4.849 < 12.72$; don't reject H_0. Comparing $(\overline{X}_1 + \overline{X}_3)/2$ to $\overline{X}_2$, $F = 18.694 > 12.72$; reject H_0. Therefore, average sales are different in the front and middle locations; average sales are different in the middle location as compared to the average of the other locations.

CHAPTER 13

13.6 **(a)** There seems to be a fairly linear relationship between mileage and weight except for one observation.
(b) $b_0 = 52.153$, $b_1 = -.0141$, $\hat{Y}_i = 52.153 - .0141X_i$.
(c) For each increase of 1 pound in weight, mileage goes down by .0141 miles/gallon (for each 100-pound increase mileage goes down by 1.41 miles/gallon).
(d) $\hat{Y}_i = 26.773$.
(e) The range of weight is from 1450 to 2340 pounds. A weight of 500 pounds is outside the range of the X variable in the model.

13.7 **(a)** The plot appears to be approximately linear.
(b) $b_0 = .5831$, $b_1 = 1.04811$, $\hat{Y}_i = .5831 + 1.04811X_i$.
(c) For $X = 5$, $\hat{Y}_i = 5.824$; for $X = 7$, $\hat{Y}_i = 7.92$; for $X = 9$, $\hat{Y}_i = 10.016$.

13.18 $S_{Y|X} = 1.555$.

13.19 $S_{Y|X} = .3643$.

13.28 **(a)** SSR = 94.416, SSE = 14.513, SST = 108.929.
(b) $r^2 = .867$; 86.7% of the variation in mileage can be explained by variation in weight.

13.29 **(a)** SSR = 58.079, SSE = 0.796, SST = 58.875.
(b) $r^2 = .986$; 98.6% of the variation in the number of beers consumed can be explained by variation in the machine reading.

13.33 **(a)** Yes, the plot seems linear; **(b)** $r = .937$.

13.38 $r = -.931$.

13.42 $t = -6.25 < -3.7074$ or $F = 39.03 > 13.75$. Reject H_0. There is evidence of a linear relationship between weight and mileage.

13.48 $t = 20.92 > 2.4469$ or $F = 436.684 > 5.99$. Reject H_0. There is evidence of a linear relationship between the machine reading and the number of beers consumed.

13.50 $t = 6.0 > 2.5706$. Reject H_0. There is evidence of a linear relationship between ice cream flavor and fat content.

13.54 **(a)** $[-.0196 \leq \beta_1 \leq -.0086]$; **(b)** $[-.988 \leq \rho \leq -.655]$; **(c)** $[22.487 \leq \mu_{Y|X} \leq 25.387]$; **(d)** $[19.864 \leq Y_I \leq 28.010]$.

13.60 $[+.62 \leq \rho \leq +.99]$.

13.67 **(a)**

Observation	$\hat{Y}$	Residual
1	5.614	.386
2	9.073	−.073
3	2.784	.216
4	8.025	−.025
5	6.767	.233
6	2.679	−.679
7	10.226	−.226
8	3.832	.168

(b) Although there are only eight observations, most of the residuals appear to be close to zero.
(c) There is one residual that is much larger in magnitude than the others.

13.72 Y = GPI, X = GMAT score.
(a) $b_0 = .3003$, $b_1 = .0048702$, $\hat{Y}_i = .3003 + .0048702\ X_i$.
(b) For each increase of 1 point on the GMAT exam, grade-point index is predicted to increase by .00487 points. b_0, equal to .3003, would represent the predicted GPI when the GMAT score is 0 if that value were plausible.
(c) $\hat{Y}_i = 3.222$.
(d) $S_{Y|X} = .1559$.
(e) $r^2 = .798$; 79.8% of the variation in GPI can be explained by variation in GMAT score.
(f) $r = +.893$.
(g) $t = 8.43 > 2.1009$. Reject H_0. There is evidence of a linear relationship between GMAT score and GPI.
(h) $[3.144 \leq \mu_{Y|X} \leq 3.30]$.

(i) $[2.807 \leq Y_I \leq 3.481]$.
(j) $[.00367 \leq \beta_1 \leq .00607]$.
(k) $[.745 \leq \rho \leq .957]$.
(l) Based on a residual analysis, the model appears adequate.

13.74 (a) $r = -.292$.
(b) $t = -.864 > -1.8595$. Do not reject H_0. No evidence of a linear relationship between price per pound and proteins per gram.
(c) $[-.73 \leq \rho \leq +.31]$.

CHAPTER 14

14.3 $U = 7 < 8$. Reject H_0. There is evidence that the sequence is not random.

14.8 $6 < U = 11 < 17$. Don't reject H_0. There is no evidence that the tendency of stores to be open is not random.

14.12 $W = 27 < 47$. Reject H_0. There is evidence that the median price is less than $700.

14.13 $25 < W = 52.5 < 95$. Don't reject H_0. There is no evidence that the median is different from $20,000.

14.16 $5 < W = 7.5 < 31$. Don't reject H_0. There is no evidence of a difference in the median percentage between Brand X and Brand Y.

14.17 $W = 36 < 53$. Don't reject H_0. There is no evidence the claim is not valid.

14.19 $Z = 1.905 > 1.645$. Reject H_0. There is evidence that the median cholesterol level is higher for users of birth control pills than for nonusers.

14.21 (a) $Z = -3.003 < -1.645$. Reject H_0. There is evidence of a difference in the median acceleration rate between German and Japanese cars.
(b) $Z = -3.003$ as compared to $t = -2.82$ in Problem 10.32. The results are highly consistent.

14.25 $H = 9.789 > 5.991$. Reject H_0. There is evidence of a difference in the median attitude score based on working environment.

14.29 $H = 12.71 < 13.277$. Don't reject H_0. There is no evidence of a difference in median yield between the five varieties.

14.31 $r_s = +0.853$, $Z = +2.83 > 1.645$. Reject H_0. There is evidence of significant positive association between social status and conformity.

14.35 $r_s = -.518$, while in Problem 13.2 $r = -.428$.

14.39 $W = 91.5 < 124$. Don't reject H_0. There is no evidence that the median claim has increased above $85.

14.43 $r_s = +0.934$, $Z = 3.37 > 1.645$. Reject H_0. There is evidence of a significant positive correlation.

Index

Q

R

S

Y

Z